# Phenolic Compounds from Medicinal Plants

Medicinal plants have been trusted for centuries as natural sources of healing. Among their many bioactive constituents, phenolic compounds stand out for their remarkable diversity and wide-ranging health benefits. *Phenolic Compounds from Medicinal Plants: Pharmaceutical and Health Benefits* illustrates the science of these compounds, showing how they connect traditional plant wisdom with modern pharmaceutical research.

The chapters explain the chemistry and structures of phenolic compounds, how they are extracted and studied, and their proven effects on human health. Topics include their antioxidant, anti-inflammatory, antimicrobial, anticancer, neuroprotective, antidiabetic, and anti-obesity activities. It also looks at their contributions to skin health, digestion, and respiratory well-being. Importantly, the book highlights challenges such as bioavailability and explores strategies to make these compounds more effective in real-world applications.

A volume in the *Exploring Medicinal Plants* series, this book is designed for researchers, health-care professionals, and students seeking to understand the growing role of phenolic compounds in medicine, nutrition, and human wellness.

## Exploring Medicinal Plants

*Series Editor: Azamal Husen, Wolaita Sodo University, Ethiopia*

Medicinal plants render a rich source of bioactive compounds used in drug formulation and development; they play a key role in traditional or indigenous health systems. As the demand for herbal medicines increases worldwide, supply is declining as most of the harvest is derived from naturally growing vegetation. Considering global interests and covering several important aspects associated with medicinal plants, the Exploring Medicinal Plants series comprises volumes valuable to academia, practitioners, and researchers interested in medicinal plants. Topics provide information on a range of subjects including diversity, conservation, propagation, cultivation, physiology, molecular biology, growth response under extreme environment, handling, storage, bioactive compounds, secondary metabolites, extraction, therapeutics, mode of action, and healthcare practices.

Led by Azamal Husen, PhD, this series is directed to a broad range of researchers and professionals consisting of topical books exploring information related to medicinal plants. It includes edited volumes, references, and textbooks available for individual print and electronic purchases.

*Exploring Traditional Wild Edible Plants*
Edited by Vibhor Agarwal, Sachidanand Singh and Rhaul Datta

*Propagation to Pharmacopeia: Modern Approaches in Medicinal Plants*
Edited by Abha Manohar K, Gopal Shukla, Saikat Gantait, A P Das, and Sumit Chakravarty

*Exploring Medicinal Orchids*
Edited by Azamal Husen and Manu Pant

*Symbiotic Association of Microorganisms with Medicinal and Herbal Plants*
Edited by Allah Ditta, Sajid Mehmood, and Azamal Husen

*Tissue Culture Techniques and Medicinal Plants: Enhancing Propagation and Production*
Edited by Azamal Husen and Manu Pant

*Genetic Improvement and Conservation Practices of Medicinal Plants*
Edited by Manu Pant, Sayed Sartaj Sohrab and Azamal Husen

*Sustainable Use of Plants and their Products in Neurodegenerative Diseases Management*
Edited by Johnson O. Oladele, Ebenezer I. O. Ajayi, María L. Flores-López, and Oluwaseyi E. Okoro

*Wild Edibles of Deserts: Food Security and Sustainability*
Salman Majeed, Muhammad Zafar, Mushtaq Ahmad, Mohd Kafeel Ahmad Ansari, Shazia Sultana, and Andrea Pieroni

*Impact of Climate Change on Medicinal and Herbal Plant micro-RNA*
Kanchanlata Tungare, Parul Johri, Sachidanand Singh, and Surojeet Das

*Exploring Medicinal Bamboos*
Edited by Manu Pant and Azamal Husen

*Exploring Herbal Synergies for Optimal Human Health*
Edited by Art Devi, Nancy George, Zaved Ahmed Khan, and Vagish Dwibedi

*Exploring Medicinal and Commercial Uses of Flowers*
Edited by Rakesh Kumar Bachheti, Archana Bachheti, and Azamal Husen

*Phenolic Compounds from Medicinal Plants: Pharmaceutical and Health Benefits*
Edited by Rakesh Kumar Bachheti, Archana Bachheti, and Azamal Husen

For more information about this series, please visit: www.routledge.com/Exploring-Medicinal-Plants/book-series/CRCEMP

# Phenolic Compounds from Medicinal Plants

## Pharmaceutical and Health Benefits

Edited by
Rakesh Kumar Bachheti,
Archana Bachheti
and Azamal Husen

CRC Press
Taylor & Francis Group
Boca Raton London New York

CRC Press is an imprint of the
Taylor & Francis Group, an **informa** business

Designed cover image: Shutterstock

First edition published 2026
by CRC Press
2385 NW Executive Center Drive, Suite 320, Boca Raton FL 33431

and by CRC Press
4 Park Square, Milton Park, Abingdon, Oxon, OX14 4RN



CRC Press is an imprint of Taylor & Francis Group, LLC



**Library of Congress Cataloging-in-Publication Data**
Names: Bachheti, Rakesh Kumar editor | Bachheti, Archana editor | Husen, Azamal editor
Title: Phenolic compounds from medicinal plants : pharmaceutical and health benefits / edited by Rakesh Kumar Bachheti, Archana Bachheti, and Azamal Husen.
Other titles: Pharmaceutical & health benefits
Description: First edition. | Boca Raton, FL : CRC Press, 2026. | Series: Exploring medicinal plants | Includes bibliographical references and index. | Summary: "Medicinal plants have been trusted for centuries as natural sources of healing. Among their many bioactive constituents, phenolic compounds stand out for their remarkable diversity and wide-ranging health benefits. Phenolic Compounds from Medicinal Plants: Pharmaceutical and Health Benefits illustrates the science of these compounds, showing how they connect traditional plant wisdom with modern pharmaceutical research. The chapters explain the chemistry and structures of phenolic compounds, how they are extracted and studied, and their proven effects on human health. Topics include their antioxidant, anti-inflammatory, antimicrobial, anticancer, neuroprotective, antidiabetic, and anti-obesity activities. It also looks at their contributions to skin health, digestion, and respiratory well-being. Importantly, the book highlights challenges such as bioavailability, and explores strategies to make these compounds more effective in real-world applications. A volume in the Exploring Medicinal Plants series, this book is designed for researchers, healthcare professionals, and students seeking to understand the growing role of phenolic compounds in medicine, nutrition, and human wellness"-- Provided by publisher.
Identifiers: LCCN 2025051037 (print) | LCCN 2025051038 (ebook) | ISBN 9781032854014 hbk | ISBN 9781032864679 pbk | ISBN 9781003527671 ebk
Subjects: LCSH: Phenols--Physiological effect | Medicinal plants | Materia medica, Vegetable | Botany, Medical
Classification: LCC QP801.P4 P444 2026 (print) | LCC QP801.P4 (ebook) | DDC 547/.632--dc23/eng/20260305
LC record available at https://lccn.loc.gov/2025051037
LC ebook record available at https://lccn.loc.gov/2025051038

ISBN: 978-1-032-85401-4 (hbk)
ISBN: 978-1-032-86467-9 (pbk)
ISBN: 978-1-003-52767-1 (ebk)

DOI: 10.1201/9781003527671

Typeset in Times
by KnowledgeWorks Global Ltd.

# Contents

# Preface

Since the dawn of civilization, plants have been inseparable from the human story. They have sustained life by providing food for nourishment, shelter for protection, and, most profoundly, medicine for healing. Across continents and cultures, from Ayurveda in India and Traditional Chinese Medicine to the practices of the Egyptians, Greeks, and Native Americans, medicinal plants were revered as sources of health and balance. These age-old remedies, passed down through oral traditions, rituals, and practices, carried the wisdom of centuries long before the advent of modern science.

Today, scientific inquiry has confirmed what ancient knowledge intuitively recognized. The therapeutic virtues of plants arise from a diverse range of bioactive molecules. Among them, phenolic compounds represent one of the most significant and versatile classes. In nature, they serve as protectors of plants, enabling defense against environmental stress, pathogens, and adverse conditions. In human health, they extend an extraordinary spectrum of benefits. Their pharmacological value spans antioxidant, anti-inflammatory, antimicrobial, anticancer, neuroprotective, antidiabetic, cardioprotective, and anti-obesity activities. This remarkable breadth has placed phenolic compounds at the forefront of contemporary research in medicine, nutraceuticals, and functional foods.

This volume, under the book series *Exploring Medicinal Plant*, seeks to bridge ancient wisdom and modern innovation. It tells the story of phenolic compounds, tracing their journey from natural origins in medicinal plants to their promising roles in human health and sustainable healthcare systems. The book has been carefully structured to serve both as a foundational guide for students and newcomers and as a rich, advanced resource for seasoned researchers and professionals. The first chapters provide an overview of phenolic compounds, their distribution across medicinal plants, and their recognized health benefits. This is followed by an exploration of chemical diversity and structural characteristics, highlighting how small molecular variations give rise to profound biological functions. The section on extraction and characterization techniques explains the cutting-edge methods used to unlock these compounds from plant systems and study them in detail. The narrative then shifts toward the biological and therapeutic dimensions. The book explores the celebrated antioxidant properties of phenolic compounds and their role in mitigating oxidative stress, a key contributor to chronic illnesses. Their anti-inflammatory potential is examined next, offering insights into how they regulate pathways linked to cardiovascular disease, arthritis, and metabolic syndrome. Their antimicrobial activity is another compelling aspect, positioning them as natural alternatives to synthetic antibiotics, with applications extending into food preservation. The anticancer potential of phenolic compounds is considered in depth, alongside their neuroprotective roles, particularly in the context of Alzheimer's and Parkinson's disease.

Beyond therapeutic interventions, phenolic compounds also contribute to the prevention and management of lifestyle-related conditions. Chapters on diabetes management, anti-obesity activity, and skin health underscore their role in promoting wellness and longevity. Further sections explore their benefits for gastrointestinal health and respiratory health, including their potential in asthma management. Recognizing promise alone is not sufficient; practical challenges remain. The book therefore devotes careful attention to the pharmacokinetics and bioavailability of phenolic compounds, critical factors determining their absorption, distribution, and efficacy in the human body. A further discussion on their synergistic interactions with other natural bioactives emphasizes that these compounds rarely act alone but often work in concert, producing effects greater than the sum of their parts. The final chapters focus on formulation strategies and innovative delivery systems, bridging the gap between laboratory research and real-world applications through drug delivery technologies, encapsulation, and functional food design.

This book is conceived as more than a collection of chapters. It is an invitation to engage with the profound dialogue between plants and people, between heritage and science, and between natural resilience and human health. By uniting diverse perspectives in chemistry, pharmacology, and applied sciences, it offers a panoramic understanding of the significance of phenolic compounds. For researchers and academicians, it provides a strong foundation and a stimulus for future explorations. For healthcare professionals, it presents evidence-based insights into therapeutic possibilities. For postgraduate students, it acts as a roadmap connecting fundamental science with translational relevance. And for the curious reader, it reveals the timeless wisdom of nature and its potential to shape sustainable healthcare solutions. We believe this volume will ignite curiosity, foster collaboration, and inspire innovation in the field of natural products. By advancing the study and application of phenolic compounds, we take another step toward healthcare strategies that are not only effective but also sustainable and deeply rooted in the gifts of nature. We extend our heartfelt gratitude to all the contributors who have shared their expertise, dedication, and vision. Their collaborative spirit has transformed this project into a meaningful reference work.

**Rakesh Kumar Bachheti**

**Archana Bachheti**

**Azamal Husen**

# About the Editors

**Rakesh Kumar Bachheti** graduated from Hemwati Nandan Bahuguna University, Garhwal, India, in 1996. He completed his MSc in Organic Chemistry from the same university in 1998. Later, in 2001, he underwent a one-year Post Graduate Diploma in Pulp and Paper Technology from the Forest Research Institute, Dehradun, India. Subsequently, he pursued his PhD in Organic Chemistry from Kumaun University, Nainital, India, which he obtained in 2007. Bachheti has a total research and teaching experience of 22 years. Before joining Addis Ababa Science and Technology University (AASTU) in Ethiopia, he worked as Dean Project (Assistant) at Graphic Era University in Dehradun, India. It's worth noting that Graphic Era University is accredited with an "A+" grade by the National Assessment and Accreditation Council (NAAC). Bachheti is also a research fellow at INTI International University Persiaran Perdana BBN, Putra Nilai, Nilai, Negeri Sembilan, Malaysia. Bachheti has an impressive research background and has presented papers at various international (Malaysia, Thailand, and India) and national conferences. He has also been an active member of important committees like the Internal Quality Assurance Cell (IQAC) and the Anti-Ragging Committee. His research interests primarily focus on natural products for industrial applications, biofuels and bioenergy, green synthesis of nanoparticles and their applications, and pulp and paper technology. His passion for natural products is evident in all of his research endeavors. As an academic mentor, Bachheti has successfully advised 40 MSc and 9 PhD students to completion. Additionally, numerous undergraduates have conducted research in his laboratory under his guidance. He actively contributes to curriculum development for BSc/MSc/PhD programs. Bachheti is also recognized for his editorial work, having edited 10 books and authored 185 publications on various aspects of natural product chemistry and nanotechnology. He has contributed 30 book chapters published by prestigious publishers like Springer, Elsevier, Taylor & Francis and Nova Publisher. He is listed among the Top 2% Scientists for the year 2024 and 2025 by Stanford University, USA, and Elsevier.

**Archana (Joshi) Bachheti** completed her BSc in 1997 and MSc in 1999 from Hemwati Nandan Bahuguna Garhwal University. She earned her PhD from the Forest Research Institute, Dehradun, India, in 2006. Currently, she is a Professor at Graphic Era University in Dehradun, India. With over 20 years of extensive experience, she has served in various academic roles in India and has also provided expertise internationally. Her teaching portfolio includes subjects such as ecology and environment, environmental science, freshwater ecology, disaster management, and bryophytes and pteridophytes. Professor Bachheti's research interests span a broad and interdisciplinary field of plant ecology, with a focus on ecorestoration, green chemistry—particularly the synthesis of nanomaterials—and exploring the medicinal properties of plants. Her research has involved ecological restoration of degraded lands, studying the physical and chemical properties of plant oils, and investigating plant-based nanomaterials. Throughout her career, she has been actively involved in mentoring students and researchers. Dr. Bachheti has published more than 155 research papers and edited over 8 books. She is listed among the Top 2% Scientists for the year 2024 by Stanford University, USA, and Elsevier.

**Azamal Husen** has served as a Full Professor and Head of the Department of Biology at the University of Gondar, Ethiopia. He is currently affiliated as Foreign Delegate with Wolaita Sodo University, Ethiopia, and holds Adjunct Professorship at the Department of Biotechnology, Graphic Era (Deemed to be University) in Dehradun, Uttarakhand, India. His earlier academic roles include Full Professor at Sankalchand Patel University in Visnagar, India, and Visiting Faculty at the Forest Research Institute and Doon College of Agriculture and Forest in Dehradun, India. With more than 25 years of experience in academia and research, he is highly regarded for his expertise in nanobiotechnology, plant physiology, environmental botany, medicinal plants, and plant propagation. His research has been supported by prominent national and international organizations, including the World Bank, the National Agricultural Technology Project, the Indian Council of Agricultural Research, the Indian Council of Forestry Research and Education, and the Japan Bank for International Cooperation. He is a prolific scholar, having published more than 350 academic works, including research articles, books, and book chapters. His editorial contributions are equally noteworthy; he serves on the editorial boards and peer review panels of numerous prestigious journals published by Elsevier, Frontiers Media, Springer Nature, Taylor & Francis, RSC, Oxford University Press, the Royal Society, CSIRO, MDPI, PLOS, Sciendo, John Wiley & Sons, and UPM Journals. He is also a member of the advisory board of Cambridge Scholars Publishing, United Kingdom. He is a fellow of the Plantae group under the American Society of Plant Biologists and holds membership in several scientific organizations, including the International Society of Root Research, the Asian Council of Science Editors, and the International Natural Product Sciences Taskforce. Throughout his career, he has delivered invited lectures, keynote addresses, and technical presentations at numerous international conferences, symposia, and workshops. In the field of academic development, he has made significant contributions through the design and delivery of orientation programs for university faculty, the development of educational modules, and the formulation of curricula at the undergraduate, postgraduate, and doctoral levels. In recognition of his contributions to teaching, research, and community service, he was honored with a special award from the University of Gondar. He is currently serves as Editor-in-Chief of the *American Journal of Plant Physiology* and is Series Editor for several major academic collections, including *Exploring Medicinal Plants* (Taylor & Francis, USA), *Plant Biology, Sustainability, and Climate Change* (Elsevier, USA), and *Smart Nanomaterials Technology* (Springer Nature, Singapore). He also leads the *Handbook Series on Nanotechnology and Its Applications* (Springer Nature, Singapore) and *Plant Chemistry* (Taylor & Francis, USA). His outstanding contributions to science have earned him international acclaim, including the prestigious distinction of being named among the "World's Top 2% Scientists" for three consecutive years (2020 onward) by Stanford University, a recognition published in the Elsevier Data Repository. In addition to his regular academic responsibilities, he has served as Guest Editor for several high-impact special issues, including *Advances in Plant Metabolomics: Insights, Applications, and Integration* (Plant Physiology and Biochemistry), *Exploring Nanotechnology in Medicinal Plants* (Journal of Herbal Medicine), and *Integrating Nanotechnology with Plant Compounds: Enhancing Abiotic Stress Resilience and Crop Improvement via Signaling Molecules* (Frontiers in Plant Science), among others. His work continues to shape contemporary research at the intersection of plant science, biotechnology, and sustainable development. His academic leadership, research excellence, and international collaborations have positioned him as a prominent figure in global scientific circles.

# Contributors

**Mohamed F. Abdel-Hameed**
Department of Pharmacology
Medical Research and
Clinical Studies Institute
National Research Centre
Dokki, Giza, Egypt

**Eman Y. Ahmed**
Department of Natural
Compounds Chemistry
National Research Centre
Dokki, Giza, Egypt

**Rania F. Ahmed**
Department of Natural
Compounds Chemistry
National Research Centre
Dokki, Giza, Egypt

**Mohammad A. Alshuniaber**
Department of Food Science
and Nutrition, College of Food
and Agriculture Sciences
King Saud University
Riyadh, Kingdom of
Saudi Arabia

**Shaza H. Aly**
Department of Pharmacognosy
Faculty of Pharmacy
Badr University in Cairo
(BUC)
Cairo, Egypt

**Naglaa M. Ammar**
Therapeutic Chemistry
Department National
Research Centre
Dokki, Giza, Egypt

**Tilahun Belayneh Asfaw**
Department of Chemistry,
College of Natural and
Computational Sciences
University of Gondar
Gondar, Ethiopia

**Rakesh Kumar Bachheti**
Centre of Molecular Medicine and Diagnostics,
Saveetha Dental College & Hospitals,
Saveetha Institute of Medical and Technical
Sciences, Saveetha University
Chennai, India

**Rasha Z. Batran**
Department of Natural Compounds Chemistry
National Research Centre
Dokki, Giza, Egypt

**Monica Butnariu**
University of Life Sciences "King Mihai I"
from Timisoara, Chemistry & Biochemistry
Discipline
Timisoara, Timis, Romania

**William C. Cho**
Department of Clinical Oncology, Queen
Elizabeth Hospital
Hong Kong SAR, China

**Sudeshna Shyam Choudhury**
Microbiology Department, St. Xavier's College
Kolkata, India

**Omayma A. Eldahshan**
Department of Pharmacognosy
Faculty of Pharmacy, Ain Shams University
Abbassia, Cairo, Egypt

Center of Drug Discovery Research and
Development, Faculty of Pharmacy
Ain Shams University
Abbassia, Cairo, Egypt

**Zeinab A. El-Gendy**
Department of Pharmacology
Medical Research and Clinical Studies Institute
National Research Centre
Dokki, Giza, Egypt

**Mai M. Elghonemy**
Department of Natural Compounds Chemistry
National Research Centre
Dokki, Giza, Egypt

**Asmaa S. Abd Elkarim**
Chemistry of Tanning Materials and Leather Technology Department
National Research Centre
Giza, Egypt

**Walaa A. El-Kashak**
Department of Natural Compounds Chemistry
National Research Centre
Dokki, Giza, Egypt

**Abdelsamed I. Elshamy**
Department of Natural Compounds Chemistry
National Research Centre
Dokki, Giza, Egypt

Faculty of Pharmaceutical Sciences
Tokushima Bunri University
Yamashiro-cho, Tokushima, Japan

**Sherif E. Emam**
Department of Pharmaceutics and Industrial Pharmacy Faculty of Pharmacy Zagazig University
Zagazig, Egypt

**Ahmed F. Essa**
Department of Natural Compounds Chemistry
National Research Centre
Dokki, Giza, Egypt

**Ahmed H. Gaara**
Department of Natural Compounds Chemistry
National Research Centre
Dokki, Giza, Egypt

**Mansour K. Gatasheh**
Department of Biochemistry
College of Science, King Saud University
Riyadh, Saudi Arabia

**Mohamed S. Abd El Hafeez**
Department of Pharmacy
Kut University College
Al Kut, Wasit, Iraq

Department of Pharmacognosy, Faculty of Pharmacy, Egyptian Russian University
Badr City, Egypt

**Aranya Hazari**
Haldia Institute of Technology
Haldia, Kolkata, India

**Kulsum Hashmi**
Department of Chemistry
Isabella Thoburn College
Lucknow, Uttar Pradesh, India

**Heba A. Hassan**
Department of Pharmacology
Medical Research and Clinical Studies Institute
National Research Centre
Dokki, Giza, Egypt

**Hiroshi Imagawa**
Faculty of Pharmaceutical Sciences
Tokushima Bunri University
Yamashiro-cho, Tokushima, Japan

**Sonika Kalia**
Department of Biotechnology
School of Applied and Life Sciences
Uttaranchal University
Prem Nagar, Dehradun, Uttarakhand, India

**Yusuke Kasai**
Faculty of Pharmaceutical Sciences
Tokushima Bunri University
Yamashiro-cho, Tokushima, Japan

**Abdulsalam M. Kassem**
Department of Pharmaceutics and Pharmaceutical Technology
Faculty of Pharmacy (Boys)
Al-Azhar University
Nasr City, Cairo, Egypt

**Abdul Rahman Khan**
Department of Chemistry, Integral University
Lucknow, Uttar Pradesh, India

**Tahmeena Khan**
Department of Chemistry, Integral University
Lucknow, Uttar Pradesh, India

**Rajapandiyan Krishnamoorthy**
Department of Food Science and Nutrition, College of Food and Agriculture Sciences
King Saud University
Riyadh, Kingdom of Saudi Arabia

**Tahia M. Kubacy**
Department of Natural Compounds Chemistry
National Research Centre
Dokki, Giza, Egypt

**Shankul Kumar**
Department of Pharmacognosy, Narayan Institute of Pharmacy, Gopal Narayan Singh University
Jamuhar, Sasaram, Rohtas, Bihar, India

**Arkajyoti Mahanta**
Haldia Institute of Technology
Haldia, Kolkata, India

**Anjali R. Mallya**
Department of Biotechnology
Nehru Arts and Science College
Thirumalayampalayam, Coimbatore
Tamil Nādu, India

**Deepa Mundekkad**
Department of Biotechnology
Nehru Arts and Science College
Thirumalayampalayam, Coimbatore
Tamil Nādu, India

**Ahmed M. Nagy**
Department of Animal Reproduction & AI
Veterinary Research Institute
National Research Centre
Cairo, Egypt

**Sumitra Nain**
Department of Pharmaceutical Chemistry
Faculty of Pharmacy, Banasthali Vidyapith
Newai, Tonk, Rajasthan, India

**Mahmoud I. Nassar**
Department of Natural Compounds Chemistry
National Research Centre
Dokki, Giza, Egypt

**Pradeep Singh Negi**
Department of Fruit and Vegetables Technology
CSIR-Central Food Technological Research Institute
Mysore, India

Academy of Scientific and Innovative Research (AcSIR)
Ghaziabad, India

**Omsatyam**
Narayan Institute of Pharmacy
Gopal Narayan Singh University
Jamuhar, Sasaram, Rohtas, Bihar, India

**Arvind Kumar Patel**
Department of Pharmacology,
Narayan Institute of Pharmacy, Gopal Narayan Singh University
Jamuhar, Sasaram, Rohtas, Bihar, India

**Vedant Kumar Prajapati**
Department of Pharmaceutics
Narayan Institute of Pharmacy
Gopal Narayan Singh University
Jamuhar, Sasaram, Rohtas, Bihar, India

**Indra Rautela**
Department of Biotechnology
School of Applied and Life Sciences
Uttaranchal University
Prem Nagar, Dehradun, Uttarakhand, India

**Saman Raza**
Department of Chemistry, Isabella Thoburn College
Lucknow, Uttar Pradesh, India

**Ruchi**
Department of Biotechnology
School of Applied and Life Sciences
Uttaranchal University
Prem Nagar, Dehradun, Uttarakhand, India

**Ahmed Sabt**
Department of Natural Compounds Chemistry
National Research Centre
Dokki, Giza, Egypt

**Manish Dev Sharma**
Department of Biotechnology
School of Basic and Applied Science
Shri Guru Ram Rai University
Patel Nagar, Dehradun, Uttarakhand, India

**Abdel Nasser B. Singab**
Department of Pharmacognosy Faculty of Pharmacy Ain Shams University
Abbassia, Cairo, Egypt

Center of Drug Discovery Research and Development, Faculty of Pharmacy
Ain Shams University
Abbassia, Cairo, Egypt

**Karishma Singh**
Academy of Scientific and Innovative Research (AcSIR)
Ghaziabad, India

Department of Fruit and Vegetables Technology, CSIR-Central Food Technological Research Institute
Mysore, India

**Shashi Ranjan Singh**
Department of Pharmacology
Narayan Institute of Pharmacy
Gopal Narayan Singh University
Jamuhar, Sasaram, Rohtas
Bihar, India

**Vinod Kumar Singh**
Narayan Institute of Pharmacy
Gopal Narayan Singh University
Jamuhar, Sasaram, Rohtas, Bihar, India

**Rehab F. Taher**
Department of Natural Compounds Chemistry
National Research Centre
Dokki, Giza, Egypt

**Fekade Beshah Tessema**
Department of Chemistry, College of Natural and Computational Sciences
Woldia University
Woldia, Ethiopia

**Vaibhavi Thapliyal**
Department of Biotechnology
School of Basic and Applied Science
Shri Guru Ram Rai University
Patel Nagar, Dehradun, Uttarakhand, India

**Sheetal Tyagi**
Department of Chemistry
School of Basic and Applied Science
Shri Guru Ram Rai University
Patel Nagar, Dehradun, Uttarakhand, India

# 1 An Overview of Phenolic Compounds from Medicinal Plants and Their Health Benefits

*Fekade Beshah Tessema*, Tilahun Belayneh Asfaw, and Rakesh Kumar Bachheti*

## 1.1 INTRODUCTION

Phenolic compounds are abundant, varied, and found widely in nature. These secondary metabolites are widely present in higher plants and contain aromatic rings with one or more hydroxyl substituents. These bioactive phenolic natural secondary metabolites have enormous and significant biological importance (Bachheti et al. 2023; Garde-Cerdán et al., 2017). Among the possible sources of phenolic compounds, medicinal plants are the major ones. Humans have used plants for many years as food, spice, flavoring, and medicine (Husen, 2022, 2023a,b, 2024a,b; Sun & Shahrajabian, 2023). For centuries, medicinal plants have been a vital part of traditional medical systems worldwide, providing a natural and often cost-effective alternative to conventional treatments. Both on their own and in combination with traditional medicine, medicinal plants are extremely important (Sun & Shahrajabian, 2023). They are important due to their cultural heritage, accessibility, affordability, holistic approach, efficacy, safety, potential for discoveries, and environmental benefits (Bodeker & Graz, 2017). Many medicinal plant parts, such as seeds, leaves, flowers, fruits, stems, and roots, are rich in bioactive compounds (Sun & Shahrajabian, 2023). Polyphenols are the major groups of phytochemicals significantly abundant in plants (Kisiriko et al., 2021).

Phenolic compounds have enormous health benefits (Mamari, 2022). This chapter is not meant to be a thorough survey of the literature, despite the fact that there are many original research publications and reviews on phenolic compounds. It seeks to supplement current publications and reviews by providing a quick reference on phenolic compounds' classification, origins, effects, and safety concerns. The chapter's overall goal is to summarize phenolic compounds found in medicinal plants and their advantages for health.

## 1.2 CLASSIFICATION OF PHENOLIC COMPOUNDS

The two metabolic routes that biogenetically produce phenolic compounds are the acetic acid pathways, which largely produces simple phenols, and the shikimic acid pathway, which mostly forms phenylpropanoids (Giada & Additional, 2013). There are two main categories of phenolic compounds: simple phenolics and polyphenolics. As substituted phenols, simple phenolics are identified by their ortho-, meta-, and para substitution patterns, which correspond to 1,2-, 1,3-, and 1,4-substitution patterns of the benzene ring, respectively. The hydroxyl group is one of the functional groups among the substituted groups (Vermerris & Nicholson, 2009). As illustrated in Figure 1.1, examples include phloroglucinol (1,3,5-trihydroxybenzene), a meta-trihydroxylated simple phenolic, and resorcinol (1,3-dihydroxybenzene), a meta-dihydroxylated simple phenolic.

* Corresponding author: fekade.b@wldu.edu.et[1]

DOI: 10.1201/9781003527671-1

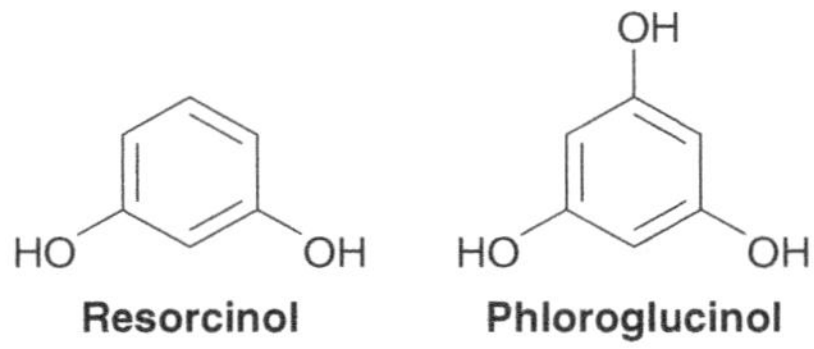


**FIGURE 1.1** Structures of simple phenolics.

Flavonoids, phenolic acids, tannins, lignans, coumarins, quinones, and xanthones are some examples of known polyphenols (Figure 1.2) (Mbaveng et al., 2014).

## 1.2.1 Flavonoids

The chemical structure of the C15 compounds, also referred to as flavonoids, is phenylbenzopyran. This structure comprises a chroman ring, sometimes called the benzopyran moiety, that is joined to a C15 (C6–C3–C6) skeleton (Figure 1.3). The fused aromatic ring is identified as the A ring, the phenyl component is labeled the B ring, and the heterocyclic benzopyran ring is designated as the C ring in this structure. There exist two forms of A rings: meta-trihydroxylated phloroglucinols and meta-dihydroxylated resorcinols. Vicinal-trihydroxylation, ortho-dihydroxylation, or monohydroxylation are the three possible states for the B ring. The core heterocycle is often one of three forms: pyran, pyrylium, or γ-pyrone. The broad structure of flavonoids makes it possible to be categorized into major classes. Two benzene rings are joined in each case by a three-carbon chain (Pereira et al., 2009). The major classes of flavonoids are the following:

- *Flavonols*
  Bearing a hydroxyl group at C-3 and as representative flavonols, quercetin and kaempferol are the most common types of flavonoids and can be found in a variety of fruits, vegetables, and beverages. These compounds are found in many foods, including tea, berries, beans, apples, broccoli, kale, Brussels sprouts, onions, and leeks. Hay fever and hive symptoms may be reduced with quercetin, an antihistamine. Its anti-inflammatory effects are another well-known characteristic. Since kaempferol and other flavonols have strong anti-inflammatory and antioxidant properties, they may help avoid chronic illnesses (Nisar, 2022).

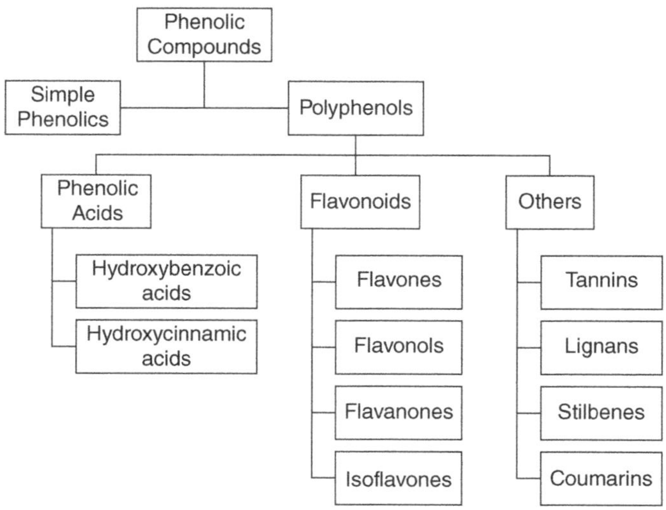


**FIGURE 1.2** Classification of phenolic compounds.

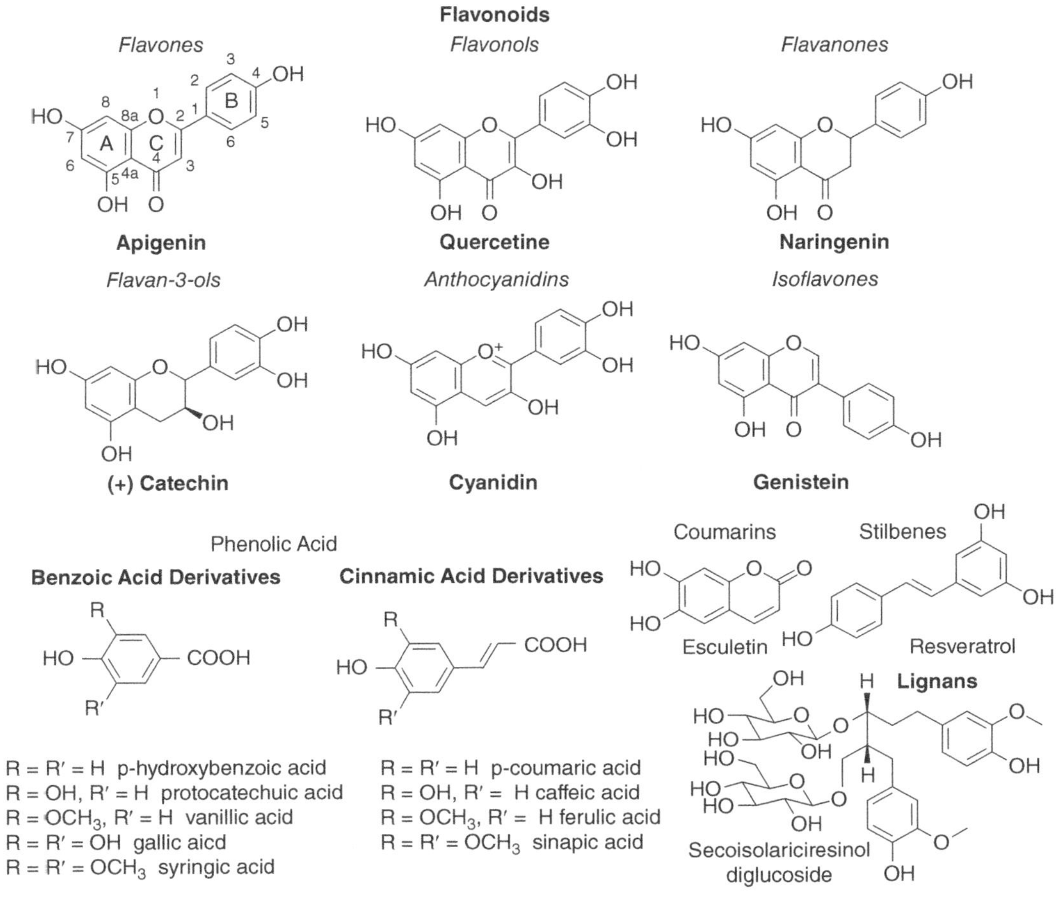


**FIGURE 1.3** Examples of major classes of phenolic compounds.

- *Flavones*
  Flavones are similar to flavonols but lack the hydroxyl group at the 3-position. Flavones have a ketone functionality at C-4, a double bond between C-2 and C-3, and a B-ring connected to C-2 (Kisiriko et al., 2021). Lutein and apigenin are common examples of flavones. Hot peppers, celery, parsley, and other plants are good sources of flavones. Flavones have been linked to delayed drug metabolization and antioxidant properties (Nisar, 2022).

- *Flavanones*
  Flavanones, which have a saturated C-ring, are commonly found in citrus fruits, tea, and red wine (Kisiriko et al., 2021). Examples of flavanones include hesperetin, eriodictyol, and naringenin. Flavanone contents are particularly high in citrus fruits. These compounds have been linked to relaxation, cardiovascular health improvement, and anti-inflammatory and antioxidant benefits (Nisar, 2022).

- *Isoflavones*
  Soybeans and other legumes are the main source of isoflavones. They have a special structure where the B-ring is joined to the C-ring in the third position. Members of this subgroup include genistein, glycitein, and daidzein. The isoflavone content in legumes, soy products, and soybeans is high. These compounds imitate the estrogen hormone because they are

phytoestrogens. Although the results of current studies are conflicting, they may potentially lower the risk of hormonal cancers such as breast, endometrial, and prostate cancer. Numerous researches have demonstrated that isoflavones function as both oxidants and antioxidants, with varying results on their impact on cancer. They are also being researched as a potential remedy for menopausal symptoms (Nisar, 2022).

- *Anthocyanidins*
  The flavylium cation is formed by the C-ring of anthocyanidins, which is hydroxylated at C-3, lacks the keto group at C-4, and has two double bonds (Kisiriko et al., 2021). Malvidin, pelargonidin, peonidin, and cyanidin are some examples of anthocyanidins. Pomegranates, plums, red wine, red and purple grapes, and red, purple, and blueberries all contain these compounds. Anthocyanidins are linked to cardiovascular health and antioxidant activity and play a role in preventing obesity and diabetes (Vermerris & Nicholson, 2009).

- *Flavanols*
  As a subtype of dihydroflavonol, flavanols are divided into three groups: dimers, polymers, and monomers (also called catechins). Flavanols are found in foods, including red wine, fava beans, apples, berries, teas, cocoa, and grapes. Black tea contains dimers, which have been linked to lowering cholesterol, but green and white teas are rich in catechins (Singla et al., 2019). Researchers think catechins could help reduce the symptoms of chronic fatigue syndrome. Furthermore, catechins have been associated with neurological and cardiovascular health (Nisar, 2022).

### 1.2.2 Phenolic Acids

Aromatic compounds with a phenolic ring and a carboxyl functional group are called phenolic acids or phenol carboxylic acids (Garde-Cerdán et al., 2017). These compounds have a carboxyl group, a hydroxyl group, and an aromatic ring. One of the most prevalent phenolic acids is salicylic acid. The two primary families of phenolic acids, which are significant naturally occurring compounds, are hydroxycinnamic and hydroxybenzoic acids (Figure 1.3) (Nisar, 2022).

- *Hydroxybenzoic acids*
  Non-phenolic benzoic acid molecules are the precursors of hydroxybenzoic acids. These are benzoic acids with a hydroxyl group added as substitute. An alternative way to think of them would be as phenols with a carboxylic acid functional group attached directly to the phenol ring. For instance, salicylic acid contains hydroxyl groups that are at ortho (o), meta (m), or para (p) positions (Figure 1.4). Benzoic acids with two hydroxyl groups added are known as dihydroxybenzoic acids. The two hydroxyl groups are generally located in 2,3-, 2,4-, 2,5-, 2,6-, 3,4-, and 3,5-positions, as shown in Figure 1.5. Trihydroxybenzoic acids are benzoic acids with three hydroxyl groups incorporated. Examples of 2,4,6-trihydroxybenzoic acid and 3,4,5-trihydroxybenzoic acid (gallic acid) are displayed in Figure 1.6 (Mamari, 2022).

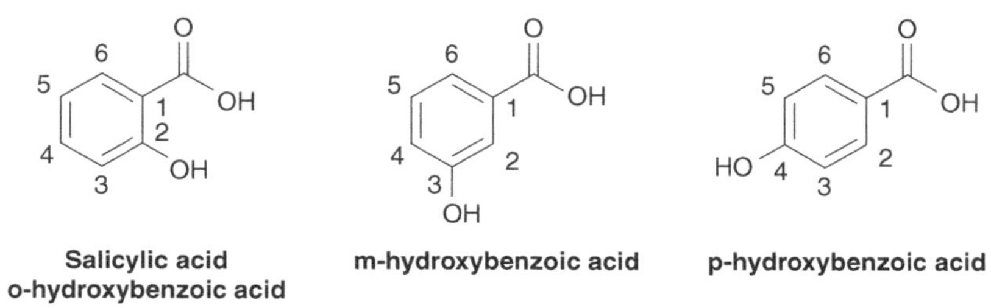


**FIGURE 1.4** Structures of hydroxybenzoic acids.

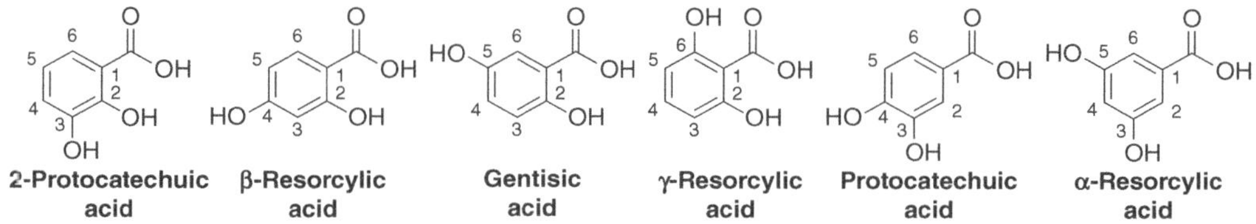


**FIGURE 1.5** Structures of main dihydroxybenzoic acids.

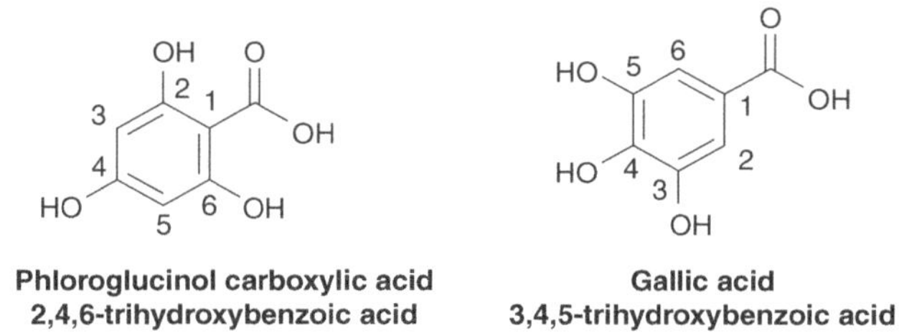


**FIGURE 1.6** Structures of main trihydroxybenzoic acids.

- *Hydroxycinnamic acids*
  Cinnamic acid molecules that are not phenolic are the source of hydroxycinnamic acids. Phenolic acids are called hydroxycinnamic acids when the carboxylic acid functional group is isolated from the phenol ring by a C=C link. 2-, 3-, and 4-hydroxycinnamic acids are a few types of hydroxycinnamic acids (Figure 1.7). Other typical cinnamic acid examples are sinapic acid, ferulic acid, and caffeic acid (Figure 1.8) (Mamari, 2022).

### 1.2.3 Other Phenolic Compounds

- *Stilbenes*
  Stilbene compounds in plants, known as phenolic compounds, play a crucial role in helping plants adapt and survive in response to sudden and ongoing stress. Polyphenols control cell growth, pollen fertility, differentiation, pollen fertility, and nodulation, indicating their essential role in maintaining plant health (Mamari, 2022).

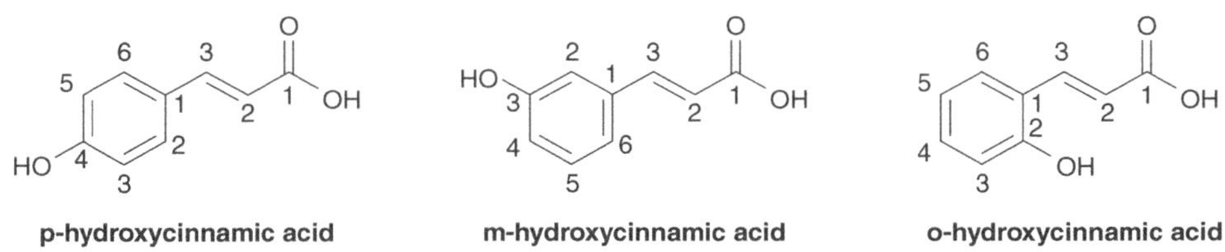


**FIGURE 1.7** Examples of hydroxycinnamic acids.

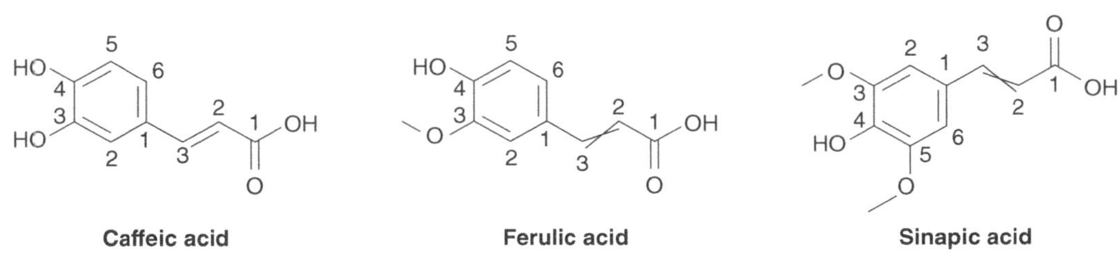


**FIGURE 1.8** Common examples of hydroxycinnamic acids.

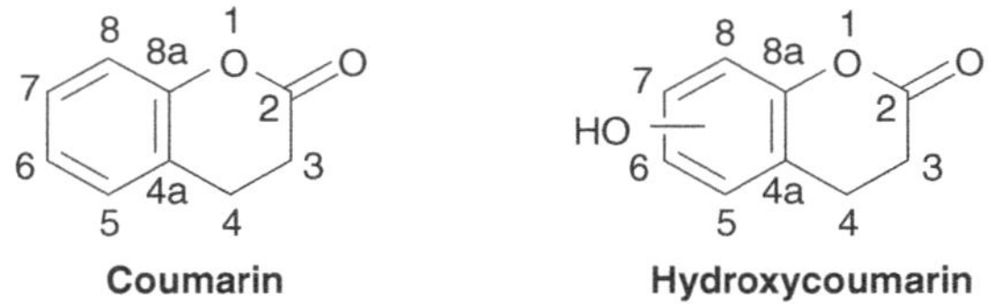


**FIGURE 1.9** General structures of hydroxy coumarins.

- *Tannins*
  A class of compounds known as tannins have many different structures, but they can all bind and precipitate proteins. They get their name from the tanning method used to create leather from animal hides, which dates back to the prehistoric era when animal fat and brain tissue were applied. The collagen chains in the hide were cross-linked due to this treatment. Plant tannins were used in the tanning process for a large portion of history until the last century when minerals like chromium and aluminum took their place (Vermerris & Nicholson, 2009).

- *Lignans*
  Flax and sesame seeds include lignans, which are chemical compounds from plants that are also present in smaller amounts in grains, seeds, fruits, and vegetables. Human gut bacteria create enterolignans, often referred to as mammalian lignans, as metabolites of dietary lignans. These compounds have been found in human plasma and urine. Although their estrogenic and other biochemical properties are moderate, they may have nutritional value in preventing cardiovascular and other chronic diseases (Nisar, 2022).

- *Coumarins*
  Hydroxy coumarins are coumarins with hydroxyl groups (see Figure 1.9). They belong to the group of phenolic compounds.
  Scopoletin and graphene are examples of hydroxy coumarins (Figure 1.10) (Mamari, 2022).

## 1.3 SOURCES OF PHENOLIC COMPOUNDS

Phenolic compounds are found in various plant tissues, including fruits, seeds, leaves, stems, and roots. You can find the occurrence of phenolic compounds in both raw and processed foods on Phenol-Explorer (http://phenol-explorer.eu/) and the USDA Database for the Flavonoid Content of Selected Foods (https://www.ars.usda.gov/) (Rosa et al., 2019). These bioactive compounds are also present in industrial waste (Garde-Cerdán et al., 2017). At least one aromatic ring containing a hydroxyl group is present in all phenolic compounds. Flavonoids and nonflavonoids are the two primary categories into which more than 8000 distinct plant phenolic chemicals can be grouped. About two-thirds of dietary phenolic compounds are flavonoids, which are also the most bioactive components. Flavonoids are the most prevalent phenolic compounds found in fruits and vegetables. Two phenyl rings (A and B) connected by a heterocyclic

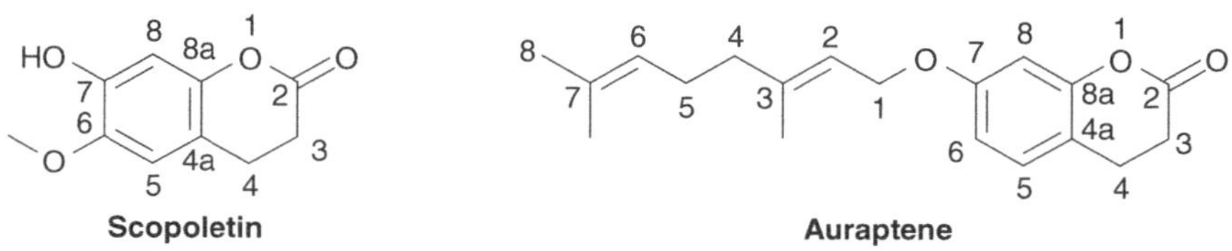


**FIGURE 1.10** Examples of hydroxy coumarins.

pyran ring (ring C, see Figure 1.3) make up their phenyl benzopyran skeleton. Based on variations in the pyran ring, flavonoids can be categorized into six categories or families. Each family's individual compounds vary in how they hydroxylate and methylate rings A and B (Rosa et al., 2019).

Next, we will consider the rich sources of known phenolic compounds among the frequently used medicinal plants.

### 1.3.1 Medicinal Plants Rich in Phenolic Compounds

- *Green tea (Camellia sinensis L.)*
  Green tea is made by drying the leaves without fermenting. Tea is well known for its stimulating properties mainly due to caffeine, which may be present from 2 to 5%. Other important constituents are the polyphenols (30–42%), and the flavonoids (–)-Epicatechin, (–)-Epigallocatechin, and theaflavin-3-O-gallate (Ben-Erik & Michael, 2017; Duke, 1997). More uses of tea include treating arthritis, asthma, diarrhea, motion sickness, and gas (Duke, 1997).

- *Turmeric (Curcuma longa L.)*
  Curcumin, monodesmethoxycurcumin, and bisdesmethoxycurcumin are examples of curcuminoids, which are yellow, non-volatile pigments that make up 3–5% of the dry weight of *C. longa*. Bisabolane, which is present in high concentrations in *C. longa*, guaiane, and a number of germacrane-type sesquiterpenes, including α- and β-turmerone, curlone, and zingiberene, are abundant in the essential oil (2–7% dry weight). Ukonan A is a polysaccharide (arabinogalactan) from turmeric rhizome (Ben-Erik & Michael, 2017). The Nigerian-harvested rhizome of *C. longa* (*Zingiberaceae*) contained the phenolic compound curcumin, which prevented rats' testicles from damage caused by di-n-butyl phthalate (DBP) (Mbaveng et al., 2014).

- *Ginkgo (Ginkgo biloba L.)*
  Diterpene lactones (sometimes referred to as ginkgolides A, B, C, J, and M), flavonoids (mostly flavonol glycosides and non-glycosidic biflavonoids), and a sesquiterpenoid (bilobalide) are among the active components. Ginkgolic acids are found in very small levels in the ginkgo leaf extracts (Ben-Erik & Michael, 2017). By lowering oxidative damage, neuroinflammation, and microglial activation, *G. biloba* extract has been found to protect dopaminergic neurons in an animal model of Parkinson's disease (Garde-Cerdán et al., 2017).

- *Garlic (Allium sativum L.)*
  The phytochemicals found in the solvent extract of the bulb of *A. sativum* include carbohydrates, total proteins, alkaloids, saponins, flavonoids, tannins, and steroids. It works against a variety of dental infections with a wide range of activity (Bin et al., 2020). Several sulfur-containing compounds in garlic and garlic oil have biological effects. Alliin, the main component of whole garlic, is broken down by the enzyme alliinase. This procedure initially forms an unstable intermediate, followed by allicin, frequently the main ingredient in processed garlic. Another metabolite that can prevent platelet aggregation is ajoene (in its Z and E forms). The three-dimensional structures of proteins can be altered by these sulfides' ability to form disulfides with their SH-groups. The sulfides can affect various proteins and enzymes due to these structural alterations (Ben-Erik & Michael, 2017).

- *Grape vine (Vitis vinifera L.)*
  Bioactive polyphenols, particularly flavonoids, stilbenes, and phenolic acids, are common in grapes and wines (Garde-Cerdán et al., 2017). Nonhydrolyzable or condensed tannins,

sometimes referred to by a variety of other names (such as procyanidins, proanthocyanidins, leucoanthocyanidins, polyphenols, or pycnogenols), are the active components of grape seed oil. Red grapes and red wine also contain anthocyanins (mostly 3-glycosides of cyanidin and peonidin), organic acids, tannins, and other phenolic compounds (Ben-Erik & Michael, 2017).

- *Ginger (Zingiber officinale Roscoe)*
  Many bioactive compounds, including phenolics and terpenes, have been identified in ginger. Gingerols, shogaols, and paradols are the major phenolic compounds that contribute to the various biological activities of ginger (Mao et al., 2019). Gingerol's phenolic component is widely recognized for its anti-inflammatory and anti-nausea properties. The major flavonoids responsible for ginger's pharmacological effects include zingerone compounds, phenolic compounds, 6-gingerol, 6-shogaol, and others (Mahmudati et al., 2019).

- *Rosemary (Rosmarinus officinalis L.)*
  Rosemary is a highly promising source of phenolic compounds found in essential oil extracted through hydro-distillation and polyphenol recovery (Kheiria et al., 2021). Rosemary phenolic compounds include a variety of chemical families, including diterpenes, flavonoids, and phenolic acids. The most prevalent phenolic acid in methanolic extracts of rosemary leaves is rosmarinic acid (Irakli et al., 2023). Even the leftover residue after recovering essential oils, which is currently thrown away as garbage, might be recovered for producing natural extracts with strong antioxidant properties and phenolic compounds (Kheiria et al., 2021). *R. officinalis* extracts were previously reported to contain luteolin, apigenin, caffeic acid, ferulic acid, carnosic acid, carnosol, and rosmarinic acid.

- *Cinnamon (Cinnamomum verum J. Presl, Syn Cinnamomum zeylanicum)*
  Cinnamon is rich in cinnamaldehyde, a phenolic compound with antioxidant and antidiabetic properties and a significant amount of phenolic acids and flavonoids (Silva et al., 2023). The two isolated chemical classes from *C. zeylanicum* are polyphenols and volatile phenols. In addition to polyphenols, cinnamon mainly contains ferulic acids, caffeic, gallic, vanillic, protocatechuic, and p-coumaric acids (Kumar et al., 2020).

- *Oregano (Origanum vulgare L.)*
  There are many different chemical components found in oregano's aerial parts, such as phenolics, terpenes, flavonoids, glucosides, sterols, tannins, resins, and mucilages (Khorsand et al., 2022). It contains phenolic compounds such as carvacrol and thymol, which have antimicrobial and antioxidant properties. Rosmarinic acid is also a potent antioxidant and has been identified as the main phenolic compound in oregano. Other phenolic compounds found in oregano include caffeic acid, eriodyctiol, luteolin, naringenin, and apigenin (Kruma et al., 2008).

- *Lavender (Lavandula angustifolia Mill.)*
  Lavender contains linalool, a phenolic compound with sedative and anti-inflammatory properties (Kruma et al., 2008). Flavonoids (such as apigenin and luteolin glycosides, catechin, naringenin, and vanillin), hydroxybenzoic acids (such as p-hydroxybenzoic acid, protocatechuic acid, vanillic acid, gentisic acid, and gallic acid), and hydroxycinnamic acids (such as rosmarinic acid, caffeic acid, p-coumaric acid, ferulic acid, chlorogenic acid, sinapic acid, cinnamic acid, 4-O-caffeoylquinic, and 5-O-caffeoylquinic) are among the many phenolic compounds found in lavender flowers (Dobros et al., 2022).

As major sources of phenolic compounds, medicinal plants are found to be rich in such groups of phytochemicals, as summarized in Table 1.1. Images of some of the common medicinal plants and/or parts are shown in Figure 1.11.

**TABLE 1.1**
**Important Medicinal Plants as Major Sources of Phenolic Compounds, Their Major Phytochemical Constituents, and Their Health Activities**

| Botanical Name of Medicinal Plant | Phytochemical Constituents | Health Activities | Key References |
|---|---|---|---|
| | **Fruits and Vegetables** | | |
| *Morus alba* L. | Phenolic acid, flavonoids, anthocyanins, pectins, sugars, artocarpin | Antimicrobial, immunological, anti-inflammatory, antioxidant, antidiabetic, anti-cancer, and protective properties | Batiha et al. (2023) |
| *Vaccinium corymbosum* L. *Vaccinium myrtillus* L. | Phenolic acid, gentisic acid, kaempferol, and quercetin | Antioxidant, anticancer, anti-inflammatory, and anti-atherosclerotic | Kopystecka et al. (2023) |
| *Casimiroa edulis* La. | Pyrogallol, p-hydroxybenzoic acid, luteolin 6-arabinose-8-glucose, acacetin | Antioxidant, antidepressant, antihypertensive, anticancer, and anti-inflammatory activities | Elkady et al. (2017) and Satheesh (2015) |
| *Dovyalis abyssinica* (A. Rich.) Warb. | Flavonoid and simple phenols, ascorbic acid | Antioxidant and antihyperglycemic effects | Baylie et al. (2024) |
| *Malus sylvestris* Miller. *Malus domestica* | Pectin, sucrose, amines, vitamins, mineral salts, polyphenols, and fruit acids (malic acid) | Antiproliferative activity. Antioxidants, cancer-preventative effects | Vuran et al. (2023) |
| *Psidium guajava* L. | Gallic acid, chlorogenic acid, epicatechin, avicularin, catechin, apigenin, guaijaverin, kaempferol, hyperin, myricetin, epigallocatechin gallate, and quercetin | Hepatoprotective, lipid-lowering, antibacterial, anticancer, antidiabetic, antioxidant, and antidiarrheal properties | Kumar et al. (2021) |
| *Mentha spicata* L. | Flavonoids, phenolic acids, and terpenes, essential oil with carvone, limonene | Antioxidant, anticancer, antiparasitic, antimicrobial, and antidiabetic effects | Menyiy et al. (2022) |
| *Vitis vinifera* L. | Anthocyanins, catechin derivatives, flavonoids (such as quercetin and kaempferol), and stilbenoids (like trans-resveratrol and trans-ε-viniferin) | Strong cosmetic characteristics, including skin-whitening, anti-aging, and anti-inflammatory properties | Sharafan et al. (2023) |
| *Allium cepa* L. | Protocatechuic acid, flavonoids, including quercetin-40-monoglucoside, isorhamnetin 3,40-diglucoside, quercetin-3,40-diglucoside, quercetin 3-glycosides, anthocyanins, kaempferol, ferulic acid, and quercetin | Neuroprotective, anti-inflammatory, and antiparasitic properties, as well as anticancer, antidiabetic, antioxidant, antiplatelet, antihypertensive, and antidepressant effects | Chakraborty et al. (2022) |
| *Brassica oleracea* var. italica | β-Carotene, a number flavonoids, and vitamins C and E | Antioxidant, anti-inflammatory, and anticancer | Syed et al. (2023) |
| | **Herbs and Spices** | | |
| *Rosmarinus officinalis* L. | Rosmarinic acid and caffeic acid, apigenin, luteolin, genkwanin and hesperidin | Antioxidant, antifungal, antibacterial, anti-inflammatory | Irakli et al. (2023) |
| *Origanum vulgare* L. | Quercetin, apigenin, luteolin, naringenin, kaempferol, coumarin, rosmarinic, chlorogenic, cinnamic, caffeic, syringic, benzoic, vanillic, gallic, chicoric, and 2,4-dihydroxybenzoic acids | High antioxidant activity, antibacterial, antifungal, antiviral, spasmolytic and anti-inflammatory | Khorsand et al. (2022) |

*(Continued)*

**TABLE 1.1** *(Continued)*
**Important Medicinal Plants as Major Sources of Phenolic Compounds, Their Major Phytochemical Constituents, and Their Health Activities**

| Botanical Name of Medicinal Plant | Phytochemical Constituents | Health Activities | Key References |
|---|---|---|---|
| *Curcuma longa* L. | Curcuminoids: monodesmethoxycurcumin, bisdesmethoxycurcumin, and curcumin. The essential oil (2–7% dry weight) contains a lot of guaiane, bisabolane, and several sesquiterpenes of the germacrane type. | Anti-inflammatory, anti-oxidative, antimicrobial, and cytotoxic to tumor cells; prevention and therapy of a wide range of disorders, including cancer, autoimmune, neurological, cardiovascular, and diabetic conditions | Rosa et al. (2019) |
| *Zingiber officinale* Roscoe | Flavonoids, phenolic compounds, 6-gingerol, 6-shogaol, and zingerone compounds | Antioxidant, antibacterial, antifungal, molluscicidal, anti-parasitic and anthelmintic. | Mahmudati et al. (2019) |
| *Cinnamomum verum* J. Presl | Its bark essential oil contains cinnamon aldehyde and trans-cinnamaldehyde (Cin), eugenol (leaf), and camphor (root). | Astringent, carminative, warming stimulant, blood purifier, digestive, antiseptic, antibacterial, antifungal, and antiviral properties | Kumar et al. (2020) |
| *Thea sinensis* L. *Camellia sinensis* (L.) Kuntze | Gallic acid (GA), coumaric acid, chlorogenic acid, caffeic acid, and its quinic acid ester, together with flavanols primarily comprised of myricetin, quercetin, and kaempferol | Antimicrobial, anticancer, antioxidant, and anti-hypercholesterolemic properties | Aboulwafa et al. (2019) |
| Coffee: coffee beans contain phenolic compounds. *Coffea arabica* L. | Purine alkaloids (mainly caffeine) Chlorogenic acid and other phenolic acid | Stimulant, diuretic; important commercial product anti-fibrotic and anticancer effects | Nwafor et al. (2022) |

### 1.3.2 Other Sources of Phenolic Compounds

Besides the foods and drinks we consume daily, waste products are also rich in phenolic compounds. Various industrial and agricultural wastes contain bioactive compounds (Sorrenti et al., 2023) and reusing these compounds could be the initial step in converting waste into a source of natural compounds with beneficial properties for pharmaceutical, food, or agricultural industries (Garde-Cerdán et al., 2017).

## 1.4 FACTORS INFLUENCING PHENOLIC COMPOUND CONTENT IN MEDICINAL PLANTS

The content of phenolic compounds in medicinal plants can vary significantly due to several factors. Some of the major factors are discussed below:

- *Environmental Factors*
  Various environmental factors such as sunlight, temperature, water availability, and soil composition can impact the levels of phenolic compounds in plants. Generally, increased sunlight exposure and optimal temperatures tend to result in higher levels of phenolic compounds,

**FIGURE 1.11** Image of major medicinal plant which are good source of phenolic compounds.

particularly flavonoids, as a defense mechanism against UV radiation. On the other hand, extreme temperatures and excessive water can negatively affect both plant growth and phenolic synthesis. In regions with higher solar irradiation, there are typically higher contents of flavanols and flavonols. Furthermore, altitude is associated with hydroxybenzoic acid concentration, promoting phenolic acid synthesis. On the other hand, dihydrochalcone concentration and annual rainfall are negatively correlated, most likely as a result of hydric stress (Solar et al., 2022). It is important to remember that the nutrient content and pH of the soil can have an impact on both plant growth and the production of secondary metabolites. Specific phenolic compounds can be synthesized under certain soil conditions. The geographical regions of origin of Mentha samples collected from various Algerian localities have shown an impact on their total phenolics (TPC), total flavonoids (TFC), rosmarinic acid (RAC) concentrations, and antioxidant activity when all other parameters are held essentially constant (Brahmi et al., 2022). The ethanolic root extracts of *Valeriana carnosa* and *Valeriana clarionifolia* from Patagonian populations were subjected to phytochemical analysis, which showed variation in antioxidant activity and phenolic component content (both quantitatively and qualitatively) (Nagahama et al., 2020).

- *Plant Factors*
  The genetic makeup of a plant species determines its inherent capacity to produce phenolic compounds. Different cultivars or varieties within the same species can exhibit varying levels of phenolic content. Phenolic compound levels often increase as plants mature. The phenolic component content and antioxidant activity of the samples are strongly influenced by their annual growing conditions (Pérez-Ochoa et al., 2023). A plant's leaves, stems, roots, and flowers can all have different concentrations of phenolic compounds (Tungmunnithum et al., 2018).

- *Agronomic Practices*
  Agronomic practices, including the use of fertilizers, pruning, and harvesting time, affect plant growth and the synthesis of secondary metabolites, which in turn affects the phenolic composition differently. The cultivar and specific agricultural technique employed to determine how agricultural activities affect the chemical composition of raspberries. This demonstrates that organic farming methods can impact the composition of polyphenols in both favorable and negative ways (Anjos et al., 2020). Agronomic practices, in particular plant density, can alter the plant's total phenolics content and antioxidant activity, which in turn can affect the protective properties of palm-tree kale in cells (Danesi et al., 2014). In another study, *Passiflora quadrangularis* L. plants treated with organic methods showed increased antioxidant properties and improved profiles of secondary metabolites (Devi et al., 2021).

- *Processing and Storage*
  The methods used for drying, storing, and processing can affect the yield and composition of phenolic compounds. In the fresh blend, pasteurization produces higher concentrations of chlorogenic acid (+188%), carotenoids (+16–42%), and catechins (+98–132%) than untreated samples, while pascalization preserves vitamin C (−0.3%), quercetin-3-glucoside (+22%), and sulforaphane (+4.3%) better than untreated samples (Debenedictis et al., 2023). The total phenolic and flavonoid contents and antioxidant activity of okra samples preserved using the same approach during storage were marginally affected by the various processing and preservation techniques (Al-dabbas et al., 2023). By understanding these factors, it is possible to optimize cultivation practices and processing methods to maximize the phenolic compound content in medicinal plants, thereby enhancing their potential health benefits.

## 1.5 HEALTH BENEFITS OF PHENOLIC COMPOUNDS FROM MEDICINAL PLANTS

As a type of secondary metabolite present in plants, phenolic compounds have been linked to a number of health advantages. It is crucial to remember that the precise health advantages of phenolic compounds can differ based on the kind of phenolic compounds, the plant from which they come, and the health status of the individual (Lin et al., 2016). Notwithstanding encouraging findings, additional research is required to completely comprehend the potential advantages of these compounds. Some of the main health advantages linked to the phenolic compounds found in medicinal plants are listed below:

### 1.5.1 Antioxidant Properties

Antioxidants are abundant in medicinal plants, which can help lower oxidative stress in cells and treat conditions, including cancer, heart disease, and inflammatory diseases. The following compounds are primarily responsible for the antioxidant properties of medicinal plants (Xu et al., 2017):

- *Phenolics*: These include lignans, anthocyanins, flavonoids, phenolic acids, and stilbenes.
- *Terpenes*: These are mostly present in medicinal plants' essential oils.
- *Carotenoids*: These consist of β-carotene, lutein, and lycopene.
- *Tocopherols and tocotrienols*: These are found in tinctures and ethanolic extracts, these compounds are soluble in polar organic solvents.
- Ascorbic acid, sometimes referred to as vitamin C, is one of the most prevalent antioxidants found in plants.

By donating electrons to neutralize free radicals, phenolic compounds can protect cells from oxidative damage. The chemical structure of these substances has a significant impact on their antioxidant action (Giada & Additional, 2013). Medicinal herbs have a dose-dependent antioxidant effect since most low-molecular-weight antioxidants can act as prooxidants at high concentrations (Ziyatdinova & Kalmykova, 2023). A lower risk of chronic illnesses like cancer, heart disease, and neurological diseases has been linked to lavender's antioxidant activity. The main medical benefits of lavender, which have been used for a long time, are its antibacterial, antifungal, anticholinesterase, sedative, antidepressant, spasmolytic, and antioxidant properties (Dobros et al., 2023). About 90% of all fruits' antioxidant activity comes from anthocyanins, compared to 10% from flavonols, flavan-3-ols, and phenolic acids (Jakobek et al., 2009). Organic cultivation increased *P. quadrangularis*' antioxidant activities due to its vitamins and secondary metabolites (Devi et al., 2021). Recent attempts to enhance the antioxidant activity of individual plant extracts involve formulating by combining essential oil extracts from various medicinal plants with antioxidant properties (Elbouzidi et al., 2024).

### 1.5.2 Anti-Inflammatory Effects

The body uses inflammation to combat infections and promote healing, but it can occasionally get out of control and persist longer than is necessary. Anti-inflammatory agents are medications or compounds that lessen inflammation by preventing the body from producing specific compounds. Because of their shown anti-inflammatory properties, medicinal plants and their extracts are regarded as a secure and reliable substitute for synthetic anti-inflammatory medications (Radovanović et al., 2023). Phytochemicals found in medicinal plants can aid in the treatment of inflammatory disorders. Medicinal plants that affect inflammation include sage, ginger, black pepper, chamomile, peppermint, turmeric, John's wort, and white willow. Berberis vulgaris, which contains berberine, and Datura stramonium, which includes hyoscine, both have anti-inflammatory properties (Gonfa et al., 2023). Plant foods' vital minerals, phytochemicals, and antioxidants can also help lower inflammation. These nutrients can help reduce inflammation and counteract toxins from processed foods, bacteria, viruses, and pollutants. Inflammatory diseases can be controlled with the use of phytochemicals derived from medicinal plants, including flavonoids, phenolic acids, sterols, terpenoids, sesquiterpenes, and tannins (Yatoo et al., 2018).

The four components of the inflammatory process are effectors, mediators, sensors, and inducers. Drugs' anti-inflammatory effects often include the following: modifying receptors and signaling pathways, controlling target tissues' reaction to inflammatory mediators, reversing the medium's effect on the target tissue, and generating anti-inflammatory mediators. By controlling inflammatory signaling pathways and preventing the synthesis of inflammatory mediators, for example, curcumin has anti-inflammatory properties (Yu et al., 2021). Extracts of medicinal plants are considered excellent traditional remedies for inflammatory conditions globally. They are abundant in phytochemicals used in herbal medicine and serve as a valuable foundation for developing new anti-inflammatory medications (Gonfa et al., 2023). Ginger (Mao et al., 2019) and Lavender (Dobros et al., 2023) are among the common medicinal plants that possess multiple biological activities, including antioxidant and anti-inflammatory.

### 1.5.3 Anticancer Potential

Cancer is a leading cause of death, with approximately 9.6 million reported cases in 2018. Various research studies have shown that natural products like fruits and medicinal plants have anti-cancer properties. The potential anticancer effects of medicinal plants include the following:

- *Antioxidants*: The antioxidant properties of many plant components can decrease oxidative stress and prevent cancer formation (Vijayalakshmi et al., 2024).
- Inducing apoptosis, stopping the cell cycle, and blocking the AKT/mTOR signaling pathway are just a few of the ways phytoconstituents might target cancer cells (Vijayalakshmi et al., 2024).
- *Cell cycle arrest*: Certain plants can stop the cell cycle at particular stages, such as G2/M (Elshnoudy et al., 2024).
- Certain plants can cause cancer cells to undergo apoptosis (Huang et al., 2024).
- Certain plants can stop the spread of cancer by blocking particular signaling pathways (Vijayalakshmi et al., 2024).

Ginger's anti-cancer properties have recently been the subject of much research on a variety of cancer types, including prostate, colorectal, breast, and cervical cancer (Mao et al., 2019). Phenolic compounds are able to trigger apoptosis (programmed cell death) in cancer cells (Sun & Shahrajabian, 2023). They may also prevent the formation of new blood vessels that supply tumors, a process known as angiogenesis (Mbaveng et al., 2014). Anti-inflammatory, strong antioxidant, anticancer, and anticarcinogenic properties have all been documented for grapes (Colitis, 2014).

Here are some more examples of therapeutic plants with anticancer properties:

- In oral carcinoma models, an extract from *Oroxylum indicum* can cause apoptosis and prevent the spread of the malignancy (Vijayalakshmi et al., 2024).
- Banana lectin, one of the bioactive substances found in *Musa paradisiaca*, can induce apoptosis and halt the cell cycle at G2/M (Malviya et al., 2023).
- *Colchicum autumnale's* colchicine has the ability to disturb the production of microtubules, which can prevent cell division and trigger apoptosis (Khan et al., 2020).
- The alkaloids vincristine and vinblastine, extracted from *Catharanthus roseus*, can suppress microtubule dynamics, resulting in apoptosis and cell cycle arrest (Vijayalakshmi et al., 2024).
- The plant *Psidium guajava*, also known as guava, can block the AKT/mTOR signaling pathway, which is critical for the survival and growth of cancer cells (Vijayalakshmi et al., 2024).
- *Annona atemoya*/muricate's fruit contains bullatacin, an acetogenin that has the ability to cause apoptosis and tumor cell condensation (Malviya et al., 2023).
- The plant *Dracaena loureiri Gagnep* has flavonoids and phenolics that may have anti-inflammatory and anticancer effects (Huang et al., 2024).
- Curcumin is a compound that can cause breast cancer cells to undergo apoptosis and stop the growth of colon cancer cells (Khan et al., 2020).

### 1.5.4 Cardiovascular Health

Heart failure, coronary heart disease (CHD), stroke, myocardial infarction, and hypertension are among the cardiovascular diseases (CVD) that are a leading cause of death globally (Michel et al., 2020). Historically, medicinal plants have been used to treat a wide range of illnesses, including heart problems (Emeje et al., 2024). The following are a few medicinal plants that could improve cardiovascular health:

- Curcumin, one of the bioactive components found in turmeric (*C. longa*), may help lower cholesterol, lessen inflammation, and prevent atherosclerosis (Emeje et al., 2024).
- Herbs like hawthorn may help treat heart problems like high blood pressure, chest discomfort, irregular heartbeats, and artery hardening (Michel et al., 2020).
- A novel medication for hypertension is presently being developed using the medicinal herb *Marrubium vulgare* L (Michel et al., 2020).

Phenolic compounds have been shown to help regulate blood pressure (Mamari, 2022). To improve the lipid profile, they might also aid in raising HDL ("good") cholesterol levels and lowering LDL ("bad") cholesterol levels. Anti-inflammatory and antioxidant properties, relaxation, and cardiovascular health are linked to the majority of flavonoid groups (Nisar, 2022). Phenolics can decrease inflammation, which plays a crucial role in the development of heart disease (Gonfa et al., 2023). Eating fruits, vegetables, and cereals is linked to a lower risk of diseases associated with oxidative stress, including cardiovascular issues, cancer, osteoporosis, and aging-related disorders (Mahmudati et al., 2019).

### 1.5.5 Neuroprotective Effects

Phenolics can decrease brain inflammation, which is linked to neurodegenerative disorders such as Alzheimer's and Parkinson's (Nisar, 2022). Their impact on improving cognitive function is also considered important (Garde-Cerdán et al., 2017). Several studies indicate that the cinnamon leaf extract has anti-inflammatory and neuroprotective properties (Silva et al., 2023). Recent studies show that resveratrol promotes the health of nerves and important brain functions, such as cognitive processes (Nisar, 2022).

### 1.5.6 Other Potential Benefits

Some phenolic compounds possess antibacterial properties that can aid in the defense against ailments. Plant responses to various forms of stress are directly influenced by a large number of polyphenols, especially phenolic acids, which possess antibacterial properties and help lignify damaged areas to promote recovery. Phenolics can potentially lessen allergic reactions and assist control of blood sugar levels (Nisar, 2022).

## 1.6 SAFETY CONSIDERATIONS

Though phenolic compounds offer numerous health benefits, it's essential to be aware of potential safety concerns. Here are some factors to consider concerning safety: excess consumption, individual tolerance, and interactions with medications. Information on the acute and long-term effects of toxicants is necessary for chemical safety evaluation. A series of animal studies have historically provided such data carried out over varying time periods, from a single dose with effects observed over a few days to repeated daily dosing with observations made over a period of months (Macko et al., 2021).

### 1.6.1 Toxicity Concerns

Some phenolic compounds have a potential for toxicity at higher concentrations.

It is important to avoid consuming excessive amounts of phenolic compounds as it can lead to adverse effects. Individual differences exist in their tolerance to phenolic compounds, as do the amount and potency of these compounds in various plant sources. Before substantially increasing the consumption of phenolic compounds, it is advisable to consult a healthcare professional. Standardized supplements or products might be used to guarantee consistent ingestion. Constipation and diarrhea are two digestive problems that can occasionally result from consuming too many phenolic chemicals. Effects might be more noticeable to those with delicate digestive systems.

Some low-molecular-weight phenolic compounds, such as thymol, are used as antiseptics in medicine because of their toxicity (Giada & Additional, 2013). As allelochemicals, phenolic compounds, and extracts can suppress noxious weeds that compete with plants. They demonstrate toxicity toward weeds by inhibiting physiological functions like photosynthesis (Kisiriko et al., 2021). The water that is pumped out of coal beds to remove methane contains phenolics as well; these compounds are primarily responsible for the acute toxicity of the water (Fini et al., 2022).

A key factor in flavonoids' ability to treat ailments is their solubility. Acute harmful effects from flavonoid ingestion are rare in humans, with the exception of allergies, due to the poor solubility of flavonoid aglycones in water, their brief time of presence in the intestine, and their decreased absorption (Kumar & Pandey, 2013). The presence of phenolic compounds in soil would prevent seed germination and the growth of native plants because they are phytotoxins. In surface waters, phenolic compounds can harm fish in quantities as low as 1 mg/L (1 ppm). At levels below 1 mg/L, phenolic chemicals harm other aquatic life and devastate natural ecosystems (Fini et al., 2022).

### 1.6.2 Drug Interactions

Some phenolic compounds can potentially worsen bleeding risks when used with blood thinners. Certain drugs may interact with phenols, reducing their efficacy or producing adverse side effects. Grapefruit juice contains bergapten and some prenylated furanocoumarins (such as bergamottin), which have been shown to have unfavorable interactions with a number of drugs (Rosa et al., 2019). Phenolic chemicals can cause rare but potentially allergic reactions in certain individuals.

## 1.7 CONCLUSION

In traditional medicine, medicinal plants high in phenolic compounds were often employed. Recent studies support the potential relationships between the content of phenolic compounds and health benefits. We anticipate seeing an increasing number of phenolic compounds used in several healthcare domains as research advances. We can maximize their potential to enhance human health and well-being by comprehending their modes of action and refining their delivery. By investigating the chemistry, pharmacology, and therapeutic potential of phenolic compounds derived from medicinal plants, scientists can make a valuable contribution to advancing innovative approaches toward the prevention, treatment, and general promotion of human health.

## COMPETING INTERESTS

The authors declare that they have no competing interest.

## DECLARATION OF GENERATIVE AI AND AI-ASSISTED TECHNOLOGIES IN THE WRITING PROCESS

During the preparation of this work, the book chapter author(s) have not taken any assistance from AI, and the book chapter author(s) reviewed and edited the content as needed and take(s) full responsibility for the content of the publication.

## REFERENCES

Aboulwafa, M. M., Youssef, F. S., Gad, H. A., Altyar, A. E., Al-Azizi, M. M., & Ashour, M. L. (2019). A comprehensive insight on the health benefits and phytoconstituents of *Camellia sinensis* and recent approaches for its quality control. *Antioxidants*, *8*(10). https://doi.org/10.3390/antiox8100455

Al-dabbas, M. M., Moumneh, M., Hamad, H. J., Abughoush, M., Abuawad, B., Al-nawasrah, B. A., Al-jaloudi, R., & Iqbal, S. (2023). Impact of processing and preservation methods and storage on total phenolics, flavonoids, and antioxidant activities of okra (*Abelmoschus esculentus* L.). *Foods*, *12*, 1–10. https://doi.org/10.3390/foods12193711

Anjos, R., Cosme, F., Gonçalves, A., Nunes, F. M., & Vilela, A. (2020). Effect of agricultural practices, conventional vs organic, on the phytochemical composition of 'Kweli 'and 'Tulameen 'raspberries (*Rubus idaeus* L.). *Food Chemistry*, *328*(April), 126833. https://doi.org/10.1016/j.foodchem.2020.126833

Bachheti, R. K., & Bachheti, A. (Eds.). (2023). Secondary Metabolites from Medicinal Plants: Nanoparticles Synthesis and their Applications. Boca Raton, CRC Press. https://doi.org/10.1201/9781003213727

Batiha, G. E. S., Al-Snafi, A. E., Thuwaini, M. M., Teibo, J. O., Shaheen, H. M., Akomolafe, A. P., Teibo, T. K. A., Al-kuraishy, H. M., Al-Garbeeb, A. I., Alexiou, A., & Papadakis, M. (2023). *Morus alba*: A comprehensive phytochemical and pharmacological review. *Naunyn-Schmiedeberg's Archives of Pharmacology*, *396*(7), 1399–1413. https://doi.org/10.1007/s00210-023-02434-4

Baylie, T., Tsega, W., Getinet, M., Abebaw, D., Azanaw, G., Adugna, A., & Jemal, M. (2024). Evaluation of antioxidant and antihyperglycemic effects *Dovyalis abyssinica* (A. Rich). *Metabolism Open*, *22*(April), 100286. https://doi.org/10.1016/j.metop.2024.100286

Ben-Erik, V. W., & Michael, W. (2017). *Medicinal Plants of the World: An Illustrated Scientific Guide to Important Medicinal Plants and Their Uses* (2nd ed.). Briza Publications.

Bin, C., Al-Dhabi, N. A., Esmail, G. A., Arokiyaraj, S., & Arasu, M. V. (2020). Potential effect of *Allium sativum* bulb for the treatment of biofilm forming clinical pathogens recovered from periodontal and dental caries. *Saudi Journal of Biological Sciences*, *27*(6), 1428–1434. https://doi.org/10.1016/j.sjbs.2020.03.025

Bodeker, G., & Graz, B. (2017). Traditional Medicine. In *Hunter's Tropical Medicine and Emerging Infectious Disease* (Tenth Edit, pp. 194–199). Elsevier Inc. https://doi.org/10.1016/B978-0-323-55512-8.00025-9

Brahmi, F., Lounis, N., Mebarakou, S., & Guendouze, N. (2022). Impact of growth sites on the phenolic contents and antioxidant activities of three Algerian Mentha species (*M. pulegium* L., *M. rotundifolia* (L.) Huds., and *M. spicata* L.). *Frontiers in Pharmacology*, *13*(June), 1–13. https://doi.org/10.3389/fphar.2022.886337

Chakraborty, A. J., Uddin, T. M., Matin Zidan, B. M. R., Mitra, S., Das, R., Nainu, F., Dhama, K., Roy, A., Hossain, M. J., Khusro, A., & Emran, T. B. (2022). *Allium cepa*: A treasure of bioactive phytochemicals with prospective health benefits. *Evidence-Based Complementary and Alternative Medicine*, *2022*. https://doi.org/10.1155/2022/4586318

Colitis, U. (2014). Polyphenols in the Prevention of Ulcerative Colitis. In *Polyphenols in Human Health and Disease*. Elsevier Inc. https://doi.org/10.1016/B978-0-12-398456-2.00050-5

Danesi, F., Valli, V., Elementi, S., & Antuono, L. F. D. (2014). The agronomic techniques as determinants of the phenolic content and the biological antioxidant effect of palm-tree Kale. *Food and Nutrition Sciences*, *5*(January), 1–7. https://doi.org/10.4236/fns.2014.51001

Debenedictis, J. N., Kok, T. M., De, & Breda, S. G. V. (2023). Impact of processing method and storage time on phytochemical concentrations in an antioxidant-rich food mixture. *Antioxidants*, *12*. https://doi.org/10.3390/antiox12061252

Devi, S., Id, R., Lee, H. H., Xiao, Y. J., Shahbani, S., Zakaria, M. H., & Bujang, J. S. (2021). Organic cultivation practices enhanced antioxidant activities and secondary metabolites in giant granadilla (*Passiflora quadrangularis* L.). *PLoS ONE*, *16*(7), 1–17. https://doi.org/10.1371/journal.pone.0255059

Dobros, N., Zawada, K., & Paradowska, K. (2022). Phytochemical profile and antioxidant activity of *Lavandula angustifolia* and *Lavandula x intermedia* cultivars extracted with different methods. *Antioxidants*, *11*. https://doi.org/10.3390/antiox11040711

Dobros, N., Zawada, K. D., & Paradowska, K. (2023). Phytochemical profiling, antioxidant and anti-inflammatory activity of plants belonging to the Lavandula genus. *Molecules*, *28*. https://doi.org/10.3390/molecules28010256

Duke, J. A. (1997). *The Green Pharmacy*. Rodale Press Inc.

Elbouzidi, A., Taibi, M., El Hachlafi, N., Haddou, M., Jeddi, M., Baraich, A., Aouraghe, A., Bellaouchi, R., Mothana, R. A., Hawwal, M. F., Mesnard, F., Hano, C., Asehraou, A., Chaabane, K., El Guerrouj, B., & Addi, M. (2024). Formulation of a three-component essential oil mixture from *Lavandula dentata, Rosmarinus officinalis,* and M*yrtus communis* for improved antioxidant activity. *Pharmaceuticals*, *17*(8). https://doi.org/10.3390/ph17081071

Elkady, W. M., Ibrahim, E. A., Gonaid, M. H., & El-Baz, F. K. (2017). Chemical profile and biological activity of *Casimiroa edulis* non-edible fruit's parts. *Advanced Pharmaceutical Bulletin*, *7*(4), 655–660. https://doi.org/10.15171/apb.2017.079

Elshnoudy, I. A., Elkhouly, A. M., Masoud, M., Rabea, H. A., & Mansour, F. R. (2024). Medicinal Plants Cultivated in Egypt With Anticancer Potential; A Systematic Review. In *Phytochemistry Reviews* (Vol. 5). Springer Netherlands. https://doi.org/10.1007/s11101-024-09957-5

Emeje, M., Ogu, E., Chidebe, I., Roko, G., Abdullahi, M., & Sule Bibinu, D. (2024). Therapeutic Potential of Medicinal Plants: Current Situation and Outlook. In *Medicinal Plants - Harnessing the Healing Power of Plants for Health [Working Title]* (pp. 1–12). IntechOpen. https://doi.org/10.5772/intechopen.1005828

Fini, E. H., Ayat, S., & Pahlavan, F. (2022). Phenolic Compounds in the Built Environment. In F. A. Badria & A. Surguchov (Eds.), *Phenolic Compounds - Chemistry, Synthesis, Diversity, Non-Conventional Industrial, Pharmaceutical and Therapeutic Applications.* DOI:10.5772/intechopen.98757

Garde-Cerdán, T., Gonzalo-Diago, A., & Pérez-Álvarez, E (Eds.). (2017). *Phenolic Compounds: Types, Effects and Research.* Nova Science Publishers, Inc..

Giada, M. de L. R., & Additional. (2013). Food Phenolic Compounds: Main Classes, Sources and Their Antioxidant Power. In *Oxidative Stress and Chronic Degenerative Diseases - A Role for Antioxidants Figure.* IntechOpen. DOI:10.5772/51687

Gonfa, Y. H., Tessema, F. B., Bachheti, A., Rai, N., Tadesse, M. G., Nasser Singab, A., Chaubey, K. K., & Bachheti, R. K. (2023). Anti-inflammatory activity of phytochemicals from medicinal plants and their nanoparticles: A review. *Current Research in Biotechnology*, *6*(July), 100152. https://doi.org/10.1016/j.crbiot.2023.100152

Huang, X., Arjsri, P., Srisawad, K., Yodkeeree, S., & Dejkriengkraikul, P. (2024). Exploring the anticancer potential of traditional Thai medicinal plants: A focus on *Dracaena loureiri* and its effects on non-small-cell lung cancer. *Plants*, *13*(2). https://doi.org/10.3390/plants13020290

Husen, A. (2022) Herbs, Shrubs and Trees of Potential Medicinal Benefits. CRC Press, Boca Raton, FL. https://doi.org/10.1201/9781003205067

Husen, A. (2023a) Exploring Poisonous Plants: Medicinal Values, Toxicity Response and Other Potential Uses. CRC Press, Boca Raton, FL. https://doi.org/10.1201/b23017

Husen, A. (2023b) Antidiabetic Medicinal Plants and Herbal Treatments. CRC Press, Boca Raton, FL. https://doi.org/10.1201/b23347

Husen, A. (2024a) Antimalarial Medicinal Plants. Taylor & Francis Group, LLC. NW ISBN: 9781032457154 https://doi.org/10.1201/9781003378396

Husen, A. (2024b) Medicinal Spices and Condiment Crops. Taylor & Francis Group, LLC. NW ISBN: 9781032469478 https://doi.org/10.1201/9781003387046

Irakli, M., Skendi, A., Bouloumpasi, E., Christaki, S., Biliaderis, C. G., & Chatzopoulou, P. (2023). Sustainable recovery of phenolic compounds from distilled rosemary by-product using green extraction methods: Optimization, comparison, and antioxidant activity. *Molecules*, *28*. https://doi.org/10.3390/molecules28186669

Jakobek, L., Eruga, M. S., Eruga, B. S., Novak, V., & Medvidovic-Kosanovi, M. (2009). Phenolic compound composition and antioxidant activity of fruits of *Rubus* and *Prunus* species from Croatia. *Nternational Journal of Food Science and Technology*, *44*, 860–868. https://doi.org/10.1111/j.1365-2621.2009.01920.x

Khan, T., Ali, M., Khan, A., Nisar, P., Jan, S. A., Afridi, S., & Shinwari, Z. K. (2020). Anticancer plants: A review of the active phytochemicals, applications in animal models, and regulatory aspects. *Biomolecules*, *10*(1). https://doi.org/10.3390/biom10010047

Kheiria, H., Mounir, A., María, Q., José, J. M., & Bouzid, S. (2021). Total Phenolic Content and Polyphenolic Profile of Tunisian Rosemary (*Rosmarinus officinalis*) Residues. In *Natural Drugs from Plants.* IntechOpen. DOI: 10.5772/intechopen.97762

Khorsand, G. J., Morshedloo, M. R., & Mumivand, H. (2022). Natural diversity in phenolic components and antioxidant properties of Oregano (*Origanum vulgare* L.) accessions, grown under the same conditions. *Scientific Reports*, *12*, 1–9. https://doi.org/10.1038/s41598-022-09742-4

Kisiriko, M., Anastasiadi, M., Terry, L. A., Yasri, A., Beale, M. H., & Ward, J. L. (2021). Phenolics from medicinal and aromatic plants : Characterisation and potential as biostimulants and bioprotectants. *Molecules*, *26*. https://doi.org/10.3390/molecules26216343

Kopystecka, A., Kozioł, I., Radomska, D., Bielawski, K., Bielawska, A., & Wujec, M. (2023). *Vaccinium uliginosum* and *Vaccinium myrtillus*—Two species—One used as a functional food. *Nutrients*, *15*. https://doi.org/10.3390/nu15194119 Academic

Kruma, Z., Andjelkovic, M., Verhe, R., & Kreicbergs, V. (2008). Phenolic compounds in basil, oregano and thyme. *Foodbalt*, 99–103.

Kumar, M., Tomar, M., Amarowicz, R., Saurabh, V., Nair, M. S., Maheshwari, C., Sasi, M., Prajapati, U., Hasan, M., Singh, S., Changan, S., Prajapat, R. K., Berwal, M. K., & Satankar, V. (2021). Guava (*Psidium guajava* L.) Leaves: Nutritional composition phytochemical profile, and health-promoting bioactivities. *Foods*, *10*(752), 1–20.

Kumar, S., Kumari, R., & Mishra, S. (2020). Pharmacological properties and their medicinal uses of Cinnamomum : A review. *Journal of Pharmacy and Pharmacology*, *71*(October 2019), 1735–1761. https://doi.org/10.1111/jphp.13173

Kumar, S., & Pandey, A. (2013). Chemistry and biological activities of flavonoids: An overview. *The Scientific World Journal*. https://doi.org/10.1155/2013/162750

Lin, D., Xiao, M., Zhao, J., Li, Z., Xing, B., Li, X., Kong, M., Li, L., Zhang, Q., Liu, Y., Chen, H., Qin, W., Wu, H., & Chen, S. (2016). An overview of plant phenolic compounds and their importance in human nutrition and management of type 2 diabetes. *Molecules*, *21*. https://doi.org/10.3390/molecules21101374

Macko, P., Palosaari, T., & Whelan, M. (2021). Extrapolating from acute to chronic toxicity in vitro. *Toxicology in Vitro*, *76*(May), 105206. https://doi.org/10.1016/j.tiv.2021.105206

Mahmudati, N., Wahyono, P., & Djunaedi, D. (2019). Antioxidant activity and total phenolic content of three varieties of Ginger (*Zingiber officinale*) in decoction and infusion extraction method. *6th International Conference on Mathematics, Science, and Education*. https://doi.org/10.1088/1742-6596/1567/2/022028

Malviya, S., Malviya, N., Joshi, A., Johariya, V., & Saxena, R. (2023). Medicinal plants and cancer chemoprevention. *Medicinal Plants and Cancer Chemoprevention*, *9*(7), 1–232. https://doi.org/10.1201/9781003251712

Mamari, H. H. A. (2022). Phenolic Compounds : Classification, Chemistry, and Updated Techniques of Analysis and Synthesis. In F. A. Badria & A. Surguchov (Eds.), *Phenolic Compounds - Chemistry, Synthesis, Diversity, Non-Conventional Industrial, Pharmaceutical and Therapeutic Applications*. IntechOpen. DOI: 10.5772/intechopen.98958

Mao, Q.-Q., Xu, X.-Y., Cao, S.-Y., Gan, R.-Y., Corke, H., Beta, T., & Li, H.-B. (2019). Bioactive compounds and bioactivities of ginger (*Zingiber officinale* Roscoe). *Foods*, *8*, 1–21. https://doi.org/10.3390/foods8060185

Mbaveng, A. T., Zhao, Q., & Kuete, V. (2014). Harmful and Protective Effects of Phenolic Compounds from African Medicinal Plants. In *Toxicological Survey of African Medicinal Plants*. Elsevier Inc. https://doi.org/10.1016/B978-0-12-800018-2.00020-0

Menyiy, N. E., Mrabti, H. N., Omari, N. E., Bakili, A. E. I., Bakrim, S., Mekkaoui, M., Balahbib, A., Amiri-ardekani, E., Ullah, R., Alqahtani, A. S., Shahat, A. A., & Bouyahya, A. (2022). Medicinal uses, phytochemistry, pharmacology, and toxicology of *Mentha spicata*. *Hindawi Evidence-Based Complementary and Alternative Medicine*, *2022*. https://doi.org/10.1155/2022/7990508

Michel, J., Abd Rani, N. Z., & Husain, K. (2020). A review on the potential use of medicinal plants from asteraceae and lamiaceae plant family in cardiovascular diseases. *Frontiers in Pharmacology*, *11*(June), 1–26. https://doi.org/10.3389/fphar.2020.00852

Nagahama, N., Gastaldi, B., Clifford, M. N., Manifesto, M. M., & Fortunato, R. H. (2020). The influence of environmental variations on the phenolic compound profiles and antioxidant activity of two medicinal Patagonian valerians (*Valeriana carnosa* Sm. And *V. clarionifolia* Phil.). *AIMS Agriculture and Food*, *6*(1), 106–124. https://doi.org/10.3934/agrfood.2021007

Nisar, A. (2022). Medicinal Plants and Phenolic Compounds. In F. A. Badria & A. Surguchov (Eds.), *Phenolic Compounds – Chemistry, Synthesis, Diversity, Non-Conventional Industrial, Pharmaceutical and Therapeutic Applications*. IntechOpen. 10.5772/intechopen.94825

Nwafor, E., Lu, P., Zhang, Y., Liu, R., Peng, H., & Xing, B. (2022). Chlorogenic acid: Potential source of natural drugs for the therapeutics of fibrosis and cancer. *Translational Oncology*, *15*(1). https://doi.org/10.1016/j.tranon.2021.101294

Pereira, D. M., Valentão, P., Pereira, J. A., & Andrade, P. B. (2009). Phenolics: From chemistry to biology. *Molecules*, *14*, 2202–2211. https://doi.org/10.3390/molecules14062202

Pérez-Ochoa, M. L., Vera-Guzmán, A. M., Mondragón-Chaparro, D. M., Sandoval-Torres, S., Carrillo-Rodríguez, J. C., Hernández-Delgado, S., & Chávez-Servia, J. L. (2023). Plant and growth condition interactions on the phenolic compound contents and antioxidant activity in *Salvia circinata* cav., a medicinal sage. *Diversity*, *15*. https://doi.org/10.3390/d15050656

Radovanović, K., Gavarić, N., & Aćimović, M. (2023). Anti-inflammatory properties of plants from Serbian traditional medicine. *Life*, *13*(4), 1–17. https://doi.org/10.3390/life13040874

Rosa, L. A., De, Moreno-escamilla, J. O., Rodrigo-garcía, J., & Alvarez-parrilla, E. (2019). Phenolic Compounds. In *Postharvest Physiology and Biochemistry of Fruits and Vegetables*. Elsevier Inc. https://doi.org/10.1016/B978-0-12-813278-4.00012-9

Satheesh, N. (2015). Review on distribution, nutritional and medicinal values of *Casimiroa edulus* Llave—An underutilized fruit in Ethiopia. *American-Eurasian Journal of Agriculture and Environmental Sciences, 15*(January 2015), 1574–1583. DOI:10.5829/idosi.aejaes.2015.15.8.9585

Sharafan, M., Malinowska, M. A., Ekiert, H., Kwaśniak, B., Sikora, E., & Szopa, A. (2023). *Vitis vinifera* (vine grape) as a valuable cosmetic raw material. *Pharmaceutics, 15*(5), 1–21. https://doi.org/10.3390/pharmaceutics15051372

Silva, A., Wang, Y. C., Wang, V., & Chen, B. H. (2023). Analysis of bioactive compounds in cinnamon leaves and preparation of nanoemulsion and byproducts for improving Parkinson's disease in rats. *Frontiers in Nutrition, August*. https://doi.org/10.3389/fnut.2023.1229192

Singla, R. K., Dubey, S. K., Garg, R., Sharma, R. K., Fiorino, M., Ameen, S. M., Haddad, M. A., & Al-Hiary, M. (2019). Natural polyphenols: Chemical classification, definition of classes, subcategories, and structures. *Journal of AOAC International, 102*, 1397–1400. DOI: 10.5740/jaoacint.19-0133

Solar, A., Medic, A., Slatnar, A., Mikulic-petkovsek, M., Botta, R., Rovira, M., Sarraquigne, J., Silva, A. P., Veberic, R., Stampar, F., Hudina, M., & Bacchetta, L. (2022). The effects of the cultivar and environment on the phenolic contents of hazelnut kernels. *Plants, 1–15*. https://doi.org/10.3390/plants11223051

Sorrenti, V., Buro, I., Consoli, V., & Vanella, L. (2023). Recent advances in health benefits of bioactive compounds from food wastes and by-products: Biochemical aspects. *International Journal of Molecular Sciences, 24*. https://doi.org/10.3390/ijms24032019

Sun, W., & Shahrajabian, M. H. (2023). Therapeutic potential of phenolic compounds in medicinal plants — Natural health products for human health. *Molecules, 28*. https://doi.org/10.3390/molecules28041845

Syed, R. U., Moni, S. S., Break, M. K., Bin, Khojali, W., Jafar, A., Alshammari, M., Abdelsalam, M. D., Taymour, K., Alreshidi, S., Elhassan Taha, K. S. M., & Mohan, M. M. (2023). Broccoli: A multi-faceted vegetable for health: An in-depth review of its nutritional attributes, antimicrobial abilities, and anti-inflammatory properties. *Antibiotics, 12*(7), 1157. https://doi.org/10.3390/antibiotics12071157

Tungmunnithum, D., Thongboonyou, A., Pholboon, A., & Yangsabai, A. (2018). Flavonoids and other phenolic compounds from medicinal plants for pharmaceutical and medical aspects: An overview. *Medicines, 5*. https://doi.org/10.3390/medicines5030093

Vermerris, D. W., & Nicholson, R. (2009). *Phenolic Compound Biochemistry*. Springer Science+Business Media.

Vijayalakshmi, M., Meganathan, S., Surendhar, S. K., & Umamaheswari, A. (2024). Exploring the systematic anticancer mechanism in selected medicinal plants: A review. *Oncology Advances, 2*(3), 141–147. https://doi.org/10.14218/OnA.2024.00012

Vuran, N. E., Kartal, D., & Çelik, İ (2023). Antiproliferative activity of *Malus sylvestris* Miller against HepG2 cell line with their antioxidant properties and phenolic composition. *Turkish Journal of Biochemistry, 48*(3), 272–281. https://doi.org/10.1515/tjb-2022-0186

Xu, D. P., Li, Y., Meng, X., Zhou, T., Zhou, Y., Zheng, J., Zhang, J. J., & Li, H. B. (2017). Natural antioxidants in foods and medicinal plants: Extraction, assessment and resources. *International Journal of Molecular Sciences, 18*(1), 20–31. https://doi.org/10.3390/ijms18010096

Yatoo, M. I., Gopalakrishnan, A., Saxena, A., Parray, O. R., Tufani, N. A., Chakraborty, S., Tiwari, R., Dhama, K., & Iqbal, H. M. N. (2018). Anti-inflammatory drugs and herbs with special emphasis on herbal medicines for countering inflammatory diseases and disorders – A review. *Recent Patents on Inflammation & Allergy Drug Discovery, 12*(1), 39–58. https://doi.org/10.2174/1872213x12666180115153635

Yu, L., Chen, Z., & Hu, C. (2021). Anti-inflammatory effects of curcumin in the inflammatory diseases: Status, limitations and countermeasures. *Drug Design, Development and Therapy, 15*, 4503–4525. https://doi.org/10.2147/DDDT.S327378

Ziyatdinova, G., & Kalmykova, A. (2023). Electrochemical characterization of the antioxidant properties of medicinal plants and products: A review. *Molecules, 28*(5). https://doi.org/10.3390/molecules28052308

# 2 Chemical Diversity and Structural Characteristics of Phenolic Compounds

*Monica Butnariu*[*]

## 2.1 INTRODUCTION

Aromatic compounds, which include one or several OH (hydroxyl) substitutes connected to the carbon atoms of the aromatic ring, are called polyphenols. The deep study of indigenous phenolic substances is particularly advantageous. These substances, which have been used for centuries as active principles in vegetable organisms, are used in engineering as indigenous colorants for textile articles, in the seasoning of leather tanning, ink, etc. This cluster of indigenous substances – vegetable colorants, tannins, lignins – requires an intricated examination both from the point of view of their physiological features in the vegetable organisms, as well as their skeleton, biosynthesis, various quality, and use [Le Goff & Ouazzani, 2014].

Vegetable organisms' phenolic compounds present a motley (inhomogeneous) cluster of very diverse organic compounds according to their chemical skeleton. This structural diversity conditions the fact that indigenous compounds are divided into various substitutes of related compounds in their systematics, and as a result, the general and integral characteristic of this significant cluster of indigenous compounds are lost. That is why phenolic hydroxyl is the basic criterion by which these compounds can be systematically distinguished from other indigenous substances. In this cluster of aromatic substances with free or bound phenolic hydroxyl content, vegetable phenolic substances are present. These substances may include one, two, or more phenolic substitutes in the benzene ring. Depending on the number of phenolic dehydroxyls, they can be considered monohydroxy derivatives of benzene, dihydroxy derivatives – with two OH substitutes, usually in the ortho position (pyrocatechin type), less often in the para position (hydroquinone type) and meta (resorcinol type); from trihydroxy derivatives, compounds with hydroxyls in positions 1,2,3 (pyrogallol type) prevail, less often in positions 1,3,5 (Phloroglucinol type) and very rarely in positions 1,2,4 (hydroxyquinone type) [Sui et al., 2024].

The molecules of vegetable organisms' phenolic compounds frequently include the C-type carbon skeleton with two benzene rings, in which more than two hydroxy phenols are included. Therefore, many of them can be called polyhydroxyphenols if they are present as monomers; the polymers will be called phenolic compounds.

Vegetable organisms, unlike microorganisms and animals, are characterized by an extremely large diversity of metabolic processes, resulting from which substances with a very varied chemical skeleton are formed. Organic substances that are not straightforwardly mixed in the growth, evolution, and multiplication of vegetable organisms have been named specialized products. A comprehensive series of secondary compounds are biologically active substances that have a distinct technicality of activity on microorganisms, animals, and humans [Schober et al., 2023].

This property is the basis for the use of specialized products in medicine and pharmaceuticals.

Phenolic compounds are widely assigned in vegetable systems. There may be 4000 compounds containing multiple phenolic functionalities in vegetable organisms, and only a few of these compounds (vitamin E and vitamin C) are familiar and investigated. Vegetable organisms' tissues can include up to a minority of grams per kilogram.

* Corresponding author: monicabutnariu@yahoo.com

DOI: 10.1201/9781003527671-2

This category of compounds can act to compose an opponent advantage between concurrent categories, such as toxic or lethal.

Metabolites of vegetable organisms' origin can be indigenous models for the synthesis of chemical compounds useful to humans [Brezani et al., 2018]. An example in this regard is salicylic acid, isolated from the bark of willow (*Salix* L.) and other vegetable organisms' category.

On the basis of salicylic acid, the drug aspirin (acetylsalicylic acid) was created.

The active substances of medicinal vegetable organisms have a very complex skeleton. Although there have been notable successes in the synthesis of organic compounds, more than one-third of medicinal seasonings are currently obtained from vegetable organisms.

The biologically active substances present in medicinal vegetable organisms that exert a therapeutic effect on the animal and human body are, as a rule, secondary products of metabolism. The reactions and chemical compounds of the primary metabolism have a general character for all living organisms. At the same time, there are a large number of metabolic pathways, which lead to the formation of chemical compounds specific only to certain vegetable organisms' categories. The more chemical reactions are necessary for the formation of a secondary metabolite, the more limited its spread [Brenna et al., 2004].

Secondary metabolism is a particularity of differentiated cells and vegetable organisms' tissues. It is characteristic only of specialized organs and only in certain phases of vegetable organisms' development. Specialized products have a taxonomic specificity, in other words each metabolite is formed within a specific taxon (family, genus, category).

Specialized outcomes have the following general characteristics:

- it has a low molecular weight (with the exception of polyisoprenides rubber, gutta-percha);
- they are not present in every organism (some metabolites are widespread, for example, phenylpropanoids are found in almost all vegetable organisms);
- they are, as a rule, biologically active substances (SBA);
- it is synthesized from primary metabolites.

The fundamental cluster of secondary compounds are phenolic compounds conformable to Table 2.1.

It should be noted the existence of a connection between primary and secondary metabolism: as a rule, the precursors of specialized outcomes are formed within primary metabolism [Sun et al., 2024].

**TABLE 2.1**
**The Most Fundamental Categories of Phenolics Substances in Vegetable Organisms**

| Basic Skeleton | No. of Carbon Atoms | Categories of Substances |
|---|---|---|
| $C_6$ | 6 | Simple Phenols, Benzoquinone |
| $C_6$–$C_1$ | 7 | Phenolic Acid |
| $C_6$–$C_2$ | 8 | Acetophenones, Phenylacetic Acid |
| $C_6$–$C_3$ | 9 | Hydroxycinnamic Acid, Polypropylenes, Coumarins, Isocoumarins |
| $C_6$–$C_4$ | 10 | Naphthoquinone |
| $C_6$–$C_1$–$C_6$ | 13 | Xanthones |
| $C_6$–$C_2$–$C_6$ | 14 | Stilbenes, Anthraquinones |
| $C_6$–$C_3$–$C_6$ | 15 | Flavonoids, Isoflavonoids |
| $(C_6$–$C_3)_2$ | 18 | Lignans, Neolignans |
| $(C_6$–$C_3$–$C_6)_2$ | 30 | Biflavonoids |
| $(C_6$–$C_3)n$ | n | Lignins |
| $(C_6)n$ | | Catecholmelanins |
| $(C_6$–$C_3$–$C_6)n$ | | (Condensed Tannins) |

Polyphenolic substances are specialized products of vegetable organisms, constituting significant causes representative of the sensorial and beneficial attributes of vegetables, fruits, and miscellaneous vegetable organisms. Many of these phenolic substances are liable for the colors of leaves, fruits, and flowers. Being a large cluster of biologically active metabolites, they show various biological features. Polyphenolic substances act as antioxidants, phytoalexins, pollinator attractants, and protective factors against ultraviolet light, among others. All these biologically active qualities make these substances perform a significant contribution to the growth and multiplication of vegetable organisms and release substantial safeguards versus pests and pathogens, causing the color and sensory features of vegetables and fruits. Phenolic substances can be classified into distinct substitutes conformable to function on the number of phenolic cores/rings they include and on the basis of some structural elements that link these cores together. The general skeletons of the main classes of phenolic substances are flavonoids (anthocyanins, flavanols, flavanones, flavones, and flavonols), coumarins, stilbenes, phenolic acids, and lignans [Wang et al., 2019].

Acyclic phenolic compounds are organic molecules featuring a hydroxyl group attached to an aromatic (benzene) ring, but unlike many common phenolics (like flavonoids), they lack a ring structure in their side chain, often appearing as open-chain structures, like phenylpropanoids (e.g., caffeic acid, ferulic acid) or simple phenols with unbranched side chains, acting as powerful antioxidants and having diverse roles in plant defense, flavor, and human health (anti-inflammatory, anti-diabetic).

Acyclic phenolic compounds are genetically connected to acylglycerols or monoacylglycerols (monoglycerides), from which they are synthesized, resulting from reduction reactions. Widespread in the vegetable kingdom are cyclic phenolic compounds, respectively, cyclitols, which are synthesized from cyclohexane. The most significant examples of cyclitols are cyclic (6-OH) hexols, respectively, quercitols. The cyclic (4-OH) tetrols and analogs are less ordinary in the vegetable kingdom. A series of cyclitol analogs is found in vegetable organisms.

Cyclitols are classified conformable to the number of (OH) hydroxyl substitutes in the framework into tetracyclitols, pentacyclitols, and hexacyclitols.

Except for tetracycline, the only polyphenolic compound with a double bond, these compounds (cyclitols) contain only single bonds, all of which are saturated.

The majority of cyclitols can be converted by opening the ring into analogues of mono-carbohydrates, and conversely, cyclization can lead to cyclitols from them.

The cyclitols are also connected to phenolic substances, in which they can be transformed by aromatization, and conversely, from phenolic substances, cyclitols can be synthesized by catalytic hydrogenation [Merchant & Richter, 2011].

## 2.2 PHENOLIC COMPOUNDS

Phenolic substances are vegetable organic substances that include an aromatic nucleus (benzene) in the framework, to which one or more hydroxyl substitutes are attached to the carbon atoms. The simplest representatives of this category phenol – $C_6H_5OH$ (Figure 2.1), hydroquinone and pyrocatechin – $C_6H_6O_2$, pyrogallol – $C_6H_6O_3$ and phloroglucinol – $C_6H_8OH_3 + 2H_2O$ (Figure 2.2) differ from each other by the number of hydroxyl substitutes. Phenol $C_6H_5OH$ is a colorless, crystalline substance. In the presence of oxygen, phenol oxidizes and acquires a pink color. It is soluble in

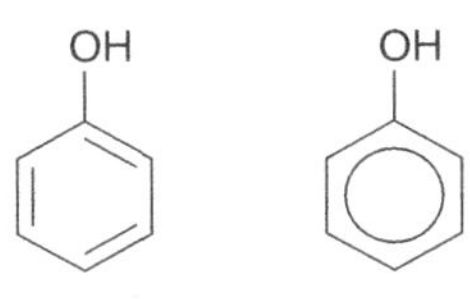


**FIGURE 2.1** The chemical structures of the Phenolic substances; organic substances that include an aromatic nucleus (benzene) in the framework, to which one or more hydroxyl substitutes are attached to the carbon atoms.

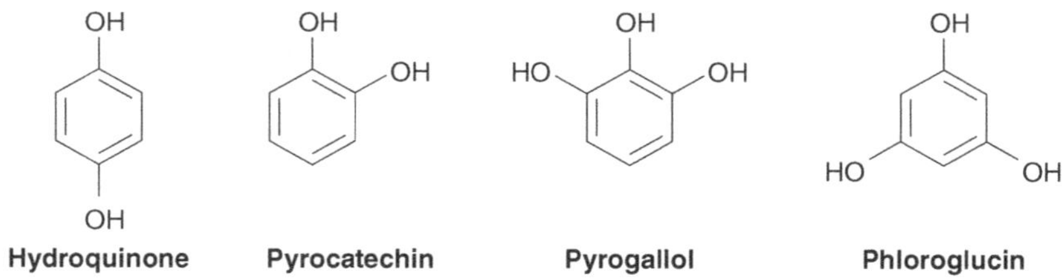


**FIGURE 2.2** The chemical structures of Hydroquinone, Pyrocatechin, Pyrogallol and Phloroglucin.

water, and at a temperature greater than 66°C, it mixes with water. It is toxic, burns the skin, and is an antiseptic.

Phenol represents the end of the side chain of the amino acid tyrosine and is part of protein frameworks. Phenols represent polar substances (dipoles) [Girón-Elola et al., 2023].

The benzene ring is the negatively charged dipole component, and the –OH cluster possesses positive charge. Phenols are weakly acidic substances. Phenol and its analogs are rarely found in the free state. Thus, phenol is found in pine cones and needles, pyrocatechol in onion peels and grapefruit fruits, and phloroglucinol in sequoia cones.

Under indigenous conditions, phenolic substances are formed following the oxidation and chemical transformation of organic substances and represent intermediate products of cellular metabolism.

Phenolic substances are classified into two large classes:

- monomeric phenolic substances;
- polymeric phenolic substances [Tinikul et al., 2018].

## 2.3 MONOMERIC PHENOLIC SUBSTANCES

Monomer phenolic substances, depending on the C (carbon) skeleton (the number of benzene cores and the number of carbon atoms in the carbon skeleton), are classified into three main classes (Table 2.2).

The three categories of monomeric phenolic substances, depending on the C (carbon) skeleton, are mentioned in the following:

1. C6–C1 substances;
2. C6–C3 substances;
3. C6–C3–C6 substances.

### 2.3.1 The Category of C6–C1 Substances

The category of C6–C1 substances includes phenolic (oxybenzoic) acids and phenolic aldehydes [Roche et al., 2017].

**TABLE 2.2**
**The Main Substitutes of Monomeric Phenolic Substances**

| Category | The Carbon Skeleton | No. Benzene Rings | No. C Atoms |
|---|---|---|---|
| Phenolic acids, phenolic aldehydes | C6–C1 | 1 | 7 |
| Hydroxycinnamic acids, coumarins | C6–C3 | 1 | 9 |
| Flavonoids | C6–C3–C6 | 2 | 15 |

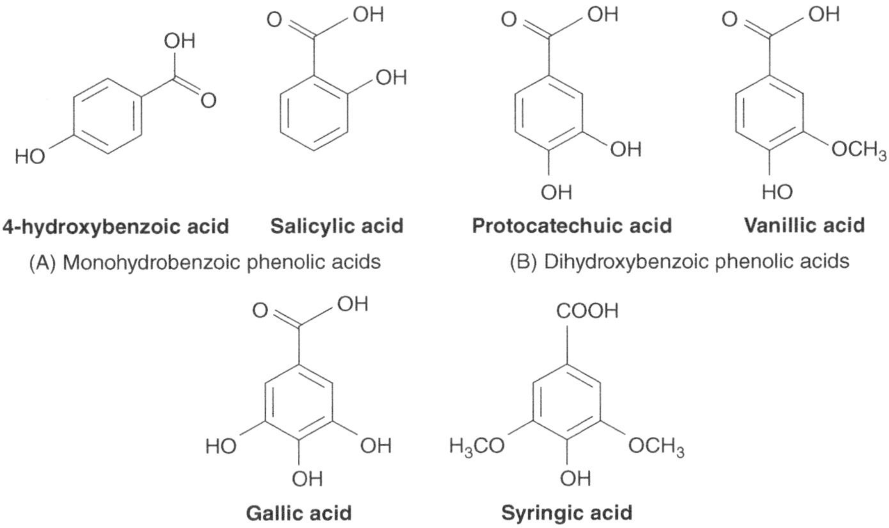


**FIGURE 2.3** Chemical Diversity and Structural Characteristics of Phenolic Compounds.

#### 2.3.1.1 Phenolic Acids

Phenolic acids include 4-hydroxybenzoic acid ($C_7H_6O_3$), salicylic acid ($C_6H_4(OH)COOH$), protocatechuic acid (PCA) ($C_7H_6O_4$), vanillic acid ($C_8H_8O_4$), gallic acid ($C_7H_6O_5$), and sirenic acid ($C_9H_{10}O_5$) (Figure 2.3.A and B).

Oxybenzoic acids are widely found in vegetable organisms. They are generally attendant in bound conformation and released by hydrolysis. 4-Hydroxybenzoic acid is found in coconuts (*Cocos nucifera*) and acai fruits (*Euterpe oleracea*), has been identified in wine, and is present in vanilla.

The $C_6H_4(OH)COOH$ (salicylic acid) [SA] is a phenolic phytohormone that contributes to growth, development, photosynthesis, transpiration, absorption, and ion transport in vegetable organisms. PCA extracted from hibiscus (*Hibiscus sabdariffa*) is a substance with an antioxidant and anti-inflammatory effect. The $C_8H_8O_4$ (vanillic acid) is a phenolic substance used in the food consumed as a flavoring factor. Great content of vanillic acid was identified in the roots of *Angelica sinensis*, a herbaceous vegetable organisms originating in China, used in traditional Chinese medicine.

Sirenic acid is found in acai fruits and in ether conformation, together with vanillic acid, are components of lignin in woody vegetable organisms [Wang et al., 2024]. Gallic acid was isolated from oak bark, tea leaves, sweet vinegar, or sumac (*Rhus aromatica*). It is found in vegetable organisms in a free state and in the conformation of a dimer – metadigallic acid (Figure 2.4).

The ester bond formed by the hydroxyl cluster of a phenolic acid and the carboxyl cluster of another phenolic acid is called a depsidic bond. Substances with a depsidic bond are called depsides, components of hydrolyzable tannins [Chen, Wei, & Matsuda, 2022].

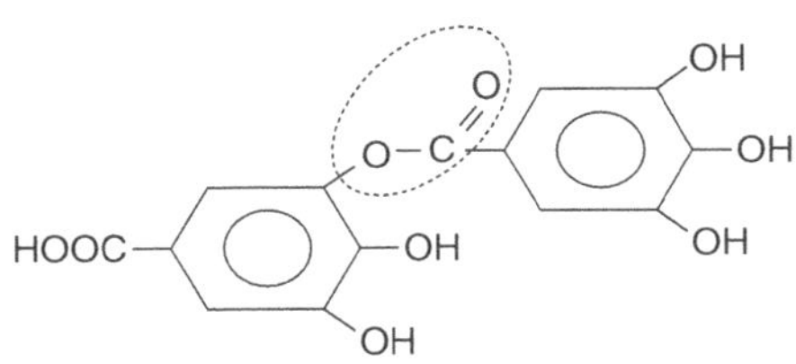


**FIGURE 2.4** The chemical structure of Gallic acid; that in a free state are in the conformation of a dimer – metadigallic acid.

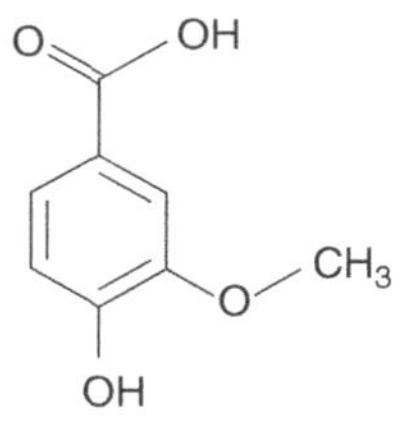

Vanillin

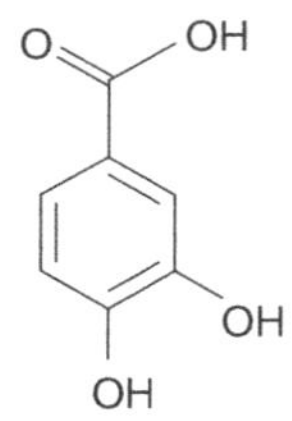

Protocatechin aldehyde

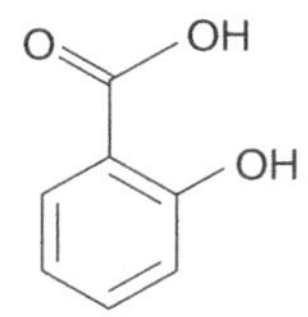

Salicylic aldehyde

**FIGURE 2.5** The chemical structures of Vanillin, Protocatechin aldehyde and Salicylic aldehyde.

#### 2.3.1.2 Phenolic Aldehydes

Phenolic aldehydes include vanillin ($C_8H_8O_3$), protocatechuic aldehyde ($C_7H_6O_3$), and salicylic aldehyde ($C_7H_6O_2$) (Figure 2.5).

The best-known phenolic aldehyde is vanillin extracted from vanilla (*Vanilla planifolia*), a vegetable organism with a pleasant smell from the category of spices, originally from Mexico. It is used in the food industry, being an essential ingredient for many recipes of cakes, puddings, cakes, chocolates, drinks, sauces, and pastries. Ripe grape brandies have a pleasant, expressive bouquet with hints of vanilla. In the maturation process of grape brandies, vanillin is synthetized because of the oxidation of coniferyl alcohol included in the oak wood of the barrels [Buitimea-Cantúa, Welti-Chanes, & Escobedo-Avellaneda, 2022]. Protocatechin aldehyde is released in the wine when the cork is made from the wood of the cork oak (*Quercus suber*). Salicylic aldehyde is a characteristic component of the aroma of the buckwheat vegetable organisms (*Fagopyrum esculentum*). It is used in perfumery [Cerezo et al., 2014].

### 2.3.2 The Category of C6–C3 Substances

The category of C6–C3 substances includes hydroxycinnamic and coumarin acids.

#### 2.3.2.1 Hydroxycinnamic (Coumaric) Acids

The hydroxycinnamic (coumaric) acids are a category of aromatic or phenylpropanoid acids; hydroxylic analogues of cinnamic acid are present in vegetables both in a free state and in a bound state. The main representatives of this category are caffeic acid, coumaric acid, sinapic acid, and ferulic acid (Figure 2.6).

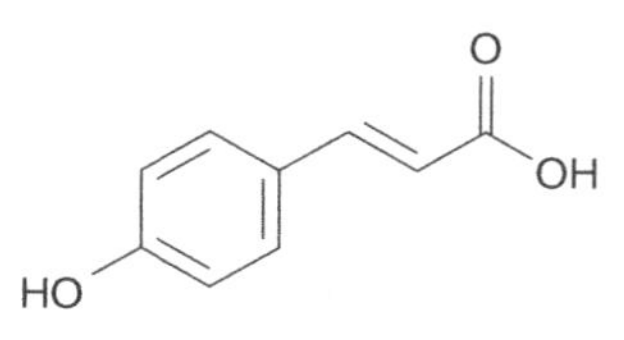

p-coumaric acid

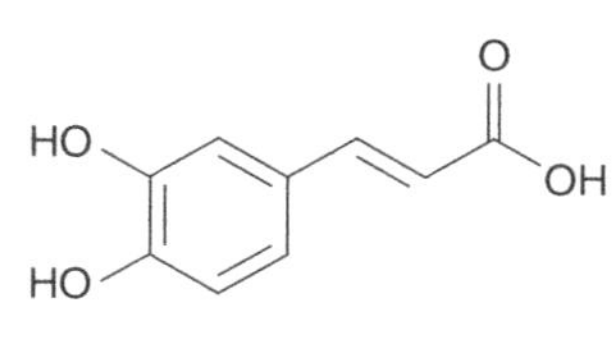

Caffeic acid

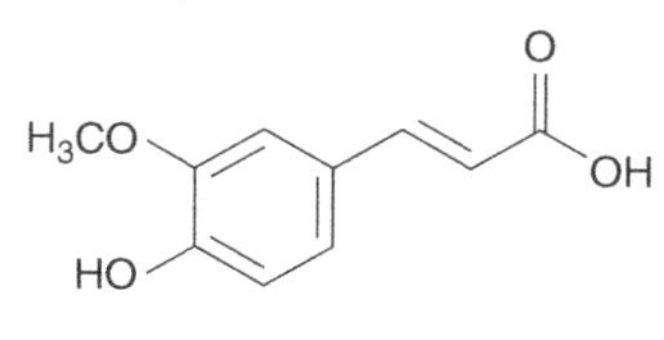

Ferulic acid

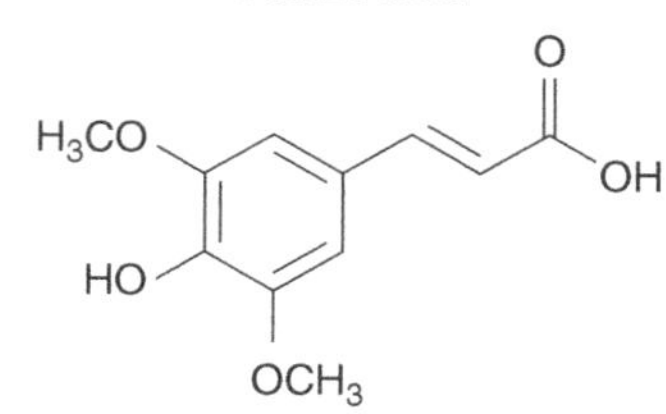

Sinapic acid

**FIGURE 2.6** The chemical structures of caffeic acid, coumaric acid, sinapic acid, and ferulic acid.

p-Coumaric acid ($C_9H_8O_3$) is found in tomatoes, carrots, garlic, and peanuts. It has been identified in wine, vinegar, and barley grains.

The $C_9H_8O_3$ in pollen is a constituent of bee honey [Li et al., 2020].

Caffeic acid ($C_9H_8O_4$) is a substance isolated from coffee beans and barley, from argan oil produced from the seeds of argan fruit (*Argania spinosa* L.). Ferulic acid ($C_{10}H_{10}O_4$) is found in the seeds of coffee, apples, peanuts, and oranges, as well as in the shells of rice grains, wheat, oats, and barley. Ferulic acid, like many other indigenous phenolic substances, has antibacterial and antioxidant quality. Sinapic acid ($C_{11}H_{12}O_5$) can be a conformation dimer with itself and with ferulic acid in the cell walls of cereals. It can be isolated from wine and vinegar [Li et al., 2021].

### 2.3.2.2 Coumarins

The coumarins are a category of indigenous aromatic organic substances. The main element of coumarins is coumarin ($C_9H_6O_2$ – Figure 2.7).

Coumarin ($C_9H_6O_2$ – Figure 2.7) is the simplest compound in that category, with a pleasant smell of freshly cut hay.

Pure coumarin and sulfine flowers (*Melilotus officinalis*) are used as flavorings in the production of some varieties of tobacco and in perfumery [Genovese et al., 2019].

Esculetin ($C_9H_6O_4$ – Figure 2.7), a hydroxylated coumarin derivative, is used in medicine to strengthen capillaries.

Bergaptol ($C_{11}H_6O_4$ – Figure 2.7) is a furanocoumarin included in bergamot oil extracted from bergamot fruits (*Citrus bergamia*). It is used in the food industry to flavor tea.

They are indigenous (natural) substances, whose chemical skeleton is structured on coumarin or isocoumarin. These substances have been identified in many vegetable organisms from the *Umbelliferae* groups: bitter gourd, angelica (*Angelica archangelica*), oranges, stag's-horn clubmoss, bear's foot, parsnip, chamomile, etc.

These substances (coumarins) are known for their anti-inflammatory, antiedema, and anti-cancer properties. The perfume industry uses coumarins for their characteristic sweet smell. Coumarins are also used in essences and in flavoring tobaccos.

Coumarins are used in medicine, representing compounds for the synthesis of anticoagulant substances (e.g., warfarin). In vegetable organisms, these substances can be determined in fruits, roots, seeds, leaves, flowers, and stems, even though, most of the time, the flowers, respectively, the fruits, contain the largest quantities.

Their contribution to vegetable organisms is mainly in the sphere of defense/protection, considering their antimicrobial and germination inhibitory quality.

The skeleton of a coumarin can be simply certified by 7-hydroxy-coumarin [$C_9H_6O_3$].

Other vegetable organisms like coumarins comprise psoralen (linear furanocoumarin), angelicin (angular furanocoumarin), seseline (pyranocoumarin), and 4-hydroxycoumarin (pyrone-substituted coumarin) [Luo et al., 2025].

*Capsicum annuum* (pepper) includes capsaicin, a strong insecticide (Figure 2.8).

Parsley (*Petroselinum sativum*) includes myristicin, α-phellandrene, α-terpinene, terpinolene, and p-cineole.

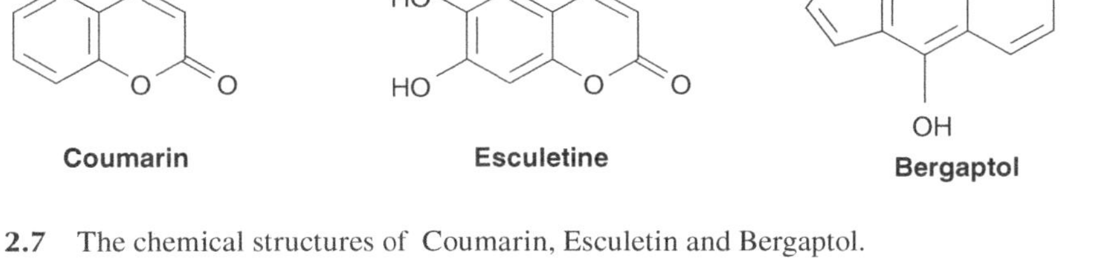


**FIGURE 2.7** The chemical structures of Coumarin, Esculetin and Bergaptol.

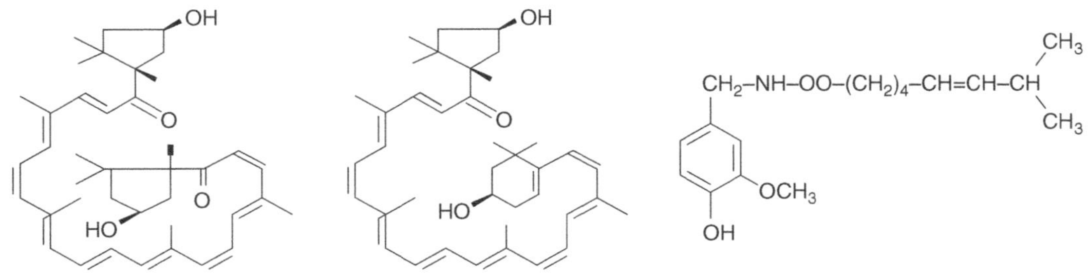


Capsorubin Capsanthin Capsaicin

**FIGURE 2.8** The chemical structures of Capsorubin Capsanthin and Capsaicin.

Comparatively to the indigenous outcomes produced via flavonoids, cinnamoyl-CoA or acetate analogues (malonyl-CoA), condensation answers/reactions occur in certain vegetable organism's category through which the adequate stilbenes, styrylpyrones, and arylpyrones are formed.

The stilbenes are attendant in bryophytes, gymnosperms, pteridophytes, and angiosperms, and over 300 distinct substances are acquainted.

Stilbenes make a significant contribution to vegetable organisms, especially in defense/protection, pharmacology, and human health.

Vegetable organisms are implicated in both constitutive and inducible protection/defense mechanisms.

The stilbenes are compounds with strong antibacterial characteristics and antifungal repercussions. They inhibit fungal growth or spore germination, cause sleepiness, and inhibit the growth of vegetable organisms.

Certain stilbenes, furthermore to features toxic to insects or other diverse organisms, have mammalian antifeedant and nematocidal qualities.

Related to their pharmacological characteristics, some stilbenes have properties that contribute to the prevention of tumor formation (e.g., resveratrol in red wine).

The biochemical, chemical, or molecular presentation of how vegetable organisms produce different metabolic substances is indispensable for understanding biodiversity and vegetable organisms' lives.

This presentation has economic involvements, bidding new possibilities for the business-like improvement of commercial vegetable organisms or the imprinting of special traits, which can establish advantages [Kaur et al., 2024].

### 2.3.3 The Category of C6–C3–C6 Substances

The category of C6–C3–C6 substances is characterized by great diversity. More than 5000 flavonoids have been separated and investigated from distinct vegetable organisms.

The substances in this category are also called flavonoids (from the Latin flavus – yellow).

Flavonoids are substances derived from flavone ($C_{15}H_{10}O_2$ – Figure 2.9). They have a skeleton with 15 carbon atoms, two benzene cores (A and B), and a heterocyclic or (pyranic C) ring.

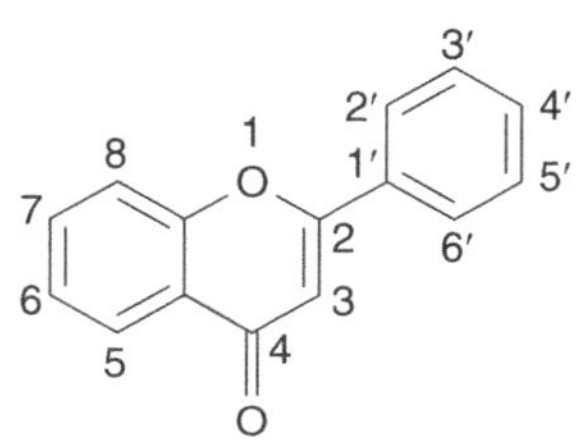


**FIGURE 2.9** The chemical structure of Flavonoids, substances derived from flavone.

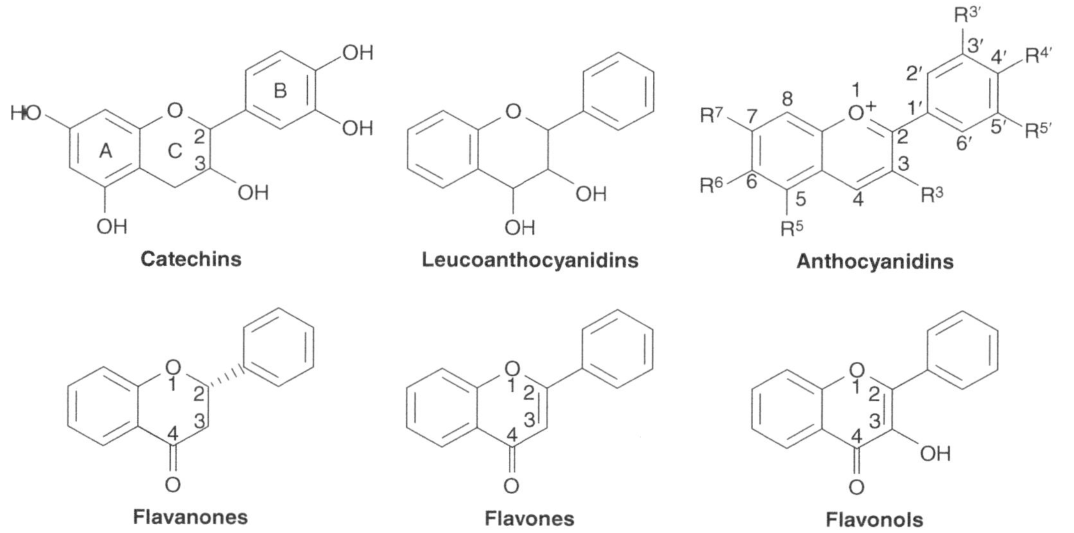


**FIGURE 2.10** The chemical structure of catechins, leucoanthocyanidins (flavan-3,4-diol), anthocyanidins and anthocyanins, flavanones, flavones, and flavanols (3-hydroxyflavones).

The pyranic heterocycle condenses with a benzene ring (A) and associates with another benzene ring (B).

Depending on the degree of oxidation or reduction of the heterocyclic segment, flavonoids are classified into several substitutes: catechins, leucoanthocyanidins (flavan-3,4-diol), anthocyanidins and anthocyanins, flavanones, flavones, and flavanols (3-hydroxyflavones) (Figure 2.10).

Catechins are the most reduced conformation of flavonoids.

The name catechin comes from catechu, an extract obtained from the catechu tree (*Acacia catechu*) used as a food additive.

The catechin framework has two benzene cores (A and B) and a dihydropyranic heterocycle (C) with a hydroxyl (–OH) hydroxyl cluster at the carbon atom in position 3 [Yang et al., 2013].

Catechins are colorless, crystalline substances soluble in alcohol and water, and they have an astringent taste.

It oxidizes easily when heated by the activity of sunlight and enzymes in a basic environment; the browning of vegetables and fruits during their mechanical and thermal treatment is caused by the oxidation of catechins.

During the polymerization of catechins, condensed tanning substances are formed.

Catechins have the following qualities:

- are antimicrobials;
- contribute to a more efficient use of ascorbic acid by the body;
- are biologically active substances;
- ensure elasticity and mechanical resistance of capillaries and regulate their permeability;
- they are antioxidant and antitumor substances; they defend the body from the action of free radicals;
- strengthen the immune system.

Flavonoids have feverish meaningful advantage because of their antioxidant or biochemical repercussion with potentially beneficial repercussion on human health.

Most of the scientific results obtained in the laboratory have demonstrated that flavonoids have several biological repercussions such as free radical scavenging action, anti-inflammatory action, antimutagenic action, anticancer, anti-HIV, antiallergic, anti-platelet, and antioxidant action.

Flavonoids are also very unstable substances, subject to numerous chemical or enzymatic ripostes during post-harvest depositing or processing, adding to the complexity of the polyphenolic composition [Goyal et al., 2023].

The skeleton of flavonoids includes a skeleton of 15 carbon atoms (C6–C3–C6), consisting of two aromatic cores and a heterocycle with oxygen.

Flavonoids are clustered into subclasses, structured on their chemical skeleton; the best-known classifications are flavonols, flavonoids, flavones, isoflavones, flavanones, anthocyanins, and chalcones.

Among these substances, flavanols, in addition to condensed tannins (proanthocyanidins), are one of the most widespread substitutes in vegetable organisms, including phenolic substances.

They are recognized to have a strong free radical elimination action, antioxidant action, and anti-tumor-promoting repercussion.

Considering the significance of proanthocyanidins in the organoleptic quality of food outcomes, as well as their significance in human health, a better comprehension of their skeleton, chemical quality, and generation workings is indispensable for predicting and controlling the quality of food outcomes.

Proanthocyanidones are generically blends of oligomers and/or polymers with flavan-3-ol units, connected either by C–C (carbon–carbon) obligation (bonds) or by C–O (carbon–oxygen) obligation [Ma et al., 2017].

The most common flavan-3-ol substitutes, implicated in the formation of proanthocyanidins, are analogues of catechin 1a, epicatechin 1b, gallocatechin 1c, or epigallocatechin 1d (Figure 2.11).

The synthesis of proanthocyanidins 2 is generically accomplished by connection the C4 atom of an electrophilic flavanyl cluster to C-6/C-8 of a nucleophilic flavanyl cluster.

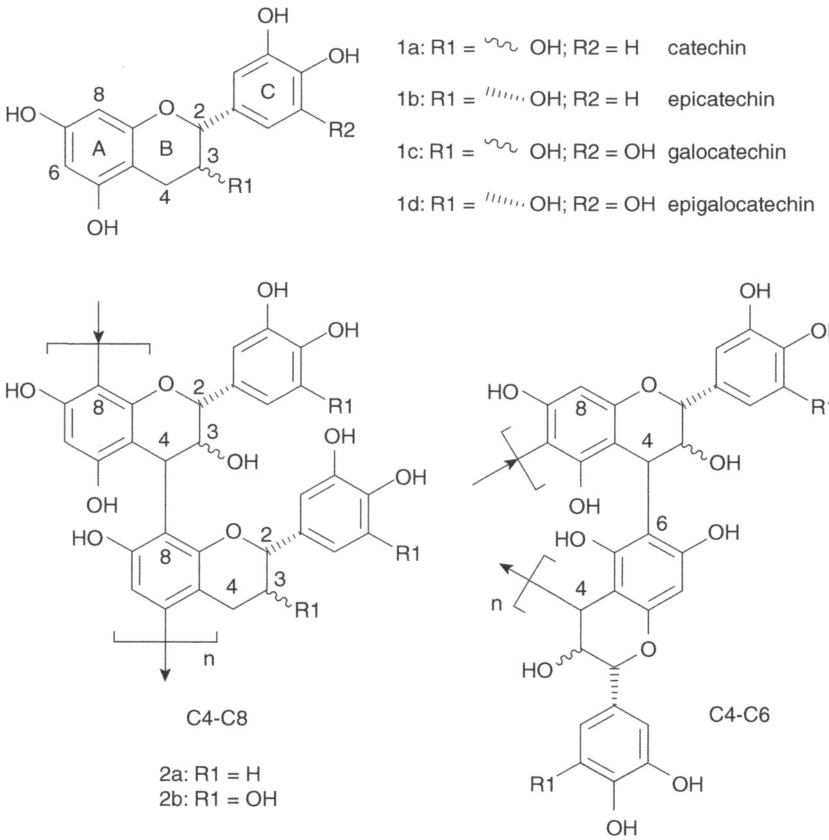


**FIGURE 2.11** The chemical structures of flavan-3-ol substitutes, implicated in the formation of proanthocyanidins, are analogues of catechin 1a, epicatechin 1b, gallocatechin 1c, or epigallocatechin 1d.

The nucleophilicity of the aromatic ring A can perform a conclusive contribution in these connection riposte, during which a new bond is formed. The stability of proanthocyanidins is mainly dependent on that of this newly created bond, which is sensitive to acidity and alkalinity conditions and can also be, a priori, converted by the type of flavanol unit.

Also, the reactivity of the electrophilic/nucleophilic category implicated in the connection riposte can be converted by the type of flavanol cluster.

More to the modification that involves the chemical riposte of flavonoids, the enzymatic oxidation of polyphenols is very implicated in many biological workings [Ferreira, Marais, & Slade, 2005].

Since the radicals formed during these oxidation processes are relatively stable, they can intervene in several connection riposte.

These connection ripostes are of major advantage in the biosynthesis of tannin, lignin, and melanin.

These ripostes are also liable for the evolution of the polyphenolic profile during the seasoning of foods and drinks structured on vegetable organisms and fruits. Indeed, when we talk about the oxidation of polyphenols that induce browning of food outcomes, we usually refer to the negative influence.

Some enzymatic browning ripostes are nevertheless very beneficial for the general acceptability of food outcomes. Black and green tea are structured on enzymatic browning ripostes to develop color and aroma.

Some data refer to the collaboration of enzymatic browning to color development during coffee processing. Similarly, color development in cocoa is facilitated by the oxidation action of polyphenols during fermentation.

Polyphenol oxidases are also responsible for the distinguishing golden-brown color of dried fruits such as grapes, plums, and figs, and bleaching is generally necessary to inactivate the enzymes after color development to minimize discoloration.

The catechins are found in large quantities in the fruits of apples, pears, peaches, apricots, raspberries, grapes, quinces, citrus fruits, forest apples, cocoa beans, in the bark of pine, fir, in the shoots and leaves of tea, and in the leaves of parsley [Bahramsoltani et al., 2024].

The shoots of the tea vegetable organisms (*Camelia sinensis*) include up to 30% of catechins in the dry mass.

## 2.4 POLYMERIC PHENOLIC SUBSTANCES

Phenolic polymer substances are heterogeneous organic substances derived from polyphenols that are divided into three categories (Table 2.3):

1. tannins (tannins);
2. lignin;
3. melanin.

**TABLE 2.3**
**Phenolic Polymers**

| | | | |
|---|---|---|---|
| The base unit | 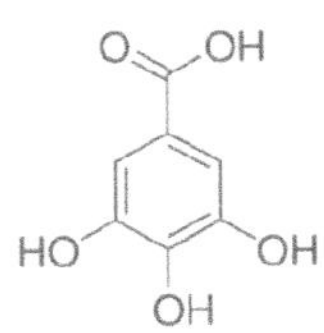 Gallic acid | 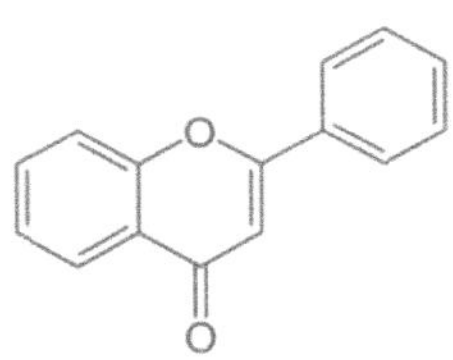 Flavonol | 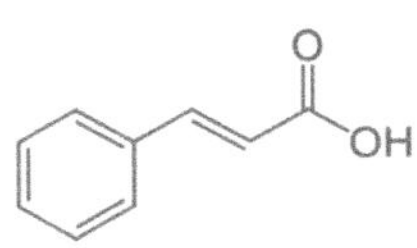 Cinnamic acid |
| Polymer category | Hydrolyzable tannins | Flavonoids, condensed tannins | Lignin |

Conformable to a recent definition accepted in the specialized literature White–Bate-Smith–Swain–Haslam (WBSSH), the cluster of polyphenols includes substances with the following features:

- in general, they are substances moderately soluble in water;
- have a molecular mass of 500–4000 Da;
- possess more than 12 hydroxyl substitutes; have in their composition 5–7 benzene cores [Luca et al., 2020].

### 2.4.1 Tannins

Tannins (from Old German tanna-oak, fir) are heterogeneous organic substances derived from polyphenols, which can form complexes with proteins, amino acids, polysaccharides, and alkaloids.

Tannins are vegetable substances with a complex chemical skeleton, polyphenolic (which includes many phenolic hydroxyl (–OH) substitutes, but also carboxylic substitutes), with an astringent taste, capable of precipitating the proteins from the raw skin, proteins with which they form insoluble, putrefactive, impermeable precipitates (tanned leather).

Tannins belong together with alkaloids in various specialized outcomes that accumulate in vegetable organisms and their organs.

The term tannins appeared in the context of the use of tanning substances from the wood of trees to transform raw animal skin into tanned leather.

At the base of the leather tanning process is the property of tannins to interact with collagen (the fibrillar protein of the skin) and to conformation a stable intercalated skeleton.

Tannins have a molecular mass that varies in a range of 500–3000 D (esters of gallic acid).

In terms of physical-chemical quality, tannins are in most cases amorphous, white or yellowish substances with an astringent or bitter taste, soluble in warm water, ethanol, and glycerin and insoluble in organic solvents; in water, they conform to colloidal solutions [Jing et al., 2022].

Tannins are weakly acidic and have a reducing quality; they oxidize easily and brown.

This phenomenon is observed on the surface of a freshly cut apple or a mechanically damaged tea leaf.

Tannins are very widespread substances in the vegetable plants. They are found in the bark of trees, in leaves, fruits, and roots. Tannins determine the food and taste value of some food outcomes (wine, tea, coffee, and cocoa).

From a chemical point of view, tannins are esters (R–O–R) of gallic acid or other phenolic acids with glucides/carbohydrates or condensation outcomes of 3-hydroxyflavones (catechins).

Tannins can be hydrolyzable, in which case they decompose by hydrolysis into a carbohydrate and phenolic acid or condensed (catechin analogues) when they do not include carbohydrates and can only be decomposed by alkaline melting.

Tannins (tanning substances) are a variety of organic substances with a heterogeneous polyphenolic skeleton. In vegetable organisms, tannins accumulate in tissues, fruits, and leaves and have exceptional physiological action.

The vegetable organisms richest in tannins are those from the families *Amentaceae*, *Fabaceae*, *Gerantaceae*, *Legumes*, *Myrtaceae*, *Rosaceae*, and *Rubiaceae*.

The tannins are polyphenolic substances (analogues of 3-hydroxy-flavan or polyesters of gallic acid) soluble in water, with an astringent taste, which attendant ripostes distinguishing of phenols and which precipitate with alkaloids and proteins and establish impermeable and putrefactive combinations (features of the leather tanning process).

Tannins are widespread in nature.

They are found in very small quantities in most vegetable organisms, and in large quantities in the bark, wood, and leaves of oak, in the bark of alder, spruce, poplar, walnut, blueberry, crowns, and mangroves, and in the fruits of some varieties, such as *Terminalia chebula* and *Caesalpinia brevifolia*.

They are astringent substances and used as antidiarrheals [Chandrasekhar et al., 2018].

The ellagic ones act on immune workings (stimulate phagocytosis), an effect also evident for epicatechol gallate and epigallocatechol gallate [Uchida et al., 1995].

Physiologically the tannins are envisaging vegetable organisms defense factors against bacterial and viral infections (they precipitate the proteins of microorganisms), antioxidants (they are strong reducing factors), and hydrogen carriers (participate in cellular redox systems); some authors consider tannins to be reserve substances for vegetable organisms as they disappear from seeds during germination.

There are no vegetable organisms that do not include tannins.

Tannins, vegetable substances with a complex chemical skeleton (comprises many phenolic hydroxyls (–OH) hydroxyl substitutes, but also carboxylic substitutes), capable of precipitating the proteins from the raw skin (with which they conformation insoluble, putrefactive, impermeable precipitates – tanned leather).

Tannins are part of the variety of specialized outcomes that accumulate in vegetable organisms and their organs [Windley et al., 2022].

In industry, tannins are used in leather tanning, ink outcome, as a food additive to give some outcomes an astringent taste, and as food coloring E181. Conformable to the hydrolysis riposte and chemical skeleton, tannins are classified into two large substitutes: hydrolysable tannins and condensed tannins (catechin tannins) [Barbehenn & Peter Constabel, 2011].

***Hydrolyzable tannins*** are indigenous esters of glucose with gallic acid (polyphenolic monomer) or condensation outcomes of gallic acid (m-digallic acid, ellagic acid, etc.).

They include from 3 to 12 gallic acid residues in their framework (Table 2.2).

Hydrolysable tannins can be extracted from edible chestnut (*Castanea sativa*), oak bark (*Quercus robur*, *Quercus petraea*, and *Quercus alba*), bird of paradise shrub (*Caesalpinia spinosa*), haritaki (*T. chebula*), and sumac (*Rhus coriaria*).

They are also obtained in large quantities from crowned galls that conform on the leaves of the following variety: *Quercus infectoria*, *Rhus semialata*, *Andricus kollari*, etc., or from rhizome cones that conforma] from the transformation of leaf buds of the variety *Quercus lusitanica* var. infectious disease.

Depending on the chemical skeleton, gallic tannins (gallotannins) and ellagic tannins are distinguished. Gallic tannins are polymers formed from gallic acid, esterified, and connected to the hydroxyl (–OH) hydroxyl cluster of glucose.

Gallotannins are easily hydrolyzed into glucose and gallic acid in an acidic environment or under the action of specific tannase enzymes, produced by the fungi *Aspergillus niger* and *Penicillium glaucum*. The main component of gallotannin is pentagalloylglucose, to which gallic acid residues are associated through depsidic association. Ellagic tannins differ from gallic tannins in that, upon hydrolysis, ellagic acid is formed, whose residues are associated with pentagalloylglucose through C–C association [Buyse et al., 2024].

***Condensed tannins*** are polymers formed by the condensation of flavans – catechins or leucoanthocyanins.

Condensed tannin frameworks do not include carbohydrate residues and can be linear (with 4→8 bonds/association) or branched (with 4→6 association) (Figure 2.12).

The trading emergence of condensed tannins is quebracho wood (*Schinopsis lorentzii*), pine bark (*Pinus sylvestris*), spruce (*Picea abies*), acacia (*Acacia mollissima*), and grapevine seeds (*Vitis vinifera*).

***Chinese tannin*** (Figure 2.13) is obtained from the galls (rice balls) [Wang et al., 2020].

***Turkish tannin*** is obtained from the galls on the branches of the *Q. infectoria* oak.

The chebulinic acid was the first crystallized gallotamine obtained from the fruits of *T. chebula*. Subject to hydrolysis, chebulinic acid splits into four frameworks of gallic acid, one framework of D-glucose, and one framework of chebulic acid. A very similar skeleton to chebulinic acid also composes chebulagic acid. Through total hydrolysis, one mole of D-glucose, gallic, ellagic, and

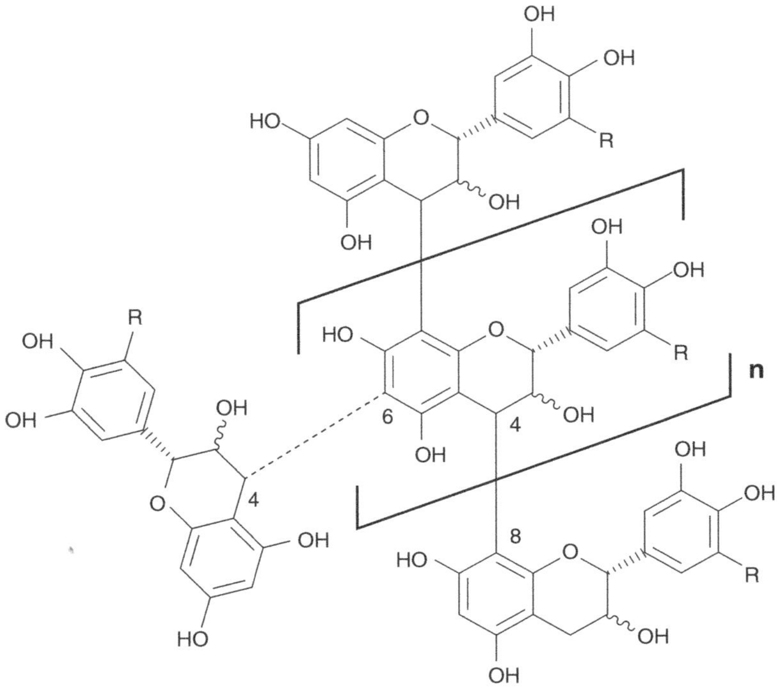


FIGURE 2.12 The chemical structure of Condensed tannin.

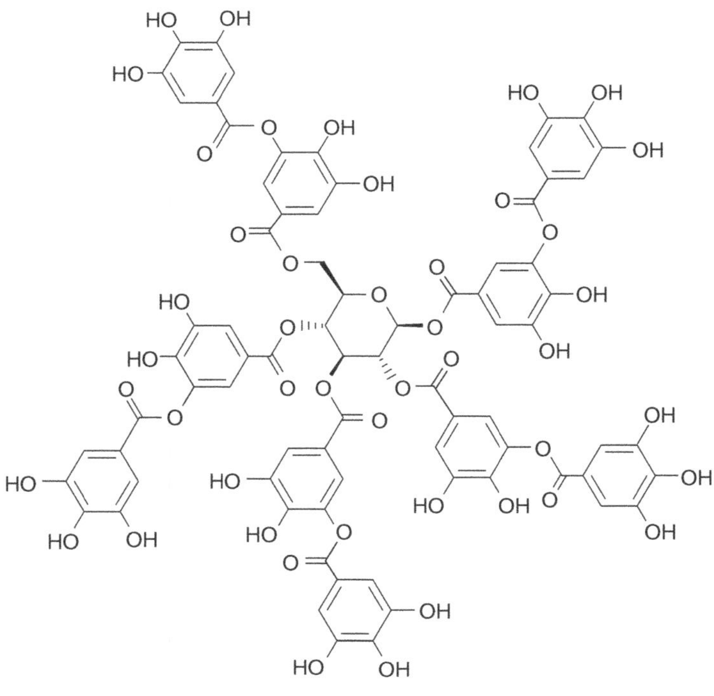


FIGURE 2.13 The chemical structure of Chinese tannin.

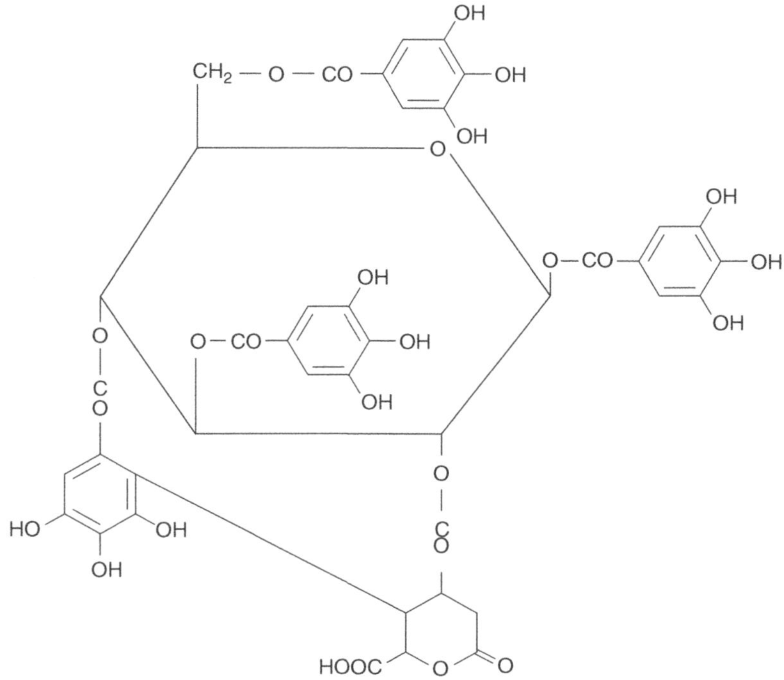


FIGURE 2.14 The chemical structure of chebulinic acid.

chebulic acid is obtained. Unlike chebulinic acid (Figure 2.14), chebulagic acid includes a hexahydroxydiphenic acid residue at carbon atoms C3 and C6 of D-glucose instead of two gallic acid residues [Niemetz & Gross, 2005].

***Elagotanins***. In the dried pods of *Caesalpinia cariora* and *Caesalpinia tinctoria*, which grow in Central America and India, there is an elagotannin, called corilagin (Figure 2.15).

Through acid hydrolysis, corilagin breaks down into one framework each of D-glucose, gallic acid, and ellagic acid.

Corilagin includes hexahydroxydiphenic acid bound to D-glucose, which, upon acid hydrolysis of corilagin, transforms immediately after release into ellagic acid [Mao et al., 2016].

An isomer of corilagin is juglanin, separated from walnut shells.

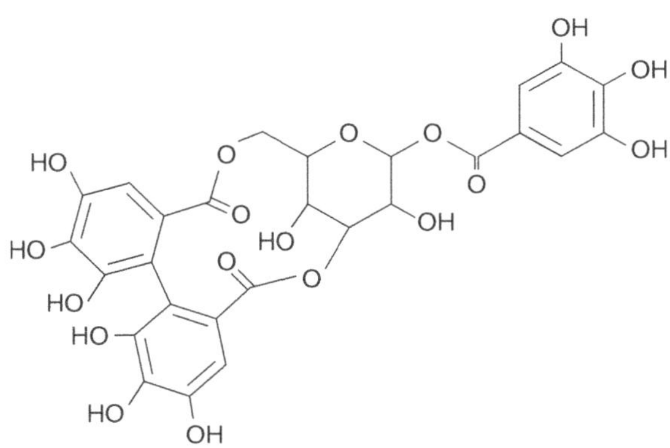


FIGURE 2.15 The chemical structure of Elagotanins.

Juglanin differs from corilagin in that it is still in an unspecified position in which the residues of gallic acid and hexahydroxydiphenic acid are connected to D-glucose.

From the pods of *C. brevifolia*, two crystalline elagotannins, yellow in color, called brevilagin I and II, were separated.

Through the hydrolysis of brevilagins, D-glucose, hexahydroxydiphenic acid and dehydrohexahydroxydiphenic acid are obtained.

Two crystallized ellagotamines, called castalagin and vescalagin, were separated from the wood of *Castanea vesca* and *Q. robur.*

A distinct skeleton from the other hydrolysable tannins has theogallin, separated from tea leaves, which is a gallic ester of quinic acid.

More than hydrogen and oxygen, tannins can also include nitrogen, phosphorus, and traces of organic ions.

These substances are extracted from vegetable organisms by using aqueous solutions; the tannins appear, for the most part, uncrystallized and have a regular shape [Valanciene & Malys, 2022].

They tend to be colloidally dispersed.

Tannin solutions have significant astringent quality; in fact they unite the collagen fibers of the gelatinous skin, reacting with them and reducing their water content.

The glycosides of these aglycones contribute to the formation of the pronounced bitter taste of citrus fruits [Mhlongo, Kenyon, & Nsahlai, 2024].

In food outcomes of vegetable origin, the tannins (tannins) accompany carbohydrates and other nutritional principles. They were separated especially from fruits and vegetables, but also from hops, tea leaves, etc.

In certain cases, the excess of iron is blamed for producing a blackening of the wines. Some catechins (e.g., epicatechin) give foods and drinks (especially wines) a slightly astringent character. The collaboration of catechinotannins in protecting blood capillaries and ensuring their permeability was also noted, acting similarly to vitamin P [Oliveira et al., 2023].

The significant collaboration of catechin–tannins in tea leaves is specified.

In epicatechin-trimeric molecular skeleton, during the seasoning of the teas, an exceptional brown coloring occurs because of the oxidation of the tannins in the green tea leaves.

Ellagic acid has been investigated more in the last decades. It is present in raspberries – 1500 μg/g wet substance; strawberries – 630 μg/g; pecans – 330 μg/g; and blueberries – 120 μg/g. From a pharmacological point of view, it was found that tannins, in moderate doses, interact with proteins to form tannin–protein complexes [Arapitsas, 2012].

Such complexes ensure anti-infective protection (contributing to the destruction of microorganisms through coagulation repercussion); they have antidiarrheal action, limit or stop fermentation processes, intervene in the coagulation processes of bloody wounds in the intestine, and produce a decrease in intestinal secretions. In the case of high doses, tannins have laxative and emetic repercussion [Melone et al., 2013].

Research carried out on laboratory animals has highlighted the fact that ellagitannins are precursors that hydrolytically release ellagic acid with repercussions in the remission of neoplasms, detoxification repercussions in the colon, and limiting the proliferation of abnormal cells. It was found experimentally that ellagic acid in laboratory animals inhibits malignancy in esophageal, lingual, lung, colon, liver, and skin cancers [Maugeri et al., 2022].

The mechanism of oncostatic action showed that ellagic acid:

- inhibits the transformation of some carcinogenic substances (e.g.: aromatic polycyclic hydrocarbons, food additives – preservatives, aflatoxins;
- conformations adduct with some carcinogenic substances that are thus inactivated;
- conformations adduct with DNA thus occupying the receptors specific to malignancy.

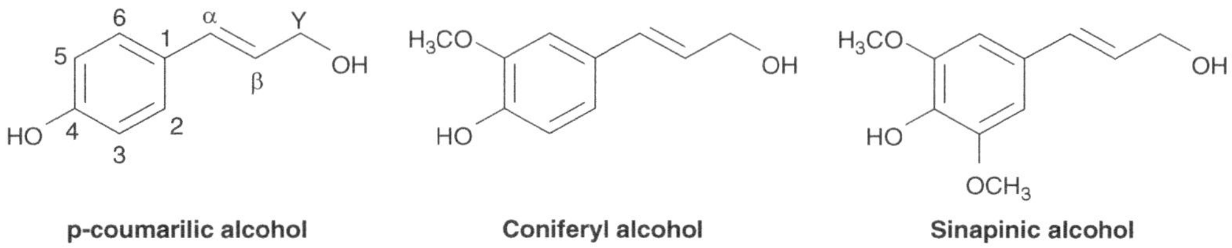


**FIGURE 2.16** The chemical structure of monomeric substances include p-coumaryl alcohol, coniferyl alcohol, and sinapic alcohol.

Remarkable is the observation regarding the fact that in small quantities elagotannins slow down the multiplication of cancer cells and in large quantities induce their tnatocytosis.

The apoptosis–necrosis relationship is currently intensively investigated for specific molecular biology and pathobiochemistry considerations.

Also, there are studies on the antimutagenic action of ellagic acid, tested *in vitro*, in relation to mutagenic substances attendant in some foods [Lakhani et al., 2019].

### 2.4.2 Lignin

Lignin (from lat. lignum – tree, wood) is a macromolecular compound, a complex polymer of aromatic alcohols (monolignols). Monomeric substances include p-coumaryl alcohol, coniferyl alcohol, and sinapic alcohol (Figure 2.16).

Lignins are establish by the condensation of include p-coumaryl alcohol, coniferyl alcohol and sinapic alcohol. Lignins from distinct vegetable organisms are not identical; they differ in the number and type of structural units.

Thus, sinapinic alcohol predominates in deciduous lignins, and coniferyl alcohol dominates in those of conifers. The molecular weight of lignins varies between 700 and 100,000 Da.

They are amorphous, dark-colored substances, insoluble in water, diluted acids, and alkalis. They are soluble in strong concentrated acids and concentrated alkalis [Rais & Zibek, 2019].

Lignins are significant collaborators in the formation of highly stable humus because they resist the action of chemical factors and are hard for microorganisms to attack.

They are used to obtain synthetic plastics, carbon black, and as substitutes for vegetable tannins [Kamimura et al., 2019].

Lignin was first mentioned by the Swiss botanist Augustin de Candolle in 1813. It is a widely distributed polymer in the vegetable kingdom, occupying second place after cellulose. Lignin is the aromatic, non-hydrolysable part of wood. It is also part of the secondary cell walls of vegetable organisms and some algae.

The wood of deciduous varieties includes 18–24% lignin, and the wood of conifers includes 27–30% lignin. The cell walls of vegetable organisms can be compared to reinforced concrete: the metal shell is cellulose, and the concrete is lignin.

Lignins are formed after the appearance of cellulose fibers by interweaving them. This gives them greater mechanical resistance and an increased resistance to the action of water, but it reduces their elasticity.

Both physical and chemical (etheric) associations are established between lignin and cellulose.

Lignins from distinct vegetable organisms differ in the number and type of structural units. In deciduous lignins, sinapic acid predominates, and in those of conifers, coniferylic acid predominates.

Lignins are amorphous, dark-colored substances insoluble in water and dilute acids and alkaline bases.

They are soluble in strong acids and concentrated alkaline bases, from which solutions they precipitate by dilution [Novaes et al., 2010].

When oxidized with nitrobenzene in a basic environment, lignin splits and establishes aromatic aldehydes (Figure 2.17).

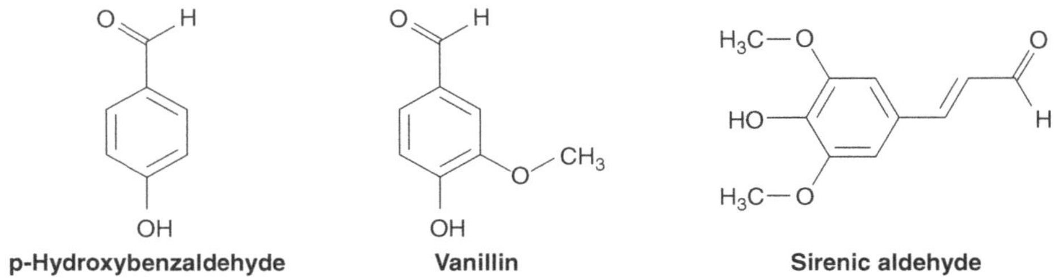


**FIGURE 2.17** The chemical structure of aromatic aldehydes, that establishes from oxidized with nitrobenzene in a basic environment, lignin splits in p-Hydroxybenzaldehyde, Vanillin and Sirenic aldehyde.

Lignin is the main component responsible for the vanilla flavor of old books. Over time, lignin is subjected to oxidative processes, giving old books a pleasant aroma.

Lignin is extracted from wood with hydrosulfite and sulfuric acid during paper manufacturing. Lignin waste from vegetables after hydrolysis is used for pressed articles, plastics, synthetic resins, and activated carbon [Wang et al., 2018].

### 2.4.3 Melanins

Melanins (from the Greek molasses – black, dark) elaborate a cluster of indigenous black nitrogen pigments, which are found in higher vegetable organisms, vertebrates, the cuticle of insects, and butterflies.

In vertebrate animals, melanins are stored in specialized cells (melanocytes) and determine the color of hair, skin, fur, etc.

Melanins are quinoid polymers and oxidation outcomes of the amino acid tyrosine.

Melanin is the generic name of a variety of polymer substances derived from tyrosine. It exists in living organisms, animals, and vegetable organisms and represents a main distinguishing feature of pigmentation and defense against solar radiation.

Melanin is what, for example, gives color to the skin, eyes, and hair of mammals.

The melanin nitrogen pigments acquainted under the generic name of "melanins" are formed from aromatic amino acids, i.e. tyrosine and hydroxyphenylamine (tyrosine).

In vegetable organisms, these pigments cause browning to blackening after heavy rainfall.

In animals, melanin is arranged in cells called "melanocytes" [Manivasagan et al., 2013].

Melanin pigments give different colors to the environment: yellow, red, and brown to black.

Melanins are particularly implicated in the expression of hair and skin color.

In animals, melanins are located in certain cells called, with a generic term, "melanocytes".

Melanins give tissues different colors, e.g.: yellow, red, brown, and black.

The black color (melanos – black) is dominant. In vegetable organisms' tissues, this becomes evident quite quickly, through the action of water stress (after prolonged precipitation).

In animal tissues, they were separated – in invertebrates – from the cuticle that covers the body of insects and in vertebrates from hair and skin [Singh et al., 2021].

This distribution is also distinguishing for humans, knowing that the absence of melanins due to the lack of the enzyme tyrosinase, which transforms tyrosine into melanin, causes albinism.

This is a condition with genetic determinism, acquainted with pathobiochemistry.

The biomodification of phenylalanine and tyrosine are carried out through oxidation, cyclization and polymerization ripostes, as a result of which substances of the type dihydroxyphenylalanine (DOPA), its quinone are formed and finally – in the presence of cysteine – phenolmelanins result.

Melanins are hydrophobic, electrically charged pigments with high molecular weights obtained by the oxidative polymerization of some precursors with aromatic nuclei. They are widespread in the living world, from bacteria and fungi to vegetable organisms and animals.

The pigment is not necessary for growth and development, but it increases the chances of survival and competitiveness of the variety in hostile conditions.

UV and electromagnetic radiation, dryness, and extreme temperatures, most likely thanks to its quality as a neutralizer of free radicals, behave like a "sponge" for radicals generated by the cell in response to environmental stress.

*In vitro* studies have shown that melanized fungi are more resistant than non-melanized ones to UV, high temperatures, hydrolytic enzymes, high concentrations of heavy metals, and antimicrobial substances [Carletti, Nervo, & Cattivelli, 2014].

Apart from this, melanins have other biological repercussions, such as thermoregulation, photoprotection, antimicrobial, antiviral, cytotoxic, anti-inflammatory, radioprotective, and immunomodulatory action.

Its quality makes melanin a biomolecule with very many biotechnological applications in cosmetics, the pharmaceutical industry, electronics, and food processing.

Despite their spread and ubiquity in nature, melanins are far from being elucidated as a molecular skeleton due to their insolubility in water, the difficulty of purification (all protocols require aggressive hydrolysis steps), and especially the lack of order of the frameworks (polymer amorphous), which prevents crystallization.

Currently, the available formulas are theoretical, resulting from mathematical models, and the presentations of the substances supposed to be melanins are done indirectly.

The problem is all the more complicated as there are many types of melanins, and many organisms, especially microorganisms, can synthesize several types of melanin depending on the availability of precursors, the maturity of the organism, or the function of the cell that stores the pigment [Leksin et al., 2024].

### Types of melanins

Eumelanins (black or brown pigments) are produced by the oxidation of tyrosine and/or phenylalanine to o-dihydroxyphenylalanine (DOPA) and dopaquionone, which undergo further cyclization to 5,6-dihydroxyindole (DHI) or 5,6-dihydroxyindole-2-carboxylic acid (DHICA).

Pheomelanins (yellow-reddish) are initially synthesized as eumelanins, but DOPA undergoes cysteinylation, directly or sub-mediated by glutathione. The outcome of this riposte, cysteinyl DOPA, further polymerizes into different benzothiazine analogs.

Alomelanins – the least investigated and conformation the most heterogeneous cluster of these polymers – are born through the oxidation/polymerization of di-(DHN) or tetrahydroxynaphthalene, via the pentaketide pathway, leading to differently colored substances (DHN-melanins), homogentisic acid (pyomelanins), γ-glutaminyl-4-hydroxybenzene, catechols, as well as to acid 4-hydroxyphenylacetic [d'Ischia et al., 2013].

### 2.4.3.1 Melanins Synthesized by Lower Filamentous Fungi (Molds)

Most melanins in fungi are derived from the precursor 1,8-dihydroxynaphthalene (DHN) and are acquainted as DHN melanins; the biosynthesis pathway for this type of melanin was called the pentaketide pathway and was intensively investigated.

Especially the last steps, in which 1,3,6,8-tetrahydroxynaphthalene (4HN) is converted, in order, to scitalon, vermelon, and DHN, are well understood.

The tyrosine ring represents a second (secondary) source of precursor frameworks, either by itself (as such) or as L-3,4-dihydroxyphenylalanine (L-DOPA) and, since dopaquinone (the oxidation outcome of DOPA), is capable of forming a cycle, and transforming into the 5,6-dihydroxyindolic ring, melanin starting from tyrosine or DOPA, typically includes indolic cores [d'Ischia et al., 2015].

***Applications of melanin.*** Due to its electronic quality, melanin is a highly sought-after biological outcome with many applications in distinct spheres: medical (nano-particles that protect healthy organs in radiotherapy), cosmetic (sunscreen for skin and hair), advanced materials (semiconductors organic, bio-friendly electronic devices, pigments for glass, or contact lenses), and food (dyes for liquors, tea).

Despite its many uses, melanin remains an expensive pigment, and many studies currently focus on the efficiency of the production processes (either chemical or biological) [Stepień et al., 2009].

Melanins are connected to proteins, and the melanin–protein complex is located in the cells in the so-called melanin granules, which are located in the cytoplasm.

Melanin can reduce ammoniacal silver nitrate solutions to metallic silver without using an external reducer. This method is far from being specific for melanins, yet it demonstrates that it has a similar reducing quality as cellular argentafin granules and lipofuchsins (lipochromes) [Sajjan et al., 2010].

## 2.5 THE BIOLOGICAL ROLE OF PHENOLIC SUBSTANCES

Phenolic substances have an antioxidant effect. The antioxidant action of phenolic substances is manifested by binding:

- heavy metal ions;
- free radicals that are formed as a result of the oxidation of organic substances.

Phenolic substances are biologically active substances.

Melanins are black pigments that protect animal skin from ultraviolet rays.

Polyphenols are a group of substances found in fruits, vegetables, nuts, legumes, coffee, tea, and cocoa.

They collaborate significantly in supporting health, so it is indispensable to consider them, especially in the case of polycystic ovary syndrome, endometriosis, or insulin resistance.

They can be divided into substitutes conformable to the number of phenolic cores and the structural elements that connect these cores, and thus, we distinguish phenolic acids, lignans, stilbenes, and flavonoids.

The latter are divided into isoflavones, flavones, flavonols, flavanones, anthocyanidins, and flavonols [Herrera-Pool et al., 2021].

Flavonoids are a significant variety of polyphenolic substances found in a wide variety of vegetable organisms. These substances are responsible for the vibrant colors of many flowers, fruits, and vegetables and are considered beneficial for health due to their antioxidant and anti-inflammatory qualities. There are several subcategories of flavonoids.

- Flavones, such as apigenin and luteolin, which are found in foods such as parsley, celery, and some teas;
- Flavonols, such as quercetin, kaempferol, and rutin, which are found in green leafy vegetables, onions, apples, and grapes;
- Flavanones, such as hesperidin and naringenin, which are found in citrus fruits;
- Anthocyanins are found in blueberries, blackberries, and red grapes.

Apart from flavonoids, there are other types of polyphenols, namely phenolic acids. They are found not only in coffee, tea, grapes, red wine, kiwi, plums, apples, and cherries, but also in corn and wheat.

Stilbenes are found in red wine, cranberries, and hazelnuts.

A well-acquainted example of stilbene is resveratrol, which has been associated with cardiovascular health benefits.

Lignans are precursors to phytoestrogens and may have benefits for hormonal health.

These substances are mainly found in seeds, especially flax and sesame seeds.

Examples of polyphenols are quercetin, resveratrol, kaempferol, apigenin, naringenin, catechins, epicatechins, proanthocyanidins, curcumin, genistein, chlorogenic acid, and many others, and each of them has valuable quality for the body. Although some names are difficult to pronounce, polyphenols are found in many outcomes.

Indigenous polyphenols derived from vegetable organisms (resveratrol, flavonoids, phenolic acids, and anthocyanins) have gained much attention in the last two decades due to their effective action in preventing and ameliorating several chronic health problems, such as cardiovascular disease, diabetes, and related diseases, as well as several cancer forms [Zhang et al., 2023].

Directing attention to anthocyanins, they have become the object of very many studies carried out on *in vivo* and *in vitro* models, used either as a daily food supplement or in synergy with standard therapies, giving significant results.

## 2.6 CONCLUSION AND FUTURE REMARKS

Medicinal vegetable organisms, structured on the ever-increasing understanding of their chemical composition and pharmacodynamic quality, with the help of high-performance analytical investigation methods, are currently one of the main sources of raw materials for the seasoning of necessary indigenous products maintaining health. Also, the therapeutic value of medicinal vegetable organisms is structured on the relationship between the chemical skeleton of the active substances and their pharmacological activity on the reactive elements of the human body. The multiple pharmacodynamic quality of the same vegetable organisms can be explained by the fact that most medicinal vegetable organisms have a complex chemical composition, starting from two to three substances up to 30–40 biologically active substances identified in some vegetable organisms; biologically active substances, although in very small concentrations, combine their activity and act through synergism. In this sense, the "miracle of nature" that knew how to create in vegetable organisms the ideal combinations of active substances with beneficial repercussions for human health must be looked at and admired. Instead, the very long way travelled by mankind was and is conditioned by the evolution of thinking and the scientific methodology of investigation. Only now can we claim that we are truly on this path, but it will take years, decades, if not centuries, to unravel all the secrets that vegetable organisms hide. Modern phytotherapy considers the entire human knowledge, particularly rich, in the sphere of phytotherapy, but greatly increasing the requirements regarding the control of the quality of vegetable organisms by modern means and their chemical composition, possible unwanted side repercussion, and especially the development of formulas and pharmaceutical forms as appropriate as bioavailability and effectiveness for health. All these aspects, of particular importance to producing pharmaceutical products, were the basis of the research in this paper to obtain new products with anti-inflammatory and hepatoprotective activity and some products used to improve cardiovascular diseases.

## COMPETING INTERESTS

The authors declare that they have no competing interest.

## DECLARATION OF GENERATIVE AI AND AI-ASSISTED TECHNOLOGIES IN THE WRITING PROCESS

During the preparation of this work, the book chapter author(s) have not taken any assistance from AI, and the book chapter author(s) reviewed and edited the content as needed and take(s) full responsibility for the content of the publication.

## REFERENCES

Arapitsas, P. (2012). Hydrolyzable tannin analysis in food. *Food Chemistry, 135*(3), 1708–1717. https://doi.org/10.1016/j.foodchem.2012.05.096

Bahramsoltani, R., Ahmadian, R., Daglia, M., & Rahimi, R. (2024). *Petroselinum crispum* (Mill.) fuss (Parsley): An updated review of the traditional uses, phytochemistry, and pharmacology. *Journal of agricultural and food chemistry, 72*(2), 956–972. https://doi.org/10.1021/acs.jafc.3c06429

Barbehenn, R. V., & Peter Constabel, C. (2011). Tannins in plant-herbivore interactions. *Phytochemistry*, *72*(13), 1551–1565. https://doi.org/10.1016/j.phytochem.2011.01.040

Brenna, E., Fronza, G., Fuganti, C., Gatti, F. G., Pinciroli, M., & Serra, S. (2004). Differentiation of extractive and synthetic salicin. The 2H aromatic pattern of natural 2-hydroxybenzyl alcohol. *Journal of Agricultural and Food Chemistry*, *52*(26), 7747–7751. https://doi.org/10.1021/jf0489320

Brezani, V., Smejkal, K., Hosek, J., & Tomasova, V. (2018). Anti-inflammatory natural prenylated phenolic compounds – Potential lead substances. *Current Medicinal Chemistry*, *25*(10), 1094–1159. https://doi.org/10.2174/0929867324666170810161157

Buitimea-Cantúa, G. V., Welti-Chanes, J., & Escobedo-Avellaneda, Z. (2022). Metabolite transformation and β-d-glucosidase activity during the high hydrostatic pressure assisted curing of vanilla beans (*Vanilla planifolia*) to improve phenolic compounds formation. *Food Chemistry*, *384*, 132497. https://doi.org/10.1016/j.foodchem.2022.132497

Buyse, K., Delezie, E., Goethals, L., Van Noten, N., Van Poucke, C., Devreese, M., Antonissen, G., Janssens, G. P. J., & Lourenço, M. (2024). Chestnut wood tannins in broiler diets: Pharmacokinetics, serum levels during rearing, and intestinal absorption pattern of gallic acid. *Journal of Agricultural and Food Chemistry*, *72*(5), 2648–2656. https://doi.org/10.1021/acs.jafc.3c09881

Carletti, G., Nervo, G., & Cattivelli, L. (2014). Flavonoids and Melanins: A common strategy across two kingdoms. *International Journal of Biological Sciences*, *10*(10), 1159–1170. https://doi.org/10.7150/ijbs.9672

Cerezo, A. B., Álvarez-Fernández, M. A., Hornedo-Ortega, R., Troncoso, A. M., & García-Parrilla, M. C. (2014). Phenolic composition of vinegars over an accelerated aging process using different wood species (acacia, cherry, chestnut, and oak): Effect of wood toasting. *Journal of Agricultural and Food Chemistry*, *62*(19), 4369–4376. https://doi.org/10.1021/jf500654d

Chandrasekhar, Y., Phani Kumar, G., Navya, K., Ramya, E. M., & Anilakumar, K. R. (2018). Tannins from *Terminalia chebula* fruits attenuates GABA antagonist-induced anxiety-like behaviour via modulation of neurotransmitters. *The Journal of Pharmacy and Pharmacology*, *70*(12), 1662–1674. https://doi.org/10.1111/jphp.13007

Chen, L., Wei, X., & Matsuda, Y. (2022). Depside bond formation by the starter-unit acyltransferase domain of a fungal polyketide synthase. *Journal of the American Chemical Society*, *144*(42), 19225–19230. https://doi.org/10.1021/jacs.2c08585

d'Ischia, M., Wakamatsu, K., Cicoira, F., Di Mauro, E., Garcia-Borron, J. C., Commo, S., Galván, I., Ghanem, G., Kenzo, K., Meredith, P., Pezzella, A., Santato, C., Sarna, T., Simon, J. D., Zecca, L., Zucca, F. A., Napolitano, A., & Ito, S. (2015). Melanins and melanogenesis: From pigment cells to human health and technological applications. *Pigment Cell & Melanoma Research*, *28*(5), 520–544. https://doi.org/10.1111/pcmr.12393

d'Ischia, M., Wakamatsu, K., Napolitano, A., Briganti, S., Garcia-Borron, J. C., Kovacs, D., Meredith, P., Pezzella, A., Picardo, M., Sarna, T., Simon, J. D., & Ito, S. (2013). Melanins and melanogenesis: Methods, standards, protocols. *Pigment Cell & Melanoma Research*, *26*(5), 616–633. https://doi.org/10.1111/pcmr.12121

Ferreira, D., Marais, J. P., & Slade, D. (2005). Heterogeneity of the interflavanyl bond in proanthocyanidins from natural sources lacking C-4 (C-ring) deoxy flavonoid nucleophiles. *Phytochemistry*, *66*(18), 2216–2237. https://doi.org/10.1016/j.phytochem.2005.01.014

Genovese, S., Epifano, F., Medina, P., Caron, N., Rives, A., Poirot, M., Silvente-Poirot, S., & Fiorito, S. (2019). Natural and semisynthetic oxyprenylated aromatic compounds as stimulators or inhibitors of melanogenesis. *Bioorganic Chemistry*, *87*, 181–190. https://doi.org/10.1016/j.bioorg.2019.03.026

Girón-Elola, C., Sasiain, I., Sánchez-Fernández, R., Pazos, E., & Correa, A. (2023). Site-selective C-H amination of phenol-containing biomolecules. *Organic Letters*, *25*(23), 4383–4387. https://doi.org/10.1021/acs.orglett.3c01560

Goyal, N., Do, C., Hill-Odom, M., Beamon, T., Ponnapakkam, T., Liu, J., Sridhar, J., Huckaba, T., & Foroozesh, M. (2023). Study of ceramide-flavone analogs showing self-fluorescence and anti-proliferation activities. *Journal of Oncology Research and Therapy*, *8*(2), 10172. https://doi.org/10.29011/2574-710x.10172

Herrera-Pool, E., Ramos-Díaz, A. L., Lizardi-Jiménez, M. A., Pech-Cohuo, S., Ayora-Talavera, T., Cuevas-Bernardino, J. C., García-Cruz, U., & Pacheco, N. (2021). Effect of solvent polarity on the ultrasound assisted extraction and antioxidant activity of phenolic compounds from habanero pepper leaves (*Capsicum chinense*) and its identification by UPLC-PDA-ESI-MS/MS. *Ultrasonics Sonochemistry*, *76*, 105658. https://doi.org/10.1016/j.ultsonch.2021.105658

Jing, W., Xiaolan, C., Yu, C., Feng, Q., & Haifeng, Y. (2022). Pharmacological effects and mechanisms of tannic acid. *Biomedicine & Pharmacotherapy=Biomedecine & pharmacotherapie*, *154*, 113561. https://doi.org/10.1016/j.biopha.2022.113561.

Kamimura, N., Sakamoto, S., Mitsuda, N., Masai, E., & Kajita, S. (2019). Advances in microbial lignin degradation and its applications. *Current Opinion in Biotechnology*, *56*, 179–186. https://doi.org/10.1016/j.copbio.2018.11.011

Kaur, G., Kaur, R., Sodhi, G. K., George, N., Rath, S. K., Walia, H. K., Dwibedi, V., & Saxena, S. (2024). Stilbenes: A journey from folklore to pharmaceutical innovation. *Archives of Microbiology*, *206*(5), 229. https://doi.org/10.1007/s00203-024-03939-z

Lakhani, N., Kamra, D. N., Lakhani, P., & Alhussien, M. N. (2019). Immune status and haemato-biochemical profile of buffalo calves supplemented with phytogenic feed additives rich in tannins, saponins and essential oils. *Tropical animal Health and Production*, *51*(3), 565–573. https://doi.org/10.1007/s11250-018-1727-z

Le Goff, G., & Ouazzani, J. (2014). Natural hydrazine-containing compounds: Biosynthesis, isolation, biological activities and synthesis. *Bioorganic & Medicinal Chemistry*, *22*(23), 6529–6544. https://doi.org/10.1016/j.bmc.2014.10.011

Leksin, I., Shelyakin, M., Zakhozhiy, I., Kozlova, O., Beckett, R., & Minibayeva, F. (2024). Ultraviolet-induced melanisation in lichens: Physiological traits and transcriptome profile. *Physiologia Plantarum*, *176*(5), e14512. https://doi.org/10.1111/ppl.14512

Li, L., Cai, Y., Sun, X., Du, X., Jiang, Z., Ni, H., Yang, Y., & Chen, F. (2020). Tyrosinase inhibition by *p*-coumaric acid ethyl ester identified from camellia pollen. *Food science & nutrition*, *9*(1), 389–400. https://doi.org/10.1002/fsn3.2004

Li, M., Zhao, X., Sun, Y., Yang, Z., Han, G., & Yang, X. (2021). Evaluation of anthocyanin profile and color in sweet cherry wine: Effect of sinapic acid and grape tannins during aging. *Molecules (Basel, Switzerland)*, *26*(10), 2923. https://doi.org/10.3390/molecules26102923

Luca, S. V., Macovei, I., Bujor, A., Miron, A., Skalicka-Woźniak, K., Aprotosoaie, A. C., & Trifan, A. (2020). Bioactivity of dietary polyphenols: The role of metabolites. *Critical Reviews in Food Science and Nutrition*, *60*(4), 626–659. https://doi.org/10.1080/10408398.2018.1546669

Luo, L., Chang, Y., Zhang, W., Liu, X., Ge, J., Chen, J., Li, Y., Zhang, D., & Sheng, L. (2025). 7-Hydroxycoumarin and its conjugated metabolites interact with organic anion transporters 1 and 3 *in vitro* and *in vivo*. *Chemico-Biological Interactions*, *405*, 111293. https://doi.org/10.1016/j.cbi.2024.111293

Ma, X., Yang, W., Laaksonen, O., Nylander, M., Kallio, H., & Yang, B. (2017). Role of flavonols and proanthocyanidins in the sensory quality of sea buckthorn (*Hippophaë rhamnoides* L.) Berries. *Journal of Agricultural and Food Chemistry*, *65*(45), 9871–9879. https://doi.org/10.1021/acs.jafc.7b04156

Manivasagan, P., Venkatesan, J., Sivakumar, K., & Kim, S. K. (2013). Actinobacterial melanins: Current status and perspective for the future. *World Journal of Microbiology & Biotechnology*, *29*(10), 1737–1750. https://doi.org/10.1007/s11274-013-1352-y

Mao, X., Wu, L. F., Zhao, H. J., Liang, W. Y., Chen, W. J., Han, S. X., Qi, Q., Cui, Y. P., Li, S., Yang, G. H., Shao, Y. Y., Zhu, D., Wang, R. F., You, Y., & Zhang, L. Z. (2016). Transport of corilagin, gallic acid, and ellagic acid from fructus phyllanthi tannin fraction in caco-2 cell monolayers. *Evidence-Based Complementary and Alternative medicine: eCAM*, *2016*, 9205379. https://doi.org/10.1155/2016/9205379

Maugeri, A., Lombardo, G. E., Cirmi, S., Süntar, I., Barreca, D., Laganà, G., & Navarra, M. (2022). Pharmacology and toxicology of tannins. *Archives of Toxicology*, *96*(5), 1257–1277. https://doi.org/10.1007/s00204-022-03250-0

Melone, F., Saladino, R., Lange, H., & Crestini, C. (2013). Tannin structural elucidation and quantitative $^{31}$P NMR analysis. 2. Hydrolyzable tannins and proanthocyanidins. *Journal of agricultural and food chemistry*, *61*(39), 9316–9324. https://doi.org/10.1021/jf401664a

Merchant, A., & Richter, A. A. (2011). Polyols as biomarkers and bioindicators for 21st century plant breeding. *Functional Plant Biology: FPB*, *38*(12), 934–940. https://doi.org/10.1071/FP11105

Mhlongo, L. C., Kenyon, P., & Nsahlai, I. V. (2024). Effect of *Acacia mearnsii* forage or tannin extract on rumen dry matter and crude protein degradation. *Journal of Animal Physiology and Animal Nutrition*, *10*, 1111–14033. Advance online publication. https://doi.org/10.1111/jpn.14033

Niemetz, R., & Gross, G. G. (2005). Enzymology of gallotannin and ellagitannin biosynthesis. *Phytochemistry*, *66*(17), 2001–2011. https://doi.org/10.1016/j.phytochem.2005.01.009

Novaes, E., Kirst, M., Chiang, V., Winter-Sederoff, H., & Sederoff, R. (2010). Lignin and biomass: A negative correlation for wood formation and lignin content in trees. *Plant Physiology*, *154*(2), 555–561. https://doi.org/10.1104/pp.110.161281

Oliveira, L. N., Pereira, M. A. N., Oliveira, C. D. S., Oliveira, C. C., Silva, R. B., Pereira, R. A. N., DeVries, T. J., & Pereira, M. N. (2023). Effect of low dietary concentrations of *Acacia mearnsii* tannin extract on chewing, ruminal fermentation, digestibility, nitrogen partition, and performance of dairy cows. *Journal of Dairy Science*, *106*(5), 3203–3216. https://doi.org/10.3168/jds.2022-22521

Rais, D., & Zibek, S. (2019). Biotechnological and biochemical utilization of lignin. *Advances in Biochemical Engineering/Biotechnology*, *166*, 469–518. https://doi.org/10.1007/10_2017_6

Roche, A., Ross, E., Walsh, N., O'Donnell, K., Williams, A., Klapp, M., Fullard, N., & Edelstein, S. (2017). Representative literature on the phytonutrients category: Phenolic acids. *Critical Reviews in Food Science and Nutrition*, *57*(6), 1089–1096. https://doi.org/10.1080/10408398.2013.865589

Sajjan, S., Kulkarni, G., Yaligara, V., Kyoung, L., & Karegoudar, T. B. (2010). Purification and physiochemical characterization of melanin pigment from Klebsiella sp. GSK. Journal of microbiology and biotechnology, 20(11), 1513–1520. DOI: 10.4014/jmb.1002.02006

Schober, J. D., Burdsall, A. C., Searcy, T., Hart, J., Shade, M., & Harper, W. F. Jr (2023). Hydroxyl radical-driven transformations of bisphenol A and 2,4-dinitroanisole: Experimental and computational analysis. *Water Environment Research: A Research Publication of the Water Environment Federation*, *95*(11), e10954. https://doi.org/10.1002/wer.10954

Singh, S., Nimse, S. B., Mathew, D. E., Dhimmar, A., Sahastrabudhe, H., Gajjar, A., Ghadge, V. A., Kumar, P., & Shinde, P. B. (2021). Microbial melanin: Recent advances in biosynthesis, extraction, characterization, and applications. *Biotechnology Advances*, *53*, 107773. https://doi.org/10.1016/j.biotechadv.2021.107773

Stepień, K., Dzierzega-Lecznar, A., Kurkiewicz, S., & Tam, I. (2009). Melanin from epidermal human melanocytes: Study by pyrolytic GC/MS. *Journal of the American Society for Mass Spectrometry*, *20*(3), 464–468. https://doi.org/10.1016/j.jasms.2008.11.003

Sui, X., Wang, J., Zhao, Z., Liu, B., Liu, M., Liu, M., Shi, C., Feng, X., Fu, Y., Shi, D., Li, S., Qi, Q., Xian, M., & Zhao, G. (2024). Phenolic compounds induce ferroptosis-like death by promoting hydroxyl radical generation in the Fenton reaction. *Communications biology*, *7*(1), 199. https://doi.org/10.1038/s42003-024-05903-5

Sun, P., Xu, S., Tian, Y., Chen, P., Wu, D., & Zheng, P. (2024). 4-Hydroxyphenylacetate 3-hydroxylase (4HPA3H): A vigorous monooxygenase for versatile *O*-hydroxylation applications in the biosynthesis of phenolic derivatives. *International Journal of Molecular Sciences*, *25*(2), 1222. https://doi.org/10.3390/ijms25021222

Tinikul, R., Chenprakhon, P., Maenpuen, S., & Chaiyen, P. (2018). Biotransformation of plant-derived phenolic acids. *Biotechnology Journal*, *13*(6), e1700632. https://doi.org/10.1002/biot.201700632

Uchida, S., Ozaki, M., Akashi, T., Yamashita, K., Niwa, M., & Taniyama, K. (1995). Effects of (–)-epigallocatechin-3-O-gallate (green tea tannin) on the life span of stroke-prone spontaneously hypertensive rats. *Clinical and Experimental Pharmacology & Physiology. Supplement*, *22*(1), S302–S303. https://doi.org/10.1111/j.1440-1681.1995.tb02928.x

Valanciene, E., & Malys, N. (2022). Advances in production of hydroxycinnamoyl-quinic acids: From natural sources to biotechnology. *Antioxidants (Basel, Switzerland)*, *11*(12), 2427.

Wang, J. P., Matthews, M. L., Williams, C. M., Shi, R., Yang, C., Tunlaya-Anukit, S., Chen, H. C., Li, Q., Liu, J., Lin, C. Y., Naik, P., Sun, Y. H., Loziuk, P. L., Yeh, T. F., Kim, H., Gjersing, E., Shollenberger, T., Shuford, C. M., Song, J., Miller, Z., … & Chiang, V. L. (2018). Improving wood properties for wood utilization through multi-omics integration in lignin biosynthesis. *Nature Communications*, *9*(1), 1579. https://doi.org/10.1038/s41467-018-03863-z

Wang, M., Wang, H., Gao, C., Wei, W., Liu, J., Chen, X., Hu, G., Song, W., Wu, J., Zhang, F., & Liu, L. (2024). Efficient production of protocatechuic acid using systems engineering of *Escherichia coli*. *Metabolic Engineering*, *82*, 134–146. https://doi.org/10.1016/j.ymben.2024.02.003

Wang, X., Chen, Q., Zhang, Z., He, H., Ma, X., Liu, Z., Ge, B., & Huang, F. (2019). Novel Enteromorpha Prolifera based carbon dots: Probing the radical scavenging of natural phenolic compounds. *Colloids and Surfaces. B, Biointerfaces*, *174*, 161–167. https://doi.org/10.1016/j.colsurfb.2018.11.019

Wang, Z., Wang, B., Yan, C., Yuan, S., & Cao, L. (2020). Neighborhood effects on the tannin-related foraging decisions of two rodent species under semi-natural conditions. *Integrative Zoology*, *15*(6), 569–577. https://doi.org/10.1111/1749-4877.12473

Windley, H. R., Starrs, D., Stalenberg, E., Rothman, J. M., Ganzhorn, J. U., & Foley, W. J. (2022). Plant secondary metabolites and primate food choices: A meta-analysis and future directions. *American Journal of Primatology*, *84*(8), e23397. https://doi.org/10.1002/ajp.23397.

Yang, H., Zhong, H. J., Leung, K. H., Chan, D. S., Ma, V. P., Fu, W. C., Nanjunda, R., Wilson, W. D., Ma, D. L., & Leung, C. H. (2013). Structure-based design of flavone derivatives as c-myc oncogene down-regulators. *European Journal of Pharmaceutical Sciences: Official Journal of the European Federation for Pharmaceutical Sciences*, *48*(1–2), 130–141. https://doi.org/10.1016/j.ejps.2012.10.010

Zhang, Y., Li, Y., Ren, X., Zhang, X., Wu, Z., & Liu, L. (2023). The positive correlation of antioxidant activity and prebiotic effect about oat phenolic compounds. *Food Chemistry*, *402*, 134231. https://doi.org/10.1016/j.foodchem.2022.134231

# 3 Extraction and Characterization Techniques for Phenolic Compounds from Medicinal Plants

*Shankul Kumar, Vedant Kumar Prajapati, Arvind Kumar Patel, Shashi Ranjan Singh, and Sumitra Nain**

## 3.1 INTRODUCTION

Phenolic compounds represent a class of metabolites that are synthesized through the secondary metabolic pathways of medicinal plants (Table 3.1). The categorization of plant phenolics is anchored in the term 'phenol', which refers to a chemical structure characterized by a phenyl ring that possesses one or more hydroxyl groups. Various classes of phenolic compounds have been systematically classified based on their fundamental structural framework (1). The spectrum of phenolic compounds encompasses a range from simple to relatively low molecular weight entities, including those with a singular aromatic ring, as well as more intricate and sizable tannins, along with their derivative polyphenols. Polyphenols comprise a wide range of compounds, including flavonoids, phenolic acids, tannins, lignans, and coumarins, which occur naturally in a variety of plant-derived foods such as fruits, vegetables, cereals, roots, and leaves (2). Phenolic compounds exhibit numerous activities, including antimicrobial (3) and anti-inflammatory effects, and can be beneficial in the treatment of diseases such as obesity (4), cancer (5), and diabetes (6), as well as possess antioxidant properties (7–9). The power of phenolics to donate hydrogen atoms or electrons in order to neutralize reactive oxygen species, or ROS, and reactive nitrogen species (RNS) and thus avoid cellular damage forms the foundation of their antioxidant potential (10). Due to their diverse and specific biological activities, secondary metabolites are considered a valuable source of lead molecules for the development of new drugs. Therefore, they are continuously being studied and explored for their potential in drug development.

Extraction of phenolic compounds from plant matrices is a complex process that is influenced by various factors, including the type of plant material, solvent polarity, temperature, extraction time, and the method employed. Because they are easy to use and inexpensive, conventional procedures such as maceration, decoction, and Soxhlet extraction are widely employed; however, they often require longer extraction times and greater solvent usage. Conversely, newer extraction methods, such as subcritical CO2 extractions require less pressure and a lower temperature than supercritical ones. The process is longer and less efficient, and the typical yield is smaller. here i have mention Supercritical fluid extraction, yield is more. (11,12).

The quality and bioactivity of phenolic compounds derived from medicinal plants are significantly influenced by the selection of the extraction method, in addition to yield. It is crucial to carefully optimize extraction conditions and choose the right solvent to guarantee good recovery and maintain the integrity of these bioactive elements (13,14). This chapter compares and contrasts classical and contemporary extraction techniques for phenolic compound isolation, describing their

* Corresponding author: nainsumitra@gmail.com

DOI: 10.1201/9781003527671-3

**TABLE 3.1**
**Medicinal Plants, Their Major Phenolic Compounds, and Therapeutic Uses**

| Medicinal Plant | Major Phenolic Compounds | Therapeutic Uses | Ref. |
|---|---|---|---|
| Green tea (*Camellia sinensis*) | Catechins (EGCG, EGC, ECG) | Antioxidant, cardioprotective, anti-inflammatory | (19) |
| Turmeric (*Curcuma longa*) | Curcumin | Anti-inflammatory, antioxidant, anti-cancer | (20) |
| Grape (*Vitis vinifera*) | Resveratrol, quercetin | Cardioprotective, anti-aging, neuroprotective | (21) |
| Pomegranate (*Punica granatum*) | Ellagic acid, punicalagin | Anti-inflammatory, anti-cancer, antimicrobial | (22) |
| Olive (*Olea europaea*) | Oleuropein, hydroxytyrosol | Antioxidant, hypolipidemic, antihypertensive | (23) |
| Cinnamon (*Cinnamomum verum*) | Cinnamic acid, catechins | Antidiabetic, antimicrobial, digestive aid | (24) |
| Clove (*Syzygium aromaticum*) | Eugenol, gallic acid | Analgesic, antimicrobial, antioxidant | (25) |
| Rosemary (*Rosmarinus officinalis*) | Rosmarinic acid, caffeic acid | Neuroprotective, antioxidant, hepatoprotective | (26) |
| Basil (*Ocimum sanctum*) | Rosmarinic acid, caffeic acid | Adaptogenic, anti-inflammatory, antimicrobial | (27) |
| Amla (*Phyllanthus emblica*) | Gallic acid, ellagic acid | Antioxidant, hypolipidemic, immune-enhancing | (28) |
| *Moringa oleifera* | Syringic acid and gallic acid | Anti-cancer activity | (29) |
| *Peganum harmala* | Gallic acid, vanillic acid, 4-hydroxybenzoic acid, and syringic acid | Antioxidant and anti-inflammatory | (30) |
| *Origanum vulgare* L. | Caffeic acid and *p*-coumaric acid | Antioxidant | (31) |
| Thymus vulgaris | Vanillin | Antioxidant | (32) |
| *Phyllanthus amarus* | Phyllanthin, niranthin, hypophyllanthin | Antioxidant and anti-inflammatory | (33) |
| Rheum palmatum | Rhein, physcion, chrysophanol, emodin, and aloe-emodin | Anti-coagulation and anti-thrombosis | (34) |

fundamental principles, benefits, drawbacks, and suitability for subsequent procedures such as compound isolation, quantitative analysis, and biological activity assessment.

## 3.2 CLASSIFICATION OF PHENOLIC COMPOUNDS

Flavonoids, phenolic acids, tannins, and lignans are common forms of phenolic chemicals found in medicinal plants (Table 3.1). These substances can be found in the leaves, stems, roots, fruits, and seeds of plants, as well as in other parts of the plant. For example:

1. *Flavonoids*: These are the most common polyphenols in diets (Figure 3.1). The skeleton flavan nucleus is the flavan nucleus, which consists of 15 carbon atoms linked together in three rings (C6–C3–C6), designated as A, B, and C. They are known for their potent anti-inflammatory and antioxidant activities. Flavonoids occur in *Ginkgo biloba* and *Camellia sinensis* (green tea) (15). Flavonoids are classified into six subclasses: flavones, flavonols, flavanols, flavanones, isoflavones, and anthocyanins, based on the oxidation state of the central C ring.
2. *Phenolic Acids*: Phenolic acids can be divided into two classes: derivatives of benzoic acid, such as gallic acid, and derivatives of cinnamic acid, such as coumaric, caffeic, and ferulic acid (Figure 3.2). These compounds are prevalent in plants such as *Rosmarinus officinalis* (rosemary) and *Punica granatum* (pomegranate) (16).
3. *Tannins*: Tannins constitute another substantial category of polyphenols in our diet and are typically further divided into two categories: (1) hydrolyzable tannins and (2) condensed tannins. Hydrolyzable tannins are substances with a core of glucose or other polyol esterified

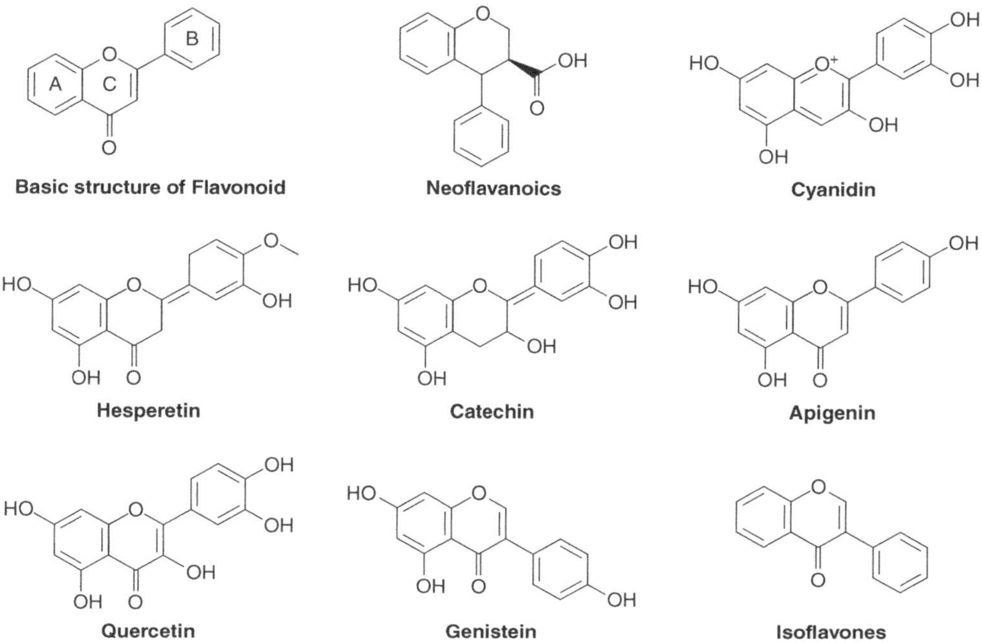


**FIGURE 3.1** Chemical structures of flavonoids and related subclasses.

either with gallic acid, also referred to as gallotannins, or with hexahydroxy diphenic acid, also known as ellagitannins. Condensed tannins are flavan-3-ol oligomers or polymers joined by an interflavan carbon bond. They are also known as proanthocyanidins, as they break down to anthocyanidins through an acid-catalyzed oxidation reaction (Figure 3.3) when heated in acidic alcohol solutions. It is present in plants such as *Terminalia chebula*, and tannins are linked with antimicrobial and astringent activities.

4. *Lignans*: Found in seeds like flaxseed, these compounds are known for their antioxidant and hormone-modulating properties (Figure 3.4) (17).
5. Stilbenes, such as resveratrol from grapes and peanuts, are a notable example with strong antioxidant potential (18).

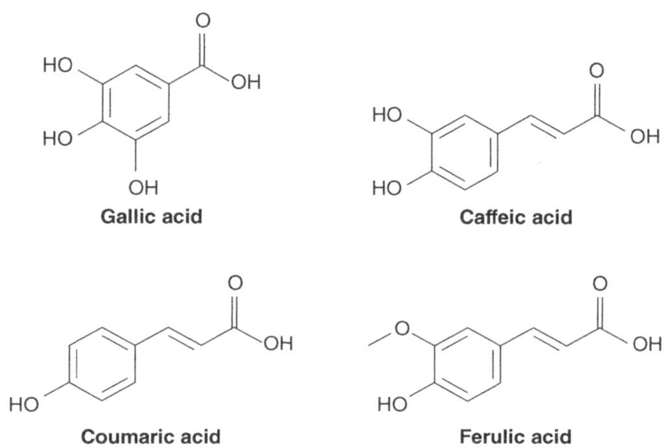


**FIGURE 3.2** Chemical structures of Phenolic acids.

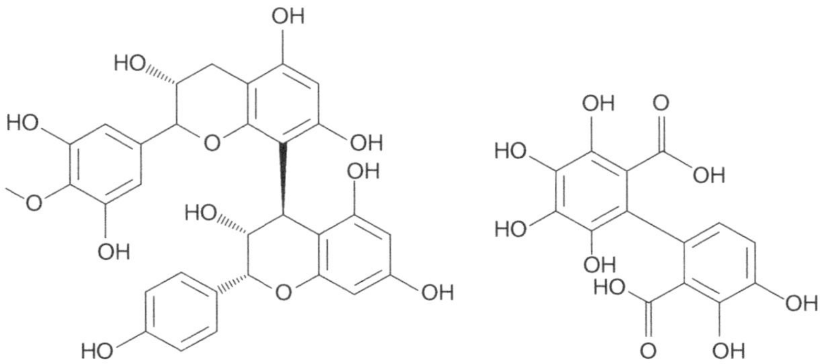

Proanthocyanidin Hexahydroxydiphenic acid

FIGURE 3.3 Chemical structures of Tannins.

## 3.3 EXTRACTION OF PHENOLIC COMPOUNDS

The initial step in the use of phytochemicals is the extraction of bioactive compounds from plant material. The phenolic compounds may be extracted from fresh, frozen, or dried plant samples. Plant samples are typically processed by treatment with milling, grinding, and homogenization, which can be followed by air-drying or freeze-drying. Overall, freeze-drying preserves greater levels of phenolic content in plant materials than air-drying (35). Phenolic compounds have been extracted from natural products using a variety of methods, including maceration, percolation, Soxhlet extraction, decoction, reflux extraction (36), UAE, pulsed electric field (PEF) extraction, enzyme-assisted extraction (EAE) (37,38), MAE, homogenate-assisted extraction (HAE), and subcritical water extraction (SWE) (39–41).

### 3.3.1 Maceration

The classical method of maceration extraction has been widely used to extract phenolic compounds from medicinal plants. Conventional extraction is an approach that moves target compounds into the solvent with the aid of solubility laws and intermolecular forces (such as dissolution). Ethanol or methanol is often selected as a solvent for phytochemical analysis in maceration extraction to obtain the organic phenolic constituents from the plant material. A shaker incubator, which maximizes the contact surface between samples and solvent, is often used to shake and control temperature, thereby hastening the extraction process.

The time duration for extraction in a given range of time sets the effectiveness of traditional extraction. More time spent extracting means more effective extractions; however, this will not affect the extraction process once solute equilibrium is achieved between the solid material inside and outside. A higher solvent-to-solid ratio is required to achieve a higher extraction yield; however,

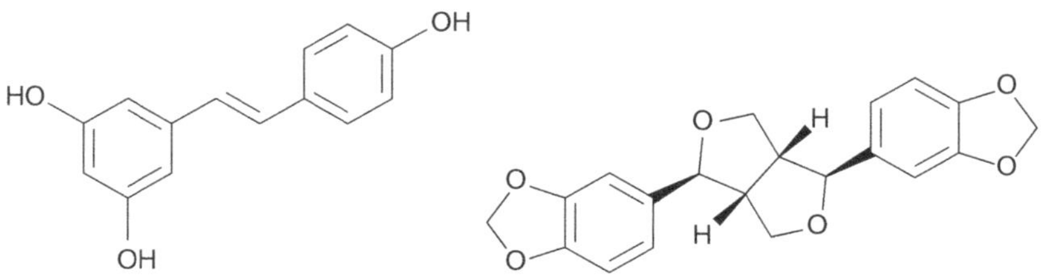

Resveratrol Sesamin

FIGURE 3.4 chemical structures of Lignans and Stilbenes.

using an excessively high ratio will result in excessive solvent consumption and a prolonged extraction time. Organic solvent maceration, then, which requires a high amount of solvents and a long extraction time, and is stable yet too inefficient, is generally employed in traditional extraction methods. Granulometry is another important factor that determines the efficiency of extraction (42).

### 3.3.2 Percolation

The crushed material is placed in a closed system, and the solvent is progressively dropped from the inlet toward the outlet in the percolation technique, which is comparable to the maceration method (43). Filtration is not necessary in this case since the percolator devices include filters that only let solvent containing the extract to flow through. The percolation approach has the same issues as the maceration method (long extraction times, high solvent volumes), but it also has issues with sample size, extraction time, and polyphenol solubility.

### 3.3.3 Soxhlet Extraction

In Soxhlet extraction, the ground-up materials are put in cellulose-based thimbles and set in the extraction chamber directly above the collection flask beneath a reflux condenser. After the solvent has been heated in the heating bottle, the vapor is released and condenses beneath cool running water before returning to the thimbles containing the sample. After repeatedly maintaining the reflux, the aqueous extract is eventually recovered from the heating flask. As a continuous procedure, Soxhlet extraction has the benefit of requiring less solvent and time than percolation and maceration techniques (43).

### 3.3.4 Decoction

The decoction approach involves either boiling the plant samples for a shorter period or covering them with boiling water and allowing the mixture to stand for a specified amount of time. This process works well with water-soluble and heat-stable phytochemicals from crude medications (44).

### 3.3.5 Reflux Extraction

The two parts of reflux extraction are the concentration tank and the extraction tank. During the extraction process, the solvent from the extraction tank is pumped into the concentration tank. The solvent is warmed in the concentration tank until it evaporates, condenses, and is pumped back into the extraction tank. The process will be repeated until the extracts in the solvent are built up. Some advantages of reflux extraction compared to conventional extraction include a minimized extraction time due to an enhanced mass transfer driving force, lower solvent costs due to solvent reuse, and a smaller fixed investment since no storage tank is required before concentration (45).

### 3.3.6 Ultrasound-Assisted Extraction (UAE)

Ultrasound is used in the UAE to mitigate cavitation caused by solvents, which can result in significant shear forces. When ultrasonics are utilized, cavitation bubbles implode, causing micro-jetting, which, along with macro-turbulence and micro-mixing, results in surface peeling, erosion, and particle breakup (46). Consequently, it speeds up heat transmission as well as the solute's solubility and diffusion. The UAE uses less solvent yet takes less time to extract than traditional extraction methods.

However, heavier ultrasonic treatment of the materials will result in a higher shear force, which might hasten the transformations. However, ultrasonic applications in the food industry are often optimized to produce the most significant outcome with the least amount of energy due to economic concerns (47). Temperature management, lowering the moisture content of food items, and increasing ultrasonic power are frequently considered to achieve maximum extraction efficiency. For instance, UAE has been reported to be more effective in extracting rosmarinic and carnosic acids than conventional extraction techniques (48).

### 3.3.7 Pulsed Electric Fieled Extraction (PEF)

The yield of extraction is enhanced by PEF extraction. PEF is a nonthermal extraction method that utilizes short, high-voltage pulses to disrupt membrane structures and release the content. Temperature, the number of pulses, specific energy input, and field intensity are some of the parameters that can affect the efficiency of PEF extraction.

The intact cytomembrane of plant cells serves as a semipermeable barrier, controlling what enters and exits the cells. As such, PEF treatment is a method that can disintegrate the cell membrane and further increase the cell wall's permeability, allowing more bioactive compounds to be released into the solvent. More material can be extracted because the selective permeability characteristics will be eliminated as a result of the breakdown of the cell's cytomembrane structure.

The research demonstrates that red onion phenolic compounds were extracted at an electric field intensity between 0.7 and 2.1 kV/cm and treatment times between 30 and 90 pulses at 25°C. Consequently, extractions of phenolic compounds and flavonoid compounds were significantly enhanced with increasing electric field intensity and treatment time (49).

### 3.3.8 Enzyme-Assisted Extraction (EAE)

The cell membrane components may be hydrolyzed by enzymes, disrupting their selective permeability and discharging the contents to accelerate extraction. Macromolecules such as polysaccharides and proteins constitute most of the cell membrane's structure. Proteins denature at high temperatures, which will lower the rate of extraction. Cellulase and other enzymes are employed in extraction as a nontoxic and nonthermal treatment to enhance extraction efficiency. Hai et al. experimented on the extraction of aged tea leaves and found that the total polyphenol content in the untreated group was less than that in samples treated with cellulase. Additionally, they found that the total polyphenol content increased with the increasing concentration of the enzyme, reaching its highest value at 74.45 mg GAE/g in a 2% v/w enzyme concentration (50).

### 3.3.9 Microwave-Assisted Extraction (MAE)

Analytes are displaced from the sample matrix into the solvent using microwave radiation through a process called MAE. In comparison to conventional extraction techniques, the major advantage is the reduction in extraction time and solvent volume (51). A number of small-molecule phenolic compounds have been extracted with it, such as trans-resveratrol, isoflavone, quercetin, and phenolic acids (such as gallic and ellagic acids) (52–55). These compounds are stable when subjected to heating using a microwave for 20 minutes at temperatures up to 100°C. Phenolic compounds that are thermally labile (such as anthocyanins) or contain more hydroxyl-type substituents (such as tannins) would not be the best fit for MAE extraction, as they will be decomposed under the conditions of MAE extraction (56).

### 3.3.10 Homogenate-Assisted Extraction (HAE)

HAE is a sophisticated and effective method that is increasingly being utilized to extract phenolic compounds from medicinal plants. By mechanically breaking down plant tissues using high-speed homogenizers, this technique significantly enhances the release of intracellular materials, including bioactive phenolics. Through the application of strong shear pressures, the homogenizer breaks down the cell walls, increasing solvent penetration and enhancing mass transfer, which in turn speeds up the extraction process. Usually, plant material is mixed with an appropriate solvent, such as methanol, ethanol, or aqueous solutions, and then homogenized at a speed of 10,000–25,000 rpm. This technique works exceptionally well with soft or semi-dry plant tissues. Its main benefits include shorter extraction times, reduced solvent use, and better preservation of thermolabile phenolic components, as it produces less heat during processing. Polyphenols from various medicinal plants have been effectively extracted using HAE, which offers high efficiency and compatibility with downstream analytical methods, including LC–MS, HPLC, and high-performance thin-layer chromatography (HPTLC) (57).

### 3.3.11 Subcritical Fluid Extraction (SFE)

SFE is a process of extraction that utilizes supercritical fluid (SF) as a solvent for extraction. SF is a solvent that exhibits liquid-like solubility and gas-like diffusivity, yet can dissolve a wide variety of natural materials. SFE significantly reduces oxidation and degradation processes compared to other extraction methods, as it operates without light or air. Supercritical carbon dioxide (S-$CO_2$) is a common SF used in extraction and has been utilized extensively. The low critical temperature (31°C), selectivity, inertness, and non-toxicity in extracting chemicals that are susceptible to high temperatures are among the benefits of employing S-$CO_2$. Due to its low polarity, S-$CO_2$ is an ideal solvent for extracting non-polar compounds, such as lipids. By adding a modifier, its solvating capabilities can be improved. However, co-solvents are frequently required to increase the solubility of phenolic compounds (particularly phenolic acids, such as gallic acid, methyl gallate, and caffeic acid) in the solvent, enabling their extraction using SFE (58).

## 3.4 CHARACTERIZATION OF PHENOLIC COMPOUNDS

The characterization of phenolic compounds extracted from medicinal plants is vital for determining their structure, concentration, purity, and potential biological activity. These compounds exist in complex matrices, and their identification requires a combination of spectroscopic, chromatographic, and hyphenated analytical techniques (59).

### 3.4.1 Spectrophotometric Method

Spectrophotometric methods are widely used for characterizing and quantifying phenolic compounds due to their simplicity, cost-effectiveness, and suitability for routine analysis. These methods are particularly useful in assessing total phenolic content (TPC) in plant extracts, food samples, and herbal formulations, especially when advanced instruments like HPLC or LC–MS are not readily available (60).

The most widely used spectrophotometric method is the Folin–Ciocalteu assay, which involves the redox reaction of phenolic compounds with the Folin–Ciocalteu reagent, a mixture of phosphomolybdic and phosphotungstic acids (61). Under alkaline conditions, phenolic hydroxyl groups reduce the reagent to form a blue-colored complex, which can be quantified at 765 nm using a UV–Visible spectrophotometer. The intensity of the blue color is proportional to the concentration of total phenolics in the sample (62).

To run the assay, an extract volume is combined with the Folin–Ciocalteu reagent and sodium carbonate, incubated for a set period (typically 30–60 minutes), and then spectrophotometrically measured (61). A standard phenolic compound, such as gallic acid, is used to create a calibration curve, and the results are expressed as milligrams of gallic acid equivalents (GAE) per gram of extract or sample. Other spectrophotometric assays to detect specific groups of phenolics are as follows:

Aluminum chloride colorimetric assay for total flavonoid content (read at ~415 nm), vanillin–HCl assay for condensed tannins, and DMACA (p-dimethylaminocinnamaldehyde) reagent for particular flavanols such as catechins (63).

These methods, while not specific to individual phenolic compounds, provide valuable information on the overall phenolic profile and antioxidant potential of samples. However, results can be influenced by interfering substances, such as ascorbic acid, sugars, and other reducing agents.

### 3.4.2 UV–Visible Spectroscopy

UV–Visible spectroscopy is based on the absorption of ultraviolet or visible light, which results in electronic transitions, primarily $\pi \rightarrow \pi^*$ and $n \rightarrow \pi^*$, within the conjugated systems typical of phenolics. These compounds typically exhibit strong absorbance in the 200–400 nm range, with simple phenols showing absorption at around 270 nm (64). Polyphenolic compounds, such as flavonoids, show two characteristic absorption bands: Band I in the range of 300–380 nm, associated with the B-ring's cinnamoyl system, and Band II in the range of 240–280 nm, associated with the benzoyl system of the A-ring. This spectral data helps determine the degree of conjugation, aromaticity, and functional groups present, and is frequently used to estimate the TPC using reagents such as the Folin–Ciocalteu method. Results are expressed in GAE and are based on absorbance measurements, typically taken at 765 nm. It is appropriate for high-throughput screening and is easy to use and cost-effective (60)

### 3.4.3 Infra-Red Spectroscopy

In contrast, when exposed to infrared (IR) radiation, IR spectroscopy uses molecular vibrations to identify functional groups. In phenolic compounds, a broad absorption band around 3200–3550 $cm^{-1}$ indicates O–H stretching due to hydrogen bonding, while peaks around 1450–1600 $cm^{-1}$ confirm aromatic C=C stretching vibrations (Table 3.2). The C–O stretching of phenolic hydroxyl groups typically appears between 1150–1250 $cm^{-1}$, and O–H bending vibrations can be seen near 1310–1410 $cm^{-1}$. These IR spectral features are essential for verifying the presence of hydroxyl groups, aromatic rings, and their bonding environment. When combined, UV–Visible and IR spectroscopy provide a reliable and non-destructive method for the initial characterization of phenolic compounds, facilitating the identification of these compounds, evaluation of their purity, and subsequent structural investigations (65).

Fourier transform infrared (FTIR) spectroscopy, a more advanced form of IR spectroscopy, provides **fingerprint information** about functional groups. It is used for quality assessment, functional group identification, and confirmation of phenolic presence in complex matrices (66).

### 3.4.4 Thin-Layer Chromatography (TLC)

Thin-layer chromatography (TLC) is a simple, rapid, and cost-effective analytical method commonly used for the initial identification and characterization of phenolic compounds, particularly in plant extracts and herbal preparations. In TLC, a minimal amount of sample extract is applied as a

**TABLE 3.2**
**Characteristic IR Absorption Peaks for Phenolic Compounds**

| Vibrational Mode | Wavenumber ($cm^{-1}$) | Assignment |
|---|---|---|
| O–H stretching (phenolic) | 3200–3550 (broad) | Due to hydrogen bonding |
| C–H stretching (aromatic) | ~3000–3100 | Aromatic ring hydrogens |
| C=C stretching (aromatic) | 1450–1600 | Aromatic ring vibrations |
| C–O stretching (phenolic) | 1150–1250 | Phenolic C–O bond |
| O–H bending | 1310–1410 | Especially if free O–H is present |

spot onto a TLC plate having a thin layer of a stationary phase, typically silica gel. The plate is then left in a developing chamber with a mobile phase (solvent system), which travels up the plate via capillary action. Phenolic compounds travel at varying speeds based on their polarity and affinities with the stationary and mobile phases, creating various spots of different heights on the plate. Every compound can be described in terms of retention factor (Rf value), which is determined as the ratio of the distance traveled by the compound to the distance traveled by the solvent front (67).

After development, the spots corresponding to phenolic compounds are visualized under UV light (254 nm or 366 nm) or by spraying with specific reagents such as ferric chloride, which forms colored complexes with phenolics, or Folin–Ciocalteu reagent for more specific detection. Common solvent systems used for phenolic separation include mixtures of ethyl acetate, formic acid, acetic acid, and water, depending on the nature of the phenolics being analyzed. It is beneficial in differentiating between types of phenolic compounds such as flavonoids, phenolic acids, and tannins (68).

### 3.4.5 High-Performance Thin-Layer Chromatography (HPTLC)

HPTLC is a sophisticated and optimized form of conventional TLC with improved resolution, sensitivity, and reproducibility. It is widely applied to the qualitative and quantitative analysis of phenolic substances in plant extracts, medicinal herbs, and foodstuffs. HPTLC has emerged as a highly effective tool in phytochemical analysis due to its speed, economy, and ability to process multiple samples in parallel (69).

In HPTLC, pre-coated plates with a uniform and thin layer of stationary phase (generally silica gel) are employed, along with automated sample applicators and development chambers to provide precision and accuracy. Phenolic compounds are resolved according to their polarity and interaction with mobile and stationary phases. The Rf value (retention factor) is determined for each compound, and reference standards such as gallic acid, quercetin, or caffeic acid are used for comparison (70).

Visualization of the phenolic spots is affected under UV light (254 or 366 nm) or with specific derivatizing reagents like ferric chloride, which gives a reaction with the phenolic hydroxyl groups to form characteristic colors. Following development and visualization, densitometric scanning is employed to evaluate the compounds by measuring the intensity of each spot, yielding precise and reproducible results.

HPTLC is particularly suitable for profiling complex mixtures of phenolic compounds, such as flavonoids, phenolic acids, and tannins. It is widely used for the quality control of herbal formulations, the detection of adulterants, and the authentication of raw materials. The method supports the analysis of multiple samples simultaneously and is therefore ideal for routine screening (71).

### 3.4.6 High-Performance Liquid Chromatography (HPLC) and LC–MS

One of the widely used analytical techniques to identify phenolic chemicals is HPLC, which is known for its high resolution, sensitivity, and ability to resolve complex mixtures. Phenolic compounds like phenolic acids, tannins, and flavonoids that are prevalently present in plant extracts exhibit broad polarities and chemical structures (72). HPLC can easily separate, identify, and quantify them. Analyte separation, achieved by a mobile phase (typically a combination of water and an organic solvent, such as methanol or acetonitrile) and a stationary phase (typically an RP-C18 column), is the basis of the technique. Improved peak shape and resolution are achieved by acidifying the mobile phase with phosphoric or formic acid. A common approach to separate phenolic compounds of varying polarity is gradient elution (14).

Phenolics are usually detected in HPLC using UV–Visible or diode-array detectors (DAD). Common detection wavelengths for flavonoids are 360–370 nm, while for simple phenols and phenolic acids, they are 280 nm. More sophisticated detection methods, including HPLC combined with mass spectrometry (HPLC–MS), provide accurate structural elucidation and molecular weight determination, which is particularly helpful for small or unknown components. To prepare a sample, plant material must be extracted, filtered, and possibly cleaned using solid-phase extraction (SPE) to remove any contaminants. Usually, injection volumes fall between 10 and 20 μL (68).

The identification of individual phenolic compounds by comparison with reference standards, the measurement of TPC, quality control of herbal and nutraceutical products, and pharmacokinetic studies of bioactive phenolics are just a few of the many uses for HPLC in natural product research. For instance, HPLC can efficiently separate and measure catechins, such as epicatechin, epigallocatechin gallate (EGCG), and gallic acid, in the examination of green tea extracts (73). All things considered, HPLC is a strong and crucial instrument for characterizing phenolic compounds, as it provides precision, repeatability, and the ability to work with complex matrices.

Although HPLC is an excellent tool for quantitative analysis, it might not be able to identify unknown chemicals with certainty. LC–MS overcomes this constraint by combining the molecular selectivity of mass spectrometry with the separation power of HPLC. The molecular weight, fragmentation patterns, and structural characteristics of phenolic compounds are all described by LC–MS. It is beneficial for complicated plant matrices when it is necessary to identify unknown or trace-level phenolics. LC–MS is ideal for metabolite profiling, structural elucidation, and untargeted phenolic screening, as it can identify compounds without the need for reference standards (74).

HPLC and LC–MS work together to provide a comprehensive platform for studying phenolic compounds. While LC–MS improves the capacity to detect novel or low-abundance compounds based on their mass-to-charge ratios, HPLC guarantees precise quantification and separation of recognized chemicals. Studies of phenolic bioavailability and metabolism in biological systems, nutritional analysis of foods, quality control of herbal medications, and phytochemical research all make substantial use of this combination method (75).

### 3.4.7 Gas Chromatography–Mass Spectrometry (GC–MS)

Gas chromatography–mass spectrometry (GC–MS) is a valuable analytical method employed to identify and structurally elucidate volatile and semi-volatile phenolic compounds, especially those of low molecular weight. GC–MS combines the excellent separation capability of gas chromatography (GC) with the accurate molecular identification capability of mass spectrometry (MS), making it a valuable analytical tool in phytochemical analysis, environmental research, and food science (76).

Phenolic compounds are generally non-volatile and thermally unstable in their native form. Therefore, before GC–MS analysis, they are often subjected to derivatization, typically via silylation (using reagents like BSTFA or MSTFA), to convert hydroxyl groups into more volatile and thermally stable trimethylsilyl (TMS) ethers. This step is crucial for ensuring efficient vaporization and improved chromatographic behavior (77).

In the GC component, derivatized phenolic compounds are vaporized and carried by an inert gas (e.g., helium) through a capillary column coated with a non-polar or moderately polar stationary phase. Compounds are separated based on their volatility and interaction with the column. As they elute, they enter the mass spectrometer, where they are ionized (usually by electron impact, EI), fragmented, and detected based on their mass-to-charge (m/z) ratios. Each compound produces a unique fragmentation pattern (mass spectrum), which can be compared with reference libraries (e.g., NIST) for identification (78).

GC–MS is particularly suited for analyzing simple phenolic acids (such as benzoic acid, ferulic acid, and vanillic acid), phenolic aldehydes, and phenolic alcohols, which are often found in essential oils, fruits, beverages, and plant extracts. It is also widely used in metabolomics and the profiling of phenolic degradation products in processed foods or herbal preparations (79).

### 3.4.8 Nuclear Magnetic Resonance (NMR) Spectroscopy

Nuclear magnetic resonance (NMR) spectroscopy is one of the most powerful and informative techniques for the structural characterization of phenolic compounds, particularly in their pure and isolated forms. Unlike chromatographic methods that focus on separation and quantification, NMR provides direct insight into the molecular structure, functional groups, and chemical environment of atoms within a compound (80).

In phenolic compounds, NMR spectroscopy is helpful in identifying the aromatic ring system, hydroxyl groups, and the positions of substituents. The most widely employed forms of NMR for phenolic analysis are $^{1}H$ NMR (proton NMR) and $^{13}C$ NMR (carbon-13 NMR). In $^{1}H$ NMR spectra, phenolic protons (–OH) generally occur as broad singlets between $\delta$ 4 and 7 ppm, whereas aromatic protons occur in the range of $\delta$ 6–8 ppm, depending on the nature of the substitution. Aromatic carbons are found in $^{13}C$ NMR between $\delta$ 100 and 160 ppm, while the C–OH carbons are typically downfield ($\delta > 150$ ppm) as a result of the hydroxyl group's electron-withdrawing effect (81).

Sophisticated 2D-NMR methods, such as COSY (correlation spectroscopy), HSQC (heteronuclear single quantum coherence), and HMBC (heteronuclear multiple bond correlation), provide extensive connectivity and structural information. These are especially important for the identification of complex polyphenols, such as flavonoids, tannins, and lignans, which contain multiple aromatic rings and hydroxyl substitutions. NMR is also important in ascertaining the purity of isolated compounds, detecting isomerism, and establishing stereochemistry in chiral phenolic molecules (81).

One major advantage of NMR is that it requires minimal sample preparation and provides non-destructive analysis, making it an ideal tool for the structural elucidation of both natural and synthetic phenolic compounds. However, NMR typically requires larger quantities of pure compounds (usually in the milligram range) and is less sensitive compared to MS-based techniques (82).

## 3.5 CONCLUSION

The successful characterization and extraction of phenolic compounds requires a strategic blending of methodologies tailored to the sample type and analytical objectives. Extraction procedures, including maceration, Soxhlet extraction, UAE, and HAE, play a crucial role in the separation of phenolics from plant matrices. The yield and phenolic composition are greatly affected by the solvent used, temperature, and extraction time. Green and advanced extraction methods, such as UAE and HAE, are also gaining popularity due to their cost-efficiency, speed, and eco-friendliness (83).

For characterization, UV–Visible spectroscopy provides a rapid and cost-effective method to detect the presence of conjugated phenolic systems and quantify TPC, commonly using the Folin–Ciocalteu reagent. IR and FTIR spectroscopy help identify characteristic functional groups, confirming phenolic structures based on O–H and aromatic C=C vibrations. HPLC is the most widely used technique for the accurate separation and quantification of individual phenolic compounds. At the same time, LC–MS adds powerful structural elucidation capabilities by combining chromatographic separation with mass-based identification, even for compounds present in trace amounts (84).

GC–MS is particularly suited for volatile or derivatized phenolics, providing detailed fragmentation patterns essential for the identification of low-molecular-weight compounds (85). NMR spectroscopy, although requiring pure samples and larger quantities, remains unmatched for detailed structural analysis, offering insights into molecular frameworks, connectivity, and stereochemistry (80). TLC and HPTLC are valuable for preliminary screening and comparison of phenolic profiles, with HPTLC providing enhanced resolution and quantitative capabilities through densitometry (85).

Spectrophotometric assays, especially the Folin–Ciocalteu method, are widely used for estimating TPC and are essential for routine screening and assessing antioxidant potential (86).

In summary, an integrated approach combining appropriate extraction techniques with complementary analytical methods—ranging from simple spectrophotometric assays to advanced chromatographic and spectroscopic tools—is essential for the comprehensive study of phenolic compounds. This holistic strategy ensures the accurate extraction, identification, quantification, and quality evaluation of compounds, supporting research, product development, and quality control in the pharmaceutical, nutraceutical, and food industries.

## REFERENCES


1. Tsao R. Chemistry and biochemistry of dietary polyphenols. Nutrients. 2010 Dec;2(12):1231–46.
2. Rana A, Samtiya M, Dhewa T, Mishra V, Aluko RE. Health benefits of polyphenols: A concise review. J Food Biochem. 2022 Oct;46(10):e14264.
3. Kabir F, Sultana MS, Kurnianta H. Antimicrobial activities of grape (*Vitis vinifera* L.) pomace polyphenols as a source of naturally occurring bioactive components. African J Biotechnol. 2015;14(26): 2157–61.
4. Jiao X, Wang Y, Lin Y, Lang Y, Li E, Zhang X, et al. Blueberry polyphenols extract as a potential prebiotic with anti-obesity effects on C57BL/6 J mice by modulating the gut microbiota. J Nutr Biochem. 2019;64:88–100.
5. Matulja D, Vranješević F, Kolympadi Markovic M, Pavelić SK, Marković D. Anticancer activities of marine-derived phenolic compounds and their derivatives. Molecules. 2022;27(4):1449.
6. Praparatana R, Maliyam P, Barrows LR, Puttarak P. Flavonoids and phenols, the potential anti-diabetic compounds from *Bauhinia strychnifolia* Craib. Stem. Molecules. 2022;27(8):2393.
7. Oldoni TLC, Merlin N, Bicas TC, Prasniewski A, Carpes ST, Ascari J, et al. Antihyperglycemic activity of crude extract and isolation of phenolic compounds with antioxidant activity from *Moringa oleifera* Lam. leaves grown in Southern Brazil. Food Res Int. 2021;141:110082.
8. Hwang IW, Chung SK. Isolation and identification of myricitrin, an antioxidant flavonoid, from Daebong persimmon peel. Prev Nutr Food Sci. 2018;23(4):341.
9. Ruiz-Ruiz JC, Matus-Basto AJ, Acereto-Escoffié P, Segura-Campos MR. Antioxidant and anti-inflammatory activities of phenolic compounds isolated from *Melipona beecheii* honey. Food Agric Immunol. 2017;28(6):1424–37.
10. Zhu X, Li N, Wang Y, Ding L, Chen H, Yu Y, et al. Protective effects of quercetin on UVB irradiation-induced cytotoxicity through ROS clearance in keratinocyte cells. Oncol Rep. 2017;37(1):209–18.
11. Zhang QW, Lin LG, Ye WC. Techniques for extraction and isolation of natural products: A comprehensive review. Chin Med. 2018;13(1):26.
12. Dai J, Mumper RJ. Plant phenolics: Extraction, analysis and their antioxidant and anticancer properties. Molecules. 2010;15(10):7313–52.
13. Azmir J, Zaidul ISM, Rahman MM, Sharif KM, Mohamed A, Sahena F, et al. Techniques for extraction of bioactive compounds from plant materials: A review. J Food Eng. 2013;117(4):426–36.
14. Ignat I, Volf I, Popa VI. A critical review of methods for characterisation of polyphenolic compounds in fruits and vegetables. Food Chem. 2011;126(4):1821–35.
15. Lobiuc A, Pavăl NE, Mangalagiu II, Gheorghiță R, Teliban GC, Amăriucăi-Mantu D, et al. Future antimicrobials: Natural and functionalized phenolics. Molecules. 2023 Jan;28(3).
16. Cushnie TPT, Lamb AJ. Antimicrobial activity of flavonoids. Int J Antimicrob Agents. 2005 Nov;26(5):343–56.
17. Shamsudin NF, Ahmed QU, Mahmood S, Ali Shah SA, Khatib A, Mukhtar S, et al. Antibacterial effects of flavonoids and their structure-activity relationship study: A comparative interpretation. Molecules. 2022 Feb;27(4).
18. Hadidi M, Liñán-Atero R, Tarahi M, Christodoulou MC, Aghababaei F. The potential health benefits of gallic acid: Therapeutic and food applications. Antioxidants. 2024;13.
19. Cabrera C, Artacho R, Giménez R. Beneficial effects of green tea—A review. J Am Coll Nutr. 2006;25(2):79–99.
20. Aggarwal BB, Gupta SC, Sung B. Curcumin: An orally bioavailable blocker of TNF and other pro-inflammatory biomarkers. Br J Pharmacol. 2013;169(8):1672–92.
21. Baur JA, Sinclair DA. Therapeutic potential of resveratrol: The in vivo evidence. Nat Rev Drug Discov. 2006 Jun;5(6):493–506.
22. Lansky EP, Newman RA. *Punica granatum* (pomegranate) and its potential for prevention and treatment of inflammation and cancer. J Ethnopharmacol. 2007;109(2):177–206.
23. Omar SH. Oleuropein in olive and its pharmacological effects. Sci Pharm. 2010;78(2):133–54.
24. Ranasinghe P, Pigera S, Premakumara GAS, Galappaththy P, Constantine GR, Katulanda P. Medicinal properties of 'true' cinnamon (*Cinnamomum zeylanicum*): A systematic review. BMC Complement Altern Med. 2013;13(1):10.
25. Chaieb K, Hajlaoui H, Zmantar T, Kahla-Nakbi AB, Rouabhia M, Mahdouani K, et al. The chemical composition and biological activity of clove essential oil, *Eugenia caryophyllata* (*Syzigium aromaticum* L. Myrtaceae): A short review. Phyther Res An Int J Devoted to Pharmacol Toxicol Eval Nat Prod Deriv. 2007;21(6):501–6.

26. Bozin B, Mimica-Dukic N, Samojlik I, Jovin E. Antimicrobial and antioxidant properties of rosemary and sage (*Rosmarinus officinalis* L. And *Salvia officinalis* L., Lamiaceae) essential oils. J Agric Food Chem. 2007;55(19):7879–85.
27. Mondal S, Mirdha BR, Mahapatra SC. The science behind sacredness of Tulsi (*Ocimum sanctum* Linn.). Indian J Physiol Pharmacol. 2009;53(4):291–306.
28. Scartezzini P, Speroni E. Review on some plants of Indian traditional medicine with antioxidant activity. J Ethnopharmacol. 2000;71(1–2):23–43.
29. Manguro LOA, Lemmen P. Phenolics of *Moringa oleifera* leaves. Nat Prod Res. 2007;21(1):56–68.
30. Sodaeizadeh H, Rafieiolhossaini M, Havlík J, Van Damme P. Allelopathic activity of different plant parts of *Peganum harmala* L. and identification of their growth inhibitors substances. Plant Growth Regul. 2009;59:227–36.
31. Koksal E., Bursal E., Gulcin İ., Korkmaz M., Caglayan C., Goren A. C., Alwasel S. H. Antioxidant activity and polyphenol content of Turkish thyme (*Thymus vulgaris*) monitored by liquid chromatography and tandem mass spectrometry. International Journal of Food Properties,2017; 20(3), 514–525. https://doi.org/10.1080/10942912.2016.1168438
32. Shan B, Cai YZ, Sun M, Corke H. Antioxidant capacity of 26 spice extracts and characterization of their phenolic constituents. J Agric Food Chem. 2005;53(20):7749–59.
33. Singh M, Tiwari N, Shanker K, Verma RK, Gupta AK, Gupta MM. Two new lignans from *Phyllanthus amarus*. J Asian Nat Prod Res. 2009;11(6):562–8.
34. Tian K, Zhang H, Chen X, Hu Z. Determination of five anthraquinones in medicinal plants by capillary zone electrophoresis with β-cyclodextrin addition. J Chromatogr A. 2006;1123(1):134–7.
35. Abascal K, Ganora L, Yarnell E. The effect of freeze-drying and its implications for botanical medicine: A review. Phyther Res An Int J Devoted to Pharmacol Toxicol Eval Nat Prod Deriv. 2005;19(8):655–60.
36. Arora S, Itankar P. Extraction, isolation and identification of flavonoid from *Chenopodium album* aerial parts. J Tradit Complement Med. 2018;8(4):476–82.
37. Hao J, Wang Z, Jia Y, Sun L, Fu Z, Zhao M, et al. Optimization of ultrasonic-assisted extraction of flavonoids from *Lactuca indica* L. cv. Mengzao and their antioxidant properties. Front Nutr. 2023;10:1065662.
38. Liu R, Chu X, Su J, Fu X, Kan Q, Wang X, et al. Enzyme-assisted ultrasonic extraction of total flavonoids from *Acanthopanax senticosus* and their enrichment and antioxidant properties. Processes. 2021;9(10):1708.
39. Choommongkol V, Punturee K, Klumphu P, Rattanaburi P, Meepowpan P, Suttiarporn P. Microwave-assisted extraction of anticancer flavonoid, 2′, 4′-dihydroxy-6′-methoxy-3′, 5′-dimethyl chalcone (DMC), rich extract from *Syzygium nervosum* fruits. Molecules. 2022;27(4):1397.
40. Yang J, Li N, Wang C, Chang T, Jiang H. Ultrasound-homogenization-assisted extraction of polyphenols from coconut mesocarp: Optimization study. Ultrason Sonochem. 2021;78:105739.
41. Radovanović K, Gavarić N, Švarc-Gajić J, Brezo-Borjan T, Zlatković B, Lončar B, et al. Subcritical water extraction as an effective technique for the isolation of phenolic compounds of *Achillea* species. Processes. 2022;11(1):86.
42. Shi L, Zhao W, Yang Z, Subbiah V, Suleria HAR. Extraction and characterization of phenolic compounds and their potential antioxidant activities. Environ Sci Pollut Res. 2022;29(54):81112–29.
43. Azwanida NN. A review on the extraction methods use in medicinal plants, principle, strength and limitation. Med Aromat Plants. 2015; 4(3):196, doi:10.4172/2167-0412.1000196.
44. Sticher O. Natural product isolation. Nat Prod Rep. 2008;25(3):517–54.
45. Zhao YL, Hou QC, Li L, Guo JR, Cui YJ, Li P, et al. Preparation of polysaccharides from *Grifola frondosa* using dynamic hot reflux extraction. J Cent South Univ (Science Technol. 2011;42:38–45.
46. Chemat F, Rombaut N, Sicaire AG, Meullemiestre A, Fabiano-Tixier AS, Abert-Vian M. Ultrasound assisted extraction of food and natural products. Mechanisms, techniques, combinations, protocols and applications. A review. Ultrason Sonochem. 2017;34:540–60.
47. Bermúdez-Aguirre D, Barbosa-Cánovas GV. An update on high hydrostatic pressure, from the laboratory to industrial applications. Food Eng Rev. 2011;3:44–61.
48. Zu G, Zhang R, Yang L, Ma C, Zu Y, Wang W, et al. Ultrasound-assisted extraction of carnosic acid and rosmarinic acid using ionic liquid solution from *Rosmarinus officinalis*. Int J Mol Sci. 2012;13(9):11027–43.
49. Liu Z, Zeng X, Ngadi M. Enhanced extraction of phenolic compounds from onion by pulsed electric field (PEF). J Food Process Preserv. 2018;42(9):e13755.
50. Hai TC, Nam ND, Hong Anh LT, Vu TA, Man PV. Enzyme assisted extraction of polyphenols from the old tea leaves. J Nutr Heal Sci. 2016;3(4):404–10.

51. Eskilsson CS, Björklund E. Analytical-scale microwave-assisted extraction. J Chromatogr A. 2000;902(1):227–50.
52. Li H, Chen B, Nie L, Yao S. Solvent effects on focused microwave assisted extraction of polyphenolic acids from *Eucommia ulmodies*. Phytochem Anal An Int J Plant Chem Biochem Tech. 2004;15(5):306–12.
53. Huang J, Zhang Z. Microwave-assisted extraction of quercetin and acid degradation of its glycosides in *Psidium guajava* leaves. Anal Sci. 2004;20(2):395–7.
54. Rostagno MA, Palma M, Barroso CG. Microwave assisted extraction of soy isoflavones. Anal Chim Acta. 2007;588(2):274–82.
55. Du FY, Xiao XH, Li GK. Application of ionic liquids in the microwave-assisted extraction of trans-resveratrol from *Rhizma Polygoni Cuspidati*. J Chromatogr A. 2007;1140(1–2):56–62.
56. Liazid A, Palma M, Brigui J, Barroso CG. Investigation on phenolic compounds stability during microwave-assisted extraction. J Chromatogr A. 2007;1140(1–2):29–34.
57. Chemat F, Abert Vian M, Ravi HK, Khadhraoui B, Hilali S, Perino S, et al. Review of alternative solvents for green extraction of food and natural products: Panorama, principles, applications and prospects. Molecules. 2019;24(16):3007.
58. Radzali SA, Markom M, Saleh NM. Co-solvent selection for supercritical fluid extraction (SFE) of phenolic compounds from *Labisia pumila*. Molecules. 2020;25(24):5859.
59. Khoddami A, Wilkes MA, Roberts TH. Techniques for analysis of plant phenolic compounds. Molecules. 2013;18:2328–75.
60. Ainsworth EA, Gillespie KM. Estimation of total phenolic content and other oxidation substrates in plant tissues using Folin–Ciocalteu reagent. Nat Protoc [Internet]. 2007;2(4):875–7. Available from: https://doi.org/10.1038/nprot.2007.102
61. Singleton VL, Orthofer R, Lamuela-Raventós RM. [14] Analysis of total phenols and other oxidation substrates and antioxidants by means of Folin-Ciocalteu reagent. In: Oxidants and Antioxidants Part A [Internet]. Academic Press; 1999. p. 152–78. Available from: https://www.sciencedirect.com/science/article/pii/S0076687999990171
62. Everette JD, Bryant QM, Green AM, Abbey YA, Wangila GW, Walker RB. Thorough study of reactivity of various compound classes toward the Folin–Ciocalteu reagent. J Agric Food Chem. 2010;58(14):8139–44.
63. Chang CC, Yang MH, Wen HM, Chern JC. Estimation of total flavonoid content in propolis by two complementary colometric methods. J Food Drug Anal. 2002;10(3):3.
64. Silverstein RM, Webster FX, Kiemle DJ. Chapter 3, Proton NMR spectrometry. In: Spectrometric identification of organic compounds (7th ed). Hoboken John Wiley Sons, Inc; 2005. p. 127–203.
65. Harborne AJ. Phytochemical methods a guide to modern techniques of plant analysis. Springer Science & Business Media; Springer Dordrecht. 1998, 3: p. 37–89.
66. Skoog DA, Holler FJ, Crouch SR. Textbook "principles of instrumental analysis." Cengage Learn. 2019.
67. Sasidharan S, Chen Y, Saravanan D, Sundram KM, Latha LY. Extraction, isolation and characterization of bioactive compounds from plants' extracts. African J Tradit Complement Altern Med. 2011;8(1).
68. Stalikas CD. Extraction, separation, and detection methods for phenolic acids and flavonoids. J Sep Sci. 2007;30(18):3268–95.
69. El-Haggar M, El-Hosseiny L, Ghazy NM, El-Fiky FK, El-Hawiet A. Phytochemical investigation, antimicrobial and cytotoxic activities of suspension cultures of *Lepidium sativum* L South African J Bot [Internet]. 2021;138:500–5. Available from: https://www.sciencedirect.com/science/article/pii/S0254629921000028
70. Cho, Ei Mon, "Authentication of andrographis paniculata and herbal products using dna barcoding and high-resolution melting analysis" (2022). Chulalongkorn University Theses and Dissertations (Chula ETD). 12367. https://digital.car.chula.ac.th/chulaetd/12367
71. Srivastava M. High-Performance Thin-Layer Chromatography (HPTLC). Springer Science & Business Media; 2010.
72. Adamczak A, Ożarowski M, Karpiński TM. Antibacterial activity of some flavonoids and organic acids widely distributed in plants. J Clin Med. 2019;9(1):109.
73. Jakabová S, Árvay J, Šnirc M, Lakatošová J, Ondejčíková A, Golian J. HPLC-DAD method for simultaneous determination of gallic acid, catechins, and methylxanthines and its application in quantitative analysis and origin identification of green tea. Heliyon. 2024;10(16).
74. Kılınc H, D'Urso G, Paolillo A, Alankus O, Piacente S, Masullo M. LC-MS and NMR based plant metabolomics: A comprehensive phytochemical investigation of *Symphytum anatolicum*. Metabolites. 2023;13.

75. Muller M, de Villiers A. A detailed evaluation of the advantages and limitations of online RP-LC×HILIC compared to HILIC×RP-LC for phenolic analysis. J Chromatogr A. 2023 Mar;1692:463843.
76. Tran TK, Ha PTT, Henry RJ, Nguyen DNT, Tuyen PT, Liem NT. Polyphenol contents, gas chromatography-mass spectrometry (GC-MS) and antibacterial activity of methanol extract and fractions of *Sonneratia caseolaris* fruits from Ben Tre Province in Vietnam. J Microbiol Biotechnol. 2024 Jan;34(1):94–102.
77. Santa-Cruz-Mérida GV, Otomo JI, Araoz-Prado DR, Rodrigues EA, de Andrade DA, Bustillos OV. Advanced analytical approaches for phenolic compounds in groundwater: A PRISMA systematic review. Water. 2025;17.
78. Maji S, Roy C, Sinha SK. Gas chromatography-mass spectrometry (GC-MS): A comprehensive review of synergistic combinations and their applications in the past two decades. J Anal Sci Appl Biotechnol. 2023 Jul 30;5:72–85.
79. Garcia-Viñola V, Ruiz-de-Villa C, Gombau J, Poblet M, Bordons A, Reguant C, et al. Simultaneous analysis of organic acids, glycerol and phenolic acids in wines using gas chromatography-mass spectrometry. Foods. 2024;13.
80. Kisiriko M, Anastasiadi M, Terry L, Yasri A, Beale M, Ward J. Phenolics from medicinal and aromatic plants: Characterisation and potential as biostimulants and bioprotectants. Molecules. 2021 Oct 20;26:6343.
81. Krivdin LB. Recent advances in liquid-phase NMR of the coal-derived products. Magn Reson Chem. 2024;62(11):775–802.
82. Schild K, Sönnichsen FD, Martin D, Garamus VM, Van der Goot AJ, Schwarz K, et al. Unraveling the effects of low protein-phenol binding affinity on the structural properties of beta-lactoglobulin. Food Chem [Internet]. 2023;426:136496. Available from: https://www.sciencedirect.com/science/article/pii/S0308814623011147
83. Sulejmanović M, Milić N, Mourtzinos I, Nastić N, Kyriakoudi A, Drljača J, et al. Ultrasound-assisted and subcritical water extraction techniques for maximal recovery of phenolic compounds from raw ginger herbal dust toward in vitro biological activity investigation. Food Chem. 2024 Mar;437(Pt 1):137774.
84. Wongsa P, Phatikulrungsun P, Prathumthong S. FT-IR characteristics, phenolic profiles and inhibitory potential against digestive enzymes of 25 herbal infusions. Sci Rep. 2022 Apr;12(1):6631.
85. Custodio-Mendoza JA, Pokorski P, Aktaş H, Napiórkowska A, Kurek MA. Advances in chromatographic analysis of phenolic phytochemicals in foods: Bridging gaps and exploring new horizons. Foods. 2024;13.
86. Pérez M, Dominguez-López I, Lamuela-Raventós RM. The chemistry behind the Folin-Ciocalteu method for the estimation of (poly)phenol content in food: Total phenolic intake in a Mediterranean dietary pattern. J Agric Food Chem. 2023 Nov;71(46):17543–53.

# 4 Antioxidant Properties of Phenolic Compounds from Medicinal Plants

*Shaza H. Aly, Mohamed S. Abd El Hafeez, Omayma A. Eldahshan, and Abdel Nasser B. Singab*

## 4.1 INTRODUCTION

The increased popularity of antioxidants and phenolic compounds is due to their major importance as tools to counteract oxidative stress and prevent or treat chronic diseases (Sharifi-Rad et al. 2020). This chapter provides a foundational understanding of these compounds, their sources, and their mechanisms of action, making it an essential starting point for delving deeper into their biological significance and applications. Antioxidants are molecules that neutralize free radicals, thereby safeguarding cellular components such as lipids, proteins, and DNA from oxidative damage (Rahman et al. 2021); primarily, reactive oxygen species (ROS) and reactive nitrogen species (RNS) are highly reactive substances generated through physiological metabolism and exposure to external factors, including pollution, radiation, and toxins. While modest levels of ROS are crucial to cell communication and immune responses, their excessive accumulation induces oxidative stress, leading to aging and the onset of related illnesses, including cancer, cardiovascular disorders, diabetes, and neurodegeneration (Asif 2015). Natural antioxidants obtained from diet or endogenous metabolic cycles are essential for the preservation of oxidative status (Mileo & Miccadei 2016). They pair synergistically to detoxify free radicals, chelate transition metals, restore oxidative damage, and promote endogenous antioxidant gene expression. Phenolic compounds are significant natural antioxidants due to their high prevalence in plant-based foods and potent bioactive characteristics (Dixit et al. 2025; Lin et al. 2016). Phenolic compounds constitute a diverse category of secondary metabolites generated by plants, distinguished by the presence of one or more hydroxyl functional groups attached to an aromatic ring (Lutz et al. 2019). They include flavonoids, phenolic acids, tannins, lignans, and stilbenes. Plants produce these compounds as a defense against biotic and abiotic stressors, such as pathogens, ultraviolet radiation, and oxidative stress (Kruk et al. 2022).

Phenolic substances exhibit a range of complexity, encompassing small molecules such as phenolic acids and complicated structures like tannins (Rajendran et al. 2014). This structural diversity strengthens their broad biological activity, encompassing antioxidant, anti-inflammatory, antibacterial, and anticancer properties. Phenolic compounds primarily demonstrate antioxidant properties by donating hydrogen atoms or electrons, aiding in the neutralization of free radicals and their conversion into less reactive forms (Demirci-Cekic et al. 2022). Because of its conjugated aromatic systems and hydroxyl groups, it is an excellent radical scavenger and oxidatively damaging chain reaction inhibitor (Juszczyk et al. 2021). This review explores the current comprehension of the possible effectiveness of polyphenols from medicinal plants on various biological systems and their efficacy in vitro and in vivo studies, along with understanding their versatile structures and mechanisms of action against oxidative stress.

### 4.1.1 Importance of Medicinal Plants in Antioxidant Research

Berries, citrus fruits, apples, and grapes are all good sources of flavonoids and phenolic acids. Vegetables, leafy greens, broccoli, onions, and tomatoes, contain a wide array of phenolics with


DOI: 10.1201/9781003527671-4

antioxidant activity. Catechins are prevalent in green tea, while resveratrol and flavonoids subsidize the antioxidant properties of red wine (Padmaja et al. 2011). Some of the traditional medicinal plants contain phenolic compounds as reserves. As an example, turmeric harbors curcumin, a polyphenol with strong antioxidative and anti-inflammatory activities (Krishnaiah et al. 2011). In particular, *Ginkgo biloba*, green tea, and rosemary are known for their phenolic content and pharmacological effects (Pammi et al. 2023). Purple-fleshed sweet potato (*Ipomoea batatas*) cultivars were demonstrated to possess phenolic profile associated with radical scavenging activity, chlorogenic acid and anthocyanins being the main radical scavengers. Leaf extracts of *Bauhinia variegata*, a traditional medicinal plant, showed antioxidant properties through radical neutralization and iron-binding capabilities. Also, Spanish *Opuntia ficus-indica* cultivars revealed significant antioxidant activity due to their elevated concentration of phenolic components (Zafar et al. 2023).

Additional research on *Aesculus indica* revealed that mandelic acid and quercetin are two significant bioactive compounds with strong antioxidant activities. Rhizome extracts of *Polygonatum verticillatum* exhibited antioxidant properties that are related to their phenolic contents (Sultana et al. 2023). The wild tea *Camellia fangchengensis* was identified as having high levels of flavan-3-ol-type, bioactive, antioxidant compounds (Škrovánková et al. 2012). Gymnosperms like various *Pinus* species, including Turkish pine and Scots pine, also revealed high antioxidant activity in bark and needle extracts, with bark providing a richer source of phenolics (Hassan et al. 2017).

## 4.2 TYPES OF PHENOLICS IN MEDICINAL PLANTS

Phenolics constitute the key constituents of medicinal plants with various structures and functions that contribute to their well-known antioxidant, anti-inflammatory, and therapeutic actions (Palipoch 2013). The main categories of phenolic compounds, including flavonoids, phenolic acids, tannins, and lignans, are recognized for their bioactivity and significance in pharmacology and nutrition (Figure 4.1). Knowledge of their chemical structure, origins, how they work, and possible health benefits improves our capacity to exploit these molecules in contemporary medicine (Atawodi 2005).

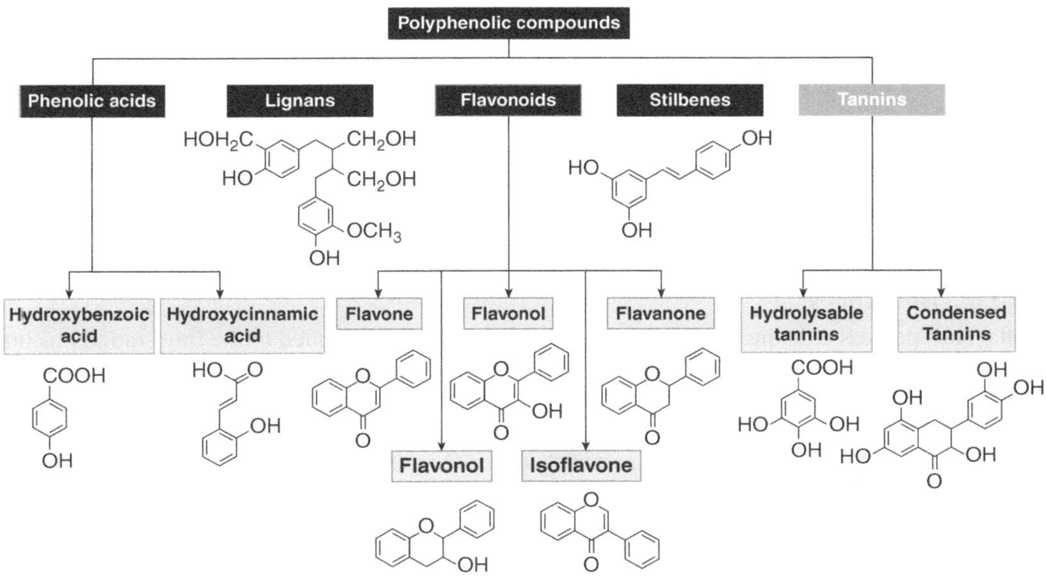


**FIGURE 4.1** The general chemical structure of common polyphenolic compounds.

### 4.2.1 Flavonoids

This class is defined by a 15-carbon structure and two aromatic rings (A and B) connected by a three-carbon bridge (C-ring), and it is categorized into flavonols, flavones, flavanones, flavan-3-ols, anthocyanins, and isoflavones (Panche et al. 2016). According to Harborne and Mabry (2013), flavonoids constitute a vast and varied category of secondary metabolites present in plants, esteemed for their beneficial health effects and their role in plant defense mechanisms. There are six primary types based on variations in their chemical structures and the extent of oxidation of the core pyran ring (Pietta 2000). Additionally, the source materials, hallmark molecules, and therapeutic effects of flavonoids vary (Pietta 2000).

#### 4.2.1.1 Flavones

Flavones are distinguished by the presence of a ketone group and a double bond in the ring between the second and third positions. These compounds occur in high concentrations in parsley, celery, chamomile, and citrus (sic). Notable flavones encompass apigenin and luteolin, which are recognized for their significant antioxidant, anti-inflammatory, and anticancer characteristics (Samanta et al. 2011).

#### 4.2.1.2 Flavonols

Flavonols are structurally related to flavones with one more hydroxyl group at the 3-position of the core ring (Harborne 2013). They are also prevalent in onions, kale, broccoli, apples, and berries. Flavonols, including quercetin, kaempferol, and myricetin, are known as strong antioxidants. These compounds are particularly valuable for preventing cardiovascular disease and strengthening immunological function (Kumar & Pandey 2013).

#### 4.2.1.3 Flavanones

Flavanones differ from flavones by having a saturated central ring, with no double bond between the second and third carbon atoms (Iwashina 2000). These flavonoids are especially concentrated in citric fruits, such as oranges, lemons, and grapefruits (Kesarkar et al., 2009). Hesperidin, naringenin, and eriodictyol are some super flavanones, and they have got great antioxidant powers as well as knocking back inflammation and some pretty impressive antibacterial punch too. They also show a role in promoting skin health and reducing inflammation (Guven et al. 2019).

#### 4.2.1.4 Flavanols (or Flavan-3-Ols)

Flavanols are unique since they do not contain the central ring double bond, and they are highly hydroxylated (Ullah et al. 2020). They are highly prevalent in green tea, cocoa, grapes, and apples. Catechin, epicatechin, and epigallocatechin gallate (EGCG) are well-known flavanols (Terahara 2015). These compounds are recognized for their capacity to enhance heart health, stimulate brain function, and lower oxidative stress, making them crucial in nutritional foods like tea and dark chocolate (Tapas et al. 2008).

#### 4.2.1.5 Isoflavones

Isoflavones possess a distinctive structure wherein the B-ring is bonded to the third radial position of the central ring rather than the second (Kumar & Pandey 2013). These ingredients are usually abundant in legumes like soy, chickpeas, and other beans. Genistein and daidzein are potent isoflavones with phytoestrogenic activity, useful for the treatment of hormonal dysfunctions and for preventing hormone-related cancer, e.g., breast cancer (Nijveldt et al. 2001).

#### 4.2.1.6 Anthocyanins

Anthocyanins are intensely colored flavonoids that give red, blue, and purple shades to a variety of fruits and vegetables. They are numerous in blueberries, blackberries, cherries, and red cabbage. Cyanidin, delphinidin, and malvidin belong to the most important anthocyanins exhibiting antioxidant and anti-inflammatory properties. These compounds contribute to cardiovascular health, vision, and neurodegenerative disease prevention (Harborne & Williams 2001).

Additionally, there are other flavonoid subtypes worth noting. Chalcones, flavonoid precursors, are present in tomatoes and licorice and possess anti-inflammatory and antimicrobial activity. The hepatoprotective properties of flavonolignans become evident from their mixture of flavonoids and lignans (Harborne & Williams 2001) as seen in the silymarin substance from milk thistle. The unique protective mechanisms that each flavonoid provides for both plants and humans explain why they are essential dietary components and promising research targets for medical and pharmaceutical work (Harborne & Williams 2001).

### 4.2.2 Phenolic Acids

The plant-derived metabolite group known as phenolic acids contains a phenolic ring structure and a carboxylic acid functional group (Robbins 2003). Natural environments contain these compounds abundantly in fruits and vegetables, cereals, and medicinal plants because of their potent protection against oxidation (Kumar & Goel 2019). The prevention and treatment of multiple persistent diseases hinge on phenolic acids, while these plant compounds play multiple defense roles in both plant and human health (Kumar & Goel 2019).

The chemical group of hydroxybenzoic acids includes gallic acid as well as vanillic acid, and protocatechuic acid since these compounds possess a benzene ring structure with hydroxyl groups along with a single carboxylic acid group, according to Croft (1998). These compounds occurring in berries, nuts, and tea can scavenge free radicals while defending against oxidative stress (Rashmi & Negi 2020). The hydroxycinnamic acid group consists of caffeic, ferulic, sinapic, and p-coumaric acids. These acids exist in higher concentrations than hydroxybenzoic acids and primarily appear in coffee, whole grains, and vegetables. The antioxidant capabilities of hydroxycinnamic acids stand out as prominent while they simultaneously demonstrate anti-inflammatory behavior and bactericidal properties (Khadem & Marles 2010).

Phenolic acids function as antioxidants because they remove ROS and RNS byproducts produced through metabolic processes (Heleno et al. 2015). Oxidative stress occurs when ROS and RNS measurements are outside normal ranges, damaging cells and triggering chronic diseases such as cardiovascular disease, neurodegenerative disorders, and cancer (Soares 2002). These effects are mitigated by phenolic acids, which bind to free radicals and chelate metal ions to protect cells and tissues from oxidative damage (Saibabu et al. 2015).

In addition to their antioxidant properties, phenolic acids show a large variety of biological activities. This has been known to have anti-inflammatory capabilities that lead to the modulation of inflammatory pathways and inhibition of proinflammatory cytokines (Abotaleb et al. 2020; Saibabu et al. 2015). As a result, they are helpful in the treatment of arthritis and other inflammatory bowel diseases. In addition, many phenolic acids, such as ferulic acid and caffeic acid, have antifungal and antibacterial effects, preventing bacterial, fungal, and viral growth (Valanciene et al. 2020).

Phenolic acids contribute to the prevention of chronic diseases. For instance, ferulic acid is examined for its cardioprotective benefits, as it aids in reducing serum cholesterol and improves vascular function. Caffeic acid is associated with neuroprotection and has the potential to avert age-related cognitive decline and neurodegenerative disorders, including Alzheimer's disease. Gallic acid has been studied for its anticancer properties, as it induces apoptosis in cancer cells while sparing healthy cells (Valanciene et al. 2020).

The type and content of phenolic acids in plants may be different owing to plant species, growth conditions, and processing treatment. Phenolic acids are commonly enriched in the outer parts of seed, grain, and fruit, and as such whole foods are highly enriched sources of these phytochemicals. The bioavailability of phenolic acids in the human body is also influenced by their structural features and interaction with the gut microbiota, which could metabolize phenolic acids into bioactive forms (Khan et al. 2022).

### 4.2.3 Tannins

Tannins are high-molecular-weight polyphenolic compounds that can bind to proteins and other macromolecules. They are classified as hydrolyzable tannins (e.g., gallotannins and ellagitannins) and condensed tannins (proanthocyanidins) (Chung et al. 1998). Tannins are a family of naturally occurring polyphenolic chemicals present at high levels in plants. They are essentially best known for protein binding and precipitation, a trait that leads to a characteristic astringency taste (Crozier et al. 2006). Tannins are found in various fruits, seeds, bark, leaves, and wood, all of which are integral to the human diet and traditional medicine. They perform crucial roles in plant defense mechanisms and possess numerous attributes within the human body (Hassanpour et al. 2011).

*Hydrolysable tannins and condensed tannins*: Hydrolysable tannins are esters derived from gallic acid (gallotannins) or ellagic acid (ellagitannins) and a polyol, commonly glucose (Schofield et al. 2001). They are water-soluble and readily hydrolyzed to their building blocks in acidic or enzyme conditions (De Bruyne et al. 1999). These tannins are present in plants, such as pomegranates, chestnuts, and oak bark (De Bruyne et al. 1999). Condensed tannins, or proanthocyanidins, are polyphenolic substances originating from flavan-3-ol or flavan-3,4-diol. They have superior stability compared to hydrolyzable tannins and are distinctive from chocolate, tea, grapes, and berries (Barbehenn & Constabel 2011). Both kinds of tannins have been well-researched in their antioxidant, antimicrobial, and health benefits (Hagerman et al. 1999).

The antioxidant activity of tannins is a defining biological property. Tannins neutralize ROS and metal ions, therefore preventing oxidative injury to cells and tissues. This preventive effect has lowered the threat of the development of common diseases (i.e., cardiovascular disease, cancer, and neurodegenerative disease) (Koleckar et al. 2008). In particular, proanthocyanidins are strong antioxidants associated with positive vascular function by improving endothelial function and decreasing blood pressure (Hagerman et al. 1998).

Tannins exhibit notable antibacterial activity. They impede the proliferation of bacteria, fungi, and viruses by breaking microbial membranes, reducing enzymatic activity, or binding to microbial proteins (Hagerman et al. 1998). For example, oral bacteria that lead to dental caries and gum disease are suppressed by tannins present in tea and wine. Their antimicrobial activity also contributes to preserving food and beverages, a property traditionally exploited in winemaking and leather tanning (Hässig et al. 1999).

In ancient medicine, plants with tannin content have served as the treatment for multiple disorders. Because of their astringent properties, they are beneficial for treating both diarrhea and intestinal disorders by promoting protein precipitation in the intestinal wall and forming a protective layer. Tannins also exhibit anti-inflammatory effects, making them valuable in the treatment of conditions such as sore throats, wounds, and skin irritation. Additionally, hydrolyzable tannins like ellagitannins have been investigated for their anticancer potential, as they promote apoptosis and suppress the growth of cancer cells (Ghosh 2015).

Although moderate tannin intake is beneficial, high levels of tannin intake can have negative consequences, such as decreased bioavailability of essential nutrients, such as iron and zinc, due to their high binding ability to minerals. This antinutritive characteristic is an issue for condensed tannins and needs to be taken into account in areas where tannin-rich foods will be eaten as a staple diet (Macáková et al. 2014).

Tannins serve as a natural defense mechanism in plants. They reduce the palatability of plant tissues and disrupt digestive enzymes, giving insect herbivores a strong repellent effect. Tannins also protect plants from microorganism attacks and abiotic stresses, including ultraviolet radiation and oxidative stress.

### 4.2.4 Lignans

Lignans are a class of diphenolic compounds formed through the dimerization of phenylpropanoid units. They serve as precursors of enterolignans, bioactive compounds synthesized by gut bacteria

(Fauré et al. 1990). Lignans are a category of plant-derived polyphenolics that belong to a broader class of phytoestrogens. Plant seeds, grains, vegetables, and fruits are rich in lignans, flaxseeds, sesame seeds, and whole grains in particular (Touré & Xueming 2010). These compounds have attracted interest in their large contribution to human health, especially their antioxidant, anti-inflammatory, and hormone-regulating activities (Touré & Xueming 2010). Lignans are structurally formed by the combination of two phenylpropanoid units. Lignans serve a protective function in plants, safeguarding against infections, UV radiation, and oxidative stress. Upon consumption, lignans are converted by intestinal microbiota in the human gut into mammalian lignans, specifically enterodiol, and enterolactone, which exhibit estrogenic and antiestrogenic properties. The synergistic effect of lignans allows them to mitigate hormone-mediated occurrences, making them beneficial for conditions having a hormonal foundation, including hormonal imbalances and hormone-associated malignancies (Durazzo et al. 2013). The antioxidant activity of lignans constitutes a significant health-protective attribute of these compounds. The lignans' oxygen neutralizes ROS and protects cells from being attacked by oxidative damage to limit the risk of chronic diseases, including cardiovascular and neurological disorders. In a review paper (Soleymani et al. 2020), flaxseed lignans are reported to be beneficial to cardiovascular health through a reduction in blood pressure as well as a decrease in the levels of LDL cholesterol and improvements in overall vascular function. However, lignans come with many excellent benefits, and their effectiveness varies depending on factors such as gut microbiota composition, dietary habits, and genetic predisposition. The richest known dietary source of lignans is flaxseeds, which are often recommended by those who are looking to augment their intake of this compound. However, a diet rich in various foods and other sources, such as sesame seeds, whole grains, cruciferous vegetables, and berries, can also provide substantial health benefits (Soleymani et al. 2020).

## 4.3 ANTIOXIDANT MECHANISMS OF PHENOLICS

The various antioxidant modes of action of phenolic compounds include free radical scavenging, metal chelation, inhibition of oxidative enzymes, activation of antioxidant enzymes, and complexation of metal ions, all contributing to their antioxidant efficacy. The various mechanisms are illustrated in Figure 4.2.

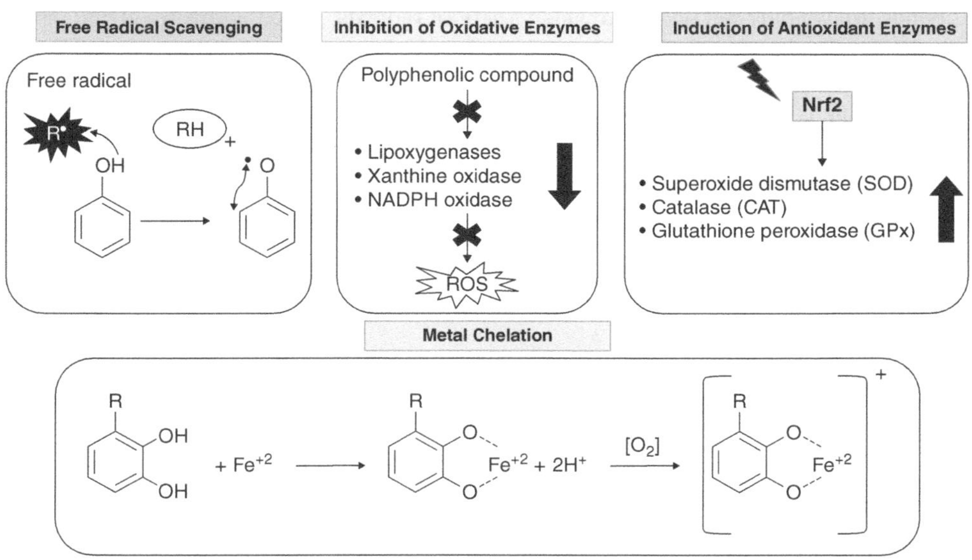


**FIGURE 4.2** Schematic representation of the different antioxidant mechanism of phenolic compounds.

### 4.3.1 Free Radical Scavenging by Phenolic Compounds

Phenolic compounds represent a diverse category of plant secondary metabolites characterized by the presence of one or more hydroxyl groups attached to an aromatic ring. These compounds are acknowledged for their substantial antioxidant activities, due to their ability to donate hydrogen to free radicals, thus transforming them into stable entities and inhibiting chain reactions (Sreejayan & Rao 1996). Phenolic compounds primarily mitigate oxidative stress through free radical scavenging, alleviating numerous chronic and degenerative diseases (Villaño et al. 2007). The processes of free radical scavenging by phenolic compounds involve the donation of hydrogen atoms, wherein phenolic compounds provide a hydrogen atom from their hydroxyl groups to free radicals, thus neutralizing their reactivity. The aromatic characteristics of phenolics impart stability to the resulting phenoxyl radical, thus preventing it from engaging in additional oxidative chain reactions. An illustrative case is the efficient utilization of phenolic acids, such as gallic acid, and flavonoids, such as quercetin (Leopoldini et al. 2004). Also, phenolics function as electron donors, neutralizing free radicals by reducing them to less reactive species. This mechanism underlies their capacity to put an end to oxidative chain reactions (Zeb 2020). Moreover, transition metals, namely iron ($Fe^{2+}$) and copper ($Cu^{2+}$), facilitate the production of very reactive radicals through reactions like the Fenton reaction. Phenolics chelate these metals, decreasing their bioavailability and thus preventing radical generation. For example, tannins and flavonoids are reputed to possess high metal-chelating properties (Shahidi et al. 1992). Certain phenolic compounds form direct contact with ROS, including hydroxyl radicals (OH•) and superoxide anions ($O_2^-$), and quench their activity (Zeb 2020).

The molecular structure of phenolic substances affects their antioxidant capability since more hydroxyl groups enhance radical scavenging capacity. For example, catechins with multiple hydroxyl groups exhibit high antioxidant activity (Seyoum et al. 2006). The conjugation of double bonds and electron delocalization within the aromatic ring or between the aromatic ring and side chain increase the stability of phenoxyl radicals (Seyoum et al. 2006). Moreover, the type of substitution as methoxy groups may modulate antioxidant activity by increasing or decreasing scavenging activity (Seyoum et al. 2006).

### 4.3.2 Examples of Phenolic Compounds with Free Radical Scavenging Activity

1. *Flavonoids*:
   These compounds, including quercetin, catechins, and anthocyanins, exhibit exceptional free radical scavenging capabilities due to their hydroxyl groups and conjugated double bonds. Flavonoids can quench a wide variety of radicals, including hydroxyl and peroxyl radicals (Gulcin & Alwasel 2023).
2. *Phenolic Acids*:
   Scavenging DPPH radicals and superoxide anions are performed well by compounds such as caffeic acid, ferulic acid, and gallic acid. With three hydroxyl groups on its aromatic ring, gallic acid is most effective in ROS scavenging (Gulcin & Alwasel 2023).
3. *Tannins*:
   Hydrolyzable tannins (e.g., ellagitannins) and condensed tannins (e.g., proanthocyanidins) are potent antioxidants. The scavenging of radicals is due to their high hydroxyl content and ability to chelate metal ions (Gulcin & Alwasel 2023).
4. *Lignans*:
   Flaxseed-derived lignans, such as secoisolariciresinol, have revealed potent radical scavenging properties. Their structural elements permit them to act energetically and efficiently on free radicals (Gulcin & Alwasel 2023).

### 4.3.3 Chelating of Metal Ions

Transition metal ions (especially $Fe^{2+}/Fe^{3+}$ and $Cu^{+}/Cu^{2+}$) cause free radical generation via Fenton and Haber–Weiss reactions. Phenolic compounds reduce oxidative stress by chelating these metals and so the formation of highly energetic hydroxyl radicals (Haider et al. 2020). The mechanism is based on the ability of phenolic compounds to interact with metal ions *via* hydroxyl and carboxyl groups to yield stable complexes which in turn inactivate the metals (Shi et al. 2022). Also, inhibition of redox cycling and chelation prevent metals from cycling between reduced and oxidized states, which is essential for free radical generation, so inhibition of redox cycling (Shi et al. 2022). For example, flavonoids with the catechol unit in quercetin and rutin have high binding affinity for iron and copper ions (Shi et al. 2022). Phenolic acids, such as gallic acid and ferulic acid, chelate iron effectively owing to their carboxylic and hydroxyl groups (Becerril-Sánchez et al. 2021). Phenolic constituents, having one or more hydroxyl substituents on a benzene ring, widely occur in the natural world and comprise flavonoids, phenolic acids, and tannins. The capacity of these compounds to transfer electrons (by the hydroxyl group) and to form stable complexes with metal ions is one of the main features responsible for their antioxidant activity (Haider et al. 2020). The mechanism of metal ion chelation by phenolics is based on electron donation (Zeb 2020), coordination sites (Gulcin & Alwasel 2022), chelate stability (Zeb 2020), and ring formation (Li et al. 2022).

### 4.3.4 Inhibition of Oxidative Enzymes

Phenolic compounds block the activity of prooxidant enzymes involved in oxidative stress, including lipoxygenases, xanthine oxidase, and NADPH oxidase, which generate ROS during metabolic processes (Zhai et al. 2022). Phenolic compounds exert enzyme inhibitory effects, thereby alleviating ROS production and associated oxidative injury (Ozougwu 2016). Phenolics bind to the catalytic sites or cofactors of enzymes and inhibit the action of these enzymes (Dwibedi et al. 2022). Or modulation of expression as few phenolic compounds affects the expression of genes coding for prooxidant enzymes, downregulating the latter over time (Dwibedi et al. 2022). The key enzymes targeted are lipoxygenases (LOX) that catalyze the oxidation of polyunsaturated fatty acids, leading to the production of inflammatory mediators. Phenolic acids, such as caffeic acid, and flavonoids, such as kaempferol, are potent inhibitors of LOX (Zhai et al. 2022). Xanthine oxidase (XO) generates superoxide radicals during purine metabolism. As XO inhibitors, quercetin and luteolin possess strong oxidizing abilities to decrease the generation of ROS (Zhai et al. 2022). Also, NADPH oxidase produces superoxide radicals as part of immune responses. Phenolics modulate NADPH oxidase activity, preventing excessive ROS formation (Dwibedi et al. 2022).

### 4.3.5 Induction of Antioxidant Enzymes

Phenolic compounds also augment the host's endogenous defense system by stimulating the expression and activity of endogenous antioxidant enzymes, including superoxide dismutase (SOD), catalase (CAT), and glutathione peroxidase (GPx), by signaling pathways, including nuclear factor erythroid 2-related factor 2 (Nrf2). This enhances the body's defense mechanisms against oxidative stress (Tomás-Barberán & Espín 2001; Zhai et al. 2022). Numerous phenolic compounds induce the Nrf2 pathway, which is a master transcription factor of antioxidant responses. Nrf2 binds to antioxidant response elements (AREs) located in the promoter of target genes and promotes production of the antioxidant enzymes (Zhai et al. 2022). Also, phenolics can affect the epigenetic regulation of antioxidant enzyme gene(s) via histone acetylation and DNA methylation (Dwibedi et al. 2022). For example, EGCG and curcumin activate the Nrf2 pathway and upregulate SOD and CAT expression (Dwibedi et al. 2022). Rosmarinic acid activates GPx and glutathione S-transferase (GST) activity, which promotes cellular detoxification (Dwibedi et al. 2022).

## 4.4 ANTIOXIDANT PROPERTIES OF PHENOLIC COMPOUNDS

Phenolic compounds are well-known for their substantial antioxidant activities, which are essential in safeguarding cells against oxidative stress and associated illnesses (Aly et al. 2019; Saber et al.

2022, 2023). These compounds are primarily found in various plants (El-Nashar et al. 2022) and are recognized for their capacity to scavenge free radicals, which may be evaluated using several *in vitro* assays as mentioned in the following section.

The antioxidant properties of phenolic compounds derived from the fruit peels and leaves of *Pyrus ussuriensis* were evaluated utilizing the DPPH radical scavenging test. Among the isolated compounds quercetin and quercetin-3-O-(3″-O-galloyl)-α-L-rhamnopyranoside exhibited potent DPPH radical scavenging activity with $IC_{50}$ values of 6.06 and 9.60 μg/mL, respectively, as compared to the control *tert*-butylhydroquinone (TBHQ) with $IC_{50}$ value of 3.88 μg/mL (Qiu et al. 2018).

Another research aimed to investigate different extracts of *Lonicera macranthoides* rich in phenolic compounds and assess their antioxidant capacity by different assays. In the DPPH assay, ethyl acetate fraction and *n*-butanol fractions exhibited the highest antioxidant activity with $IC_{50}$ values of 10.02 ± 1.35 and 12.14 ± 1.43 μg/mL (butylated hydroxytoluene (BHT) = 29.17 μg/mL). In the ABTS assay, *n*-butanol fraction and water extract showed the highest antioxidant activity with $IC_{50}$ values of 10.03 ± 1.66 and 18.91 ± 1.95 μg/mL (BHT = 60.45 μg/mL). The potent antioxidant activity of the *n*-butanol fraction is attributed to the phenolic acids and flavonoids identified using HPLC/MS analysis such as caffeoylshikimic acid, caffeoylquinic acid, and their derivatives also quercetin, luteolin, and chrysoeriol with their corresponding glycosides (Hu et al. 2016).

Another study indicates that the leaves of *Armeniaca sibirica* are rich in polyphenolic compounds, such as hydroxycinnamic acids and flavonols. Notably, quercetin-3-O-rutinoside, 5-O-caffeoylquinic acid, and 3-O-caffeoylquinic acid contribute significantly to the antioxidant capacity of the extract. The antioxidant activity of sibiriaester A and sibiriaester B was the highest among the tested compounds in the DPPH assay with $IC_{50}$ values of 11.35 and 7.43 μg/mL, respectively, although sibiriaester A was slightly better than the positive control (Trolox $IC_{50}$ = 12.25 μg/mL). In ABTS assay sibiriaester A and sibiriaester B showed $IC_{50}$ values of 5.37 and 5.61 μg/mL, respectively (Trolox $IC_{50}$ = 12.37 μg/mL). Regarding the FRAP assay, at the concentration of 0.75 mg/ml, sibiriaester A and sibiriaester B (48.21 and 57.34 mg/mL) were found to be less potent than the positive control Trolox (87.58 mg/ml). The results demonstrated that the leaf extracts possess strong antioxidant properties, which are essential for combating oxidative stress in biological systems (Wu et al. 2018).

The methanolic extract of the rhizomes of *Astilbe rivularis* showed antioxidant activity in the DPPH assay ($IC_{50}$ = 13 μg/mL); it also revealed the presence of 11-O-galloylbergenin (2), (+)-catechin (3), (–)-catechin (4), (–)-afzelechin (5), and (–)-epiafzelechin (6) that showed potent antioxidant activity with $IC_{50}$ values of 4.6, 3.9, 4.0, 6.0, and 5.7 μg/mL, respectively (Trolox = 12.2 μg/mL) (Hori et al. 2018).

The characterization of phenolic compounds in early- and late-ripening sweet cherries (*Prunus avium* L.) was investigated in a separate study that assessed phenolic content and its correlation with antioxidant activity across various sweet cherry cultivars. Both early- and late-ripening cherries had a high concentration of phenolics, primarily free epicatechin and conjugated 2,3,4-trihydroxybenzoic acids, and demonstrated antioxidant activity in DPPH and FRAP assays. The ability to combat oxidative stress through antioxidant capacity is crucial for mitigating oxidative damage and may confer numerous health benefits (Wang et al. 2017).

Another study emphasizes several phenolic components found in the native Brazilian fruits grumixama (*Eugenia brasiliensis*) and guabiju (*Myrcianthes pungens*), including gallic acid, catechin, vanillic acid, and ellagic acid. The antioxidant activity of the fruit extracts was assessed using DPPH and NO radical scavenging tests. Results indicated that extracts suppressed DPPH radicals by up to 95% and NO radicals by roughly 80% at a concentration of 2.5 mL of gallic acid equivalents. This indicates that this fruit can significantly combat oxidative stress due to its elevated concentrations of antioxidant compounds (Tischer et al. 2023).

The methanol extract of *Moricandia sinaica* leaves was examined via HPLC–ESI–MS/MS, identifying specific phenolic constituents, such as flavonoids, anthocyanins, and phenolic acids.

This extract has significant free radical scavenging activity in the DPPH and FRAP assays, with $IC_{50}$ values of 10.22 µg/mL and 20.89 µg/mL, crucial in combating oxidative stress (Aly et al. 2024).

The antioxidant ability of *Eryngium thorifolium* was evaluated by DPPH and ABTS radical scavenging tests. The findings demonstrated that both the aerial parts and roots displayed significant antioxidant activity, with the methanol and water extracts of the aerial parts exhibiting superior efficacy in DPPH and ABTS experiments. Amounts of 21.9 and 89.26 mg TE/g in DPPH and ABTS assays, respectively, can be ascribed to the presence of rosmarinic acid, myricetin glycosides, derivatives of caffeic and quinic acid, luteolin, and quercetin (Zengin et al. 2024).

Chickweed (*Stellaria media*) contains numerous bioactive compounds, including flavonoids such as quercetin, rutin, apigenin, and kaempferol derivatives, as well as phenolic acids. The antioxidant activity of various chickweed extracts was substantial, with the water extract exhibiting the highest activity, as demonstrated by DPPH and ABTS assays (13.04 ± 0.36 and 59.98 ± 0.57 mg TE/g, respectively) (Cusumano et al. 2024).

Ultrasound-assisted extraction (UAE) was utilized for the extraction of phenolic chemicals, proving effective in maximizing output. The investigation utilized UPLC–ESI–QTOF–MS/MS, enabling the identification of 11 distinct phenolic compounds, mostly including phenolic acids such as caffeic acid derivatives and hydroxycinnamic acids, as well as flavonoids, including quercetin, luteolin, and apigenin. The extracts from red corn cobs exhibited enhanced radical scavenging capacity as indicated by the findings obtained from the ABTS+ procedure with values ranging from 114.63 to 155.14 TE mg/100g in comparison to the DPPH assay (Hernández et al. 2018).

Another study demonstrated significant antioxidant properties of lemon myrtle (*Backhousia citriodora*) leaves, effectively scavenging free radicals as measured by DPPH and ABTS assays at 83.31 and 60.13%, respectively (Kang et al. 2020).

The extract of *Ruta graveolens* is rich in phenolic compounds, including rutin, syringic acid, and naringenin, quantified as 464.95, 179.74, and 109.78 µg/g, respectively. The DPPH and β-carotene tests assessed the antioxidant properties, revealing significant activity with values of 0.77 mM TE/g DW for DPPH and 0.37 mM β-CE/g DW for β-carotene. These results indicate that *R. graveolens* possesses a strong potential as a natural antioxidant source (Mokhtar et al. 2022).

A different investigation examines the *in vitro* antioxidant characteristics and phenolic composition of acid aqueous ethanol extracts derived from the seed coat of *Torreya grandis*. Identified phenolic components comprise protocatechuic acid, 2-hydroxy-2-(4-hydroxyphenylethyl) malonic acid, catechin, epigallocatechin gallate, and epicatechin gallate, which enhance the total antioxidant activity. The antioxidant capacity was assessed utilizing the DPPH radical scavenging test and the ABTS assay. The extract demonstrated scavenging abilities of 93.4 and 96.5% in the ABTS and DPPH assays, respectively, comparable to the control ascorbic acid (Quan et al. 2022).

The fruits of *Eugenia calycina* and Eugenia *stigmatosa* are identified as potential sources of antioxidant phenolic compounds, such as cyanidin-3-glucoside, rutin, gallic acid, and ellagic acid. The results indicated that both species exhibited significant antioxidant activity; particularly *E. calycina* revealed antioxidant capacity in ORAC, FRAP, and ABTS assays at 83.46, 53.29, and 40.66 µmol TE/g, respectively, while *E. stigmatosa* 44.00, 53.48, and 47.13 µmol TE/g, respectively (Silva et al. 2024).

A different investigation examines the UAE of phenolic compounds from *Phyllanthus emblica* L. and assesses their antioxidant activities. The extract exhibited significant antioxidant activity, with $IC_{50}$ values recorded at 26.00 µg/mL for DPPH radicals, 50.11 µg/mL for total antioxidant capacity, and 115.91 µg/mL for ferrous ion chelating ability (Tsai et al. 2014).

Another research identified a diverse range of phenolic compounds in the extract of *Nelumbo nucifera*, including phenolic acids, i.e. gallic acid, caffeic acid, chlorogenic acid, and coumaric acid, and flavonoids, i.e. rutin, catechin, and quercetin. Its three parts flesh, peel, and nodes showed $IC_{50}$ values of 46.00, 26.43, and 21.72 µg/mL, respectively, in the DPPH assay, while in the FRAP assay at 75.91, 87.66, and 100.43 µg TE/100 µg GAE (Yi et al. 2016).

The phenolic extract (RFE) from *Rhodiola rosea* contains high levels of epigallocatechin (EGC), epigallocatechin gallate (EGCG), and gallic acid (GA), which are responsible for antioxidant properties. It showed peroxyl radical scavenging capacity (PSC) and oxygen radical absorbance capacity (ORAC) equivalent to 342.22 ± 15.83 μM VCE/g DW and 1644.06 ±46.50 μMTE/g DW, respectively (Zhang et al. 2022a).

A study on tea seed oil, *Camellia sinensis*, displayed various assays to evaluate the antioxidant capacity. The results showed that phenolic extracts demonstrated significant proton and electron transfer capabilities, with an $IC_{50}$ value of 77.93 ± 4.80 μg/mL and cell antioxidant capacity of 101.05 ± 6.70 μmol·QE/100 g, indicating effective radical scavenging activity. Furthermore, the *in vivo* study on galactose-induced aging in rats demonstrated that *C. sinensis* significantly enhanced organ index, reduced MDA levels, and increased the activity of antioxidant enzymes, including SOD and glutathione peroxidase. The results were linked to the presence of flavonoids and phenolic acids identified through UPLC-MS/MS analysis, including naringenin, gentisic acid, hydroxytyrosol, 7-hydroxycoumarin, homovanillic acid, pyrocatechol, *p*-coumaraldehyde, and trans-cinnamic acid (Liu et al. 2022).

The antioxidant capacity of *Parthenium hysterophorus* L. flower extract was estimated using the DPPH assay, where the extract showed significant inhibition of free radicals, achieving a maximum inhibition of 76.90% at a concentration of 80 μg with an $IC_{50}$ value of 54.278 μg/μL. HPLC-DAD analysis exposed the presence of phenolics in the flower extract as well as gallic acid, chlorogenic acid, ellagic acid, and p-coumaric acid, with chlorogenic acid being most abundant (Iqbal et al. 2022).

## 4.5 HEALTH BENEFITS ASSOCIATED WITH PHENOLICS

Phenolics serve as potent antioxidants, effectively neutralizing and reducing oxidative stress. This property plays a vital role in safeguarding against cellular damage and reducing the risk of chronic diseases such as cancer, cardiovascular diseases, and neurodegenerative disorders (Figures 4.3 and 4.4). The ingesting of phenolic-rich foods has been associated with numerous health benefits.

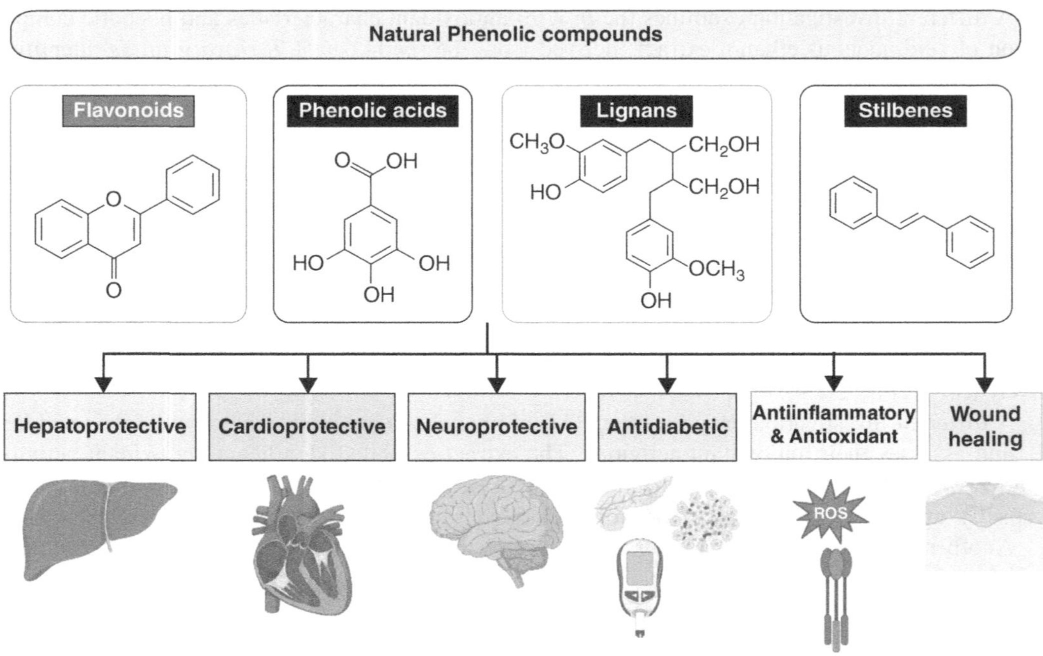


**FIGURE 4.3** Schematic depiction of different health benefits of phenolic compounds.

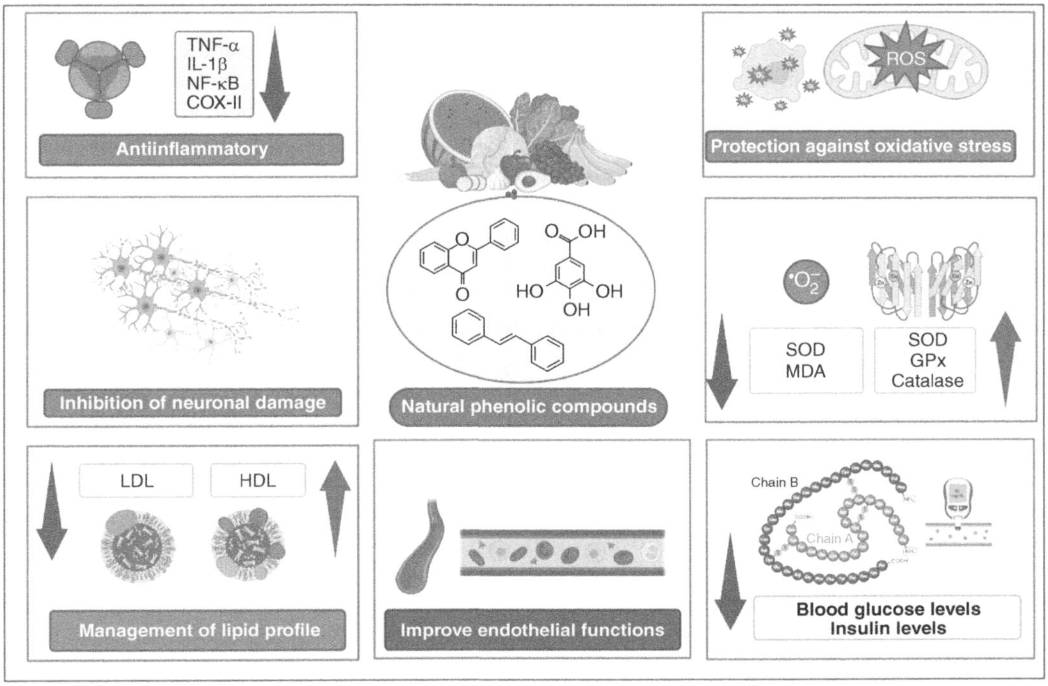


**FIGURE 4.4** Mechanism of action of natural phenolics as antioxidants in different diseases.

### 4.5.1 Hepatoprotective Properties

Phenolic compounds provide protective effects against hepatotoxic agents by modulating antioxidant parameters. For example, they can reduce the expression of MDA involved in lipid peroxidation while increasing antioxidant enzymes such as SOD and catalase, thus improving the clearance of ROS and preventing further oxidative stress in liver tissues.

The hepatoprotective potential of *M. sinaica* methanol leaf extract was evaluated in a rat model treated with paracetamol-induced liver toxicity. Among its antioxidant activities, SOD activity was increased by 3.4 folds, nonprotein sulfhydryl increased by 2.2 folds, and GSH peroxidase increased by 2.6 folds. Furthermore, lipid peroxidation levels were correspondingly reduced by 68% (Aly et al. 2024). An additional study demonstrated that the antioxidant potential of the extract of *Hyophorbe verschaffeltii* is capable of elevating the level of SOD in $CCl_4$-induced hepatic injury by 124.7% as compared to control and declined the level of MDA by 20.25% as compared to $CCl_4$ group. Therefore, the extract of *H. verschaffeltii* shows free radical scavenging activity and thus the potential to neutralize harmful free radicals in biological systems (Elgindi et al. 2016).

For the use of the defatted extract as well as the phenolic fraction of *Phlomis Tuberosa*, hepatocytes exposed to $CCl_4$ showed considerably reduced cell damage. The extracts displayed strong antioxidant activity, effectively scavenging free radicals and reducing MDA and enhancing levels of GSH. Regarding the *in vivo* mode, the extract's effects on these biochemical markers were more pronounced than those of silymarin. The major compounds identified using HPLC analysis are luteolin, quercetin, spiraeoside, apigenin-7-glucoside, and verbascoside (Kondeva-Burdina et al. 2023).

One more study investigates protective properties of poncirin, a flavonoid compound, against liver injury induced by paracetamol. Poncirin significantly improved liver function tests and reduced markers of oxidative stress and inflammation. It enhanced antioxidant enzyme activities and improved hematological parameters. Moreover, after concern treatment, immunohistochemistry revealed decreased expression of inflammatory proteins like NF-κB, JNK, and COX-2 (Ullah et al.,

2022). A study identified the secondary metabolites in *Echinops ritro* extracts, together with acylquinic acids, flavonoids, and coumarins. Key compounds such as protocatechuic acid O-hexoside, chlorogenic acid, and flavonoids such as apigenin, apigenin 7-O-glucoside, hyperoside, jaceosidin, and cirsiliol were prominent. Extracts from flowers and leaves of *E. ritro* both alleviated oxidative stress in isolated rat microsomes and hepatocytes. These extracts enhanced cultured cell survival, elevated GSH levels, and reduced LDH efflux and MDA generation, indicating a potent antioxidant function (Zheleva-Dimitrova et al. 2023).

### 4.5.2 Cardioprotective Effects

The uses of polyphenolic compounds in several trials effectively decreased coronary heart disease risk like atherosclerosis (Cheng et al. 2017; Lutz et al. 2019). Ingesting of high amounts of soy products and isoflavones was significantly associated with the reduction of myocardial infarction, for example, women consuming isoflavones at high levels (Kokubo et al. 2007). A second study showed that polyphenol-rich fruit (açaí) consumption caused a decrease in oxidative stress markers with ROS, oxidized LDL (ox-LDL), and MDA. In addition, in attendance was increase in paraoxonase 1, an antioxidant defense enzyme (Pala et al. 2018). Resveratrol was also investigated for other effects together with the effects of resveratrol on arterial stiffness, a common complication of type 2 diabetes; resveratrol was shown to reduce systolic blood pressure and cardio ankle vascular index (Imamura et al. 2017). The various studies that discussed polyphenols as potent inhibitors of LDL oxidation were supported to be involved in the progression of atherosclerosis. It has been stated that antiplatelet, anti-inflammatory, and antioxidant properties of polyphenols could protect against cardiovascular disease as polyphenols could elevate high density lipoprotein (HDL) levels and improve the endothelial functions (Ahmadi et al. 2022; Santhakumar et al. 2014).

### 4.5.3 Neuroprotective Properties

Certain studies indicate that phenolic compounds are capable of providing neuroprotection by counteracting oxidative stress and inflammation to neural tissue. This is relevant, particularly for Alzheimer's disease. In order to investigate the neuroprotective properties of *Sophora secundiflora* and *Sophora tomentosa* leaves and the isoflavonoid formononetin in scopolamine made dementia, a study was carried out. These extracts treated plated cells and reduced oxidative stress markers to a notable extent. In particular, MDA levels and lipid peroxidation markers were significantly lowered, and the levels of the antioxidant enzymes increased (Aly et al. 2021). *Schinus polygamus* leaf extract produces agathisflavone flavonoid compounds (at 1, 3, and 5 μg/L), significantly mitigating the memory deficit imposed by scopolamine in the zebrafish model. Further, it efficiently decreased oxidative stress markers in zebrafish brains. As in previous reports, we found that agathisflavone decreases levels of MDA, while increasing the activity of enzymes like SOD and catalase. In addition, we suggest the role of agathisflavone in promoting protective effects counter to oxidative damage and in mitigating the effect of oxidative damage by enhancing antioxidant defenses, thereby providing cognitive protection from neurotoxic challenges (Dumitru et al. 2019). With respect to neurotoxicity, resveratrol, curcumin, myricetin, ginsenosides, EGCG, and ginkgetin bind to and prevent the iron chelating effects which is thought to play a neuroprotective role in disorders like Huntington's, Alzheimer's, and Parkinson's; also, stopping the neurotoxic effects of the β-amyloid protein (Rana et al. 2022; Wang et al. 2020).

### 4.5.4 Wound Healing Properties

Phenolic compounds have been shown to significantly enhance tissue repair and integrity and lead to improved wound closure and healing rates. Phenolic compounds can modulate inflammatory responses, which is vital in the early stages of wound healing; these compounds help to create an ideal environment for tissue regeneration by reducing inflammation. To illustrate, a current study inspected the achieved synergistic impacts of extractive fractions high in flavonoids from leaves of *Sophora japonica* (SjL/E) and *Glycyrrhiza glabra* (GgR/E) roots related to wound mending, such as clear falls in MDA levels and increases in GSH and SOD levels in the compartment of the wound, which might propose enhanced antioxidant action in the tissue of the wound. The major phenolics found in *G. glabra* were isoliquiritigenin, neolicuroside, kanzonol Y, glabridin, and glabrol. In addition, *S. japonica* possessed the content of kaempferol, sophoraflavonoloside, sophoraflavanone G, genistein, tamarixetin, and kurarinone (Aly et al. 2023). A recent study showed that electrospun scaffolds based on *Tamarindus indica* extract not only supported cell adhesion and proliferation but also accelerated the wound healing process that would be due to the presence of flavonoids as major constituents in methanol extract, i.e. luteolin-7-*O*-hexoside, vicenin 1, and apigenin-*O*-hexoside (as well as caffeic acid) (Goher et al. 2023). Another recent study also examined the incorporation of *Syzygium cumini* extract in nanofibers, and it significantly improved wound healing outcomes with a notable reduction in wound size and enhanced tissue regeneration, noteworthy the richness of *S. cumini* extract with flavonoid glycosides, namely myricetin, kaempferol, apigenin, and isorhamnetin glycosides besides tannins and anthocyanins (Abdelazim et al. 2024).

### 4.5.5 Antidiabetic Properties

Phenolics play a role in regulating glucose metabolism and enhancing insulin sensitivity by inhibiting key enzymes involved in the breakdown of carbohydrates (e.g., α-glucosidase and α-amylase), which can aid in managing blood glucose levels and decreasing risk of diabetes. Additionally, they may enhance β-cell function and promote healthy adipose tissue metabolism. In diabetic rats induced by a high-fat diet and streptozotocin injection, therapy with *Amomum tsao-ko* extract greatly improved glucose tolerance. It reduced fasting blood glucose levels, insulin levels, and MDA, while elevating SOD levels, indicating an improved antioxidant status. Its antioxidant potential is due to the presence of flavonoid and phenolic constituents such as 6″-O-acetylgenistin, ephedrannin A, mulberrin, 12a-hydroxyrotenone, cyclomorusin, epigallocatechin 3-gallate, isoastilbin, and sec-O-glucosylhamaudol (Zhang et al. 2022b). Another study indicated that catechin may exert its protective effects through antioxidant mechanisms, reducing oxidative stress markers and inflammation associated with diabetes. It showed reduced circulatory MMP-9 levels and inhibition of neuronal damage, so it can protect against diabetic autonomic neuropathy (Addepalli & Suryavanshi 2018).

### 4.5.6 Anti-inflammatory Properties

Phenolic compounds exhibit anti-inflammatory properties by modulating inflammatory pathways. For instance, they can inhibit the activation of proinflammatory transcription factors like NF-κB, leading to decreased expression of inflammatory cytokine properties. This action is beneficial in managing conditions such as arthritis. Rhoifolin, a bioactive flavonoid, was isolated from *Teucrium polium* and loaded into PLGA nanoparticles; in a formalin-induced rat paw edema model, rhoifolin nanoparticles at 10 mg/kg significantly reduced paw thickness and mitigated histopathological changes associated with inflammation. The treatment normalized nitric oxide levels and modulated expression of proinflammatory cytokines, for example, TNF-α and IL-1β (Al-Shalabi et al. 2022).

## 4.6 CONCLUSION

The most prevalent secondary metabolites in plants are phenolic compounds, which can take many different forms, including lignans, flavonoids, phenolic acids, and stilbenes. Phenolic compounds are very potent and have broad biological effects, and various studies have shown their efficacy based on *in vitro* and *in vivo* studies. Their action mechanism would depend on the downregulation of oxidative stress markers, upregulation of antioxidant enzymes, and alterations in some signaling pathways. The health benefits linked to phenolics highlight their significance in nutrition and disease prevention. Their varied biological functions render them crucial elements of a nutritious diet, especially in mitigating the danger of chronic disorders such as diabetes, and cardiovascular and liver diseases. An ongoing investigation into the mechanisms that facilitate these advantages will clarify their significance in enhancing health and controlling ailments.

## COMPETING INTERESTS

The authors declare that they have no competing interest.

## DECLARATION OF GENERATIVE AI AND AI-ASSISTED TECHNOLOGIES IN THE WRITING PROCESS

During the preparation of this work, the book chapter author(s) have not taken any assistance from AI, and the book chapter author(s) reviewed and edited the content as needed and take(s) full responsibility for the content of the publication.

## REFERENCES

Abdelazim E, Goher S, Aly S, El-Nashar H, EL-Moslamy S et al. 2024. In vitro and in vivo studies of *Syzygium cumini* loaded electrospun PLGA/PMMA/collagen nanofibers for accelerating topical wound healing. *RSC Adv.* 14(1):101–17

Abotaleb M, Liskova A, Kubatka P, Büsselberg D. 2020. Therapeutic potential of plant phenolic acids in the treatment of cancer. *Biomolecules.* 10(2):221

Addepalli V, Suryavanshi SV. 2018. Catechin attenuates diabetic autonomic neuropathy in streptozotocin induced diabetic rats. *Biomed Pharmacother.* 108:1517–23

Ahmadi A, Jamialahmadi T, Sahebkar A. 2022. Polyphenols and atherosclerosis: A critical review of clinical effects on LDL oxidation. *Pharmacol Res.* 184:106414

Al-Shalabi E, Abusulieh S, Hammad AM, Sunoqrot S. 2022. Rhoifolin loaded in PLGA nanoparticles alleviates oxidative stress and inflammation in vitro and in vivo. *Biomater Sci.* 10(19):5504–19

Aly SH, Elissawy AM, Eldahshan OA, Elshanawany MA, Efferth T, Singab ANB. 2019. The pharmacology of the genus Sophora (Fabaceae): An updated review. *Phytomedicine.* 64:153070

Aly SH, Elissawy AM, Fayez AM, Eldahshan OA, Elshanawany MA, Singab ANB. 2021. Neuroprotective effects of *Sophora secundiflora*, *Sophora tomentosa* leaves and formononetin on scopolamine-induced dementia. *Nat Prod Res.* 35(24):5848–52

Aly SH, Elissawy AM, Mahmoud AMA, El-Tokhy FS, Mageed SSA, et al. 2023. Synergistic effect of *Sophora japonica* and *Glycyrrhiza glabra* flavonoid-rich fractions on wound healing: In vivo and molecular docking studies. *Molecules.* 28(7):2994

Aly SH, Mahmoud AMA, Abdel Mageed SS, Khaleel EF, Badi RM et al. 2024. Exploring the phytochemicals, antioxidant properties, and hepatoprotective potential of *Moricandia sinaica* leaves against paracetamol-induced toxicity: Biological evaluations and in silico insights. *PLoS One.* 19(10):e0307901

Asif M. 2015. Chemistry and antioxidant activity of plants containing some phenolic compounds. *Chemistry International.* 1(1):35–52

Atawodi SE. 2005. Antioxidant potential of African medicinal plants. *Afr J Biotechnol.* 4(2):128–33

Barbehenn RV, Constabel CP. 2011. Tannins in plant–herbivore interactions. *Phytochemistry.* 72(13):1551–65

Becerril-Sánchez AL, Quintero-Salazar B, Dublán-García O, Escalona-Buendía HB. 2021. Phenolic compounds in honey and their relationship with antioxidant activity, botanical origin, and color. *Antioxidants.* 10(11):1700

Cheng Y-C, Sheen J-M, Hu WL, Hung Y-C. 2017. Polyphenols and oxidative stress in atherosclerosis-related ischemic heart disease and stroke. *Oxid Med Cell Longev.* 2017(1):8526438

Chung K-T, Wong TY, Wei C-I, Huang Y-W, Lin Y. 1998. Tannins and human health: A review. *Crit Rev Food Sci Nutr.* 38(6):421–64

Croft KD. 1998. The chemistry and biological effects of flavonoids and phenolic acids a. *Ann N Y Acad Sci.* 854(1):435–42

Crozier A, Jaganath IB, Clifford MN. 2006. Phenols, polyphenols and tannins: An overview. *Plant Secondary Metabolites: Occurrence, Structure and Role in the Human Diet.* 1:1–25

Cusumano G, Flores GA, Cetiz MV, Kurt U, Ak G, et al. 2024. Small steps to the big picture for health-promoting applications through the use of chickweed (*Stellaria media*): In vitro, in silico, and pharmacological network approaches. *Food Sci Nutr.* 12(11):9295–313.

De Bruyne T, Pieters L, Deelstra H, Vlietinck A. 1999. Condensed vegetable tannins: Biodiversity in structure and biological activities. *Biochem Syst Ecol.* 27(4):445–59

Demirci-Cekic S, Özkan G, Avan AN, Uzunboy S, Çapanoğlu E, Apak R. 2022. Biomarkers of oxidative stress and antioxidant defense. *J Pharm Biomed Anal.* 209:114477

Dixit V, Chaubey KK, Dayal D, Chole PB, BT M, Bachheti RK, Gupta PC, Bachheti A, Worku LA. 2025. Natural products in aging: Cellular mechanisms and emerging therapeutics for age-related disorders. *Journal of Aging Research.* 2025(1):6868732

Dumitru G, El-Nashar HAS, Mostafa NM, Eldahshan OA, Boiangiu RS et al. 2019. Agathisflavone isolated from *Schinus polygamus* (Cav.) Cabrera leaves prevents scopolamine-induced memory impairment and brain oxidative stress in zebrafish (*Danio rerio*). *Phytomedicine.* 58:152889

Durazzo A, Turfani V, Azzini E, Maiani G, Carcea M. 2013. Phenols, lignans and antioxidant properties of legume and sweet chestnut flours. *Food Chem.* 140(4):666–71

Dwibedi V, Jain S, Singhal D, Mittal A, Rath SK, Saxena S. 2022. Inhibitory activities of grape bioactive compounds against enzymes linked with human diseases. *Appl Microbiol Biotechnol.* 106(4):1399–1417

Elgindi MR, Singab AE-NB, Aly SH, Mahmoud II. 2016. Phytochemical investigation and antioxidant activity of *Hyophorbe verschaffeltii* (Arecaceae). *J Pharmacogn Phytochem.* 5(2):39–46

El-Nashar HAS, Aly SH, Ahmadi A, El-Shazly M. 2022. The impact of polyphenolics in the management of breast cancer: Mechanistic aspects and recent patents. *Recent Pat Anticancer Drug Discov.* 17(4):358–79

Fauré M, Lissi E, Torres R, Videla LA. 1990. Antioxidant activities of lignans and flavonoids. *Phytochemistry.* 29(12):3773–75

Ghosh D. 2015. Tannins from foods to combat diseases. *Int J Pharm Res Rev.* 4(5):40–44

Goher SS, Aly SH, Abu-Serie MM, El-Moslamy SH, Allam AA et al. 2023. Electrospun *Tamarindus indica*-loaded antimicrobial PMMA/cellulose acetate/PEO nanofibrous scaffolds for accelerated wound healing: In-vitro and in-vivo assessments. *Int J Biol Macromol.* 128793

Gulcin İ, Alwasel SH. 2022. Metal ions, metal chelators and metal chelating assay as antioxidant method. *Processes.* 10(1):132

Gulcin İ, Alwasel SH. 2023. DPPH radical scavenging assay. *Processes.* 11(8):2248

Guven H, Arici A, Simsek O. 2019. Flavonoids in our foods: A short review. *J Basic Clin Health Sci.* 3(2):96–106

Hagerman AE, Riedl KM, Jones GA, Sovik KN, Ritchard NT et al. 1998. High molecular weight plant polyphenolics (tannins) as biological antioxidants. *J Agric Food Chem.* 46(5):1887–92

Hagerman AE, Riedl KM, Rice RE. 1999. Tannins as biological antioxidants. *Plant Polyphenols 2: Chemistry, Biology, Pharmacology, Ecology.* 495–505

Haider K, Haider MR, Neha K, Yar MS. 2020. Free radical scavengers: An overview on heterocyclic advances and medicinal prospects. *Eur J Med Chem.* 204:112607

Harborne JB. 2013. The flavonoids: Advances in research since 1980. (New York: Chapman and Hall).

Harborne JB, Mabry TJ, Markham KR, Chari VM. 1982. In JR. Harborne & TJ. Mabry (Eds.), The flavonoids. Advances in Research (pp. 52–134). London: Chapman and Hall.

Harborne JB, Williams CA. 2001. Anthocyanins and other flavonoids. *Nat Prod Rep.* 18(3):310–33

Hassan W, Noreen H, Rehman S, Gul S, Amjad Kamal M, et al. 2017. Oxidative stress and antioxidant potential of one hundred medicinal plants. *Curr Top Med Chem.* 17(12):1336–70

Hassanpour S, MaheriSis N, Eshratkhah B. 2011. Plants and secondary metabolites (Tannins): A Review

Hässig A, Linag WX, Schwabl H, Stampfli K. 1999. Flavonoids and tannins: Plant-based antioxidants with vitamin character. *Med Hypotheses.* 52(5):479–81

Heleno SA, Martins A, Queiroz MJRP, Ferreira ICFR. 2015. Bioactivity of phenolic acids: Metabolites versus parent compounds: A review. *Food Chem.* 173:501–13

Hernández M, Ventura J, Castro C, Boone V, Rojas R et al. 2018. UPLC-ESI-QTOF-MS2-based identification and antioxidant activity assessment of phenolic compounds from red corn cob (*Zea mays* L.). *Molecules.* 23(6):1425

Hori K, Wada M, Yahara S, Watanabe T, Devkota HP. 2018. Antioxidant phenolic compounds from the rhizomes of *Astilbe rivularis. Nat Prod Res.* 32(4):453–56

Hu X, Chen L, Shi S, Cai P, Liang X, Zhang S. 2016. Antioxidant capacity and phenolic compounds of *Lonicerae macranthoides* by HPLC–DAD–QTOF-MS/MS. *J Pharm Biomed Anal.* 124:254–60

Imamura H, Yamaguchi T, Nagayama D, Saiki A, Shirai K, Tatsuno I. 2017. Resveratrol ameliorates arterial stiffness assessed by cardio-ankle vascular index in patients with type 2 diabetes mellitus. *Int Heart J.* 58(4):577–83

Iqbal J, Khan AA, Aziz T, Ali W, Ahmad S et al. 2022. Phytochemical investigation, antioxidant properties and in vivo evaluation of the toxic effects of *Parthenium hysterophorus*. *Molecules*. 27(13):4189

Iwashina T. 2000. The structure and distribution of the flavonoids in plants. *J Plant Res*. 113(3):287

Juszczyk G, Mikulska J, Kasperek K, Pietrzak D, Mrozek W, Herbet M. 2021. Chronic stress and oxidative stress as common factors of the pathogenesis of depression and Alzheimer's disease: The role of antioxidants in prevention and treatment. *Antioxidants*. 10(9):1439

Kang E-J, Lee J-K, Park H-R, Kim H, Kim H-S, Park J. 2020. Antioxidant and anti-inflammatory activities of phenolic compounds extracted from lemon myrtle (*Backhousia citriodora*) leaves at various extraction conditions. *Food Sci Biotechnol*. 29:1425–32

Kesarkar S, Bhandage A, Deshmukh S, Shevkar K, Abhyankar M. 2009. Flavonoids: An overview. *J Pharm Res*. 2(6):1148–54

Khadem S, Marles RJ. 2010. Monocyclic phenolic acids; hydroxy- and polyhydroxybenzoic acids: Occurrence and recent bioactivity studies. *Molecules*. 15(11):7985–8005

Khan J, Khan MZ, Ma Y, Meng Y, Mushtaq A, et al. 2022. Overview of the composition of whole grains' phenolic acids and dietary fibre and their effect on chronic non-communicable diseases. *Int J Environ Res Public Health*. 19(5):3042

Kokubo Y, Iso H, Ishihara J, Okada K, Inoue M, Tsugane S. 2007. Association of dietary intake of soy, beans, and isoflavones with risk of cerebral and myocardial infarctions in Japanese populations: The Japan public health Center–based (JPHC) study cohort I. *Circulation*. 116(22):2553–62

Koleckar V, Kubikova K, Rehakova Z, Kuca K, Jun D et al. 2008. Condensed and hydrolysable tannins as antioxidants influencing the health. *Mini Rev Med Chem*. 8(5):436–47

Kondeva-Burdina M, Shkondrov A, Popov G, Manov V, Krasteva I. 2023. In vitro/in vivo hepatoprotective and antioxidant effects of defatted extract and a phenolic fraction obtained from *Phlomis tuberosa*. *Int J Mol Sci*. 24(13):10631

Krishnaiah D, Sarbatly R, Nithyanandam R. 2011. A review of the antioxidant potential of medicinal plant species. *Food Bioprod Process*. 89(3):217–33

Kruk J, Aboul-Enein BH, Duchnik E, Marchlewicz M. 2022. Antioxidative properties of phenolic compounds and their effect on oxidative stress induced by severe physical exercise. *J Physiol Sci*. 72(1):19

Kumar N, Goel N. 2019. Phenolic acids: Natural versatile molecules with promising therapeutic applications. *Biotechnol Rep*. 24:e00370

Kumar S, Pandey AK. 2013. Chemistry And biological activities of flavonoids: An overview. *Sci World J*. 2013(1):162750

Leopoldini M, Marino T, Russo N, Toscano M. 2004. Antioxidant properties of phenolic compounds: H-atom versus electron transfer mechanism. *J Phys Chem A*. 108(22):4916–22

Li Y, Miao Y, Yang L, Zhao Y, Wu K, et al. 2022. Recent advances in the development and antimicrobial applications of metal–phenolic networks. *Adv Sci*. 9(27):2202684

Lin D, Xiao M, Zhao J, Li Z, Xing B, et al. 2016. An overview of plant phenolic compounds and their importance in human nutrition and management of type 2 diabetes. *Molecules*. 21(10):1374

Liu G, Zhu W, Zhang J, Song D, Zhuang L et al. 2022. Antioxidant capacity of phenolic compounds separated from tea seed oil in vitro and in vivo. *Food Chem*. 371:131122

Lutz M, Fuentes E, Ávila F, Alarcón M, Palomo I. 2019. Roles of phenolic compounds in the reduction of risk factors of cardiovascular diseases. *Molecules*. 24(2):366

Macáková K, Kolečkář V, Cahlíková L, Chlebek J, Hošťálková A, et al. 2014. Tannins and their influence on health. In *Recent Advances in Medicinal Chemistry*, pp. 159–208. Elsevier

Mileo AM, Miccadei S. 2016. Polyphenols as modulator of oxidative stress in cancer disease: New therapeutic strategies. *Oxid Med Cell Longev*. 2016(1):6475624

Mokhtar M, Youcefi F, Keddari S, Saimi Y, Otsmane Elhaou S, Cacciola F. 2022. Phenolic content and in vitro antioxidant, anti-inflammatory and antimicrobial evaluation of Algerian *Ruta graveolens* L. *Chem Biodivers*. 19(9):e202200545

Nijveldt RJ, Van Nood ELS, Van Hoorn DEC, Boelens PG, Van Norren K, Van Leeuwen PAM. 2001. Flavonoids: A review of probable mechanisms of action and potential applications. *Am J Clin Nutr*. 74(4):418–25

Ozougwu JC. 2016. The role of reactive oxygen species and antioxidants in oxidative stress. *Int J Res*. 1(8):1–8

Padmaja M, Sravanthi M, Hemalatha KPJ. 2011. Evaluation of antioxidant activity of two Indian medicinal plants. *J Phytol*. 3(3)

Pala D, Barbosa PO, Silva CT, de Souza MO, Freitas FR, et al. 2018. Açai (*Euterpe oleracea* Mart.) dietary intake affects plasma lipids, apolipoproteins, cholesteryl ester transfer to high-density lipoprotein and redox metabolism: A prospective study in women. *Clin Nutr*. 37(2):618–23

Palipoch S. 2013. A review of oxidative stress in acute kidney injury: Protective role of medicinal plants-derived antioxidants. *Afr J Tradit Complement Alternat Med.* 10(4):88–93

Pammi SSS, Suresh B, Giri A. 2023. Antioxidant potential of medicinal plants. *J Crop Sci Biotechnol.* 26(1):13–26

Panche AN, Diwan AD, Chandra SR. 2016. Flavonoids: An overview. *J Nutr Sci.* 5:e47

Pietta P-G. 2000. Flavonoids as antioxidants. *J Nat Prod.* 63(7):1035–42

Qiu D, Guo J, Yu H, Yan J, Yang S et al. 2018. Antioxidant phenolic compounds isolated from wild *Pyrus ussuriensis* Maxim. fruit peels and leaves. *Food Chem.* 241:182–87

Quan W, Xu Y, Xie Y, Peng F, Lin Y. 2022. In vitro antioxidant properties and phenolic profile of acid aqueous ethanol extracts from *Torreya grandis* seed coat. *Molecules.* 27(17):5560

Rahman MM, Rahaman MS, Islam MR, Rahman F, Mithi FM, et al. 2021. Role of phenolic compounds in human disease: Current knowledge and future prospects. *Molecules.* 27(1):233

Rajendran P, Nandakumar N, Rengarajan T, Palaniswami R, Gnanadhas EN et al. 2014. Antioxidants and human diseases. *Clin Chim Acta.* 436:332–47

Rana A, Samtiya M, Dhewa T, Mishra V, Aluko RE. 2022. Health benefits of polyphenols: A concise review. *J Food Biochem.* 46(10):e14264

Rashmi HB, Negi PS. 2020. Phenolic acids from vegetables: A review on processing stability and health benefits. *Food Res Int.* 136:109298

Robbins RJ. 2003. Phenolic acids in foods: An overview of Analytical methodology. *J Agric Food Chem.* 51(10):2866–87

Saber FR, Aly SH, Khallaf MA, El-Nashar HAS, Fahmy NM, et al. 2022. *Hyphaene thebaica* (Areceaeae) as a promising functional food: Extraction, analytical techniques, bioactivity, food, and industrial applications. *Food Anal Methods.* 16(9):1447–67.

Saber FR, Munekata PES, Rizwan K, El-nashar HAS, Fahmy NM, et al. 2023. Family Myrtaceae : The treasure hidden in the complex/diverse composition. *Crit Rev Food Sci Nutr.* 64(19):6737–55.

Saibabu V, Fatima Z, Khan LA, Hameed S. 2015. Therapeutic potential of dietary phenolic acids. *Adv Pharmacol Pharm Sci.* 2015(1):823539

Samanta A, Das G, Das SK. 2011. Roles of flavonoids in plants. *Carbon N Y.* 100(6):12–35

Santhakumar AB, Bulmer AC, Singh I. 2014. A review of the mechanisms and effectiveness of dietary polyphenols in reducing oxidative stress and thrombotic risk. *J Hum Nutr Diet.* 27(1):1–21

Schofield P, Mbugua DM, Pell AN. 2001. Analysis of condensed tannins: A review. *Anim Feed Sci Technol.* 91(1–2):21–40

Seyoum A, Asres K, El-Fiky FK. 2006. Structure–radical scavenging activity relationships of flavonoids. *Phytochemistry.* 67(18):2058–70

Shahidi F, Janitha PK, Wanasundara PD. 1992. Phenolic antioxidants. *Crit Rev Food Sci Nutr.* 32(1):67–103

Sharifi-Rad M, Anil Kumar NV, Zucca P, Varoni EM, Dini L, et al. 2020. Lifestyle, oxidative stress, and antioxidants: Back and forth in the pathophysiology of chronic diseases. *Front Physiol.* 11:694

Shi L, Zhao W, Yang Z, Subbiah V, Suleria HAR. 2022. Extraction and characterization of phenolic compounds and their potential antioxidant activities. *Environmental Science and Pollution Research.* 29(54):81112–29

Silva JDR, Arruda HS, Andrade AC, Berilli P, Borsoi FT et al. 2024. *Eugenia calycina* and *Eugenia stigmatosa* as promising sources of antioxidant phenolic compounds. *Plants.* 13(15):2039

Škrovánková S, Mišurcová L, Machů L. 2012. Antioxidant activity and protecting health effects of common medicinal plants. *Adv Food Nutr Res.* 67:75–139

Soares SE. 2002. Phenolic acids as antioxidants. *REVISTA DE NUTRICAO-CAMPINAS-.* 15(1):71–82

Soleymani S, Habtemariam S, Rahimi R, Nabavi SM. 2020. The what and who of dietary lignans in human health: Special focus on prooxidant and antioxidant effects. *Trends Food Sci Technol.* 106:382–90

Sreejayan N, Rao MN. 1996. Free radical scavenging activity of curcuminoids. *Arzneimittelforschung.* 46(2):169–71

Sultana N, Saini PK, Kiran SR, Kanaka S. 2023. Exploring the antioxidant potential of medicinal plant species: A comprehensive review. *J Plant Biota.* 2(2).

Tapas AR, Sakarkar DM, Kakde RB. 2008. Flavonoids as nutraceuticals: A review. *Trop J Pharm Res.* 7(3):1089–99

Terahara N. 2015. Flavonoids in foods: A review. *Nat Prod Commun.* 10(3):1934578X1501000334

Tischer B, Pangloli P, Nieto-Veloza A, Reeder M, Dia VP. 2023. Bioactive compounds, antioxidant capacity and anti-inflammatory activity of native fruits from Brazil. *PLoS One.* 18(5):e0285625

Tomás-Barberán FA, Espín JC. 2001. Phenolic compounds and related enzymes as determinants of quality in fruits and vegetables. *J Sci Food Agric.* 81(9):853–76

Touré A, Xueming X. 2010. Flaxseed lignans: Source, biosynthesis, metabolism, antioxidant activity, bioactive components, and health benefits. *Compr Rev Food Sci Food Saf.* 9(3):261–69

Tsai C, Chou C, Liu Y, Hsieh C. 2014. Ultrasound-assisted extraction of phenolic compounds from *Phyllanthus emblica* L. and evaluation of antioxidant activities. *Int J Cosmet Sci.* 36(5):471–76

Ullah A, Munir S, Badshah SL, Khan N, Ghani L et al. 2020. Important flavonoids and their role as a therapeutic agent. *Molecules.* 25(22):5243

Ullah H, Khan A, Bibi T, Ahmad S, Shehzad O, et al. 2022. Comprehensive in vivo and in silico approaches to explore the hepatoprotective activity of poncirin against paracetamol toxicity. *Naunyn Schmiedebergs Arch Pharmacol.* 395(2):195–215.

Valanciene E, Jonuskiene I, Syrpas M, Augustiniene E, Matulis P, et al. 2020. Advances and prospects of phenolic acids production, biorefinery and analysis. *Biomolecules.* 10(6):874

Villaño D, Fernández-Pachón MS, Moyá ML, Troncoso AM, García-Parrilla MC. 2007. Radical scavenging ability of polyphenolic compounds towards DPPH free radical. *Talanta.* 71(1):230–35

Wang M, Jiang N, Wang Y, Jiang D, Feng X. 2017. Characterization of phenolic compounds from early and late ripening sweet cherries and their antioxidant and antifungal activities. *J Agric Food Chem.* 65(26):5413–20

Wang W, Wang S, Liu T, Ma Y, Huang S et al. 2020. Resveratrol: Multi-targets mechanism on neurodegenerative diseases based on network pharmacology. *Front Pharmacol.* 11:694

Wu Y-N, Wang W, Yao G-D, Jiang X-F, Zhang Y, Song S-J. 2018. Antioxidant phenolic acids from the leaves of *Armeniaca sibirica. J Asian Nat Prod Res.* 20(10):969–76

Yi Y, Sun J, Xie J, Min T, Wang L-M, Wang H-X. 2016. Phenolic profiles and antioxidant activity of lotus root varieties. *Molecules.* 21(7):863

Zafar F, Asif HM, Shaheen G, Ghauri AO, Rajpoot SR et al. 2023. A comprehensive review on medicinal plants possessing antioxidant potential. *Clin Exp Pharmacol Physiol.* 50(3):205–17

Zeb A. 2020. Concept, mechanism, and applications of phenolic antioxidants in foods. *J Food Biochem.* 44(9):e13394

Zengin G, Yagi S, Eldahshan OA, Singab AN, Selvi S et al. 2024. Decoding chemical profiles and biological activities of aerial parts and roots of *Eryngium thorifolium* Boiss by HPLC-MS/MS, GC-MS and in vitro chemical assays. *Food Biosci.* 104556

Zhai R, Hu J, Jin M. 2022. Towards efficient enzymatic saccharification of pretreated lignocellulose: Enzyme inhibition by lignin-derived phenolics and recent trends in mitigation strategies. *Biotechnol Adv.* 61:108044

Zhang S, Jiang S, Deng N, Zheng B, Li T, Liu RH. 2022a. Phytochemical profiles, antioxidant activity and antiproliferative mechanism of *Rhodiola rosea* L. phenolic extract. *Nutrients.* 14(17):3602

Zhang X-F, Tang Y-J, Guan X-X, Lu X, Li J et al. 2022b. Flavonoid constituents of *Amomum tsao-ko* Crevost et Lemarie and their antioxidant and antidiabetic effects in diabetic rats–in vitro and in vivo studies. *Food Funct.* 13(1):437–50

Zheleva-Dimitrova D, Simeonova R, Kondeva-Burdina M, Savov Y, Balabanova V et al. 2023. Antioxidant and hepatoprotective potential of *Echinops ritro* L. Extracts on induced oxidative stress in vitro/in vivo. *Int J Mol Sci.* 24(12):9999

# 5 Anti-Inflammatory Effects of Phenolic Compounds

*Rajapandiyan Krishnamoorthy**, *Mohammad A. Alshuniaber, and Mansour K. Gatasheh*

## 5.1 INTRODUCTION

A biological reaction known as inflammation occurs when a material that is physical, chemical, or biological enters the body and disturbs the homeostasis of the tissue. Certain agents include trauma, such as shock or burns; immune system reactions, such as hypersensitivity, are examples of specific agents (Medzhitov, 2008). Pathogens are bacteria, fungi, and viruses, and toxic compounds are pollutants. Depending on the type of stimulation and how well the body responds to get rid of the stimulus or damaged tissues, inflammation can be categorized as either acute or chronic leukocyte (mostly neutrophil) emigration, fluid and plasma protein exudation, and a rapid (within minutes) onset that lasts for several hours or days are the characteristics of acute inflammation (Serhan & Savill, 2005). When the immune system effectively eliminates harmful substances, the acute phase of inflammation comes to an end; if not, an inadequate reaction may cause a chronic phase. Chronic inflammation is characterized by tissue damage, fibrosis, and the development of new blood vessels, as well as the presence of macrophages and lymphocytes. More than 50% of deaths globally currently occur from chronic inflammatory illnesses, which present a major risk to public health. The reduction of inflammatory indicators by up to 30% through meals high in phenolic compounds has been demonstrated, suggesting that these diets could be utilized in addition to or as an alternative to conventional anti-inflammatory medications (Furman et al., 2019).

In 2020, research revealed that more than 80% of cardiovascular diseases were linked to chronic inflammation, underscoring the significance of anti-inflammatory treatments, such as phenolic compounds. Novel therapeutics for disorders linked to inflammation may result from more research into the processes of phenolics (Jiang & Dusting, 2003). During inflammation, some cells primarily macrophages assist in the synthesis of substances that promote inflammation. A few of the mediators are interleukin (IL)-1β, PGs, reactive oxygen species (ROS), nitric oxide (NO), tumor necrosis factor (TNF)-α, IL-6, IL-8, and NO. Owing to the oversynthesis of these mediators in chronic inflammation, atherosclerosis, diseases such as dementia, cancer, arthritis, and asthma have all been associated with the onset of chronic degenerative disorders. As a result, it became necessary to decrease inflammation, and nonsteroidal anti-inflammatory drugs (NSAIDs) are commonly used to do this.

NSAID use is associated with a number of adverse effects, including bronchospasm, fluid retention, renal failure, and gastrointestinal problems. Numerous plants and herbs have been used as alternatives in folk medicine to relieve inflammation, with little to no negative side effects (Lai & Roy, 2020).

Studies have shown interest in phenolic compounds as important components that contribute to their anti-inflammatory qualities. The presence of aromatic rings and two hydroxyl groups or more characterizes the chemical entities known as polyphenols, also referred to as phenolic compounds. Plants naturally contain phenolic compounds that protect them from UV radiation and other

* Corresponding author: moorthy@ksu.edu.sa

DOI: 10.1201/9781003527671-5

environmental stresses. Scientists are becoming increasingly interested in phenolic compounds because of their exceptional pharmacological and nutritional qualities, which include antioxidant, anti-inflammatory, antimicrobial, antibacterial, and antiviral effects (Sun & Shahrajabian, 2023).

By decreasing oxidative stress and inflammation and activating antioxidant enzymes, phenols can shield essential biological components from harm caused by reactive free radicals. In order to stop cellular damage, it can also transfer electrons to free radicals and block or sequester ROS. Due to its ability to regulate glucose metabolism, it seems to have tremendous potential for both treating and preventing diabetes. Additionally, it lowers cholesterol and blood pressure, thereby reducing the risk of cardiovascular disease (Liu et al., 2024).

Polyphenols can block the growth of cells by affecting the signaling pathways of PI3K/Akt, CDK, and Erk1/2. Polyphenols possess the capacity to enhance endogenous defense mechanisms such as glutathione peroxidase (GPX) and superoxide dismutase (SOD). Their ability to simultaneously regulate several proteins and transcription factors makes them attractive options for investigating anti-cancer treatments. This review examines the ability of phenolic compounds to reduce inflammation (Martínez & Sánchez-Reus, 2021).

## 5.2 CLASSIFICATION OF PHENOLIC COMPOUNDS

There are several groups of phenolic and polyphenolic compounds, including lignins, tannins, coumarins, flavonoids, and phenolic acids. Cereals, legumes, and other seeds are among the numerous natural sources of phenolic acids, a noteworthy class of phenolic compounds. They act as structural elements of cell wall matrices, enabling the formation of dense cell wall structures by bonding with macromolecules like cellulose, hemicellulose, and pectin. As a result, they usually do not exist in the free type but rather in many conjugated forms (Zhang & Hamauzu, 2020). The two classes of phenolic acids are hydroxybenzoic acids and hydroxycinnamic acids. P-coumaric, caffeic, ferulic, and sinapic acids are examples of hydroxycinnamic acids; p-hydroxybenzoic, protocatechuic, vanillic, syringic, and gallic acids are examples of hydroxybenzoic acids (Khan & Mukhtar, 2021).

The distinct characteristics of every phenolic acid are dictated by functional groups, specifically hydroxyl and methoxy groups, and substitution. Ingested phenolic acids are absorbed in the gastrointestinal tract and subsequently circulate in the bloodstream following methylation, sulfation, and glucuronidation in the liver (Zheng & Wang, 2020). The digestive system absorbs the hydroxycinnamic acids in wine and transforms them into conjugates of glucuronide and sulfate, which are then absorbed by the bloodstream. The hydrophilicity of phenolic compounds is increased by this conjugation process, which makes it easier for the biliary or urine systems to eliminate them (Crespy & Williamson, 2020).

Flavonoids are a class of phenolics that include flavonols, flavones, flavanonols, flavanones, anthocyanidins, and isoflavones, among many other subclasses. Flavonoids are defined by a three-ring structure in the $C_6$–$C_3$–$C_6$ configuration. These substances could show different hydroxyl and methoxy group substitution patterns. Flavonoids have significant bioactivities, such as antiviral, anticancer, and anti-inflammatory properties, and they can lower the risk of type 2 diabetes and cardiovascular disease (García-López & de la Torre, 2021). The compound 1,2-diphenylethene, also known as stilbenes, has a $C_6$–$C_3$–$C_6$ carbon structure that exhibits a variety of important bioactivities. Kobophenol-A and -B are two more tetrastilbenes that exhibit antibacterial action against *Staphylococcus aureus*. Topoisomerase II, which aids in unwinding coiled DNA during transcription in cells, is known to be potently blocked by stilbenes. One of the most common stilbenes found in grapes and wine is resveratrol, which also exhibits a variety of other bioactivities, including anti-tumorigenic, anti-inflammatory, and cardioprotective effects. Resveratrol also prevents Alzheimer's and Parkinson's disease by regulating heme oxygenase activity, as abnormal heme oxygenase function is a component of both conditions (Pérez-Jiménez et al., 2020a).

Benzopyrones, which contain coumarins, are defined by the $C_6$–$C_3$ structure and the presence of an oxygen heterocycle in the C3 component. Several plant species contain them, including mullein (*Verbascum* spp.), sweet-clover (*Melilotus* sp.), deer tongue (*Dichanthelium clandestinum*), sweet grass (*Hierochloe odorata*), tonka beans (*Dipteryx odorata*), and sweet woodruff (*Galium odoratum*) (Bae & Kim, 2021). It has been documented that bacteria, in addition to plants, can also create coumarins as byproducts. Specifically, these are species of *Aspergillus* and *Streptomyces*, which include coumermycin and novobiocin. Anti-inflammatory, antibacterial, antiviral, antioxidant, anticancer, antiasthmatic, antidepressant, anti-HIV, and anti-Alzheimer's properties are only a few of the many biological activities of coumarins. Moreover, coumarin reduces blood coagulation by preventing the hepatic manufacture of chemicals required for vitamin K-dependent coagulation. For this reason, coumarin is used to treat conditions such as thrombotic phlebitis and pulmonary embolism as an anticoagulant (Khan & Mukhtar, 2021).

Furthermore, plants also contain phenolic chemicals in polymeric form; lignins are polymers made from monolignols, which include sinapic acid and p-coumaric acid. Tannins can be classified into two categories: condensed tannins, sometimes referred to as proanthocyanidins, and hydrolyzable tannins, which include ellagitannins and gallotannins. The latter is further classified into multiple subclasses based on whether the flavonoids have a single or double connection. Rich sources of proanthocyanidins include apples, lentils, grape seeds, cocoa seeds, and other foods originating from plants (Huang & Zhan, 2022).

The structural properties of proanthocyanidins, especially those with two or three hydroxyl groups in the flavan-3-ol unit's B-ring, make them bioactive. As an example, proanthocyanidins can decrease the oxidation of low-density lipoprotein (LDL) cholesterol, mitigate mucosal damage, provide radioprotection against chromosomal damage, and have significant anti-adhesion activity against bacteria (Cao & Zhang, 2021). They also possess potent antioxidant properties, such as the ability to scavenge oxygen radicals.

## 5.3 BIOSYNTHESIS OF PHENOLICS

The endoplasmic reticulum (ER), a surface region of plant cells, is where phenols are produced. Phenylalanine and tyrosine are precursors used in the manufacture of numerous distinct phenolic compounds, including flavonoids, phenolic acids, coumarins, lignins, and tannins. The enzymes erythrose 4-phosphate, p-coumarate-3-hydroxylase, cinnamonate-4-hydroxylase, o-methyl transferase, and phenylalanine ammonia lyase (PAL) are a few that help in the synthesis of phenolics. Phenylalanine is the primary component required to create various phenolics. Both the tyrosine and, to a lesser degree, the shikimate pathways create it (Kumar & Singh, 2021).

PAL releases an ammonia group from phenylalanine, initiating the enzymatic production of phenolics. As a result, a double bond is introduced, creating trans-cinnamic acid. Cinnamate-4-hydroxylase converts trans-cinnamic acid into p-coumaric acid. Then, by adding a hydroxyl group to the aromatic ring of p-coumaric acid, p-coumarate-3-hydroxylase facilitates the synthesis of caffeic acid (Wang & Zhang, 2022). The enzymes o-methyl transferase and hydroxylase convert p-coumaric acid to ferulic acid, which is further transformed into sinapic acid. These enzymes facilitate the synthesis of sinapic acid by substituting a methoxy group. With the help of the enzyme tyrosine ammonia lyase, tyrosine, another precursor, is transformed into p-coumaric acid by removing an ammonia group. Afterward, comparable processes are followed by p-coumaric acid to produce additional phenolic acids (Jia & Lu, 2022).

Furthermore, the removal of two carbon atoms from phenylpropanoids results in the production of benzoic acid. Three molecules of p-coumaroyl Coenzyme A (CoA) and malonyl CoA combine to form the flavonoid branch pathway, which is the process of producing flavonoids. Naringenin chalcone is thus generated, which is then broken down by chalcone synthase (CHS) and ultimately transformed into flavanone by chalcone isomerase. The generated flavanone can be transformed into a variety of distinct types of flavonoids, including isoflavone, anthocyanins, flavone, flavanol, and flavanone, by the appropriate enzymatic processes. Flavanol synthase/flavanone-3-hydroxylase and flavone synthase convert flavanone into kaempferol and apigenin, while isoflavone synthase generates genistein (Zhang & Wang, 2021).

## 5.4 HEALTH BENEFITS OF PHENOLIC COMPOUNDS

Phenolic chemicals are found in a wide variety of fruits and vegetables, such as tomatoes, berries, and grapes. By lowering the prevalence of metabolic diseases, such as type 2 diabetes, phenolic substances may enhance health. The biological features of phenolic compounds are varied, although the precise methods via which they confer disease prevention remain unidentified. Antioxidants help in the removal of these oxidative wastes. For example, ROS and reactive nitrogen species (RNS) can damage vital biological macromolecules, such as DNA and proteins, and hinder mitochondrial respiration (Pérez-Jiménez et al., 2020b).

These compounds are highly reactive in a physiological setting. The anti-aging, anti-inflammatory, antiproliferative, and antioxidant properties of phenolic compounds have been shown in numerous studies. In a physiological environment, a steady stream of highly reactive oxidized molecules is produced, including superoxide, peroxide, singlet oxygen, hydroxyl radical, NO, and peroxynitrite ($OONO^-$). These molecules are known as RNS and ROS. Reduced antioxidant levels can contribute to cellular degradation, a process that is a cause of disease, aging, and cell death (Shen & Zhao, 2021).

The mitochondrial permeability transition pore (mPTP) of individual mitochondria can be activated by photo-induced ROS in intact biological systems. ROS-induced ROS release (RIRR) is a phenomenon that was discovered after the mPTP was exposed to ROS and further stimulation of ROS production. The mitochondrial response to oxidative stress is the opening of the mPTP, which raises ROS signals and can have a variety of effects depending on ROS concentrations. Apart from the consequences of ROS generated within mitochondria, ROS released into the cytosol can trigger RIRR in adjacent mitochondria and can result in a complex cellular signaling reaction (the source of the RIRR) (Giorgio & Bianchi, 2021). In this instance, the movement of ROS between mitochondria may function as a positive feedback loop, raising the level of ROS generation that permeates the cell and clearly damages the mitochondria and the cellular architecture.

Although significant, avalanche-like releases of ROS can be produced in a lab via photo-induced ROS generation, this suggests a deeper process involving light-independent spontaneous redox transitions associated with the opening of the mPTP or other mitochondrial channels that could occur in various physiological or pathological circumstances. Oxidative and reductive stressors can arise from an imbalance in ROS intake, neutralization, and outflow, as well as from their related triggers in particular cell signaling pathways. This imbalance may result in a variety of diseases or even the death of biological organisms. This mechanism of anti-inflammatory activity against several diseases, which include metabolic and other inflammation-related diseases (Cai & Zhao, 2022).

## 5.5 PHENOLIC COMPOUNDS AND INFLAMMATION

A vast family of secondary metabolites created during plant metabolism are known as phenolic compounds. Many experts are interested in phenolic compounds because of their many health benefits, particularly when it comes to their presence in food. There is usually at least one aromatic ring with one or more hydroxyl groups attached to it in a phenolic compound. Phenolic compounds can be either aromatic or aliphatic. There are two types of phenolic compounds: flavonoids and non-flavonoids (Manach & Scalbert, 2021).

### 5.5.1 Flavonoids

These are heterocyclic substances that have an oxygen atom separating the two aromatic rings. Flavonoids can be divided into four groups according to their hydrogenation state and type of heterocyclic substitution: flavones, flavanols, anthocyanins, and isoflavones. Glycosides are a common characteristic of flavonoids (García-Viguera & Gil-Izquierdo, 2021).

### 5.5.2 Non-Flavonoids

Benzoic and cinnamic acids, sometimes known as phenolic acids, are two of the most unique non-flavonoid substances. Phenolic acids yield lignins, tannins, and stilbenes. Based on epidemiological statistics, food products containing phenolic compounds (hydroquinone, resorcinol, and catechol) have been linked to a lower risk of developing chronic illnesses, such as diabetes mellitus, heart disease, Parkinson's disease, Alzheimer's disease, and infections. It is believed that phenolic compounds are responsible for these positive outcomes. As mentioned, the formation of chronic problems has been linked to long-lasting and severe diseases (Khan & Miah, 2022).

Since it has been demonstrated that glycosides increase absorption while decreasing the efficacy of anti-inflammatory medications, their usefulness is a matter of debate. In other specialized applications, the relationship between pro-inflammatory mediators and the phenolic system has been demonstrated. Three structural characteristics of flavonoids have the potential to inhibit NO: (A) the double bond C2=C3; (B) a large substituent group that may significantly reduce inhibition (aglycones have a stronger inhibitory effect than glycosides); and (C) the presence of 7 and 4′ hydroxyl groups, which also affect inhibition, albeit less so than the previously mentioned characteristics. Therefore, invasive events that alter the inflammatory cascade associated with the evolution of the disease may be helpful targets for disease prevention (Chakraborty & Basak, 2022).

Some phenolic substances have anti-inflammatory properties. A decrease in exaggerated reactions is correlated with increased absorption of these drugs, even if the exact mechanisms behind these anti-inflammatory effects are still unknown. A set of standards based on interactions with numerous inflammatory targets was developed to establish the connection between phenolic complexes and their anti-inflammatory capabilities (Rafique & Ali, 2023).

## 5.6 ANTI-INFLAMMATORY EFFECTS OF PHENOLS

An essential biological response to tissue damage is inflammation, which is necessary for survival. The immune system releases pro-inflammatory cytokines in response to various stimuli, including sickness, damage, or irritation. Adults with elevated levels of pro-inflammatory cytokines TNF-α, IL-6, and IL-1b have severe health problems such as asthma, allergies, atherosclerosis, and cancer. Limiting diseases linked to inflammation requires limiting the overproduction of pro-inflammatory cytokines (Miller & Raison, 2021).

Plant-based compositions containing phytochemicals have been utilized for the treatment of inflammation and related disorders since ancient times. Among the known phytochemicals, phenols are essential for reducing inflammation; recent research has shown that they possess potent anti-inflammatory properties (Li & Zhang, 2022).

Phenolic substances generally cooperate with NSAIDs. However, some phenolic substances, such as those that inhibit cyclooxygenase (COX), can suppress the expression of genes that support inflammation. Phenolic substances can alter transcriptional elements associated with antioxidant pathways, including nuclear factor-kappa B (NF-κB) and nuclear erythroid-related factor 2 (Nrf-2). These effects were caused mainly by phenolic compounds' antioxidant properties (Pérez-Jiménez & Neveu, 2021).

Changes in gene expression or the suppressive effects of pro-inflammatory mediators are partially responsible for the effects of phenolic substances. While synthetic drugs often only affect one pathway, higher amounts of phenolic compounds, or a range of phenolics, can have anti-inflammatory effects across multiple pathways. Phenolic ligands can facilitate the formation of covalent bonds between flavonoids and macromolecules. It is thought that phenolic compounds decrease

**TABLE 5.1**
**Anti-Inflammatory Mechanism of Phenolic Compounds**

| Phenolic Compound/Source | Key Mechanisms | Anti-Inflammatory Effects | Research/Study Highlights |
|---|---|---|---|
| **General phenolic compounds** | Inhibit NF-κB, Nrf-2 pathways; modulate gene expression | Reduce COX activity and inhibit pro-inflammatory cytokines (IL-6, IL-1 beta, and TNF-α). | Synergistic effects of NSAIDs, reduce inflammation-related diseases |
| **Flavonoids (e.g., cocoa, tea)** | Modulate inflammatory mediators (IL-6) affect eicosanoid synthesis | Dose-dependent decrease of IL-6 and LDL levels reduces vascular inflammation | Beneficial effects on cardiovascular health: lower risk of heart disease |
| **Procyanidins (grape phenolic extracts)** | Suppress inflammatory mediators; reduce ROS and NO production | Decrease nitric oxide, prostaglandin E2, and ROS | Studied in both in vitro and in vivo settings |
| **p-Coumaric acid** | Inhibits TNF-α expression | Decreases TNF-α levels in arthritic rats | Significant anti-inflammatory effect in adjuvant-induced arthritis model |
| **Catechols** | Inhibit COX-2, NO synthase | Affect nucleophilic additions and binding of pro-inflammatory mediators | Depend on the B-ring configuration and electrophilic domain formation |
| **Phenolic ligands** | Establish covalent bonds with flavonoids | Inhibit the binding of pro-inflammatory mediators | Modulate multiple components, unlike synthetic drugs targeting single components. |

active resistance units, modify eicosanoid synthesis, obstruct pro-inflammatory mediator binding, and suppress NO synthase and COX-2 activity by inhibiting NF-κB (Bai & Wu, 2023).

The anti-inflammatory properties of phenolic compounds can be significantly influenced by their structures (Table 5.1). Furthermore, the creation of a double bond between C2 and C3 results in coplanarity between the A and C rings, which improves the interaction between flavonoids and artificial active sites. Catechols can limit enzymatic activity only if the B ring is configured in a way that requires nucleophilic additions and is impacted by the development of electrophilic domains (Wang et al., 2023).

Dietary flavonoids, such as the flavones found in tea and cocoa, have an impact on inflammatory mediators like IL-6 and show a dose-response connection with bloodstream IL-6 concentrations. A small number of people showed improvement in the decrease of inflammatory markers. However, there is evidence that the assimilation of the phenolic components of cocoa leads to the upregulation of inflammatory markers (TNF-α, IL-6, and IL-1), as well as a reduction in LDL. This may lead to decreased levels of platelet aggregation inhibition, NO, and vascular inflammation in addition to oxidative stress, hence reducing the risk of cardiovascular illnesses (Alonso-Magdalena & Quesada, 2022).

Numerous investigations detailing the anti-infective properties of phenolic compounds have been conducted. Procyanidins have been demonstrated to have the ability to suppress inflammatory mediators, hence reducing NO, prostaglandin E2, and ROS. In order to investigate the anti-inflammatory properties of p-coumaric acid, Pragasam et al. tracked the release of TNF-α in the synovial tissue of rats undergoing adjuvant-induced arthritis treatment. Studies showed that by reducing TNF-α expression, p-coumaric acid had a strong anti-inflammatory effect (Pragasam et al., 2023). Table 5.1 shows the anti-inflammatory mechanism of phenolic compounds.

## 5.7 PHENOLS AND THEIR MECHANISMS OF ACTION AS ANTIOXIDANTS

Owing to the deleterious consequences of ROS, such as peroxyl ($•RO_2^-$), hydroxyl (•OH), and superoxide ($•O_2^-$), on the skin, numerous studies have examined antioxidant compounds to strengthen the body's defense mechanisms against oxidative stress-related illnesses. Phenolic acids, flavonoids, and other phenolics were the focus of research on natural antioxidants and their derivatives (Sharma et al., 2023).

In addition to their known role as direct antioxidants, phenolic compounds also exhibit indirect antioxidant action through the enhancement of regulatory effects on signaling pathways and the promotion of endogenous protective enzymes. Phenolic acids have antioxidant properties due to the reactivity of the phenolic molecule, namely, the hydroxyl substituent on the aromatic ring. Whilst there might be alternatives as well, radical scavenging via hydrogen atom donation is the main concept explaining the antioxidant action of phenolic acids. Studies have indicated that the antioxidant capacities of free, esterified, glycosylated, and nonglycosylated phenolics vary, as do the antioxidant activities of different phenolic acids (Zhao et al., 2023 a and b).

## 5.8 PHENOLS AS ANTIOXIDANT AGENTS AND THEIR SIGNALING PATHWAYS

### 5.8.1 Phenol-Influenced Signaling Pathways

Phenolic chemicals are found in large quantities and exhibit a variety of antioxidant mechanisms of action. Allicin enhances the NRF2 antioxidant signaling pathway at the level of the metabolic route, activating NRF2 through its electrophilic center, which acts as a target for nucleophiles such as the sulfhydryl group on KEAP1. Through the identification of binding sites within the human antioxidant response element (ARE) promoter region, NRF2 regulates the production of key antioxidants. This contains rate-limiting enzymes that shield cells from oxidative damage brought on by free radicals, such as glutamyl cysteine synthetase (GCS). Glycosides from *Tripterygium wilfordii* reduce ANGII levels by activating the ERK1/NOX2/NRF2/2 pathway, while ANGII increases antioxidant enzyme activity by ERK1/2 activation and NRF2 nuclear translocation, lowering AngII-induced oxidative stress (Zhang et al., 2023b).

By obstructing the ERK1/2-RSK-nNOS pathway, resveratrol can stop oxidation. Rats with fructose-induced hypertension can lower their blood pressure by stimulating adenosine monophosphate-activated protein kinase (AMPK) with resveratrol. It also prevents the ERK1/2-RSK-nNOS pathway, which phosphorylates Ser 1416 in nNOS and generates NO and Rac1-mediated increases in P22Phox subunit activity (Gao & Zhang, 2023).

One anti-inflammatory drug that works by blocking the Akt/PI3K signaling pathway is quercetin. By activating the AMPK/SIRT pathway, quercetin enhances the oxidative stress response. Adenine nucleotide levels can be measured by quercetin at both baseline and LPS stimulation. Additionally, it can reduce pro-inflammatory cytokines by reducing the M1 macrophage marker NOS2 both at rest and in response to LPS stimulation. By raising the AMP/ATP ratio, quercetin activates AMPK and SIRT1 expression, both of which suppress lipogenesis (Khan et al., 2023).

Through the inhibition of vital survival pathways that are activated in tumor cells, quercetin demonstrates anti-malignant effects. Examination and assessment: Research conducted on lymphoid cancer cells, both in vivo and in vitro, demonstrates quercetin's cytotoxic properties. Ferry et al. conducted a Phase I clinical trial to examine the antiproliferative properties of quercetin in ovarian cancer that was resistant to cisplatin. One patient's CA 125 levels dropped from 295 to 55 units/m$^2$ after receiving 420 mg/m$^2$. It was discovered that the serum alpha-fetoprotein levels of another patient had dropped. Consequently, it was determined that intravenous quercetin had an anticancer impact by inhibiting lymphocyte tyrosine kinase activity. A range of enzymes and cytokines, such as the cyclooxygenases COX-1 and COX-2, which transform arachidonic acid into prostaglandins that influence pain and stimulate inflammatory processes, are also necessary for anti-inflammatory and inflammatory activities (Ferry et al., 2022). Table 5.2 shows phenols and their signaling pathways as antioxidant and anti-inflammatory agents.

**TABLE 5.2**
**Phenols and Their Signaling Pathways as Antioxidant and Anti-Inflammatory Agents**

| Phenolic Compound | Signaling Pathway/Target | Antioxidant/Anti-inflammatory Action | Other Key Effects |
|---|---|---|---|
| **Allicin** | NRF2-ARE pathway | Activates NRF2, enhances antioxidant response by increasing expression of enzymes like GCS. | Reduces oxidative damage |
| ***Tripterygium wilfordii* glycosides** | NRF2/ERK1/2/NOX2 pathway | Activates NRF2 and ERK1/2 and reduces oxidative stress by decreasing ANGII levels. | Improves antioxidant enzyme activity |
| **Resveratrol** | ERK1/2-RSK-nNOS pathway | Inhibits oxidation, enhances nitric oxide release via phosphorylation of nNOS. | Reduces blood pressure, inhibits Rac1-mediated oxidative stress |
| **Quercetin** | AMPK/SIRT1/PI3K/Akt pathway | Improves oxidative stress response, activates AMPK and SIRT1, inhibits lipogenesis | Reduces M1 macrophage activity, has anticancer effects |
| **Curcumin** | TLR2/4, MAPK/NF-κB pathway | Inhibits NF-κB and ROS production, blocks TLR signaling | Reduces inflammation, enhances NRF2 activity |
| **Epigallocatechin 3-gallate (EGCG)** | TLR/MyD88/NF-κB pathway | Inhibits NF-κB activation, suppresses inflammatory gene products | Inhibits TLR4/9 signaling and inflammatory responses |
| **Fisetin, naringenin** | MAPK pathway | Inhibits LPS-induced inflammatory protein expression, blocks p38 and ERK pathways. | Reduces COX-2 expression |
| **Berberine** | AMPK/NOX4 pathway | Enhances antioxidant capacity by reducing ROS from NOX4 | Increases GSH and GPX levels, inhibits protein carbonylation |
| **Luteolin** | COX-2, MAPK pathway | Inhibits COX-2 enzyme activity, reduces TNF-α and nitric oxide production | Reduces prostaglandin synthesis and inflammatory cytokines |
| **Myristoin** | COX-1 and COX-2 | Inhibits both COX-1 and COX-2 enzymes | Anti-inflammatory |

### 5.8.2 Pathways Associated with NF-κB

The primary mechanism associated with NF-κB that has been extensively studied is immune receptor-activated kinase TGF-β, which in turn activates kinase 1 (TAK1). Members of the IκB kinase (IκB) family are primarily responsible for phosphorylating the trimeric IκB kinase (IKK) complex, which is initiated by TAK1. IκBα and p105 are degraded by ubiquitin (Ub)-mediated proteasome, targeting the synthesis of inflammatory mediators by facilitating the nuclear translocation of canonical NF-κB family members. This mechanism might be suppressed by curcumin (Zhang et al., 2023).

Limiting the generation of ROS via the TLR4-MAPK/NF-κB pathway can help mitigate LPS-induced inflammation. Lutein's inhibition of LPS-induced production of NF-κB, intercellular adhesion molecule-1 (ICAM-1), and TNF-α is evidence of its anti-inflammatory capabilities. Through modulation of IKK and IκB phosphorylation, resveratrol selectively inhibits LPS-induced NF-κB activation, leading to a considerable decrease in TNF-α and IL-6 levels (Zhao et al., 2023).

### 5.8.3 MAPK-Associated Pathways

ERK, JNK, and p38 are essential members of the mammalian MAPK family. Mitogens and growth factors frequently start the ERK1/2 pathway. Stress and inflammation, on the other hand, primarily activate the JNK and p38 pathways. Many phenolic substances can impede MAPKs. The MAPK signaling system, which controls the generation of cytokines and the expression of inflammatory proteins produced by LPS, is inhibited by fisetin, naringenin, and butane (Seyed-Farid et al., 2023).

Resveratrol has the potential to mitigate inflammation in cases of mastitis by inhibiting the phosphorylation of p38 and ERK. Moreover, it may reduce COX-2 levels by preventing the activation of p38 and ERK brought on by PMA. By reducing MAPK and AP-1 DNA binding, quercetin affects the synthesis of TNF-alpha and NO synthase in response to lipopolysaccharide (LPS) in RAW murine cells (Zhou et al., 2023).

### 5.8.4 TLR-Associated Pathways

TLRs are divided into two groups: intracellular kinds that identify microbial nucleic acids and external types that identify components of microbial membranes. TLR is capable of inducing inflammatory reactions, and it initiates the downstream pathway that is dependent on MyD88 upon activation. The phosphorylation of IL-1 receptor-associated kinase-4 (IRAK-4) is enhanced by MyD88. IRAK-1 is phosphorylated by this mechanism, which leads to its degradation. The transcription factor NF-κB and the traditional IKKα/β/γ complex are activated when phosphorylated IRAK-1 interacts with TRAF6. Therefore, NF-κB activation leads to the synthesis of inflammatory gene products, including cytokines (Hacker et al., 2023).

Curcumin can suppress TLR2/4 and TLR4 dimerization in extracellular medium. On cell TLR9 signaling, curcumin had contradictory effects. MyD88 promotes the phosphorylation of IL-1 receptor-associated kinase-4 (IRAK-4). This method phosphorylates IRAK-1, leading to its degradation. Phosphorylated IRAK-1 interacts with TRAF6 to activate both the transcription factor NF-κB and the classical IKKα/β/γ complex. Therefore, inflammatory gene products, such as cytokines, are synthesized as a result of NF-κB activation (Zhao et al., 2023).

Additionally, it affects the TLR pathway, which is dependent on TRIF, and inhibits the signaling pathway that is dependent on MyD88. This, in turn, controls the inflammatory response that follows TLR activation.

### 5.8.5 Regulation of Genes and Proteins by Phenols

Numerous antioxidative properties of phenolic compounds are demonstrated at the gene and protein levels by blocking the activities of enzymes such as mitochondrial succinic oxidase, microsomal monooxygenase, and NADH oxidase that are involved in the synthesis of reactive oxygen. Phenolic substances act as scavengers of free radicals. Berberine also increases AMPK expression to prevent oxidation. Higher concentrations of AMPK affect NOX, which can reduce ROS produced by NOX4 (Zhang et al., 2023).

One important pathogenic mechanism linked to oxidative stress is the accumulation of advanced glycation end-products (AGEs). Glyoxal (GO) and methylglyoxal (MGO) are the most common precursors of AGEs, and they may help create AGEs from higher lysine and arginine residues in the proteome. Lipid peroxidation causes polyphenols to reduce the amount of GO, which inhibits the formation of AGEs. The antioxidant qualities of polyphenols are closely linked to the inhibitory action of GO. Because polyphenols can compete with arginine and lysine, they can prevent the formation of AGEs. When extra virgin olive oil (EVOO) stimulates NRF2, it increases the production of glutathione S-transferase (GST) and other antioxidant enzymes (Khan et al., 2023).

A subset of phenolic substances has antioxidant activity, which is linked to genes upstream of the signaling cascade, as well as proteins directly involved in antioxidant functions. These genes cooperate and are interdependent. A large number of genes that shield cells from oxidative stress are on the NRF2 target list. ARE and kelch-like ECH-associated protein-1 (KEAP1) can regulate NRF2 (Ma et al., 2023).

Oxidative stress-induced modifications to KEAP1 interfere with the way KEAP1 and NrF2 interact, which prevents NrF2 from degrading. Consequently, NRF2 moves into the nucleus and attaches itself to ARE to initiate the activation of antioxidant enzymes.

Curcumin increases the amounts of DNA repair enzymes such as APE1 and poly ADP-ribose polymerase 1 (PARP1), which improves NRF2 and promotes ARE. By moderating ERK1/2 activation and enhancing NRF2's nuclear translocation, *T. wilfordii* extracellularly signals antioxidant enzyme activity. KEAP1 is oxidized by oxidases, which activate NRF2. Next, the transcription of several antioxidant genes is encouraged by mitogen-activated protein kinase (MAPK), which also contributes to NRF2 buildup. Through their interference with the nRF2/KEAP1/HO-1 axis, curcumin analogs mainly prevent Aβ25–35-induced oxidative damage in vitro. Increased GSH and GPX levels, together with decreased MDA and protein carbonyl levels, are how berberine enhances antioxidant capability (Wang et al., 2023).

Certain polyphenols have the potential to obstruct an enzyme's function. While resveratrol acts as a nonselective COX inhibitor, myristoin inhibits COX1 and COX2. Certain flavonoids have similar processes. While luteolin influences both COX2 enzyme activity and expression, chrysin reduces COX2 enzyme activity without altering its expression.

## 5.9 THERAPEUTIC EFFECTS OF PHENOLS

### 5.9.1 The Anti-Inflammatory Properties of Phenols

Conventionally, natural plants and their extracts have been employed as anti-inflammatory agents; considerable anti-inflammatory properties have even been demonstrated for some extracts. Inflammation can arise from various abnormal pathophysiological conditions or from the invasion and colonization of the body by harmful microorganisms. Cancer, neurological conditions, autoimmune diseases, and autoinflammatory diseases can all be brought on by inflammation. A primary class of natural chemicals with anti-inflammatory properties is phenolics. Plant-based phenolic chemicals, such as flavonoids and catechins, reduce pro-inflammatory environments and oxidative stress to help treat intestinal inflammation, whether it is acute or chronic.

Phenols, a class of naturally occurring substances, protect the heart by regulating the NLRP3 inflammatory pathway, which has a variety of anti-inflammatory actions in cardiovascular illnesses. Additionally, phenols have anti-platelet properties, but polyphenols greatly improve endothelial and platelet function in the vascular system. Phenols have several ways in which they reduce inflammation. For instance, they can demonstrate anti-inflammatory properties via MAPK, PI3K/Akt, and NF-κB, three cellular signaling pathways. On macrophages, phenolics exhibit their anti-inflammatory properties by inhibiting IL-6, IL-1-beta, and TNF-α, three important inflammatory response regulators. By blocking pro-inflammatory enzymes such as COX and lipoxygenase (LOX), they may exhibit anti-inflammatory properties (Khattak et al., 2023).

### 5.9.2 Phenols and the Regulation of Cells Associated with Inflammation

Inflammation is linked to a variety of cells, including neutrophils, monocytes, eosinophils, and macrophages. Different cells have different processes of inflammation and anti-inflammation. It is believed that macrophages play a role in several autoimmune and chronic inflammatory diseases. To eradicate invasive infections, they generate NO, TNF, and IL-1, among other mediators that activate antibacterial mechanisms, including peroxidation (Murray & Wynn, 2023).

T-cell polarization requires IL-12 and IL-23, which are generated by active macrophages. Natural killer (NK) cells are essential immune cells involved in the regulation of immune responses. They protect the body from tumors and viral infections and occasionally trigger autoimmune disorders and hypersensitive reactions. Research has shown that polyphenols possess anti-allergic properties through their inhibitory effects on cells associated with allergies, such as basophils and mast cells. Both tea polyphenols and brown algal polyphenols have shown promise as therapeutic agents by having a significant impact on cellular responses brought on by immune reactions (Sharma et al., 2023). Figure 5.1 highlights the interactions that phenolic compounds have with important immune cells and pathways to demonstrate the therapeutic benefits of these compounds on inflammation.

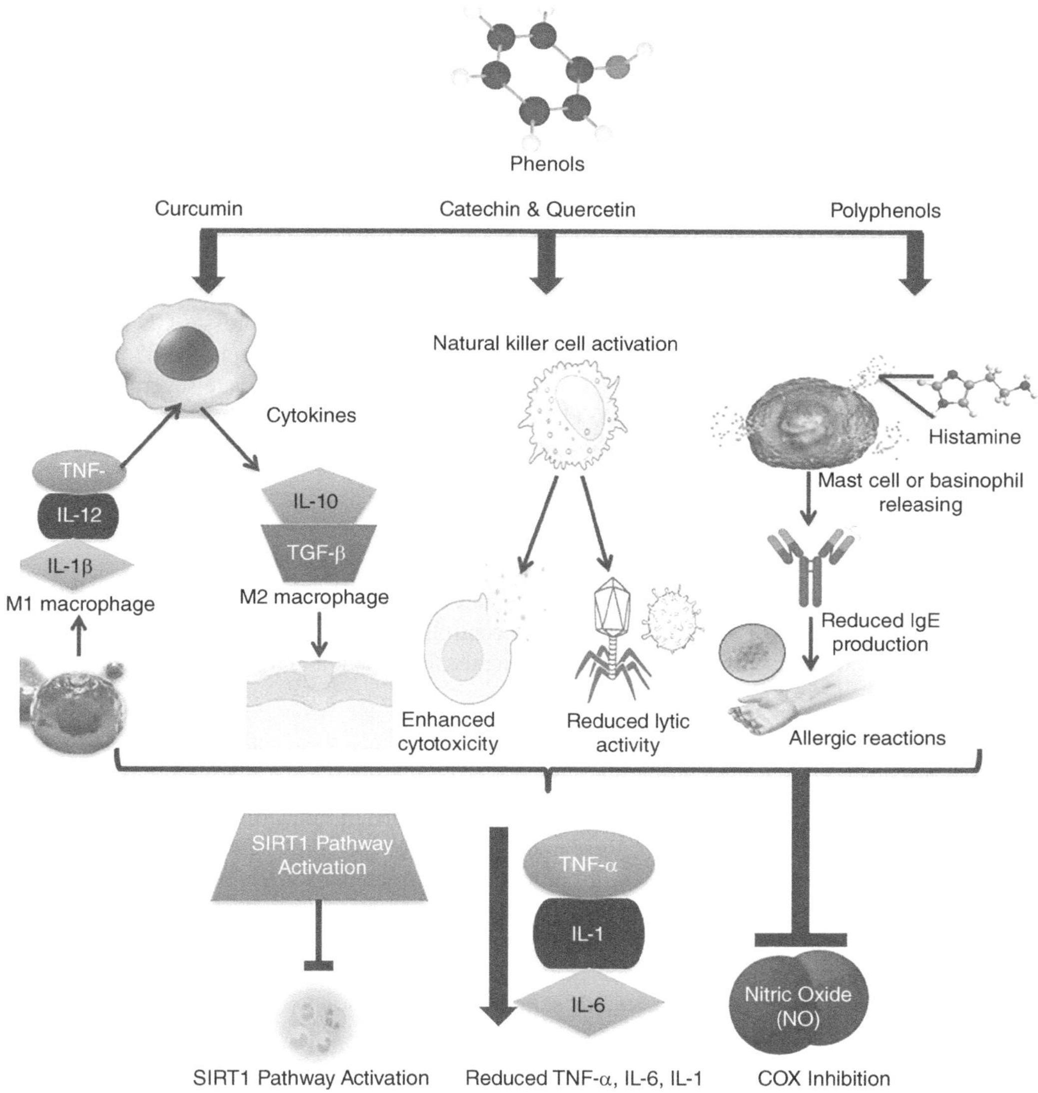


**FIGURE 5.1** Therapeutic benefits of phenolic compounds on immune cells and pathways.

### 5.9.3 Phenolics Control Macrophages

Macrophages release anti-inflammatory mediators in response to stimuli associated with infection or inflammation. These mediators include tumor, IL-23, IL-1 beta (IL-1β), IL-12, and necrosis factor-alpha (TNF-α). The continuation of the M1 phase results in detrimental tissue damage and prolongs inflammation. It has been demonstrated that M2 macrophages release significant levels of the anti-inflammatory cytokines IL-10 and TGF-beta, which support tissue remodeling, healing, and angiogenesis (Gordon & Martinez, 2023).

Curcumin prevented M1 macrophages from producing pro-inflammatory cytokines and chemokines, but it enhanced the phenotypic polarization of M2 macrophages. In LPS-stimulated macrophages, catechins reduced the production of NO and TNF. By interacting with Toll-like receptors-4 (TLR-4), which are activated in macrophages by LPS, MAPK, protein kinase B (PkB), and TNF receptor-associated factor 6 (TRAF6), RSV might influence macrophage function.

Citrus hesperidin reduced LPS-stimulated macrophages' production of TNF-α in a concentration-dependent manner. Hesperidin stopped monocytes from adhering to endothelial cells and stopped these cells from producing the TNF-stimulated adhesion protein VCAM-1.

### 5.9.4 Phenols Influence NK Cells

Naturally occurring killer cells have the ability to eliminate pathogens. NK cells are essential for the regulation of immunity because they exhibit potent cytolytic activity when stimulated by $CD4^+$ T cells' production of IFN-γ and IL-2. A phenolic substance called catechin may make NK cells more cytotoxic, and quercetin may make NK cells more lytic. Resveratrol affects both innate and adaptive immunity, activates T cells, NK cells, and macrophages, and suppresses $CD4^+$ and $CD25^+$ T cells through its interactions with other receptors, including integrin receptor αvβ3.

This phenolic molecule can inhibit COX activity, remove ROS, and activate several anti-inflammatory pathways, including SIRT1. STAT/TLR-4/NF-κB signaling is suppressed by SIRT1, which reduces immune cell production of cytokines and pro-inflammatory mediators such as PAF, TNF-α, and histamine from mast cells or macrophages (Liao et al., 2022).

### 5.9.5 Phenols Regulate Cells Associated with Allergic Reactions

Polyphenols can affect the way the allergen-IgE complex forms by preventing the complex from binding to FceRI on mast cells or basophils. They act by controlling the equilibrium between Th1 and Th2, and polyphenols have antiallergic effects by preventing the formation of antigen-specific IgE antibodies.

## 5.10 CONCLUSION

Owing to their diversity and heterogeneity, phenolic compounds can be found in many different plant-based diets and have been demonstrated to have significant anti-inflammatory properties by blocking pro-inflammatory mediators. Researchers have proposed dietary phenolic chemicals as a promising natural therapeutic option for inflammation and related disorders, with few to no adverse effects. However, most of these studies have employed an in vitro approach using phenolic extracts from various plants, overlooking the impact of human metabolism and digestion on the potential anti-inflammatory properties of phenolic compounds found in food. Therefore, it is still unclear exactly how and where phenolic compounds suppress inflammation.

## COMPETING INTERESTS

The authors declare that they have no competing interests.

## DECLARATION OF GENERATIVE AI AND AI-ASSISTED TECHNOLOGIES IN THE WRITING PROCESS

During the preparation of this work, the book chapter author(s) have not taken any assistance from AI, and the book chapter author(s) reviewed and edited the content as needed and take(s) full responsibility for the content of the publication.

## REFERENCES

Alonso-Magdalena, P., & Quesada, I. (2022). Dietary flavonoids and their role in modulating inflammatory markers and cardiovascular risk. Nutrients, 14(6), 1265.

Bae, J. H., & Kim, J. Y. (2021). Coumarin: A natural product with a wide range of biological activities. Frontiers in Pharmacology, 12, 639444.

Bai, X., & Wu, Y. (2023). Mechanisms of phenolic compounds in modulating inflammatory responses: A multi-target approach. Journal of Functional Foods, 95, 105888.
Cai, Y., & Zhao, M. (2022). Mitochondrial ROS and inflammation: A double-edged sword in disease. Cellular Signalling, 89, 110137.
Cao, Y., & Zhang, Y. (2021). Proanthocyanidins: Structural characteristics and biological activities. Molecules, 26(4), 850.
Chakraborty, S., & Basak, P. (2022). Flavonoids as anti-inflammatory agents: Structural insights and therapeutic applications. Frontiers in Pharmacology, 13, 848127.
Crespy, V., & Williamson, G. (2020). Bioavailability and metabolism of polyphenols in humans. Food & Function, 11(6), 5673–5685.
Ferry, D. R., et al. (2022). Phase I clinical trial of quercetin in ovarian cancer: A preliminary report. Cancer Chemotherapy and Pharmacology, 79(4), 607–617.
Furman, D., Campisi, J., Verdin, E., Carrera-Bastos, P., Targ, S., Franceschi, C., Ferrucci, L., & Slavich, G. M. (2019). Chronic inflammation in the etiology of disease across the life span. Nature Medicine, 25(12), 1822–1832.
Gao, J., & Zhang, J. (2023). Resveratrol and hypertension: Mechanistic insights into its role in oxidative stress and nitric oxide synthesis. Molecules, 28(5), 1892.
García-López, P. M., & de la Torre, M. C. (2021). Flavonoids: Chemical characteristics and health benefits. Frontiers in Nutrition, 8, 673902.
García-Viguera, C., & Gil-Izquierdo, A. (2021). Flavonoids: Bioactive compounds in the diet with potential health benefits. Journal of Food Science, 86(2), 295–308.
Giorgio, V., & Bianchi, K. (2021). Mitochondrial ROS and the role of mPTP in the regulation of cell death and survival. Free Radical Biology and Medicine, 164, 123–132.
Gordon, S., & Martinez, F. O. (2023). Alternative activation of macrophages: Mechanisms and functions. Annual Review of Immunology, 41, 225–250. doi:10.1146/annurev-immunol-110221-095750.
Hacker, H., et al. (2023). Toll-like receptors and the evolution of the immune system. Nature Reviews Immunology, 23(2), 107–120.
Huang, W., & Zhan, Y. (2022). Tannins: Structural characteristics, biological activities, and health benefits. Foods, 11(6), 727.
Jia, X., & Lu, Y. (2022). Biosynthetic pathways of phenolic acids in plants: A review of the role of key enzymes. Molecules, 27(1), 233.
Jiang, F., & Dusting, G. J. (2003). Natural phenolic compounds as cardiovascular therapeutics: Potential role of their antiinflammatory effects. Current Vascular Pharmacology, Jun;1 (2), 135–156.
Khan, M. F., et al. (2023). Role of polyphenols in mitigating oxidative stress and formation of advanced glycation end products. Food Chemistry, 407, 135083.
Khan, M. I., & Miah, M. M. (2022). Phenolic acids: A comprehensive review of their health benefits and role in the prevention of chronic diseases. Molecules, 27(10), 3141.
Khan, M. I., & Mukhtar, H. (2021). Phenolic acids: Structure, classification, and biological properties. International Journal of Molecular Sciences, 22(12), 6610.
Khattak, K. F., et al. (2023). The role of phenolic compounds in inflammation and immune response. Journal of Medicinal Food, 26(4), 300–309.
Kumar, V., & Singh, A. (2021). Biosynthesis and functions of phenolic compounds in plants: A review. Journal of Plant Biology, 64(2), 203–215.
Lai, P. K., & Roy, J. (2020). Antimicrobial and chemopreventive properties of herbs and spices. Current Medicinal Chemistry, 17(26), 3248–3260.
Li, X., & Zhang, X. (2022). Phytochemicals in the management of inflammation: A historical perspective and recent developments. Molecules, 27(5), 1436.
Liao, H., Xu, H., & Wang, Y. (2022). Phenolic compounds: Molecular mechanisms and clinical applications in inflammation. Phytochemistry Reviews, 21(3), 611–637.
Liu, Y., Luo, J., Peng, L., Zhang, Q., Rong, X., Luo, Y., & Li, J. (2024). Flavonoids: Potential therapeutic agents for cardiovascular disease. Heliyon, Jun 6;10(12), e32563.
Ma, Q., et al. (2023). The role of NRF2 in the regulation of antioxidant genes and cellular defense mechanisms. Antioxidants, 12(2), 353.
Manach, C., & Scalbert, A. (2021). Polyphenols and health: From dietary sources to potential health benefits. Critical Reviews in Food Science and Nutrition, 61(17), 2856–2877.
Martínez, M. A., & Sánchez-Reus, I. (2021). Polyphenols in cancer therapy: Recent advances and future challenges. Nutrients, 13(12), 4214.
Medzhitov, R. (2008). Origin and physiological roles of inflammation. Nature, 454(7203), 428–435. doi:10.1038/nature07201.

Miller, A. H., & Raison, C. L. (2021). Cytokine and immune system interactions in the pathogenesis of chronic inflammatory diseases. Nature Reviews Immunology, 21(2), 69–84.

Murray, P. J., & Wynn, T. A. (2023). Protective and pathogenic functions of macrophage subsets. Nature Reviews Immunology, 23(1), 12–27. doi:10.1038/s41577-022-00681.

Pérez-Jiménez, J., & Neveu, V. (2021). The role of phenolic compounds in modulating inflammation: Mechanisms and therapeutic potentials. Frontiers in Nutrition, 8, 757562.

Pragasam, V., et al. (2023). Anti-inflammatory effects of p-coumaric acid on TNF-α expression in adjuvant-induced arthritis in rats. Journal of Ethnopharmacology, 305, 116035.

Rafique, R., & Ali, M. (2023). Exploring the anti-inflammatory potential of phenolic compounds: Mechanisms and applications. Phytochemistry Reviews, 22(1), 235–252.

Serhan, C. N., & Savill, J. (2005). Resolution of inflammation: The beginning programs the end. Nature Immunology, 6(12), 1191–1197.

Seyed-Farid, A., et al. (2023). Naringenin modulates MAPK pathways to alleviate inflammation in LPS-stimulated macrophages. Biomedicine & Pharmacotherapy, 157, 113931.

Sharma, P., et al. (2023). Tea polyphenols as potential therapeutic agents against allergic responses. Journal of Nutritional Biochemistry, 114, 109175. doi: 10.3390/nu15224823.

Shen, J., & Zhao, Y. (2021). The role of phenolic compounds in preventing oxidative stress-related diseases: A review. Frontiers in Pharmacology, 12, 691124.

Sun, W., & Shahrajabian, M. H. (2023). Therapeutic potential of phenolic compounds in medicinal plants-natural health products for human health. Molecules, Feb 15;28(4), 1845.

Wang, L., et al. (2023). *Tripterygium wilfordii* extract induces antioxidant enzyme activity via the ERK1/2-NRF2 signaling pathway. Frontiers in Pharmacology, 14, 1067503.

Wang, Y., & Zhang, S. (2022). Enzymatic pathways for phenolic acid biosynthesis in plants: Insights and applications. Plants, 11(3), 374.

Zhang, Y., et al. (2023a). Berberine enhances AMPK activation and reduces ROS production in endothelial cells. Journal of Cardiovascular Pharmacology, 81(4), 297–307.

Zhang, Y., et al. (2023b). Curcumin suppresses inflammatory responses by inhibiting the NF-κB signaling pathway: A systematic review and meta-analysis. Journal of Inflammation Research, 16, 1201–1215.

Zhang, Y., & Hamauzu, Y. (2020). Phenolic compounds in cereals, legumes, and seeds: Structural characteristics and biological activities. Foods, 9(2), 200.

Zhao, C., et al. (2023a). Curcumin modulates TLR signaling to inhibit inflammation and cancer progression. Phytotherapy Research, 37(3), 1371–1384. doi:10.1002/ptr.7626.

Zhao, J., et al. (2023b). Lutein mitigates LPS-induced inflammation via inhibition of the NF-κB signaling pathway. Journal of Nutritional Biochemistry, 116, 109085.

Zheng, W., & Wang, S. Y. (2020). Phenolic acids: A comprehensive review on their absorption, metabolism, and biological activities. Food Chemistry, 309, 125679.

Zhou, Q., et al. (2023). Resveratrol inhibits mastitis inflammation by targeting p38 and ERK signaling pathways. Molecules, 28(4), 1624.

# 6 Antimicrobial Activities of Phenolic Compounds Derived from Medicinal Plants

*Tahmeena Khan*, Kulsum Hashmi, Saman Raza, and Abdul Rahman Khan*

## 6.1 INTRODUCTION

Antibiotic resistance has emerged as a global problem. As per the data of the World Health Organization (WHO), mortalities related to multiple drug resistance will be the leading cause of death by 2025. The United Nations Organization in 2016 also declared that the antibiotic resistance is of great public concern (López Navas et al., 2019). The European Antimicrobial Resistance Surveillance Network (EARS-Net) is a system established for the surveillance of antimicrobial resistance (AMR), with the aim of monitoring and gathering data regarding AMR, as well as conducting temporal and spatial analyses. Similarly, the National Antibiotic Resistance Plan (NARP) operates in Spain (AEMPS). In 2017, the WHO released data on global priority pathogens threatening human health. The list focusses on Gram-negative bacteria that are resistant to multiple antibiotics. Alternative strategies are needed to combat resistance, such as bacteriophages and antimicrobial peptides (Davies & Bennett, 2017).

Plant extracts play a crucial role in combating pathogenic microorganisms. The medicinal properties of plants have been known for ages. Extracts of medicinal plants possess different biological activities (Mehta et al., 2001). The compounds present in the plant extract may have antimicrobial activities, inhibiting the growth of bacteria, fungi, and viruses through various mechanisms (Abate et al., 2022; Shankar et al., 2010). A few compounds also exhibit intrinsic antibacterial activity and antibiotic resistance when combined with antibiotics. Complex compounds often exhibit significant therapeutic activity and lower side effects compared to synthetic drugs (Ruddaraju et al., 2020). The efficacy of the medicinal plant extracts against microbes is also related to the synergistic effect between the active compounds of the extract (Wagner & Ulrich-Merzenich, 2009). The synergism may be affected by enhanced bioavailability, solubility, and a reduction in toxicity (Wagner & Ulrich-Merzenich, 2009). Medicinal plants that are abundant in phytochemicals have in vitro antimicrobial activities (Cowan, 1999). Among the different classes of compounds found in plants, polyphenols are important ones. They contain one or more polyphenolic groups and are classified according to their structure. Polyphenols are secondary metabolites that take part in the scavenging of hydrogen peroxide (Supritha & Radha, 2018). They are abundantly found in plants and exhibit structural diversity. Phenolic compounds possess immunomodulatory and antiinflammatory activities (Yahfoufi et al., 2018). Phenolic compounds are also known for their antiviral activity (Tirado-Kulieva et al., 2022). Carvacrol and thymol, and diterpenes like carnosol and carnosic, hydroxybenzoic acids like p-hydroxybenzoic, gallic, and ellagic, phenylpropanoic acids like p-coumaric, caffeic, and chlorogenic, phenylpropenes like eugenol, coumarins, flavanones like naringenin, eriocitrin, and flavanols like catechin and astragalin (Sun et al., 2022) possess several

* Corresponding author: tahminakhan30@yahoo.com

DOI: 10.1201/9781003527671-6

**TABLE 6.1**
**Medicinal Properties of Plant-Based Phenolic Compounds**

| Medicinal Property | Example of a Plant | Part of the Plant | References |
|---|---|---|---|
| Antioxidant activity | *Pistacia lentiscus* L. | Fruits | Tebbi et al. (2023) |
| | *Acacia nilotica* | Pods and bark | Muddathir et al. (2017) |
| | *Vaccinium stenophyllum* (blueberry) | Fruit | Bernal-Gallardo et al. (2021) |
| | *Litsea glaucescens* (Mexican bay leaf) | Leaves | César et al. (2022) |
| | *Amsonia orientalis* (European bluestar) | Leaves | Acemi et al. (2020) |
| | Cranberry | Fruit | Vattem et al. (2005) |
| | Strawberry | Fruit | Pinto et al. (2007) |
| | *Urtica dioica* (nettle) | Root, stalk, leaves | Otles and Yalcin (2012) |
| Antiinflammatory activity | *Allium scorodoprasum* L. (sand leek) | Bulb | Demir et al. (2022) |
| | *Aurea helianthus* (a Korean medicinal herb) | Leaves | Kim et al. (2017) |
| | Green tea | Leaves | Wang et al. (2021) |
| Antifungal | Walnut | Fruits | Xu et al. (2022) |
| | *Ulmus davidiana* (elm variety) | Root bark | Alishir et al. (2021) |
| | *Vitaceae* (grapevine) | Leaves and berries | Jeandet et al. (2002) |
| Antibacterial | Lavender | Leaves and flowers | Cavanagh and Wilkinson (2002) |
| | *Rhus verniciflua* (lacquer tree) | Bark, leaves, and fruits | Jang et al. (2018) |
| | Olive | Fruit | Fei et al. (2018) |
| Anticancer | *Sedum dendroideum* (a medicinal plant) | Stem and leaves | Enriquez-Ochoa et al. (2020) |
| | *Ficus palmata* (fig) | Fruit | Al-Qahtani et al. (2023) |
| Antidiabetic | *Eleusine coracana* L. (finger millet) | Grains | Shobana et al. (2009) |

beneficial properties, including antiinflammatory, antidiabetic, antiatherogenic, antihypertensive, anticancer, antibacterial, antiviral, and antifungal effects, among others (Fraga et al., 2019). Polyphenols are also used as preservatives owing to their antimicrobial and antioxidant properties (Efenberger-Szmechtyk et al., 2021; Olszewska et al., 2020). Polyphenols exhibit antibacterial activity against various bacterial strains, and fungi as well. Polyphenols have at least one aromatic ring in their structure and one or more hydroxyl groups. Oligomerization of simple phenols also results in some polyphenols, which are capable of forming glycoconjugates and derivatives such as esters and methyl esters. Flavonoids are the most active polyphenols, along with hydrolysable tannins like gallotannins and ellagitannins. The antimicrobial activity of polyphenols is typically assessed using microdilution and disk diffusion methods, with the aid of minimum inhibitory concentration (MIC), minimum fungicidal concentration (MFC), and minimum bactericidal concentration (MBC). A few of the medicinal properties of plant-based phenolic compounds are given in Table 6.1.

## 6.2 POLYPHENOLIC COMPOUNDS DERIVED FROM PLANTS

With over 8,000 known phenolic structures, ranging from simple molecules like phenolic acids to highly polymerised substances like lignins, phenolic compounds are widely distributed throughout the plant kingdom. They are the most prevalent secondary metabolites of plants. They are an essential component of the human diet since they are found in every plant organ. Plant-based foods, i.e., fruits, vegetables, cereals, and legumes, and beverages, like tea, coffee, and wine, are abundant in phenolic compounds. In addition to serving as a defence mechanism for plants against UV rays, as well as specific diseases, parasites, and predators, they also contribute to the colour, flavour, and aroma of plant parts.

Different polyphenols isolated from plant extracts have been tested against various microbes. Against *Acinetobacter baumannii*, norwogonin and theaflavin (Betts et al., 2011; Miyasaki et al., 2013) showed activity against clinical isolates. Extracts of *Polygonum cuspidatum* (Su et al., 2015) and *Amaranthus retroflexus* (Marinaş et al., 2014) also showed activity against this strain. Norwogonin was also isolated from the extracts of *Scutellaria baicalensis*, *Rosa rugosa*, and *Terminalia chebula*, showing an MIC value of 256 mg/mL against different strains of *A. baumannii*. Along with norwogonin, ellagic acid, and a mixture of chebulagic acid, chebulinic acid, corilagin, and terchebulin were also obtained. *Pseudomonas aeruginosa* is another crucial pathogen, and polyphenols derived from different extracts have exerted activity against it. Epigallocatechin gallate (ECGC) (Betts et al., 2019), *Curcuma longa* (Álvarez et al., 2016), the ethanolic extract of *P. cuspidatum* (Su et al., 2015), and the extract of *A. retroflexus* (Marinaş et al., 2014) have shown activity against *P. aeruginosa*. A combined extract of *A. retroflexus* with trimethoprim–sulphamethoxazole, piperacillin–tazobactam, ciprofloxacin, and ticarcillin–clavulanic acid has been active against *P. aeruginosa*. Against the Multi-Drug Resistent (MDR) strains of *P. aeruginosa*, a combination of ECGC and polyphenol has been active. *C. longa* L (Zingiberaceae), containing flavonoids and tannins, also showed antibacterial activity against clinical strains of *P. aeruginosa* (Álvarez et al., 2016). Crude ethanolic extracts of *P. cuspidatum* and *A. retroflexus* have also been active against *P. aeruginosa*. Pathogens belonging to the Enterobacteriaceae family have also been treated with polyphenols. Extracts of *Acacia nilotica*, *Cinnamomum zeylanicum*, and *Syzygium aromaticum* (Khan et al., 2009), tea polyphenols (catechins, ECGC, ECG (epigallocatechin), GCG (epicatechin gallate), EC (epicatechin) (Zhang et al., 2020), extracts of *C. longa* (Álvarez et al., 2016), extracts of *Prosopis laevigata* and *Opuntia ficus-indica* (Sánchez et al., 2016), and extracts of *Lawsonia inermis* (*Klebsiella* spp.) (Gull et al., 2013) containing polyphenols have been active against *Salmonella* and *Shigella*. Catechins, EGG, ECG, and GCG are some of the important polyphenols found in the extract. Different polyphenols isolated from the extracts of *Sophora secundiflora*, *Sphaeralcea ambigua*, etc. (Sánchez et al., 2016), ethanol crude extracts of *P. cuspidatum* (Su et al., 2015), *L. inermis* extracts (Gull et al., 2013), and pomegranate fruits (*Punica granatum* L.) extracts (Pagliarulo et al., 2016) have been active against *Staphylococcus aureus*. Quercetin, catechin, protocatechuic acid ethyl ester, and caffeic acid (Miklasińska-Majdanik et al., 2018) were the active polyphenolic moieties against the same bacterial strain. Antifungal activity of polyphenols has also been reported. Polyphenol-containing extracts of *A. nilotica*, *C. zeylanicum*, and *S. aromaticum* showed antifungal properties (Khan et al., 2009). Extracts of *A. retroflexus* were also active against *Candida famata*, *Candida utilis*, and *Candida albicans* (Marinaş et al., 2014). Gallic acid, ellagic acid, and corilagin were the most active polyphenols against *Candida* species and *Cryptococcus neoformans*. Phenolic compounds are extracted through maceration, digestion, extraction through Soxhlet, percolation and fermentation, supercritical fluid extraction, etc. The water content supplied to plants and the stress exposure and the factors that influence the polyphenolic content. High-performance liquid chromatography (HPLC) and colourimetric method are usually employed to quantify phenolic compounds (Albergaria et al., 2020). Phenolic compounds also have substantial antioxidant activity which is linearly correlated with the phenolic content (Chirinos et al., 2013). Tannins show the highest activity (Barros et al., 2012; Cai et al., 2006). In *Cytisus multiflorus*, chrysin was the most abundant flavone, whereas in *Sambucus nigra* and *Malva sylvestris*, quercetin-3-O-rutinoside and kaempferol-3-O-rutinoside were prominent flavanols (Barros et al., 2012). *Nepeta italica* subsp. *cadmea* and *Teucrium sandrasicum* have also been found to be rich in phenolics (Kaska et al., 2019) having therapeutic properties. Quinic, syringic, gallic, and p-coumaric acids were found in *Pituranthos tortuosus*, *Ephedra*

**TABLE 6.2**
**Biological Activity of Plant-Derived Polyphenols**

| S. No. | Plant | Activity | References |
|---|---|---|---|
| 1 | *Hypericum empetrifolium* | Antifungal | Boga et al. (2021) |
| 2 | *Allium sativum* | Antifungal | Abirami et al. (2021) |
| 3 | *Rosa micrantha* | Antifungal | Barros et al. (2013) |
| 4 | *Ulmus davidiana* var. *japonica* | Antifungal | Alishir et al. (2021) |
| 5 | *Scirpus holoschoenus* | Antibacterial | Oussaid et al. (2017) |
| 6 | *Rhanterium adpressum* | Antibacterial | Boussoussa et al. (2016) |
| 7 | *Rhus verniciflua* | Antibacterial | Jang et al. (2018) |
| 8 | *Scrophularia ningpoensis* Hemsl. | Antibacterial | Lin et al. (2022) |
| 9 | *Isatis indigotica* | Antiviral | Lin et al. (2005) |
| 10 | *Sambucus FormosanaNakai* | Antiviral | Weng et al. (2019) |
| 11 | *Ribes nigrum* | Antiviral | Granato et al. (2022) |

*alata*, and *Retama raetam* (Benabderrahim et al., 2019). Some activities of plant-derived polyphenolic compounds are given in Table 6.2 (Manso et al., 2021).

## 6.3 CLASSES OF POLYPHENOLIC COMPOUNDS

Phenolic compounds are produced as secondary metabolites in plants. They were classified according to their basic structure, having them classified in more than ten categories. The three major classes of phenolics contain (a) a single benzene ring (C6 class), (b) one C6 ring with an attached chain having one to seven carbon atoms (C6-Cn class), and (c) complex skeleton, where carbon chain is also attached to a second benzene ring ($C_6$–Cn–$C_6$ class) (Tsimogiannis & Oreopoulou, 2019). The classification of polyphenols is depicted in Table 6.3.

**TABLE 6.3**
**Broad Classification of Polyphenols**

| S. No. | Number and Layout of Carbon Atoms | Polyphenol Class or Family (Polyphenol Subclass) |
|---|---|---|
| 1 | $C_6$ | Simple phenolics |
| 2 | $C_6$–$C_1$ | Hydroxybenzoic acids |
| 3 | $C_6$–$C_2$ | Acetophenones, phenylacetic acids |
| 4 | $C_6$–$C_3$ | Cinnamic acids, coumarins, isocoumarins |
| 5 | $C_6$–$C_4$ | Naphthoquinones |
| 6 | $C_6$–$C_1$–$C_6$ | Xanthones |
| 7 | $C_6$–$C_2$–$C_6$ | Stilbenes, anthraquinones |
| 8 | $C_6$–$C_3$–$C_6$ | Flavan-3-ols |
| | | Flavonols |
| | | Flavones |
| | | Flavanone |
| 9 | $(C_6–C_1)_n$ | Hydrolysable tannins |
| 10 | $(C_6–C_3)_2$ | Lignans, neolignans |
| 11 | $(C_6–C_3)_n$ | Lignins |
| 12 | $(C_6–C_3–C_6)_n$ | Condensed tannins |

## 6.4 SPECIFIC POLYPHENOLIC COMPOUNDS WITH ANTIMICROBIAL ACTIVITIES

### 6.4.1 Hydroxybenzoic Acids (Gallic Acid and Protocatechuic Acid)

Hydroxybenzoic acid (HBAs) are phytochemicals found in many medicinal plants and used for the treatment of a number of human diseases (Ossipov et al., 2022). 4-HBAs (4HBA) is associated with antifungal properties (Joshi et al., 2021).

#### 6.4.1.1 Gallic Acid

It is one of the most abundant polyphenols found in nature (Asdaq et al., 2021; Osamudiamen et al., 2020). Its antibacterial and antifungal properties are well known (Gong et al., 2023; Tan et al., 2022). It exhibits a significant affinity for binding and disrupting proteases, demonstrating a promising phytotherapeutic effect against fungal proteases, a virulence factor (Hasan et al., 2023). Transmission of resistance between and within individuals leads to resistance towards pathogenic microorganisms. Secondary metabolites synthesized by plants are involved in the defence mechanism. Plant-derived gallic acid has been assessed on *Escherichia coli*, *P. aeruginosa*, *S. aureus*, and *Listeria monocytogenes* in a study. Appreciable MIC and MBC were obtained. Irreversible alterations in membrane properties, causing changes in hydrophobicity and a decrease in negative surface charge, were observed, indicating that gallic acid is a sustainable source of broad-spectrum antimicrobial activities (Borges et al., 2013). Similarly, different extracts of *Caesalpinia mimosoides* Lamk were tested against pathogenic bacterial and fungal strains. Aqueous and ethanolic extracts exhibited potent activity against some of the tested microorganisms (Chanwitheesuk et al., 2007).

#### 6.4.1.2 Protocatechuic Acid

Protocatechuic acid (3,4-dihydroxybenzoic acid) is a natural phenolic compound and a chief metabolite of complex polyphenols (Erukainure et al., 2017). It is found in bran and brown rice, the scales of onion, and grapes (Dare et al., 2020; Krishna & Muraleedharan, 2023). *Veronica montana* also contains protocatechuic acid, which exhibits antibacterial activity by affecting its cytoplasmic membrane (Stojković et al., 2013). Lipid oxidation in beef tissues was found to be decreased by protocatechuic acid, which is derived from roselle calyx (*Hibiscus sabdariffa* L.) (Chao & Yin, 2009), suggesting that it may prevent contamination from *Campylobacter* and aerobes. The effect of protocatechuic acid on murine cytochrome P450 and phase II enzymes was investigated, and when administered to rodents, the mutagenicity of amine derivatives was decreased (Kuźniak et al., 2005). In another study, protocatechuic acid was tested against several Gram-positive and Gram-negative bacteria through a microdilution assay. Protocatechuic acid inhibited the bacterial growth, causing membrane lysis (Stojković et al., 2013).

### 6.4.2 Hydroxycinnamic Acids (p-Coumaric Acid, Caffeic Acid, Ferulic Acid, Sinapic Acid)

Derivatives of hydroxycinnamic acid are important polyphenols found in vegetables and fruits and are used in the human diet (Yuan et al., 2022). They are also found in plants, cereals, fruits, and coffee (Ribas-Agustí et al., 2019).

#### 6.4.2.1 p-Coumaric Acid

p-Coumaric acid (4-hydroxycinnamic acid) is known as a phytochemical as well as a nutraceutical, which is found in carrots, tomatoes, and cereals. It occurs in the cell walls of graminaceous plants. It reduces low-density lipoprotein (LDL) peroxidation and exerts antioxidant and antimicrobial activities. It is found in the peripheral tissues of kernels. It is also found in good quantities in the pericarp fractions in barley, wheat, oat, and corn (Fărcaş et al., 2013). Its antimicrobial activity is more prominent against Gram-negative bacteria than Gram-positive strains (Cetin-Karaca, 2011). p-Coumaric acid extracted from *Crocus baytopiorum* showed considerable activity against *Bacillus cereus* and *Bacillus subtilis* (Acar et al., 2010). Lou et al. (2012) also showed that p-coumaric acid exerted activity against *Salmonella typhimurium* and *Shigella dysenteriae* as well as *E. coli*. The effect was found to increase with a reduction in pH. It changes the permeability of the cell membrane and disrupts cell functions (Lou et al., 2012). Hydroxycinnamic acids have shown higher antimicrobial activity as compared to hydroxybenzoic acids having the same hydroxyl groups, and the antimicrobial activity decreases when hydroxyl groups increase (Sánchez-Maldonado et al., 2011).

#### 6.4.2.2 p-Caffeic Acid Caffeic Acid (3,4-Dihydroxycinnamic Acid)

It is a phenolic compound found in apple, pear, and coffee (Silva et al., 2020). It is biosynthesized through L-tyrosine, or through deamination and hydroxylation of L-tyrosine (Wang et al., 2023). Caffeic acid shows antimicrobial activity in combination with other drugs (Kępa et al., 2018; Palaniappan & Holley, 2010). It shows activity against harmful intestinal bacteria (Lima et al., 2016). Caffeic acid has also been more effective against *Staphylococcus epidermidis* than *Klebsiella pneumoniae* or *S. aureus* (Koike et al., 1979). β-Cyclodextrin (βCD)-coated caffeic acid was also effective against skin infections. In addition to antibacterial activity, esters of caffeic acid have also demonstrated antifungal activity (Pinho et al., 2015). Methyl, ethyl, propyl, isopropyl, and butyl caffeate derivatives of caffeic acid were effective against several *Candida* sp. The MIC was obtained through the microdilution method. Moderate antifungal activity was observed, with the best activity exhibited by methyl caffeate ester (Pukkila-Worley et al., 2009). Coffee pulp, a byproduct of coffee processing, was used to quantify polyphenols through HPLC, followed by mass spectrometry modes of the liquid chromatography–quadrupole time-of-flight (QTOF) technique. The fraction was tested for antifungal activity against *Alternaria brassicicola*, *Pestalotiopsis* sp., and *Paramyrothecium breviseta*. The results showed that caffeic acid, present in the polyphenol fraction, inhibited fungi with IC50 values of 0.09, 0.31, and 0.14, respectively (Sangta et al., 2021).

#### 6.4.2.3 Ferulic Acid (4-Hydroxy-3-Methoxycinnamic Acid)

It is a polyphenol having therapeutic utility (Bao et al., 2019). Ferulic acid is useful for the synthesis of coniferyl alcohol, diferulic acid, vanillin, curcumin, etc. (Raj & Singh, 2022). It is found in bananas, eggplant, citrus fruits, and cabbage (Ali et al., 2021). It is associated with polysaccharides in cell walls, such as those found in cereal bran, and is a primary compound of *Angelica sinensis*, *Chuanxiong rhizoma*, and *Ferula* (Rosa et al., 2018). *A. sinensis*, a perennial herb, is a source of bioactive ferulic acid (Singh et al., 2022). Ferulic acid is known for its potential therapeutic properties, exerting pharmacological effects such as antioxidant and antiinflammatory properties. Ferulic acid was isolated from the methanolic extract of *Ligusticum mutellina*, as analysed by HPLC. The extract showed moderate antifungal activity with an MIC of 1.25–2.5 mg/mL against *Micrococcus luteus*, *P. aeruginosa*, and *Candida* spp., which were the most sensitive to the extracts (Sieniawska et al., 2013). In another study, plant-derived ferulic acid was tested for its antibiofilm activity against *P. aeruginosa*, *E. coli*, *L. monocytogenes*, and *S. aureus*. Ferulic acid inhibited the swarming of *L. monocytogenes* and *E. coli* (Borges et al., 2012).

#### 6.4.2.4 Sinapic Acid

Sinapic acid is a prevalent hydroxycinnamic acid (von Danwitz & Schulz, 2020), known to protect lysosomes by preventing lysosomal dysfunction (Roy & Prince, 2012). It was obtained from *Brassica juncea* L. seed. The phenolic fraction showed antibacterial activity against *S. aureus* and *L. monocytogenes* (Engels et al., 2012). Black cumin (*Nigella sativa* L.) seed extracts were tested for the presence of phenolic compounds. The inhibitory action was tested against butyrylcholinesterase (BChE) and catalase (CAT), and their antibacterial activity was tested against several bacterial and yeast strains. The extracts were tested using the agar diffusion method. Sinapic acid was identified as the major component (Topcagic et al., 2017). A study

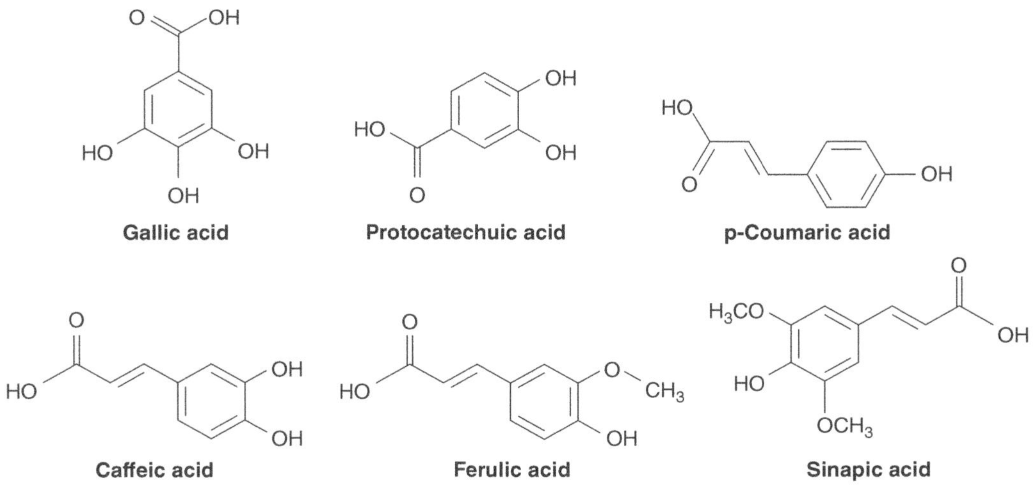


**FIGURE 6.1** Structure of some polyphenolic compounds.

was conducted using hydroxybenzoic and hydroxycinnamic acids on various bacteria and a yeast strain (*C. albicans*). The phytochemicals were ranked based on the relative inhibitory capacity index (RICI), which showed that sinapic acid and 4-hydroxybenzoic acid were the most effective. Sinapic acid exhibited a zone of inhibition ranging from 9.00 to 27.00 mm and an MIC of 18.00 to 72.00 mg/mL against microorganisms, indicating that it can act as an alternative antimicrobial agent (Gurbuzer, 2021). The general structures of some polyphenolic compounds are represented in Figure 6.1.

### 6.4.3 Coumarins (Umbelliferone, Esculetin, Scopoletin)

Coumarins (2H-chromen-2-one ring) have a molecular formula of $C_9H_6O_2$, and it is one of the important compounds used in cosmetics and food. They belong to the class of heterocyclic compounds having a benzopyrone moiety in their structure (Borah et al., 2022). Coumarin has antibacterial, anticancer, antioxidant, and antiinflammatory properties (Çakmakçı et al., 2022). Their derivatives are found in over 30 plant families such as Clusiaceae, Umbelliferae, Guttiferae, Rutaceae, Oleaceae, and Fabaceae (Dandriyal et al., 2016). Different coumarins were found in the roots of *Angelica dahurica*.

#### 6.4.3.1 Umbelliferone

It is a 7-hydroxycoumarin and an isomer of caffeic acid (Salau et al., 2020) and is associated with different pharmacological properties (Seong et al., 2019). *A. nilotica*, *Angelica decursiva*, and *Aegle marmelos* are rich sources of umbelliferone (Mazimba, 2017). It is associated with antifungal (Pan et al., 2017), antihyperglycemic (Kundu et al., 2020), and radical scavenging properties. Umbelliferone also exhibits antibacterial and antibiofilm properties and is commonly found in plants of the Rutaceae and Apiaceae families. The antibacterial and antibiofilm potential of the compound was tested against *B. cereus* and *B. subtilis*, and its mechanism was elucidated. An agar well assay revealed its efficacy against *K. pneumoniae*. The antibiofilm activity was dose-dependent. Deformation of the bacterial cell membrane and outflow of intracellular contents were observed (Mandal et al., 2025). The antimicrobial activity of umbelliferone, derived from *Rhododendron lepidotum*, was tested against various bacteria. The obtained MC was 500 µg/mL against *S. aureus* and *P. aeruginosa* (Khan et al., 2010). It also inhibited the growth of *Fusarium culmorum* (Ojala et al., 2000). Antibacterial activity was tested against *E. coli*, *Enterobacter aerogenes*, and

*K. pneumoniae* exerted by umbelliferone obtained from the stem bark of *Balsamocitrus camerunensis* (Fouotsa et al., 2013). Biofilm formation related to *E. coli* is difficult to treat as it is tolerant to physical and chemical antimicrobial treatment (Patel et al., 2011). Umbelliferone inhibited biofilm formation by *E. coli* O157:H7 at 50 µg/mL (Lee et al., 2014).

#### 6.4.3.2 Esculetin

Esculetin (6,7-dihydroxycoumarin), a natural coumarin found in herbs, exhibits various pharmacological activities (Park et al., 2008). Aesculin has been associated with the treatment of obesity, diabetes, cardiovascular diseases, cancer, and neurological disorders (Kadakol et al., 2016). It is found in *Artemisia capillaris*, *Matricaria chamomilla* L., *Artemisia scoparia*, etc. (Yang et al., 2017). Esculetin is a key component of Cortex Fraxini, a traditional herbal medicine. Esculetin exhibits antibacterial activity. It demonstrates activity through free radical scavenging or by modulating the nuclear factor erythroid 2-related factor 2 pathway, among other mechanisms (Cai & Cai, 2023). *E. coli* O157:H7 is a human pathogen that spreads by direct or indirect contact. It causes haemorrhagic colitis and does not have an effective therapy for cure. Upon treatment with aglycone esculetin, the pathogen decreased significantly in faecal samples (Duncan et al., 2004). Another study demonstrated that esculetin inhibited the Shiga-like toxin gene stx2 in *E. coli* O157:H7 and suppressed the virulent factor in *Caenorhabditis elegans* (Lee et al., 2014).

#### 6.4.3.3 Scopoletin

Scopoletin (6-methoxy-7-hydroxycoumarin) belongs to the coumarin family (Dou et al., 2013). It is known for its medicinal properties (Connell et al., 2017). It is found in *Convolvulus prostratus* Forssk. Moreover, it is known for its antimicrobial and antioxidative properties (Sushil et al., 2021). It is also found in *Morinda citrifolia* L. (Tasfiyati et al., 2022). The derivative enhances immunity by boosting the phagocytic potential of macrophages (Alkorashy et al., 2020). Scopoletin has shown good antibacterial and antifungal activities. The antimicrobial activity of scopoletin, isolated from cassava, was tested for its antimicrobial properties (Rodriguez et al., 2000). The fungicidal effect of scopoletin was also tested against *Microcyclus ulei* (Garcia et al., 1995). Scopoletin, isolated from *Mitracarpus frigidus*, a member of the family Rubiaceae, was tested against *Candida tropicalis* for its ability to inhibit biofilm formation. Fungal growth inhibition was evaluated through MIC, and the ability of scopoletin to affect *C. tropicalis* was tested through spectrophotometric and slide imaging. Scopoletin had an MIC value of 50 µg/L against *C. tropicalis* and resulted in a 91.7% reduction in cell density. The mechanism was linked to the fungal cell wall, which affected plasma membrane sterols. It also suppressed the growth of biofilm on the surface of coverslips (Lemos et al., 2020). Scopoletin (1.4 µmol/g by dry weight), isolated from the *Lasianthus lucidus* Blume species, was tested against *P. aeruginosa* and *P. aeruginosa* DMSC 37166. The highest antibacterial activity was observed at a concentration of 128 µg/mL, and morphological changes in the bacterial cells were evident. Scopoletin was found to cause cell lysis, cell wall swelling, and its sinking into the cell (Napiroon et al., 2018).

### 6.4.4 Stilbenes (Resveratrol, Piceatannol, Pterostilbene)

Stilbenes are phytophenols having good bioactivity (Oyama et al., 2023). They are commonly found in the plants of Vitaceae, Gnetaceae, Leguminosae (Leguminaceae), and Dipterocarpaceae, existing in two isomeric forms and having a $C_6$–$C_2$–$C_6$ skeleton (Avendaño-Godoy et al., 2022). They have excellent antiinflammatory, antiviral, and other properties and also find applications as colouring agents and dietary supplements (Choi et al., 2021). Stilbenes are abundantly found in grapes and wine (Fernández-Marín et al., 2012). They are synthesized in response to abiotic and biotic stress in plants. They act as phytoalexins towards pathogens or herbivore attack in their mechanism as antimicrobials (Guerrero et al., 2020).

#### 6.4.4.1 Resveratrol

Resveratrol (3,5,4-trihydroxy-trans-stilbene) is a plant polyphenol. It is a non-flavonoid polyphenol, consisting of phenolic rings linked by a double bond, giving two isomeric forms (Pinilla-Peñalver et al., 2022). The trans form, being more stable, converts into the cis form under pressure, high pH, or UV light. It is particularly found in red wine, cranberry, etc. (Shen & Ji, 2018) and also obtained from the roots of *P. cuspidatum* and *Rheum rhaponticum* (Nicolau et al., 2022). Antimicrobial activity of resveratrol against *B. cereus*, *Bacillus megaterium*, *E. coli*, *K. pneumoniae*, and fungi such as *Trichophyton mentagrophytes*, *T. tonsurans*, and *Trichophyton rubrum* (Vestergaard & Ingmer, 2019). Resveratrol has inhibited the growth of *B. cereus*, *Mycobacterium smegmatis*, *Helicobacter pylori*, *Vibrio cholerae*, and *Neisseria gonorrhoeae* at a concentration below 100 µg/mL. However, for many bacterial strains like *P. aeruginosa*, *Streptococcus pyogenes*, and *K. pneumoniae*, the activity was observed above 100 µg/mL. This may be due to its poor penetration across the outer membrane of Gram-negative bacteria (Vestergaard & Ingmer, 2019). Resveratrol extracted from *P. cuspidatum* was tested against Avian Pathogenic *E. coli* (APEC) and to determine the antibiofilm effect. The swimming and swarming effect of resveratrol was tested using a semisolid medium culture. The MIC of resveratrol was 128 µg/mL, and the swimming and swarming of APEC were reduced in a dose-dependent manner. Resveratrol also exhibited activity against APEC biofilm formation above 1 µg/mL ($p < 0.01$) (Ruan et al., 2021).

#### 6.4.4.2 Piceatannol

Piceatannol (3,4,3,5-tetrahydroxy-trans-stilbene) is a polyphenolic stilbene with therapeutic potential (Algandaby & Al-Sawahli, 2021). It is found in blueberries, grapes, and passion fruit (Takei et al., 2005). Piceatannol, obtained from macaúba epicarp, was found to increase the tolerance of *Saccharomyces cerevisiae* towards oxidative stress. The $Me_2CO:H_2O$ extract demonstrated antibacterial and antifungal activity, and piceatannol enhanced the survival rate of *Galleria mellonella* after fungal infection. The absorption, distribution, metabolism, excretion, and toxicity (ADMET) study showed low toxicity (Santos et al., 2025). Larch (*Larix decidua*) bark is used as an antiseptic and to treat dermatological conditions like eczema and psoriasis. A study reported the antibacterial activity evaluation of larch bark extract against *S. aureus*, *S. pyogenes*, *Streptococcus pneumoniae*, *K. pneumoniae*, *P. aeruginosa*, and *Haemophilus influenza*. The activity was tested using the microplate dilution method between 2 and 200 µg/mL. The extract rich in piceatannol had a growth inhibition between 10 and 40% towards majority of the strains (Faggian et al., 2021). Piceatannol was also obtained from wild grapevine (*Vitis vinifera* subsp. *sylvestris*), and the ethanolic extract was tested against *S. aureus*, methicillin-resistant *S. aureus*, and *Enterococcus faecalis*. The MIC and MBC values ranged from 125 to 250 µg/mL. The MBC/MIC value was ≤ 2, showing the bactericidal activity (Daldoul et al., 2024).

#### 6.4.4.3 Pterostilbene

It is a dimethyl ester derivative of resveratrol (Figure 6.2) and found in blueberries, grapevine, and heartwood (Azzolini et al., 2017). It exhibits antifungal activities against *Leptosphaeria maculans* and *Peronophythora litchii*, among others (Xu et al., 2022). Lignans and phenylpropanoids are involved in biotic and abiotic stresses (Samari et al., 2022). Lignans, important components with antibacterial and antiviral activities, are found in the xylem of plants (Torres-Moreno et al., 2022). It is found in *Artemisia absinthium*, *Artemisia gorgonum*, *Chrysanthemum cinerariaefolium*, *Capsicum frutescens*, *Chrysanthemum indicum*, *Diotis maritima*, *Eupatorium ageratina*, *E. fleischmannia*, *Otanthus maritimus*, *Aptosimum spinescens*, *Gmelina arborea* Roxb., *Acanthopanax senticosus*, *Eleutherococcus divaricatus*, *Asarum sieboldii*, *Aristolochia cymbifera*, *Alnus glutinosa*, and *Salicomia europaea* (Dossou et al., 2023). Sesamin could boost intestinal probiotics, modulating gut microbiota for the improvement of intestinal health (Wang et al., 2021).

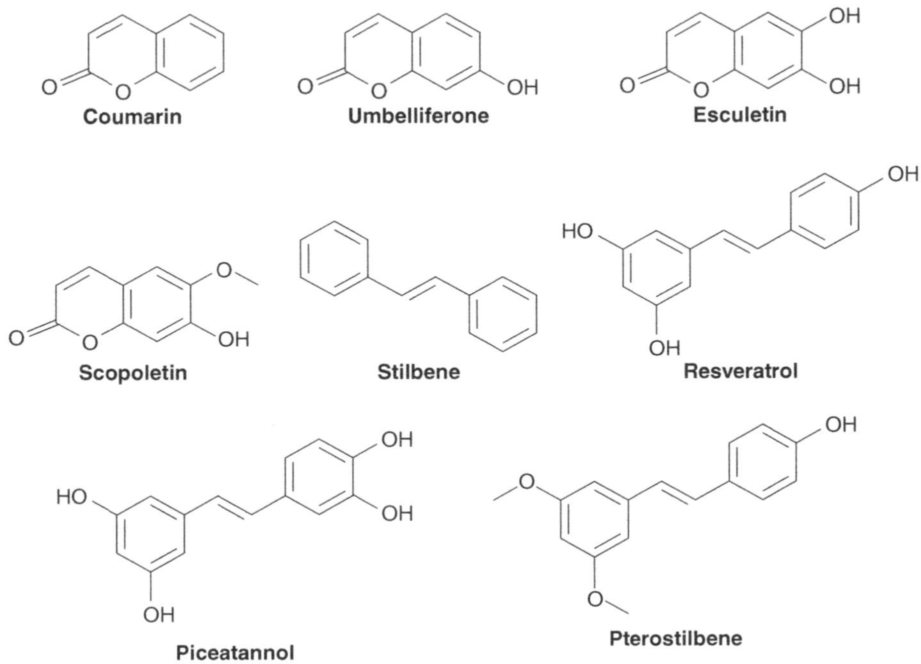


**FIGURE 6.2** Structure of polyphenolic compounds.

## 6.5 CONDENSED TANNINS OR PROANTHOCYANIDINS (PROCYANIDIN B1)

Proanthocyanidins (Figure 6.3) are the condensed tannins and are secondary metabolites and found in fruits, seeds, nuts, and bark (Abhijit et al., 2020). They were prevalent in lotus seed coats. They are of three types: procyanidins, prodelphinidins, and propelargonidins (Tao et al., 2022). Their precursors are prepared by the phenylpropanoid pathway in the cytosol and converted to the vacuole, and upon polymerization they create proanthocyanidins (Sharma & Beniwal, 2022). They are associated with various bioactivities, including antibacterial, anti-cancer, and antioxidant (Morimoto et al., 2022). Ethanolic extracts of the bark of *Dalbergia monetaria*, an Amazonian plant, were prepared. The in vitro antibacterial activity was tested against ten bacterial strains, including *S. aureus* and methicillin-resistant *P. aeruginosa*. UHPLC–HRMS/MS characterized the extrat, which was enriched with proanthocyanidins, and its subfractions had (epi)-cassiaflavan and (epi)-catechin units (Brito de Moura et al., 2020). The antibacterial, antiinflammatory, and cytoprotective activity of proanthocyanidin fraction (PACN), obtained from *Pelargonium sidoides* DC root extract, exerted a strong effect against

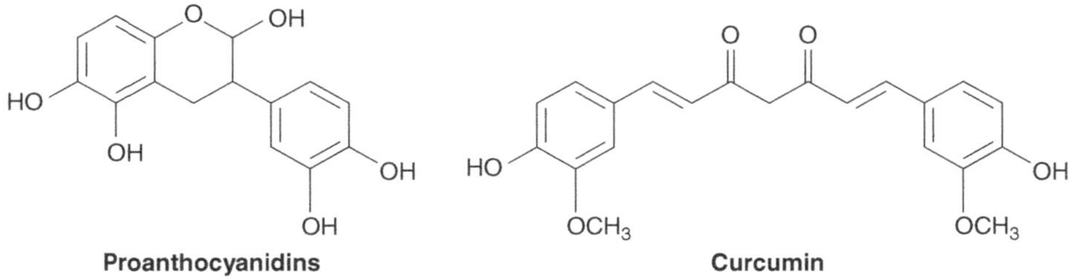


**FIGURE 6.3** Structure of Proanthocyanidins and Curcumin.

*Porphyromonas gingivalis*. The work also reported activity against *S. aureus*, *S. epidermidis*, *Aggregatibacter actinomycetemcomitans*, and *E. coli* (Jekabsone et al., 2019).

## 6.6 CURCUMINOIDS

Curcuminoids are polyphenols found in plant species of *Curcuma* like *C. longa*, *Curcuma wenyujin*, and *Curcuma zedoaria* (Wu et al., 2022). In turmeric they are obtained through the phenylpropanoid biosynthetic pathway (El-Hawaz et al., 2016). Curcumin and demethoxycurcumin have phenylmethoxy group (Wang et al., 2020).

## 6.7 CURCUMIN

A bis-unsaturated-diketone or a hydrophobic polyphenol (Figure 6.3) is found in the rhizome of the perennial herb genus *Curcuma* and species like *Curcuma amada* and *Curcuma aromatica* (Feng et al., 2022). The active functional groups of curcumin are oxidized by electron transfer and hydrogen abstraction (Priyadarsini, 2014). Curcuminoids, the main constituents of curcumin, show broad-spectrum antibacterial properties. Curcumin has demonstrated activity against microbes associated with surgical infections, primarily *S. aureus* and *E. coli*. It has pronounced activity against *H. pylori* and *Mycobacterium tuberculosis*. Curcumin also exhibits antifungal activity against various fungal strains, which cause a range of infections. *Candida* sp., which is responsible for candidemia and candidiasis, has been reported to be treated with curcumin. Life-threatening viral diseases have also been counteracted by curcumin. Combination therapies involving curcumin have been employed to combat antibiotic-resistant microbes (Hussain et al., 2022). A study reported the antimicrobial activity of curcumin against 100 pathogens. Curcumin showed enhanced sensitivity against Gram-positive bacteria compared to Gram-negative bacteria. It was effective against *S. pyogenes* (median MIC = 31.25 µg/mL), methicillin-sensitive *S. aureus* (250 µg/mL), *Acinetobacter lwoffii* (250 µg/mL), and individual strains of *E. faecalis* and *P. aeruginosa* (62.5 µg/mL) (Adamczak et al., 2020).

## 6.8 MECHANISTIC INSIGHT

The antimicrobial activity may arise due to the interference of the phytochemical with the functioning or synthesis of vital microbial components or overcoming AMR. However, bacteria develop resistance towards multiple antimicrobials or acquire it from neighbouring microbes. The mechanism of AMR to secondary metabolites is depicted in Figure 6.4.

## 6.9 MECHANISMS OF ACTION

Bacterial ribosomal subunits are effective targets to inhibit protein synthesis. Antimicrobials have worked by altering the shape of ribosomes to prevent the deciphering of mRNA codons or by interfering with the wobbly bases of tRNA anticodons (Walsh, 2003). Inhibition of DNA gyrase is another important target for antibacterial agents. DNA gyrase is known to assist in synthesis, replication, repair, and transcription of bacterial DNA (Maxwell, 1997). Similarly, bifunctional enzymes, such as transglycosylases and transpeptidases, also play important roles in the formation of the bacterial cell wall. Antimicrobials may bind to the peptidoglycan layer, preventing the enzyme's action (Schneider & Sahl, 2010). Antibiotics can also bind to the lipid component of lipopolysaccharides and alter their structure by facilitating phospholipid exchange, which leads to osmotic imbalance and ultimately bacterial death (Poirel et al., 2017). Phytochemicals derived from plants are mostly secondary metabolites and used for medicinal purposes. Secondary metabolites are produced through secondary plant metabolism and also

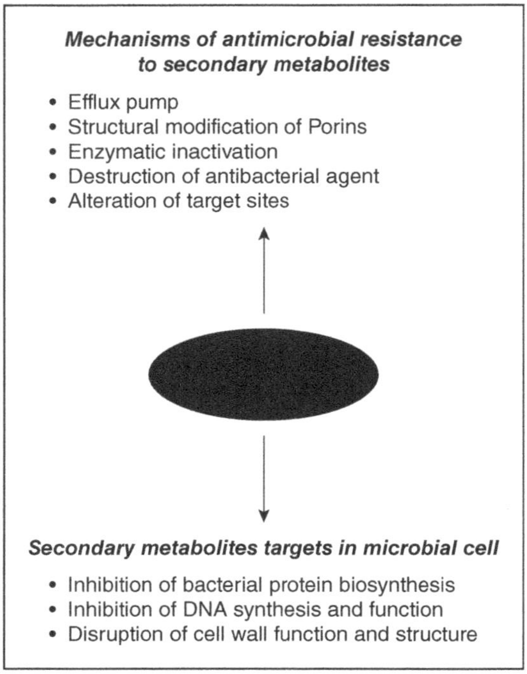


**FIGURE 6.4** Mechanism of action of antimicrobial agents (Vaou et al., 2021).

occur as intermediate or final products (Stefanovic & Comic, 2012). Based on their structure, substituents in their structure, position of the substituents, glycosidic linkages, alkylation of OH groups, and other features like the conditions and climate in which the plant is cultivated, their antimicrobial activity varies (Arima et al., 2002; Assob et al., 2011; Merkl et al., 2010). Bioactive plant extracts also contain complex mixtures of compounds, resulting in enhanced effects. The primary target of the microbes is the cytoplasmic membrane, which affects its permeability and functionality (Radulovic et al., 2013; Saleem et al., 2010). Plant extracts may also contain EP inhibitors (Savoia, 2012) and inhibit normal cell communication, especially against MDR pathogens. QS inhibitors decrease the expression of QS-controlled genes and resist metabolic and disposal processes of the host (Radulovic et al., 2013). Some phytochemicals also modify protein–protein interactions and modulate immune response (Vadhana et al., 2015). They also interfere with intermediary metabolism (Anandhi et al., 2014) and coagulate cytoplasmic constituents (Mogosanu et al., 2015). Biofilm inhibition is another mode through which antimicrobials act (Quave et al., 2012). The antiviral components present in the plant extract interfere with viral proteins responsible for their replication (Rajasekaran et al., 2014). Phenolic compounds show diverse mechanisms of action and are extensively used against pathogenic bacteria (Urs & Dunleavy, 1975). The inhibition of EP and the potential to modify cell membrane permeability are some common mechanisms, along with a shift in several intracellular functions, enzyme binding, or loss of cell wall integrity (Farhadi et al., 2019; Górniak et al., 2019). Flavones are known to disrupt the envelope of microbes (Cazarolli et al., 2008). In contrast, flavanols interact with the microbial cell wall and deactivate specific enzymes by binding to sulfhydryl groups or exhibiting non-specific interactions with proteins (Betts et al., 2011). Flavonoids form complexes with extracellular proteins and bacterial membranes. The antimicrobial property is primarily exerted by inhibiting bacterial virulence factors, disrupting the cytoplasmic membrane, deactivating microbial enzymes, and causing a deficiency of substrates such as zinc and iron, which are essential for microbial growth (Cushnie & Lamb,

2005). Quinones form complexes with amino acids present in microbial proteins, causing them to be denatured (García et al., 2012). Tannins work by disrupting the cell wall and membrane, causing inhibition of oxidative phosphorylation and disrupting microbial metabolism. They also interfere with the intercalation of DNA base pairs and hamper the action of enzymes, ultimately leading to cell death (Engels et al., 2011).

## 6.10 FUTURE PROSPECTS AND CONCLUSION

Medicinal plants are a treasure trove of beneficial phytoconstituents, of which polyphenols are an important class of compounds. Many medicinal plants remain to be explored in order to discover novel compounds as therapeutic agents (Quave, 2016). Chemical investigation of complex mixtures of natural compounds is needed. In vivo testing in animal models is also needed to correlate with the in vitro findings (Huie, 2002). Furthermore, structural alterations may also be explored to enhance both pharmacokinetics and pharmacodynamics, as well as to elucidate structure–activity relationships (Ahmad et al., 2024). The synergistic effect also needs thorough elaboration, particularly between the compounds obtained from plants with antibiotics. Target identification and mechanistic insights also need deeper understanding (Ahmad et al., 2020; Azad et al., 2023). Toxicity is another important aspect to be focussed upon to explore the role of phytoconstituents as biomedical agents. Improved methodologies must be developed, such as for MIC determination, to obtain a more accurate picture and comparison between the activity caused by a particular phytoconstituent in the plant extract and a standard antibiotic (Huie, 2002). Techniques like genomics and proteomics, among others, are also being developed, which hold promise for contributing to research related to antimicrobial agents.

## COMPETING INTERESTS

The authors declare that they have no competing interest.

## DECLARATION OF GENERATIVE AI AND AI-ASSISTED TECHNOLOGIES IN THE WRITING PROCESS

During the preparation of this work the book chapter author(s) have not taken any assistance from AI, and the book chapter author(s) reviewed and edited the content as needed and take(s) full responsibility for the content of the publication.

## REFERENCES

Abate, L., Tadesse, M. G., Bachheti, A., & Bachheti, R. K. (2022). Traditional and phytochemical bases of herbs, shrubs, climbers, and trees from Ethiopia for their anticancer response. *BioMed Research International*, *2022*. https://doi.org/10.1155/2022/1589877

Abhijit, S., Tripathi, S. J., Rao, B. S., & Devi, S. A. (2020). Grape seed proanthocyanidin extract and swimming training enhances neuronal number in dorso-medial prefrontal cortex in middle-aged male rats by alleviating oxidative stress. *Journal of Functional Foods*, *64*, 103693.

Abirami, S., Raj, B. E., Soundarya, T., Kannan, M., Sugapriya, D., Al-Dayan, N., & Mohammed, A. A. (2021). Exploring antifungal activities of acetone extract of selected Indian medicinal plants against human dermal fungal pathogens. *Saudi Journal of Biological Sciences*, *28*(4), 2180–2187.

Acar, G., Mercan Doğan, N., Duru, M. E., & Kıvrak, I. (2010). Phenolic profiles, antimicrobial and antioxidant activity of the various extracts of Crocus species in Anatolia. *African Journal of Microbiology Research*, *4*, 1154–1161.

Acemi, R. K., Acemi, A., Çakır, M., Polat, E. G., & Özen, F. (2020). Preliminary screening the antioxidant potential of in vitro-propagated *Amsonia orientalis*: An example to sustainable use of rare medicinal plants in pharmaceutical studies. *Sustainable Chemistry and Pharmacy*, *17*, 100302.

Adamczak, A., Ożarowski, M., & Karpiński, T. M. (2020). Curcumin, a natural antimicrobial agent with strain-specific activity. *Pharmaceuticals*, *13*(7), 153.

Ahmad, M. I., Veg, E., Joshi, S., Khan, A. R., & Khan, T. (2024). Thiosemicarbazone derivatives in search of potent medicinal agents: QSAR approach (a review). *Russian Journal of General Chemistry*, *94*(9), 2470–2495.

Ahmad, R., Khan, M. A., Srivastava, A. N., Gupta, A., Srivastava, A., Jafri, T. R., Siddiqui, Z., Chaubey, S., Khan, T., & Srivastava, A. K. (2020). Anticancer potential of dietary natural products: A comprehensive review. *Anti-Cancer Agents in Medicinal Chemistry (Formerly Current Medicinal Chemistry-Anti-Cancer Agents)*, *20*(2), 122–236.

Albergaria, E. T., Oliveira, A. F. M., & Albuquerque, U. P. (2020). The effect of water deficit stress on the composition of phenolic compounds in medicinal plants. *South African Journal of Botany*, *131*, 12–17.

Algandaby, M. M., & Al-Sawahli, M. M. (2021). Augmentation of anti-proliferative, pro-apoptotic and oxidant profiles induced by piceatannol in human breast carcinoma MCF-7 cells using zein nanostructures. *Biomedicine & Pharmacotherapy*, *138*, 111409.

Ali, S. A., Saifi, M. A., Pulivendala, G., Godugu, C., & Talla, V. (2021). Ferulic acid ameliorates the progression of pulmonary fibrosis via inhibition of TGF-β/smad signalling. *Food and Chemical Toxicology*, *149*, 111980.

Alishir, A., Yu, J. S., Park, M., Kim, J. C., Pang, C., Kim, J. K., Jang, T. S., Jung, W. H., & Kim, K. H. (2021). Ulmusakidian, a new coumarin glycoside and antifungal phenolic compounds from the root bark of *Ulmus davidiana* var. *japonica*. *Bioorganic & Medicinal Chemistry Letters*, *36*, 127828.

Alkorashy, A. I., Doghish, A. S., Abulsoud, A. I., Ewees, M. G., Abdelghany, T. M., Elshafey, M. M., & Elkhatib, W. F. (2020). Effect of scopoletin on phagocytic activity of U937-derived human macrophages: Insights from transcriptomic analysis. *Genomics*, *112*(5), 3518–3524.

Al-Qahtani, J., Abbasi, A., Aati, H. Y., Al-Taweel, A., Al-Abdali, A., Aati, S., Yanbawi, A. N., Khan, M. A., Ghalloo, B. A., Anwar, M., & Khan, K. U. R. (2023). Phytochemical, antimicrobial, antidiabetic, thrombolytic, anticancer activities, and in silico studies of *Ficus palmata* Forssk. *Arabian Journal of Chemistry*, *16*(2), 104455.

Álvarez, N. M., Ortíz, A. A., & Martínez, O. C. (2016). In vitro antibacterial activity of *Curcuma longa* (Zingiberaceae) against nosocomial bacteria in Montería, Colombia. Revista de Biología Tropical, 64(3), 1201–1208.

Anandhi, D., Srinivasan, P. T., Kumar, G. P., & Jagatheesh, S. (2014). DNA fragmentation induced by the glycosides and flavonoids from *C. coriaria*. *International Journal of Current Microbiology and Applied Sciences*, *3*(12), 666–673.

Arima, H., Ashida, H., & Danno, G. I. (2002). Rutin-enhanced antibacterial activities of flavonoids against *Bacillus cereus* and Salmonella enteritidis. *Bioscience, Biotechnology, and Biochemistry*, *66*(5), 1009–1014.

Asdaq, S. M. B., Alamri, A. S., Alsanie, W. F., Alhomrani, M., & Yasmin, F. (2021). Potential benefits of gallic acid as skeletal muscle relaxant in animal experimental models. *Saudi Journal of Biological Sciences*, *28*(12), 7575–7580.

Assob, J. C., Kamga, H. L., Nsagha, D. S., Njunda, A. L., Nde, P. F., Asongalem, E. A., Njouendou, A. J., Sandjon, B., & Penlap, V. B. (2011). Antimicrobial and toxicological activities of five medicinal plant species from Cameroon traditional medicine. *BMC Complementary and Alternative Medicine*, *11*(1), 11.

Avendaño-Godoy, J., Ortega, E., Urrutia, M., Escobar-Avello, D., Luengo, J., von Baer, D., Mardones, C., & Gómez-Gaete, C. (2022). Prototypes of nutraceutical products from microparticles loaded with stilbenes extracted from grape cane. *Food and Bioproducts Processing*, *134*, 19–29.

Azad, I., Khan, T., Ahmad, N., Khan, A. R., & Akhter, Y. (2023). Updates on drug designing approach through computational strategies: A review. *Future Science OA*, *9*(5), FSO862.

Azzolini, M., Mattarei, A., La Spina, M., Fanin, M., Chiodarelli, G., Romio, M., Zoratti, M., Paradisi, C., & Biasutto, L. (2017). New natural amino acid-bearing prodrugs boost pterostilbene's oral pharmacokinetic and distribution profile. *European Journal of Pharmaceutics and Biopharmaceutics*, *115*, 149–158.

Bachheti, R. K., & Bachheti, A. (Eds.). (2023). *Secondary Metabolites from Medicinal Plants: Nanoparticles Synthesis and Their Applications*. CRC Press. https://doi.org/10.1201/9781003213727

Bao, Y., Chen, Q., Xie, Y., Tao, Z., Jin, K., Chen, S., Bai, Y., Yang, J., & Shan, S. (2019). Ferulic acid attenuates oxidative DNA damage and inflammatory responses in microglia induced by benzo(a)pyrene. *International Immunopharmacology*, *77*, 105980.

Barros, L., Alves, C. T., Dueñas, M., Silva, S., Oliveira, R., Carvalho, A. M., Henriques, M., Santos-Buelga, C., & Ferreira, I. C. (2013). Characterization of phenolic compounds in wild medicinal flowers from Portugal by HPLC–DAD–ESI/MS and evaluation of antifungal properties. *Industrial Crops and Products, 44*, 104–110.

Barros, L., Dueñas, M., Carvalho, A. M., Ferreira, I. C., & Santos-Buelga, C. (2012). Characterization of phenolic compounds in flowers of wild medicinal plants from Northeastern Portugal. *Food and Chemical Toxicology, 50*(5), 1576–1582.

Benabderrahim, M. A., Yahia, Y., Bettaieb, I., Elfalleh, W., & Nagaz, K. (2019). Antioxidant activity and phenolic profile of a collection of medicinal plants from Tunisian arid and Saharan regions. *Industrial Crops and Products, 138*, 111427.

Bernal-Gallardo, J. O., Molina-Torres, J., Angoa-Pérez, M. V., Cárdenas-Valdovinos, J. G., García-Ruíz, I., Ceja-Díaz, J. A., & Mena-Violante, H. G. (2021). Phenolic compound content and the antioxidant and antimicrobial activity of wild blueberries (*Vaccinium stenophyllum* Steud.) Fruits extracts during ripening. *Horticulturae, 8*(1), 15.

Betts, J. W., Hornsey, M., Higgins, P. G., Lucassen, K., Wille, J., Salguero, F. J., Seifert, H., & La Ragione, R. M. (2019). Restoring the activity of the antibiotic aztreonam using the polyphenol epigallocatechin gallate (ECGC) against multidrug-resistant clinical isolates of *Pseudomonas aeruginosa. Journal of Medical Microbiology, 68*(10), 1552–1559.

Betts, J. W., Kelly, S. M., & Haswell, S. J. (2011). Antibacterial effects of theaflavin and synergy with epicatechin against clinical isolates of *Acinetobacter baumannii* and *Stenotrophomonas maltophilia. International Journal of Antimicrobial Agents, 38*(5), 421–425.

Boga, M., Ersoy, E., Ozkan, E. E., Cinar, E., Kara, E. M., Canturk, Y. Y., & Zengin, G. (2021). Volatile and phenolic profiling of a traditional medicinal plant, *Hypericum empetrifolium* with in vitro biological activities. *Journal of Ethnopharmacology, 272*, 113933.

Borah, B., Dwivedi, K. D., Kumar, B., & Chowhan, L. R. (2022). Recent advances in the microwave-and ultrasound-assisted green synthesis of coumarin-heterocycles. *Arabian Journal of Chemistry, 15*(3), 103654.

Borges, A., Ferreira, C., Saavedra, M. J., & Simões, M. (2013). Antibacterial activity and mode of action of ferulic and gallic acids against pathogenic bacteria. *Microbial Drug Resistance, 19*(4), 256–265.

Borges, A., Saavedra, M. J., & Simões, M. (2012). The activity of ferulic and gallic acids in biofilm prevention and control of pathogenic bacteria. *Biofouling, 28*(7), 755–767.

Boussoussa, H., Khacheba, I., Djeridane, A., Mellah, N., & Yousfi, M. (2016). Antibacterial activity from *Rhanterium adpressum* flowers extracts, depending on seasonal variations. *Industrial Crops and Products, 83*, 44–47.

Brito de Moura, P. H., Almeida, A., Porzel, D. S., Wessjohann, A., Ramos Leal, L. A., & Campos Martins, I. C. (2020). Characterization of antibacterial proanthocyanidins of *Dalbergia monetaria*, an Amazonian medicinal plant, by UHPLC-HRMS/MS. *Planta Medica, 86*(12), 858–866.

Cai, T., & Cai, B. (2023). Pharmacological activities of esculin and esculetin: A review. *Medicine, 102*(40), e35306.

Cai, Y. Z., Sun, M., Xing, J., Luo, Q., & Corke, H. (2006). Structure–radical scavenging activity relationships of phenolic compounds from traditional Chinese medicinal plants. *Life Sciences, 78*(25), 2872–2888.

Çakmakçı, E., Özdemir, M., Şen, F., Bulut, M., & Yalçın, B. (2022). Vegetable oil-based, coumarin-containing antibacterial thermosets with improved thermal stability via copper-free thermal azide-alkyne click polymerization. *Industrial Crops and Products, 182*, 114870.

Cavanagh, H. M. A., & Wilkinson, J. M. (2002). Biological activities of lavender essential oil. *Phytotherapy Research, 16*(4), 301–308.

Cazarolli, L. H., Zanatta, L., Alberton, E. H., Bonorino Figueiredo, M. S. R., Folador, P., Damazio, R. G., Pizzolatti, M. G., & Barreto Silva, F. R. M. (2008). Flavonoids: Prospective drug candidates. *Mini Reviews in Medicinal Chemistry, 8*(13), 1429–1440.

César, L. R. J., Javier, H., Fernando, A. Z. J., Carlos, V., Enrique, R. Z. R., Efrain, A., Evelin, M. B., Inocencio, H. C., Luis, O. J., Zaira, D., & Humberto, G. R. (2022). Identification of the main phenolic compounds responsible for the antioxidant activity of *Litsea glaucescens* Kunth. *South African Journal of Botany, 147*, 208–214.

Cetin-Karaca, H. (2011). Evaluation of natural antimicrobial phenolic compounds against foodborne pathogens (Doctoral dissertation, University of Kentucky Libraries).

Chanwitheesuk, A., Teerawutgulrag, A., Kilburn, J. D., & Rakariyatham, N. (2007). Antimicrobial gallic acid from *Caesalpinia mimosoides* Lamk. *Food Chemistry, 100*(3), 1044–1048.

Chao, C. Y., & Yin, M. C. (2009). Antibacterial effects of roselle calyx extracts and protocatechuic acid in ground beef and apple juice. *Foodborne Pathogens and Disease*, *6*(2), 201–206.

Chirinos, R., Pedreschi, R., Rogez, H., Larondelle, Y., & Campos, D. (2013). Phenolic compound contents and antioxidant activity in plants with nutritional and/or medicinal properties from the Peruvian Andean region. *Industrial Crops and Products*, *47*, 145–152.

Choi, Y. S., Yoon, D. H., Kim, S. Y., Kim, C. S., & Lee, K. R. (2021). Stilbene oligomers from the stems of *Parthenocissus tricuspidata* and their potential anti-neuroinflammatory and neuroprotective activity. *Tetrahedron Letters*, *71*, 153027.

Connell, B. J., Saleh, M. C., Rajagopal, D., & Saleh, T. M. (2017). UPEI-400, a conjugate of lipoic acid and scopoletin, mediates neuroprotection in a rat model of ischemia/reperfusion. *Food and Chemical Toxicology*, *100*, 175–182.

Cowan, M. M. (1999). Plant products as antimicrobial agents. *Clinical Microbiology Reviews*, *12*(4), 564–582.

Cushnie, T. T., & Lamb, A. J. (2005). Antimicrobial activity of flavonoids. *International Journal of Antimicrobial Agents*, *26*(5), 343–356.

Daldoul, S., Hanzouli, F., Fares, N., Tabbene, O., Zemni, H., Vincenzi, S., Mliki, A., & Gargouri, M. (2024). Enhancing biological activities and phenolic content of wild grapevine roots by severe drought stress. *Plant Biosystems-An International Journal Dealing with All Aspects of Plant Biology*, *158*(3), 344–353.

Dandriyal, J., Singla, R., Kumar, M., & Jaitak, V. (2016). Recent developments of C-4 substituted coumarin derivatives as anticancer agents. *European Journal of Medicinal Chemistry*, *119*, 141–168.

Dare, R. G., Oliveira, M. M., Truiti, M. C., Nakamura, C. V., Ximenes, V. F., & Lautenschlager, S. O. (2020). Abilities of protocatechuic acid and its alkyl esters, ethyl and heptyl protocatechuates, to counteract UVB-induced oxidative injuries and photoaging in fibroblasts L929 cell line. *Journal of Photochemistry and Photobiology B: Biology*, *203*, 111771.

Davies, O. L., & Bennett, S. (2017). WHO publishes list of bacteria for which new antibiotics are urgently needed. *WHO Newsletters*.

Demir, T., Akpınar, Ö, Kara, H., & Güngör, H. (2022). Phenolic profile and investigation of biological activities of *Allium scorodoprasum* L. Subsp. rotundum. *Food Bioscience*, *46*, 101548.

Dossou, S. S. K., Xu, F. T., Dossa, K., Rong, Z. H. O. U., Zhao, Y. Z., & Wang, L. H. (2023). Antioxidant lignans sesamin and sesamolin in sesame (*Sesamum indicum* l.): A comprehensive review and future prospects. *Journal of Integrative Agriculture*, 22(1), 14–30.

Dou, Y., Tong, B., Wei, Z., Li, Y., Xia, Y., & Dai, Y. (2013). Scopoletin suppresses IL-6 production from fibroblast-like synoviocytes of adjuvant arthritis rats induced by IL-1β stimulation. *International Immunopharmacology*, *17*(4), 1037–1043.

Duncan, S. H., Leitch, E. C. M., Stanley, K. N., Richardson, A. J., Laven, R. A., Flint, H. J., & Stewart, C. S. (2004). Effects of esculin and esculetin on the survival of *Escherichia coli* O157 in human faecal slurries, continuous-flow simulations of the rumen and colon and in calves. *British Journal of Nutrition*, *91*(5), 749–755.

Efenberger-Szmechtyk, M., Nowak, A., & Czyzowska, A. (2021). Plant extracts rich in polyphenols: Antibacterial agents and natural preservatives for meat and meat products. *Critical Reviews in Food Science and Nutrition*, *61*(1), 149–178.

El-Hawaz, R., Tharayil, N., Bridges, W., & Adelberg, J. (2016). Mineral nutrition of *Curcuma longa* L. in bioreactors affects subsequent development of curcuminoids following transfer to the greenhouse. *Industrial Crops and Products*, *83*, 186–193.

Engels, C., Schieber, A., & Gänzle, M. G. (2011). Inhibitory spectra and modes of antimicrobial action of gallotannins from mango kernels (*Mangifera indica* L.). *Applied and Environmental Microbiology*, *77*(7), 2215–2223.

Engels, C., Schieber, A., & Gänzle, M. G. (2012). Sinapic acid derivatives in defatted oriental mustard (*Brassica juncea* L.) seed meal extracts using UHPLC-DAD-ESI-MS n and identification of compounds with antibacterial activity. *European Food Research and Technology*, *234*, 535–542.

Enriquez-Ochoa, D., Sánchez-Trasviña, C., Hernández-Sedas, B., Mayolo-Deloisa, K., Zavala, J., Rito-Palomares, M., & Valdez-García, J. E. (2020). Aqueous two-phase extraction of phenolic compounds from *Sedum dendroideum* with antioxidant activity and anti-proliferative properties against breast cancer cells. *Separation and Purification Technology*, *251*, 117341.

Erukainure, O. L., Hafizur, R. M., Choudhary, M. I., Adhikari, A., Mesaik, A. M., Atolani, O., Banerjee, P., Preissner, R., Muhammad, A., & Islam, M. S. (2017). Anti-diabetic effect of the ethyl acetate fraction of *Clerodendrum volubile*: Protocatechuic acid suppresses phagocytic oxidative burst and modulates inflammatory cytokines. *Biomedicine & Pharmacotherapy*, *86*, 307–315.

Faggian, M., Bernabè, G., Ferrari, S., Francescato, S., Baratto, G., Castagliuolo, I., Dall'Acqua, S., & Peron, G. (2021). Polyphenol-rich *Larix decidua* bark extract with antimicrobial activity against respiratory-tract pathogens: A novel bioactive ingredient with potential pharmaceutical and nutraceutical applications. *Antibiotics*, *10*(7), 789.

Fărcaş, A., Tofană, M., Socaci, S., Scrob, S., Salanţă, L., & Borşa, A. (2013). Preliminary study on antioxidant activity and polyphenols content in discharged waste from beer production. *Journal of Agroalimentary Processes and Technologies*, *19*(3), 319–324.

Farhadi, F., Khameneh, B., Iranshahi, M., & Iranshahy, M. (2019). Antibacterial activity of flavonoids and their structure–activity relationship: An update review. *Phytotherapy Research*, *33*(1), 13–40.

Fei, P., Ali, M. A., Gong, S., Sun, Q., Bi, X., Liu, S., & Guo, L. (2018). Antimicrobial activity and mechanism of action of olive oil polyphenols extract against *Cronobacter sakazakii*. *Food Control*, *94*, 289–294.

Feng, Y., Li, B., Yang, L., & Liu, Y. (2022). Co-amorphous delivery systems based on curcumin and hydroxycinnamic acids: Stabilization, solubilization, and controlled release. *Lwt*, *170*, 114091.

Fernández-Marín, M. I., Guerrero, R. F., García-Parrilla, M. C., Puertas, B., Richard, T., Rodriguez-Werner, M. A., Winterhalter, P., Monti, J. P., & Cantos-Villar, E. (2012). Isorhapontigenin: A novel bioactive stilbene from wine grapes. *Food Chemistry*, *135*(3), 1353–1359.

Fouotsa, H., Mbaveng, A. T., Mbazoa, C. D., Nkengfack, A. E., Farzana, S., Iqbal, C. M., Marion Meyer, J. J., Lall, N., & Kuete, V. (2013). Antibacterial constituents of three Cameroonian medicinal plants: *Garcinia nobilis*, *Oricia suaveolens* and *Balsamocitrus camerunensis*. *BMC Complementary and Alternative Medicine*, *13*, 1–10.

Fraga, C. G., Croft, K. D., Kennedy, D. O., & Tomás-Barberán, F. A. (2019). The effects of polyphenols and other bioactives on human health. *Food & Function*, *10*(2), 514–528.

Garcia, D., Sanier, C., Macheix, J. J., & D'Auzac, J. (1995). Accumulation of scopoletin in *Hevea brasiliensis* infected by *Microcyclus ulei* (P. Henn.) V. ARX and evaluation of its fungitoxicity for three leaf pathogens of rubber tree. *Physiological and Molecular Plant Pathology*, *47*(4), 213–223.

García, A., Bocanegra-García, V., Palma-Nicolás, J. P., & Rivera, G. (2012). Recent advances in antitubercular natural products. *European Journal of Medicinal Chemistry*, *49*, 1–23.

Gong, W., Wang, R., Huang, H., Hou, Y., Wang, X., He, W., Gong, X., & Hu, J. (2023). Construction of double network hydrogels using agarose and gallic acid with antibacterial and antiinflammatory properties for wound healing. *International Journal of Biological Macromolecules*, *227*, 698–710.

Górniak, I., Bartoszewski, R., & Króliczewski, J. (2019). Comprehensive review of antimicrobial activities of plant flavonoids. *Phytochemistry Reviews*, *18*, 241–272.

Granato, D., Fidelis, M., Haapakoski, M., dos Santos Lima, A., Viil, J., Hellström, J., Rätsep, R., Kaldmäe, H., Bleive, U., Azevedo, L., & Marjomäki, V. (2022). Enzyme-assisted extraction of anthocyanins and other phenolic compounds from blackcurrant (*Ribes nigrum* L.) press cake: From processing to bioactivities. *Food Chemistry*, *391*, 133240.

Guerrero, R. F., Valls-Fonayet, J., Richard, T., & Cantos-Villar, E. (2020). A rapid quantification of stilbene content in wine by ultra-high pressure liquid chromatography–Mass spectrometry. *Food Control*, *108*, 106821.

Gull, I., Sohail, M., Aslam, M. S., & Athar, M. A. (2013). Phytochemical, toxicological and antimicrobial evaluation of *Lawsonia inermis* extracts against clinical isolates of pathogenic bacteria. *Annals of Clinical Microbiology and Antimicrobials*, *12*, 1–6.

Gurbuzer, A. (2021). Investigation of in vitro antimicrobial activities of some hydroxybenzoic and hydroxycinnamic acids commonly found in medicinal and aromatic plants. *International Journal of Plant Based Pharmaceuticals*, *1*(1), 42–47.

Hasan, Z., Islam, A., & Khan, L. A. (2023). Spectroscopic investigations on fungal aspartic protease as target of gallic acid. *International Journal of Biological Macromolecules*, *228*, 333–345.

Huie, C. W. (2002). A review of modern sample-preparation techniques for the extraction and analysis of medicinal plants. *Analytical and Bioanalytical Chemistry*, *373*(1), 23–30.

Hussain, Y., Alam, W., Ullah, H., Dacrema, M., Daglia, M., Khan, H., & Arciola, C. R. (2022). Antimicrobial potential of curcumin: Therapeutic potential and challenges to clinical applications. *Antibiotics*, *11*(3), 322.

Jang, J. Y., Shin, H., Lim, J. W., Ahn, J. H., Jo, Y. H., Lee, K. Y., Hwang, B. Y., Jung, S. J., Kang, S. Y., & Lee, M. K. (2018). Comparison of antibacterial activity and phenolic constituents of bark, lignum, leaves and fruit of *Rhus verniciflua*. *PLoS One*, *13*(7), e0200257.

Jeandet, P., Douillet-Breuil, A. C., Bessis, R., Debord, S., Sbaghi, M., & Adrian, M. (2002). Phytoalexins from the Vitaceae: Biosynthesis, phytoalexin Gene expression in transgenic plants, antifungal activity, and metabolism. *Journal of Agricultural and Food Chemistry*, *50*(10), 2731–2741.

Jekabsone, A., Sile, I., Cochis, A., Makrecka-Kuka, M., Laucaityte, G., Makarova, E., Rimondini, L., Bernotiene, R., Raudone, L., Vedlugaite, E., & Baniene, R. (2019). Investigation of antibacterial and antiinflammatory activities of proanthocyanidins from *Pelargonium sidoides* DC root extract. *Nutrients, 11*(11), 2829.

Joshi, A. N., Chandrakar, A. K., & Wasewar, K. L. (2021). Extractive separation of 4-hydroxybenzoic acid from aqueous solution using nontoxic and conventional solvents. *Chemical Data Collections, 36*, 100782.

Kadakol, A., Sharma, N., Kulkarni, Y. A., & Gaikwad, A. B. (2016). Esculetin: A phytochemical endeavor fortifying effect against non-communicable diseases. *Biomedicine and Pharmacotherapy, 84*, 1442–1448.

Kaska, A., Çiçek, M., & Mammadov, R. (2019). Biological activities, phenolic constituents and mineral element analysis of two endemic medicinal plants from Turkey: *Nepeta italica* subsp. cadmea and *Teucrium sandrasicum*. *South African Journal of Botany, 124*, 63–70.

Kępa, M., Miklasińska-Majdanik, M., Wojtyczka, R. D., Idzik, D., Korzeniowski, K., Smoleń-Dzirba, J., & Wąsik, T. J. (2018). Antimicrobial potential of caffeic acid against *Staphylococcus aureus* clinical strains. *BioMed Research International, 2018*(1), 7413504.

Khan, R., Bhat, K. A., Raja, A. F., Shawl, A. S., & Alam, M. S. (2010). Isolation, characterisation and antibacterial activity studies of coumarins from *Rhododendron lepidotum* Wall. ex G. Don, Ericaceae. *Revista Brasileira de Farmacognosia, 20*, 886–890.

Khan, R., Islam, B., Akram, M., Shakil, S., Ahmad, A., Ali, S. M., Siddiqui, M., & Khan, A. U. (2009). Antimicrobial activity of five herbal extracts against multi drug resistant (MDR) strains of bacteria and fungus of clinical origin. *Molecules, 14*(2), 586–597.

Kim, H. J., Park, C. G., Varghese, R., Lee, J. Y., Kim, Y., & Sung, G. H. (2017). In-vitro antioxidative, anti-inflammatory properties of *Aurea helianthus* leaf extract a Korean traditional medicinal plant. *Saudi Journal of Biological Sciences, 24*(8), 1943–1947.

Koike, S., Iizuka, T., & Mizutani, J. (1979). Determination of caffeic acid in the digestive juice of silkworm larvae and its antibacterial activity against the pathogenic streptococcus faecalis AD–4. *Agricultural and Biological Chemistry, 43*(8), 1727–1731.

Krishna, P. N., & Muraleedharan, K. (2023). Metal chelation ability of protocatechuic acid anion with 210Po84; a theoretical insight. *Computational and Theoretical Chemistry, 1220*, 113996.

Kundu, M., Chatterjee, S., Ghosh, N., Manna, P., Das, J., & Sil, P. C. (2020). Tumor targeted delivery of umbelliferone via a smart mesoporous silica nanoparticles controlled-release drug delivery system for increased anticancer efficiency. *Materials Science and Engineering: C, 116*, 111239.

Kuźniak, V. K., Szaefer, H., & Dubowska, W. B. (2005). Modulation of cytochrome P450 and phase II enzymes by protocatechuic acid in mouse liver and kidney. *Toxicology, 216*(1), 24–31.

Lee, J. H., Kim, Y. G., Cho, H. S., Ryu, S. Y., Cho, M. H., & Lee, J. (2014). Coumarins reduce biofilm formation and the virulence of *Escherichia coli* O157: H7. *Phytomedicine, 21*(8-9), 1037–1042.

Lemos, A. S., Florêncio, J. R., Pinto, N. C., Campos, L. M., Silva, T. P., Grazul, R. M., Pinto, P. F., Tavares, G. D., Scio, E., Apolônio, A., & Melo, M. (2020). Antifungal activity of the natural coumarin scopoletin against planktonic cells and biofilms from a multidrug-resistant *Candida tropicalis* strain. *Frontiers in Microbiology, 11*, 1525.

Lima, V. N., Oliveira-Tintino, C. D., Santos, E. S., Morais, L. P., Tintino, S. R., Freitas, T. S., Geraldo, Y. S., Pereira, R. L., Cruz, R. P., Menezes, I. R., & Coutinho, H. D. (2016). Antimicrobial and enhancement of the antibiotic activity by phenolic compounds: Gallic acid, caffeic acid and pyrogallol. *Microbial Pathogenesis, 99*, 56–61.

Lin, C. W., Tsai, F. J., Tsai, C. H., Lai, C. C., Wan, L., Ho, T. Y., Hsieh, C. C., & Chao, P. D. L. (2005). Anti-SARS coronavirus 3C-like protease effects of *Isatis indigotica* root and plant-derived phenolic compounds. *Antiviral Research, 68*(1), 36–42.

Lin, T., Huang, L., Cheng, N., Wang, Y., Ning, Z., Huang, S., Wu, Y., Chen, T., Su, S., & Lin, Y. (2022). The *in vitro* and *in vivo* antibacterial activities of uniflorous honey from a medicinal plant, *Scrophularia ningpoensis* Hemsl., and characterization of its chemical profile with UPLC-MS/MS. *Journal of Ethnopharmacology, 296*, 115499.

López Navas, A., Muñoz Madero, C., Aguilera Moyano, C., Alonso Herreras, M., Bueno Parralo, R., Sacristán Álvarez, S., & Toboso Escribano, S. (2019). Plan Nacional Frente a la Resistencia a los Antibióticos 2019–2021. *Agencia Española Medicam. y Prod. Sanit*, 37.

Lou, Z., Wang, H., Rao, S., Sun, J., Ma, C., & Li, J. (2012). p-Coumaric acid kills bacteria through dual damage mechanisms. *Food Control, 25*(2), 550–554.

Mandal, N., Sen, G., Pathak, A., Gauliya, K., Manjhi, M. K., Upadhyay, D. C., Khan, M. L., & Upadhyay, C. P. (2025). Comprehensive validation of bactericidal efficacy of umbelliferone against multidrug-resistant pathogens through multifaceted antimicrobial assays. *Archives of Microbiology, 207*(8), 1–18.

Manso, T., Lores, M., & de Miguel, T. (2021). Antimicrobial activity of polyphenols and natural polyphenolic extracts on clinical isolates. *Antibiotics*, *11*(1), 46.

Marinaş, I. C., Chifiriuc, C., Oprea, E., & Lazăr, V. (2014). Antimicrobial and antioxidant activities of alcoholic extracts obtained from vegetative organs of *A. retroflexus*. *Roumanian Archives of Microbiology and Immunology*, 73(1–2), 35–42.

Maxwell, A. (1997). DNA gyrase as a drug target. *Trends in Microbiology*, *5*(3), 102–109.

Mazimba, O. (2017). Umbelliferone: Sources, chemistry and bioactivities review. *Bulletin of Faculty of Pharmacy, Cairo University*, *55*(2), 223–232.

Mehta, S. R., Yusuf, S., Peters, R. J., Bertrand, M. E., Lewis, B. S., Natarajan, M. K., Malmberg, K., Rupprecht, H. J., Zhao, F., Chrolavicius, S., & Copland, I. (2001). Effects of pretreatment with clopidogrel and aspirin followed by long-term therapy in patients undergoing percutaneous coronary intervention: The PCI-CURE study. *The Lancet*, *358*(9281), 527–533.

Merkl, R., Hrádková, I., Filip, V., & Šmidrkal, J. (2010). Antimicrobial and antioxidant properties of phenolic acids alkyl esters. *Czech Journal of Food Sciences*, *28*(4), 275–279.

Miklasińska-Majdanik, M., Kępa, M., Wojtyczka, R. D., Idzik, D., & Wąsik, T. J. (2018). Phenolic compounds diminish antibiotic resistance of *Staphylococcus aureus* clinical strains. *International Journal of Environmental Research and Public Health*, *15*(10), 2321.

Miyasaki, Y., Rabenstein, J. D., Rhea, J., Crouch, M. L., Mocek, U. M., Kittell, P. E., Morgan, M. A., Nichols, W. S., Van Benschoten, M. M., Hardy, W. D., & Liu, G. Y. (2013). Isolation and characterization of antimicrobial compounds in plant extracts against multidrug-resistant *Acinetobacter baumannii*. *PLoS One*, *8*(4), e61594.

Mogosanu, G. D., Grumezescu, A. M., Huang, K. S., Bejenaru, L. E., & Bejenaru, C. (2015). Prevention of microbial communities: Novel approaches based natural products. *Current Pharmaceutical Biotechnology*, *16*(2), 94–111.

Morimoto, H., Hatanaka, T., Narusaka, M., & Narusaka, Y. (2022). Molecular investigation of proanthocyanidin from *Alpinia zerumbet* against the influenza A virus. *Fitoterapia*, *158*, 105141.

Muddathir, A. M., Yamauchi, K., Batubara, I., Mohieldin, E., & Mitsunaga, A. M. (2017). Anti-tyrosinase, total phenolic content and antioxidant activity of selected Sudanese medicinal plants. *South African Journal of Botany*, *109*, 9–15.

Napiroon, T., Bacher, M., Balslev, H., Tawaitakham, K., Santimaleeworagun, W., & Vajrodaya, S. (2018). Scopoletin from *Lasianthus lucidus* Blume (Rubiaceae): A potential antimicrobial against multidrug-resistant *Pseudomonas aeruginosa*. *Journal of Applied Pharmaceutical Science*, *8*(9), 1–6.

Nicolau, A. L. A., Peres, G. B., de Souza Silva, J., Nunes, S. H., Fortes, T. M. L., & Suffredini, I. B. (2022). Pilot project. Resveratrol intake by physical active and sedentary older adult women and blood pressure. *Experimental Gerontology*, *166*, 111883.

Ojala, T., Remes, S., Haansuu, P., Vuorela, H., Hiltunen, R., Haahtela, K., & Vuorela, P. (2000). Antimicrobial activity of some coumarin containing herbal plants growing in Finland. *Journal of Ethnopharmacology*, *73*(1–2), 299–305.

Olszewska, M. A., Gędas, A., & Simões, M. (2020). Antimicrobial polyphenol-rich extracts: Applications and limitations in the food industry. *Food Research International*, *134*, 109214.

Osamudiamen, P. M., Oluremi, B. B., Oderinlo, O. O., & Aiyelaagbe, O. O. (2020). Trans-resveratrol, piceatannol and gallic acid: Potent polyphenols isolated from *Mezoneuron benthamianum* effective as anticaries, antioxidant and cytotoxic agents. *Scientific African*, *7*, e00244.

Ossipov, V., Zubova, M., Nechaeva, T., Zagoskina, N., & Salminen, J. P. (2022). The regulating effect of light on the content of flavan-3-ols and derivatives of hydroxybenzoic acids in the callus culture of the tea plant, *Camellia sinensis* L. *Biochemical Systematics and Ecology*, *101*, 104383.

Otles, S., & Yalcin, B. (2012). Phenolic compounds analysis of root, stalk, and leaves of nettle. *The Scientific World Journal*, *2012*(1), 564367.

Oussaid, S., Chibane, M., Madani, K., Amrouche, T., Achat, S., Dahmoune, F., Houali, K., Rendueles, M., & Diaz, M. (2017). Optimization of the extraction of phenolic compounds from *Scirpus holoschoenus* using a simplex centroid design for antioxidant and antibacterial applications. *LWT*, *86*, 635–642.

Oyama, T., Yoshimori, A., Ogawa, H., Shirai, Y., Abe, H., Kamiya, T., & Tanuma, S. I. (2023). The structural differences between mushroom and human tyrosinase cleared by investigating the inhibitory activities of stilbenes. *Journal of Molecular Structure*, *1272*, 134180.

Pagliarulo, C., De Vito, V., Picariello, G., Colicchio, R., Pastore, G., Salvatore, P., & Volpe, M. G. (2016). Inhibitory effect of pomegranate (*Punica granatum* L.) polyphenol extracts on the bacterial growth and survival of clinical isolates of pathogenic *Staphylococcus aureus* and *Escherichia coli*. *Food Chemistry*, *190*, 824–831.

Palaniappan, K., & Holley, R. A. (2010). Use of natural antimicrobials to increase antibiotic susceptibility of drug resistant bacteria. *International Journal of Food Microbiology*, *140*(2–3), 164–168.

Pan, L., Li, X., Jin, H., Yang, X., & Qin, B. (2017). Antifungal activity of umbelliferone derivatives: Synthesis and structure-activity relationships. *Microbial Pathogenesis*, *104*, 110–115.

Park, C., Jin, C. Y., Kim, G. Y., Choi, I. W., Kwon, T. K., Choi, B. T., Lee, S. J., Lee, W. H., & Choi, Y. H. (2008). Induction of apoptosis by esculetin in human leukemia U937 cells through activation of JNK and ERK. *Toxicology and Applied Pharmacology*, *227*(2), 219–228.

Patel, J., Sharma, M., & Ravishakar, S. (2011). Effect of curli expression and hydrophobicity of *Escherichia coli* O157: H7 on attachment to fresh produce surfaces. *Journal of Applied Microbiology*, *110*(3), 737–745.

Pinho, E., Soares, G., & Henriques, M. (2015). Evaluation of antibacterial activity of caffeic acid encapsulated by β-cyclodextrins. *Journal of Microencapsulation*, *32*(8), 804–810.

Pinilla-Peñalver, E., García-Béjar, B., Contento, A. M., & Ríos, Á. (2022). Graphene quantum dots an efficient nanomaterial for enhancing the photostability of trans-resveratrol in food samples. *Food Chemistry*, *386*, 132766.

Pinto, M. S., Lajolo, F. M., & Genovese, M. I. (2007). Bioactive compounds and antioxidant capacity of strawberry jams. *Plant Foods for Human Nutrition*, *62*, 127–131.

Poirel, L., Jayol, A., & Nordmann, P. (2017). Polymyxins: Antibacterial activity, susceptibility testing, and resistance mechanisms encoded by plasmids or chromosomes. *Clinical Microbiology Reviews*, *30*(2), 557–596.

Priyadarsini, K. I. (2014). The chemistry of curcumin: From extraction to therapeutic agent. *Molecules*, *19*(12), 20091–20112.

Pukkila-Worley, R., Holson, E., Wagner, F., & Mylonakis, E. (2009). Antifungal drug discovery through the study of invertebrate model hosts. *Current Medicinal Chemistry*, *16*(13), 1588–1595.

Quave, C. L. (2016). *Antibiotics from Nature: Traditional Medicine as a Source of New Solutions for Combating Antimicrobial Resistance*. AMR Control, 98–102.

Quave, C. L., Estevez-Carmona, M., Compadre, C. M., Hobby, G., Hendrickson, H., Beenken, K. E., & Smeltzer, M. S. (2012). Ellagic acid derivatives from *Rubus ulmifolius* inhibit *Staphylococcus aureus* biofilm formation and improve response to antibiotics. *PLoS One*, *7*(1), e28737.

Radulovic, N. S., Blagojevic, P. D., Stojanovic-Radic, Z. Z., & Stojanovic, N. M. (2013). Antimicrobial plant metabolites: Structural diversity and mechanism of action. *Current Medicinal Chemistry*, *20*(7), 932–952.

Raj, N. D., & Singh, D. (2022). A critical appraisal on ferulic acid: Biological profile, biopharmaceutical challenges and nano formulations. *Health Sciences Review*, *5*, 100063.

Rajasekaran, D., Palombo, E. A., Yeo, T. C., Ley, D. L. S., Tu, C. L., Malherbe, F., & Grollo, L. (2014). Evidence of synergistic activity of medicinal plant extracts against neuraminidase inhibitor resistant strains of influenza viruses. *Advances in Microbiology*, *4*(16), 1260.

Ribas-Agustí, A., Martín-Belloso, O., Soliva-Fortuny, R., & Elez-Martínez, P. (2019). Enhancing hydroxycinnamic acids and flavan-3-ol contents by pulsed electric fields without affecting quality attributes of apple. *Food Research International*, *121*, 433–440.

Rodriguez, M.X., Buschmann, H., Iglesias, C. and Beeching, J.R., 2000. Production of antimicrobial compounds in cassava (*Manihot esculenta Crantz*) roots during post-harvest physiological deterioration. In Cassava Biotechnology. IVth International Scientific Meeting Cassava Biotechnology Network. Embrapa, Brasilia (pp. 320–328).

Rosa, L. D. S., Jordão, N. A., Soares, N. D. C. P., Mesquita, J. F. D., Monteiro, M., & Teodoro, A. J. (2018). Pharmacokinetic, antiproliferative and apoptotic effects of phenolic acids in human colon adenocarcinoma cells using in vitro and in silico approaches. *Molecules*, *23*(10), 2569.

Roy, S. J., & Prince, P. S. M. (2012). Protective effects of sinapic acid on lysosomal dysfunction in isoproterenol induced myocardial infarcted rats. *Food and Chemical Toxicology*, *50*(11), 3984–3989.

Ruan, X., Deng, X., Tan, M., Yu, C., Zhang, M., Sun, Y., & Jiang, N. (2021). In vitro antibiofilm activity of resveratrol against avian pathogenic *Escherichia coli*. *BMC Veterinary Research*, *17*(1), 9.

Ruddaraju, L. K., Pammi, S. V. N., sankar Guntuku, G., Padavala, V. S., & Kolapalli, V. R. M. (2020). A review on anti-bacterials to combat resistance: From ancient era of plants and metals to present and future perspectives of green nano technological combinations. *Asian Journal of Pharmaceutical Sciences*, *15*(1), 42–59.

Salau, V. F., Erukainure, O. L., Ibeji, C. U., Koorbanally, N. A., & Islam, M. S. (2020). Umbelliferone stimulates glucose uptake; modulates gluconeogenic and nucleotide-hydrolyzing enzymes activities, and dysregulated lipid metabolic pathways in isolated psoas muscle. *Journal of Functional Foods*, *67*, 103847.

Saleem, M., Nazir, M., Ali, M. S., Hussain, H., Lee, Y. S., Riaz, N., & Jabbar, A. (2010). Antimicrobial natural products: An update on future antibiotic drug candidates. *Natural Product Reports*, *27*(2), 238–254.

Samari, E., Chashmi, N. A., Ghanati, F., Sajedi, R. H., Gust, A. A., Haghdoust, F., Sharifi, M., & Fuss, E. (2022). Interactions between second messengers, SA and MAPK6 signaling pathways lead to chitosan-induced lignan production in *Linum album* cell culture. *Industrial Crops and Products, 177*, 114525.

Sánchez, E., Rivas Morales, C., Castillo, S., Leos-Rivas, C., García-Becerra, L., & Ortiz Martínez, D. M. (2016). Antibacterial and antibiofilm activity of methanolic plant extracts against nosocomial microorganisms. *Evidence-Based Complementary and Alternative Medicine, 2016*(1), 1572697.

Sánchez-Maldonado, A. F., Schieber, A., & Gänzle, M. G. (2011). Structure–function relationships of the antibacterial activity of phenolic acids and their metabolism by lactic acid bacteria. *Journal of Applied Microbiology, 111*(5), 1176–1184.

Sangta, J., Wongkaew, M., Tangpao, T., Withee, P., Haituk, S., Arjin, C., Sringarm, K., Hongsibsong, S., Sutan, K., Pusadee, T., & Sommano, S. R. (2021). Recovery of polyphenolic fraction from arabica coffee pulp and its antifungal applications. *Plants, 10*(7), 1422.

Santos, F. D. C. V., Martins, G. R., Luiz, S. R. D. S., Oliveira, I., Silva, A., Silva, L. P. D., Pereira, A. J. R. D., Lopes, M. D., Alviano, R. C., & Moreno, C. S. (2025). Exploring the epicarp potential from *Acrocomia aculeata* fruits: Chemical analysis, antioxidant and antimicrobial activities. *Antioxidants, 14*(2), 181.

Savoia, D. (2012). Plant-derived antimicrobial compounds: Alternatives to antibiotics. *Future Microbiology, 7*(8), 979–990.

Schneider, T., & Sahl, H. G. (2010). An oldie but a goodie–cell wall biosynthesis as antibiotic target pathway. *International Journal of Medical Microbiology, 300*(2–3), 161–169.

Seong, S. H., Ali, M. Y., Jung, H. A., & Choi, J. S. (2019). Umbelliferone derivatives exert neuroprotective effects by inhibiting monoamine oxidase A, self-amyloidβ aggregation, and lipid peroxidation. *Bioorganic Chemistry, 92*, 103293.

Shankar, S. R., Rangarajan, R., Sarada, D. V. L., & Kumar, C. S. (2010). Evaluation of antibacterial activity and phytochemical screening of *Wrightia tinctoria* L. *Pharmacognosy Journal, 2*(14), 19–22.

Sharma, A. K., & Beniwal, V. (2022). Biosynthesis and medicinal applications of proanthocyanidins: A recent update. *Biocatalysis and Agricultural Biotechnology, 45*, 102500.

Shen, L., & Ji, H. F. (2018). Reciprocal interactions between resveratrol and gut microbiota deepen our understanding of molecular mechanisms underlying its health benefits. *Trends in Food Science & Technology, 81*, 232–236.

Shobana, S., Sreerama, Y. N., & Malleshi, N. G. (2009). Composition and enzyme inhibitory properties of finger millet (*Eleusine coracana* L.) seed coat phenolics: Mode of inhibition of α-glucosidase and pancreatic amylase. *Food Chemistry, 115*(4), 1268–1273.

Sieniawska, E., Baj, T., Los, R., Skalicka-Wozniak, K., Malm, A., & Glowniak, K. (2013). Phenolic acids content, antioxidant and antimicrobial activity of *Ligusticum mutellina* L. *Natural Product Research, 27*(12), 1108–1110.

Silva, A. D. S., Pereira-de-Morais, L., da Silva, R. E. R., Dantas, D. D. M., Milfont, C. G. B., Gomes, M. F., Araújo, I. M., Kerntopf, M. R., Menezes, I. R. A. D., & Barbosa, R. (2020). Pharmacological screening of the phenolic compound caffeic acid using rat aorta, uterus and ileum smooth muscle. *Chemico-Biological Interactions, 332*, 109269.

Singh, S., Arthur, R., Upadhayay, S., & Kumar, P. (2022). Ferulic acid ameliorates neurodegeneration via the Nrf2/ARE signalling pathway: A review. *Pharmacological Research-Modern Chinese Medicine, 5*, 100190.

Stefanovic, O., & Comic, L. (2012). Synergistic antibacterial interaction between *Melissa officinalis* extracts and antibiotics. *Journal of Applied Pharmaceutical Science*, 01–05.

Stojković, D. S., Živković, J., Soković, M., Glamočlija, J., Ferreira, I. C., Janković, T., & Maksimović, Z. (2013). Antibacterial activity of *Veronica montana* L. extract and of protocatechuic acid incorporated in a food system. *Food and Chemical Toxicology, 55*, 209–213.

Su, P. W., Yang, C. H., Yang, J. F., Su, P. Y., & Chuang, L. Y. (2015). Antibacterial activities and antibacterial mechanism of *Polygonum cuspidatum* extracts against nosocomial drug-resistant pathogens. *Molecules. 20*(6), 11119–11130.

Sun, W., Shahrajabian, M. H., & Lin, M. (2022). Research progress of fermented functional foods and protein factory-microbial fermentation technology. *Fermentation, 8*(12), 688.

Supritha, P., & Radha, K. V. (2018). Estimation of phenolic compounds present in the plant extracts using high pressure liquid chromatography, antioxidant properties and its antibacterial activity. *Indian Journal of Pharmaceutical Education and Research, 52*(2), 321–326.

Takei, M., Umeyama, A., Arihara, S., & Matsumoto, H. (2005). Effect of piceatannol in human monocyte-derived dendritic cells in vitro. *Journal of Pharmaceutical Sciences, 94*(5), 974–982.

Tan, Q., An, X., Pan, S., Zhen, S., Hu, Y., & Hu, X. (2022). A facile and sensitive ratiometric fluorescent sensor for determination of gallic acid. *Microchemical Journal*, *172*, 106922.

Tao, W., Pan, H., Jiang, H., Wang, M., & Ye, X. (2022). Extraction and identification of proanthocyanidins from the leaves of persimmon and loquat. *Food Chemistry*, *372*, 130780.

Tasfiyati, A. N., Antika, L. D., Dewi, R. T., Septama, A. W., Sabarudin, A., & Ernawati, T. (2022). An experimental design approach for the optimization of scopoletin extraction from *Morinda citrifolia* L. using accelerated solvent extraction. *Talanta*, *238*, 123010.

Tebbi, S. O., Debbache-Benaida, N., Kadri, N., Kadi, R., & Zaidi, S. (2023). A novel strategy to improve the recovery of phenolic compounds from *Pistacia lentiscus* L. fruits using design-based statistical modeling for ultrasound-deep eutectic solvents extraction and the evaluation of their antioxidant potential. *Sustainable Chemistry and Pharmacy*, *31*, 100933.

Tirado-Kulieva, V. A., Hernández-Martínez, E., & Choque-Rivera, T. J. (2022). Phenolic compounds versus SARS-CoV-2: An update on the main findings against COVID-19. *Heliyon*, *8*(9).

Topcagic, A., Zeljkovic, S. C., Karalija, E., Galijasevic, S., & Sofic, E. (2017). Evaluation of phenolic profile, enzyme inhibitory and antimicrobial activities of *Nigella sativa* L. seed extracts. *Bosnian Journal of Basic Medical Sciences*, *17*(4), 286.

Torres-Moreno, H., López-Romero, J. C., Vidal-Gutiérrez, M., Rodríguez-Martínez, K. L., Robles-Zepeda, R. E., Vilegas, W., & Velarde-Rodríguez, G. M. (2022). Seasonality impact on the antiinflammatory, antiproliferative potential and the lignan composition of *Bursera microphylla*. *Industrial Crops and Products*, *184*, 115095.

Tsimogiannis, D., & Oreopoulou, V. (2019). Classification of Phenolic Compounds in Plants. In *Polyphenols in Plants* (pp. 263–284). Academic Press.

Urs, N. V. R., & Dunleavy, J. M. (1975). Enhancement of the bactericidal activity of a peroxidase system by phenolic compounds. *Phytopathology*, *65*(6), 686–690.

V, R. R., A, S. A., J, P. M., K, S., ST, A. S., J, D. P., B, P. G., G, M. J., N, S. (2021). Study of dynamics of genes involved in biosynthesis and accumulation of scopoletin at different growth stages of *Convolvulus prostratus* Forssk. *Phytochemistry*, *182*, 112594.

Vadhana, P., Singh, B. R., Bharadwaj, M., & Singh, S. V. (2015). Emergence of herbal antimicrobial drug resistance in clinical bacterial isolates. *Pharmaceutica Analytica Acta*, *6*(10), 434.

Vaou, N., Stavropoulou, E., Voidarou, C., Tsigalou, C., & Bezirtzoglou, E. (2021). Towards advances in medicinal plant antimicrobial activity: A review study on challenges and future perspectives. *Microorganisms*, *9*(10), 2041.

Vattem, D. A., Randhir, R., & Shetty, K. (2005). Cranberry phenolics-mediated antioxidant enzyme response in oxidatively stressed porcine muscle. *Process Biochemistry*, *40*(6), 2225–2238.

Vestergaard, M., & Ingmer, H. (2019). Antibacterial and antifungal properties of resveratrol. *International Journal of Antimicrobial Agents*, *53*(6), 716–723.

von Danwitz, A., & Schulz, C. (2020). Effects of dietary rapeseed glucosinolates, sinapic acid and phytic acid on feed intake, growth performance and fish health in turbot (*Psetta maxima* L.). *Aquaculture*, *516*, 734624.

Wagner, H., & Ulrich-Merzenich, G. (2009). Synergy research: Approaching a new generation of phytopharmaceuticals. *Phytomedicine*, *16*(2–3), 97–110.

Walsh, C. (2003). Where will new antibiotics come from? *Nature Reviews Microbiology*, *1*(1), 65–70.

Wang, H., Luo, J., Zhang, Y., He, D., Jiang, R., Xie, X., Yang, Q., Li, K., Xie, J., & Zhang, J. (2020). Phospholipid/hydroxypropyl-β-cyclodextrin supramolecular complexes are promising candidates for efficient oral delivery of curcuminoids. *International Journal of Pharmaceutics*, *582*, 119301.

Wang, S., Li, Z., Ma, Y., Liu, Y., Lin, C. C., Li, S., Zhan, J., & Ho, C. T. (2021). Immunomodulatory effects of green tea polyphenols. *Molecules*, *26*(12), 3755.

Wang, X., Liu, X., Shi, N., Zhang, Z., Chen, Y., Yan, M., & Li, Y. (2023). Response surface methodology optimization and HPLC-ESI-QTOF-MS/MS analysis on ultrasonic-assisted extraction of phenolic compounds from okra (*Abelmoschus esculentus*) and their antioxidant activity. *Food Chemistry*, *405*, 134966.

Weng, J. R., Lin, C. S., Lai, H. C., Lin, Y. P., Wang, C. Y., Tsai, Y. C., Wu, K. C., Huang, S. H., & Lin, C. W. (2019). Antiviral activity of Sambucus FormosanaNakai ethanol extract and related phenolic acid constituents against human coronavirus NL63. *Virus Research*, *273*, 197767.

Wu, J. N., Tu, Q. K., Xiang, X. L., Shi, Q. X., Chen, G. Y., Dai, M. X., Zhang, L. J., Yang, M., Song, C. W., Huang, R. Z., & Jin, S. N. (2022). Changes in curcuminoids between crude and processed turmeric based on UPLC-QTOF-MS/MS combining with multivariate statistical analysis. *Chinese Journal of Analytical Chemistry*, 50(7), 100108.

Xu, H., Wang, G., Zhang, J., Zhang, M., Fu, M., Xiang, K., Zhang, M., & Chen, X. (2022). Identification of phenolic compounds and active antifungal ingredients of walnut in response to anthracnose (*Colletotrichum gloeosporioides*). *Postharvest Biology and Technology*, *192*, 112019.

Yahfoufi, N., Alsadi, N., Jambi, M., & Matar, C. (2018). The immunomodulatory and antiinflammatory role of polyphenols. *Nutrients*, *10*(11), 1618.

Yang, J., Han, J., Li, Y., & Dong, B. (2017). Esculetin inhibits the apoptosis in H9c2 cardiomyocytes via the MAPK signaling pathway following hypoxia/reoxygenation injury. *Biomedicine & Pharmacotherapy*, *88*, 1206–1210.

Yuan, Y., Xiang, J., Zheng, B., Sun, J., Luo, D., Li, P., & Fan, J. (2022). Diversity of phenolics including hydroxycinnamic acid amide derivatives, phenolic acids contribute to antioxidant properties of proso millet. *Lwt*, *154*, 112611.

Zhang, N., Liu, W., & Qian, K. (2020). In-vitro antibacterial effect of tea polyphenols combined with common antibiotics on multidrug-resistant *Klebsiella pneumoniae*. *Minerva Medica*, *111*(6), 536–543.

# 7 Anticancer Potential of Phenolic Compounds

*Mansour K. Gatasheh, Mohammad A. Alshuniaber, and Rajapandiyan Krishnamoorthy**

## 7.1 INTRODUCTION

Cancer remains a serious public health challenge in the world, with millions of new cases and deaths each year. According to the International Agency for Research on Cancer (IARC), there were 9.7 million cancer-related deaths and about 20 million new instances of the disease in 2022. According to estimates, 1 in 5 men and women will get cancer at some point in their lives, and 1 in 9 men and 1 in 12 women will pass away from the disease (Puranik et al 2017). Chemotherapy, radiation, and surgery remain the standard cancer treatment; there are, however, many side effects that limit their usefulness.

Lately, the natural compounds have gained much interest as anticancer agents. Among them are phenolic compounds, a diverse group of phytochemicals rich in fruits, vegetables, and all plant materials, which have garnered great interest due to their possible anticancer effects. These compounds possess their structural characteristics due to aromaticity, with one or more hydroxyl groups, which confer upon them antioxidant activity and diverse biological effects, including anticancer action. The scientific community is increasingly considering these natural compounds to be a crucial means of preventing and treating cancer, as demonstrated by numerous studies on their modes of action and therapeutic potential.

Studies have shown that phenolic compounds exert their anticancer effects through a series of mechanisms, including modulation of cell signaling pathways, induction of apoptosis, and inhibition of angiogenesis and metastasis. In an excellent recent review, Bakrim et al. analyze the effects of phenolic compounds on telomerase activity and their impact on the expression of pro-metastatic oncogenes, thereby highlighting their multifunctionality in cancer therapy (Bakrim et al., 2022). Zhou et al. also describe several obligations for a natural polyphenol to interact with multiple protein targets and signaling pathways, greatly impacting its chemopreventive and chemotherapeutic activities (Braakhuis et al., 2016).

Furthermore, antioxidant activity is also associated with anticancer potential in phenolic compounds. It acts as a scavenger of free radicals, thereby preventing oxidative stress, which is one of the major etiologies for the onset of cancer (Duan et al., 2021). For example, phenolic compounds, primarily found in wine and fruits, have been linked to antioxidant properties and are involved in neutralizing reactive oxygen species (ROS), thereby protecting cells from oxidative damage and promoting overall cellular wellness (Kwon, 2018). This has drawn attention to cancer, wherein oxidative stress might result in DNA damage and thus tumorigenesis.

Although it has been indicated that polyphenol-rich foods help in reducing certain cancers, Braakhuis et al. highlighted the role of dietary polyphenolics in reducing breast cancer recurrence and suggested that the antioxidant capacity of these compounds may contribute to their protective effects against cancer (Pandya et al., 2013); furthermore, Vinson et al. reported that foods rich in antioxidants, such as purple potatoes, also retard the growth of cancer cells and induce apoptosis. Thus, the theory behind phenolic dietary compounds participating in the prevention of cancer is strengthened.

* Corresponding author: moorthy@ksu.edu.sa

 DOI: 10.1201/9781003527671-7

Apart from all the above, phenolic compounds can also exert direct cytotoxic effects on cancer cells. For example, many plant extracts contain the phenolic class of compounds, which inhibit cancer cell growth through mechanisms such as cell cycle arrest and induction of apoptosis (Vinson et al., 2012).

The anticancer potential of special phenolic acids, such as gallic acid and ellagic acid, has been well-documented, with current indications of their potential in multiple cancer models (Xia et al., 2018). Some of the scientific evidence concerning the anticancer potential provided by phenolic compounds indicates their antioxidant activity, ability to modulate cell signaling pathways, and their direct cytotoxic action against cancer cells. Multifaceted strategies against cancer that include dietary polyphenols focus on various cancer pathways such as oxidative stress, cell proliferation, apoptosis, inflammation, and angiogenesis, as well as metastasis. Forward-looking treatment strategies that utilize natural compounds in the prevention and treatment of cancer, often in conjunction with standard therapy, are focused. Relatively high availability of these agents within the common diet and low levels of toxicity make them tempting for ingredients that are supposed to be taken for longer periods to prevent cancer from further developing.

With the advancement of studies, the approach toward using phenolic compounds as deterrents to cancer will entail focusing on the bioavailability of each compound, the effective dose, and how such compounds interact with the body. The rising acceptance of the effectiveness of polyphenols in anticancer therapy provides a reasonable basis for their application in primary therapy as well as in the reduction of the burden of this disease. As research continues to uncover the complexities of these compounds, there is growing optimism regarding their potential applications in cancer prevention and therapy.

## 7.2 PHENOLIC COMPOUNDS IDENTIFIED FROM PLANTS

This is of special interest in relation to cancer, wherein oxidative stress might result in DNA damage and, thus, tumorigenesis. Although it has been indicated that polyphenol-rich foods help in reducing certain cancers, Braakhuis et al. highlighted the role of dietary polyphenolics in reducing breast cancer recurrence and suggested that the antioxidant capacity of these compounds may contribute to their protective effects against cancer; furthermore, the theory behind phenolic dietary compounds participating in the prevention of cancer is strengthened (Xiang et al., 2020).

The anticancer potential of special phenolic acids, such as gallic acid and ellagic acid, has been well-documented, with current indications of their potential in multiple cancer models (Zhou et al., 2019). Some of the scientific evidence concerning the anticancer potential provided by phenolic compounds indicates their antioxidant activity, ability to modulate cell signaling pathways, and their direct cytotoxic action against cancer cells. As the science behind the workings of these compounds becomes increasingly understood, the potential use of these in cancer prevention and treatment continues to garner greater enthusiasm.

Intisar et al. evaluated the methanol–water extract of *Polygonum bistorta*. They found that it exhibited potent cytotoxic activity against human hepatocellular carcinoma cells, reinforcing the notion that phenolic compounds can effectively target cancer cells (Intisar et al., 2012). Furthermore, the study by Freddie Bray highlighted the presence of eugenol and isoeugenol in *Balanites aegyptiaca* extracts, which were associated with antioxidant and anticancer activities. The role of phenolic compounds in modulating cancer-related pathways has also been emphasized in the literature. Such evidence suggests that specific phenolic compounds can induce apoptosis and inhibit the proliferation of cancer cells.

Shang et al. were able to identify chlorogenic acid (CGA), a well-known antioxidant, in *Ligularia fischeri*, which has been implicated in anticancer properties (Shang et al., 2010). Additionally, Anantharaju and coworkers reviewed dietary phenolics for combating oxidative stress and exhibiting potential anticancer activities, including the induction of apoptosis and cell-cycle arrest

(Anantharaju et al., 2016). This finding corresponded with Stanković's work, which also stated that phenolic compounds from *Teucrium* species exhibited antiproliferative and pro-apoptotic effects on various human cancer cell lines (Stanković et al., 2011).

Mechanisms by which phenolic compounds show inhibitory potential for cancer cell infection are multiple in nature. Iwo and others highlighted that these phenolic compounds scavenged free radicals and chelated metal ions and inhibited important enzymes such as topoisomerase II that are involved in cancer progression (Iwo et al., 2022). This has been further supported by Burhan and coworkers, who observed that the phenolic compounds present in *Morus alba* stem extracts demonstrated significant antioxidant and anticancer activities against WiDr cancer cells (Burhan et al., 2020).

Cumulative evidence suggests that phenolic compounds will, therefore, work as antioxidants and intervene directly in terms of cancer cell-growth inhibition and promotion of apoptosis. Phenolic compounds from a variety of plants hold great potential to arrest cancer depending on their antioxidant property, stimulation of apoptosis, and the modulation of cancer-related pathways. The mounting body of research underscores the significance of these natural compounds in the prevention and treatment of cancer, offering potential therapeutic avenues in oncology science.

## 7.3 PHARMACOLOGICAL ACTIONS OF PHENOLIC COMPOUNDS

Phenolic compounds are a group of phytochemicals belonging to a large group of secondary plant metabolites and are found in many plants. They have been recognized to exhibit a wide range of pharmacological effects, of which the most important is the antioxidant effect. Flavonoids and phenolic acids, as sources of phenolic compounds, were reported to have the ability to remove free radicals and decrease oxidants, which are linked with many diseases, including cancer and cardiovascular diseases.

For example, Oskoueian et al. reinforced that the phenolic compounds of *Jatropha curcas* have great potential for antioxidant activity, which can help to protect cells against death and show antineoplastic properties (Oskoueian et al., 2011). Similarly, the involvement of phenolic compounds in antioxidant practices has been attributed to their effectiveness in mitigating ROS and, subsequently, preventing cell oxidation (Kim et al., 2021). Secondly, phenolic compounds have considerable anti-inflammatory activity.

Belhadi et al. mentioned that according to their classification of phenolic compounds, flavonoids have demonstrated great anti-inflammatory and analgesic profiles mainly due to the inhibition of cyclooxygenases (COX-1 and COX-2) (Belhadi et al., 2020). It is of significant importance in anti-inflammatory and analgesic properties; therefore, flavonoids are important in the treatment of inflammation. Additionally, the study by Prabhu et al. revealed that phenolic compounds from *Cyperus rotundus* possess numerous pharmaceutical activities, such as anti-inflammatory effects, which is a breakthrough in its use (Prabhu et al., 2022).

The antibacterial effectiveness of phenolic has also been proven earlier. For instance, gallic acid, caffeic acid, and pyrogallol have been shown to significantly alter the activity of antibiotics against various microorganisms, including bacteria and fungi (Lima et al., 2016). According to Oskoueian et al., they noted that flavonoids inhibit bacterial growth by interacting with DNA gyrase and the energy metabolism of microbial cells (Oskoueian et al., 2011). This antimicrobial activity is especially important at a time when antibiotic resistance is increasing, as phenolic compounds can serve as an alternative to therapeutic drugs.

In addition, the present study has also shown that phenolic compounds possess high anticancer potential. The pharmacological activities or anticancer effects of phenolic compounds are not simple; they involve more than one biochemical pathway. These pharmacological actions underscore the possibility of phenols to act on both cancer cell multiplication and proliferation. The anticancer mechanisms of phenol compounds are somewhat complex and blend several pharmacokinetically active biological pathways.

Some of these compounds can affect specific cellular activities, such as apoptosis, cell cycle, and angiogenesis. For example, phenolic compounds have been found to induce proteins that arrest cell growth, such as p53, and inhibit signaling networks responsible for the survival and proliferation of cancer cells, including NF-κB and PI3K/Akt. In addition, phenols can act as antagonists of matrix metalloproteinases (MMPs), enzymes that participate in tumor malignancy.

Studying the effects of these compounds has revealed that they can cause cancerous cell death, slow down cell division, and affect multiple pathways implicated in cancer. For instance, Sharifi-Rad et al. have reviewed phenolic compounds as Nrf2-regulating agents, which, on the other hand, can activate phase II detoxifying enzymes and boost antioxidant protection by acting as chemopreventive agents against cancer (Sharifi-Rad et al., 2023). Additionally, Kim et al. demonstrated that phenolic compounds can serve as chemotherapeutic agents, as shown in their study (Kim et al., 2021).

The pharmacokinetics of phenolic compounds also have a major influence on their medicinal benefit. The pharmacokinetic behaviors of phenolic compounds in *Rhus verniciflua* were described by Jin et al., who noted that these compounds are widely metabolized and their availability and action may be altered (Jin et al., 2015). Pharmacokinetics of phenolic compounds is a critical knowledge in determining the potential and best uses of the compounds. In turn, there are numerous and varied pharmacological effects of phenolic compounds in the human body, including antioxidant properties, anti-inflammatory, antimicrobial, and anticancer effects. Therefore, the complex of new data on these natural compounds emphasizes their significance in modern medicine and their potential as valuable tools for the development of therapeutic agents for various diseases.

## 7.4 THE ANTI-TUMOR EFFECT OF PHENOLS

Phenolic compounds have been observed to possess a wide range of anti-tumor effectiveness in vitro. It has a cytotoxic effect on cancer cells by causing apoptosis, inhibiting mitosis in tumor cells, and blocking the most important pathways related to cellular oncogenesis. These compounds exhibit several biological activities, including antioxidant, anti-inflammatory, and anti-carcinogenic properties, making them prospective anticancer agents. Additionally, phenols are detrimental to angiogenesis, the process of forming new blood vessels, which is essential for tumor growth and thereby depriving the cancerous cells of nutrients.

Research has also shown that consumers of phenolic-rich foods are less likely to develop cancers, thus confirming the chemopreventive properties of these foods. This part presents the information gleaned from the latest research that discusses the anti-tumor properties of phenolic compounds. The first way that phenolic compounds have been shown to affect tumor development is by suppressing the growth of cancer cells and causing them to die. Anantharaju et al. state in their study that diets rich in polyphenolic compounds are beneficial in preventing diseases associated with oxidative stress, such as cancer, through the mechanisms of apoptosis and cell cycle arrest (Anantharaju et al., 2020). For example, his polo, a phenolic compound derived from *Phellinus linteus*, has been used to investigate its ability to influence the human leukemia cell cycle, specifically dividing through the G0/G1 phase, and its ability to induce apoptosis, acting as an anticancer agent (Chen et al., 2013). Likewise, Huang et al. demonstrated that his polo exerts anticancer effects on liver cancer cells, whereby it retarded its growth and the cell's invasive tendencies by preventing the phosphorylation of ERK (Huang et al., 2011).

Furthermore, phenolic compounds influence different molecular targets, which affect cancer progression and development. For instance, Falcão et al. have shown that phenolic compounds

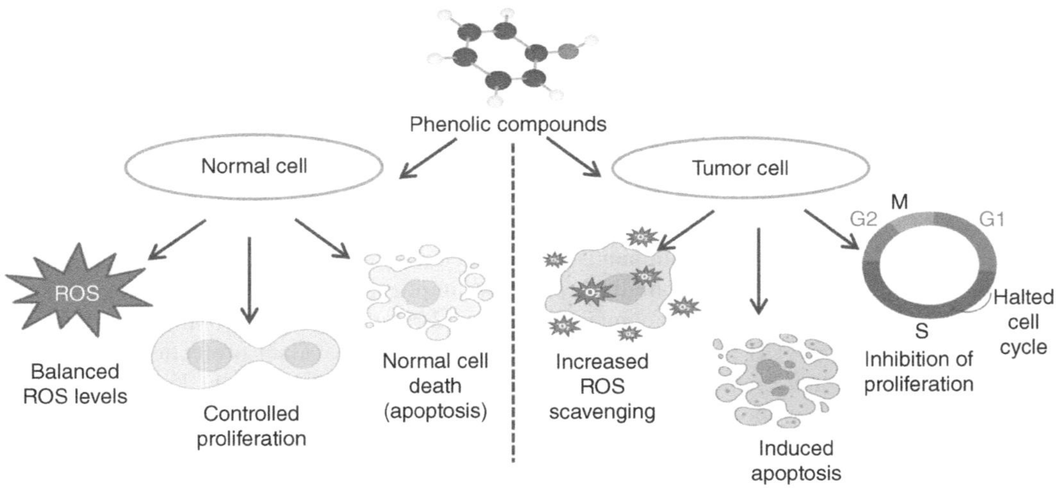


**FIGURE 7.1** Characteristics features of phenolic compounds on healthy cells and cancer cells.

extracted from Moroccan propolis are cytotoxic to human tumor cell lines by causing apoptosis, cell cycle arrest, and inhibiting MMPs that are important for tumor invasion and metastasis (Falcão et al., 2019). This dual mechanism represents the optimal strategy for cancer treatment, as it targets multiple aspects of tumor biology.

The other role of phenolic compounds is that, on an antioxidant basis, they possess anti-tumor properties. Phenolic compounds, due to their ability to scavenge free radicals and decrease oxidative stress, can help prevent changes in the DNA strands that are so important for cancer to occur. For instance, the study by Fantini et al. has revealed that the amount and efficiency of polyphenols, the major component of red wine, influence the activity of multiple targets implicated in the carcinogenesis process, targeting the growth of cancer cells (Fantini et al., 2015).

Furthermore, López et al. revealed that the phytochemicals of murta berries have antioxidant, anti-inflammatory, and anti-tumor properties. In addition, bioavailability is a significant determinant of phenolic compounds' anti-tumor potency. Muller et al. have also pointed out in their study that phenolic compounds have a very high potential in treating cancer when used in in vitro studies, the major downside being the issue of low bioavailability (Muller et al., 2019). Some methods identified include synthesizing phenolic compounds as polymeric nanoparticles that exhibit higher bioavailability and effective anticancer properties. It also affords some enhancement to the therapeutic effects of phenolic compounds in clinical practices.

These phenolic compounds possess strong anti-tumor activity by affecting signal transduction pathways, promoting apoptosis, and exerting free radical scavenging abilities. This and other related studies contribute to the growing body of evidence that suggests the use of these natural compounds in cancer prevention and management. However, mastering the challenges concerning their bioavailability will be perspective as concerns advancing their clinical perspective. Figure 7.1 shows characteristics of healthy cells and cancer cells influenced by phenolic compounds.

## 7.5 BIOAVAILABILITY

Bioavailability is the extent to which a particular compound introduced into the body gets into the circulation for activity or metabolism. Phenolic compounds and their bioavailability. As mentioned earlier, the bioavailability of phenolic compounds defines their efficacy, health effects, and therapeutic value. However, a potential drawback of many of these substances with promising pharmacological activity is their low bioavailability of phenolic compounds. However, the biological activity of most phenolic compounds remains low since they undergo rapid metabolization and have low absorbability and poor solubility. For instance, resveratrol has demonstrated high anticancer efficacy in vitro; however, due to its zero-order kinetics profile and consequently poor systemic availability in the human body, it is not very useful clinically. Efforts to increase phenol solubility and absorption include work with functional nanocarriers and chemical and concomitant administration of other agents.

Phenolic compounds exhibit differences in solubility, stability, and permeability, which are influenced by the chemical structure of the phenolics, the food matrix in which they are present, and the gastrointestinal environment. Flavonoids, in particular, are weak compounds that are only partially available in their natural structures, though the absorption varies greatly from one phenolic compound to another. For example, research has indicated that the bioavailability range of polyphenols can range from 0.3% to 43% (Rein et al., 2013). This variability is due to aspects including the effect of phenolic compounds on dietary fibers, which may polymerize or engage in similar actions and may be released or absorbed during digestion (Palafox-Carlos et al., 2011).

The food matrix has an important influence on the bioavailability of compounds because some processing techniques increase the release of these compounds from the matrix (Ribas-Agustí et al., 2017). It is also relevant that the phenolic compounds exhibit gastrointestinal behavior that is important for their bioavailability. Cañas et al. stressed that some phenolic compounds are non-absorbed and then enter the large intestine, where gut bacteria ferment them to release bioaccessible forms (Cañas et al., 2022). It contains reactions such as hydroxylation, oxidation, and glycosylation, which can enhance the solubility and absorption rates of phenolic metabolites (Cañas et al., 2022).

Besides that, the solubility and bioavailability of phenolic compounds may be further modified by the presence of other specific functions, such as methoxy or glycoside functions, as Baptista (2024) reported. Furthermore, the consumption of various phenolic compounds simultaneously improves their bioavailability. For instance, Dudonné et al. showed that enhancement of the absorption of phenolic compounds from strawberry and cranberry extracts in rats was attributed to the presence of quercetin-rich onion extract (Dudonné et al., 2016). This implies that some phenolic compounds interact competitively with efflux transport proteins to promote the absorption of other compounds found in foods. The effect that results from food processing on phenolic compounds' bioavailability is also of primary concern.

Processing and fermentation can alter the chemical composition of phenolic compounds, potentially enhancing their bioaccessibility and bioavailability (Ribas-Agustí et al., 2017). For example, the analysis of mango "Ataulfo", processed into juice, revealed that it can maintain the bioavailability and antioxidant capacity of its phenolic content (Quirós-Sauceda et al., 2017). This suggests that food processing techniques can either improve or harm the health effects of phenolic compounds.

Consequently, the bioavailability presented by phenolic compounds depends on chemical characteristics, gastrointestinal profiles, and the food matrix, amongst others, of the foods in question. This revision makes sense within the context of discussing the benefits of phenolic compounds in health-related applications (Ribas-Agustí et al., 2017). For instance, the processing of "Ataulfo" mango into juice was shown to preserve the bioavailability and antioxidant capacity of its phenolic compounds (Quirós-Sauceda et al., 2017). This indicates that appropriate food processing strategies can enhance the health benefits associated with phenolic compounds.

Thus, the bioavailability of phenolic compounds is influenced by a complex interplay of factors, including their chemical structure, the food matrix, gastrointestinal behavior, and food processing methods (Table 7.1). Understanding these factors is essential for optimizing the health benefits

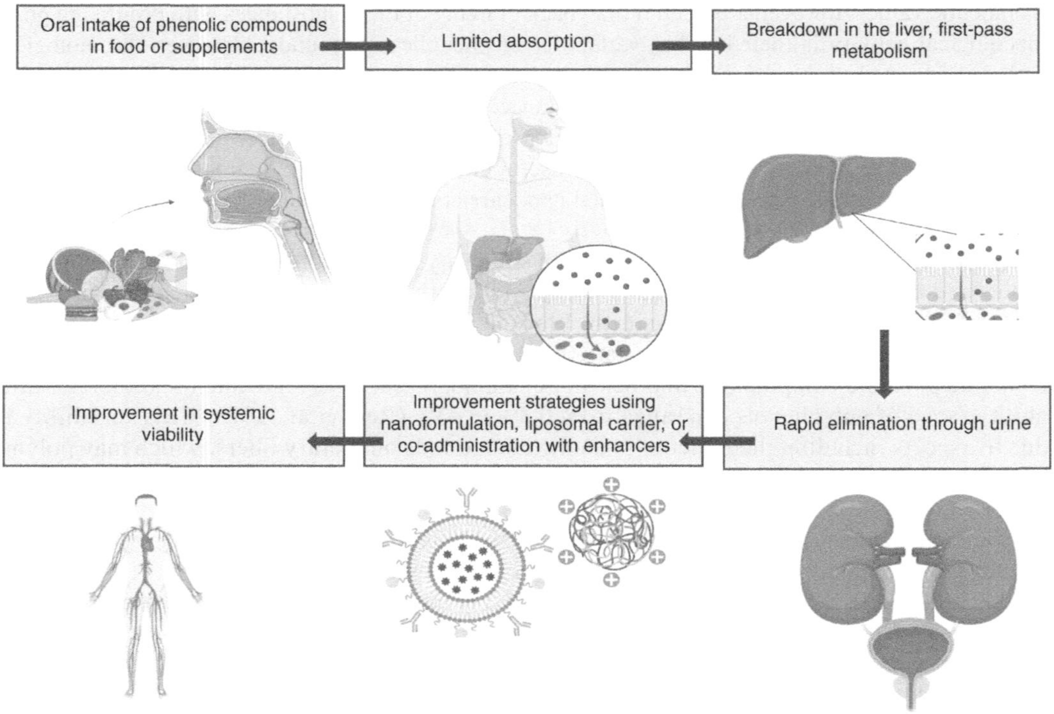


**FIGURE 7.2** Bioavailability pathways of phenolic compounds.

of phenolic compounds and their application in functional foods, therapeutics, and nutraceuticals. Figures 7.2 and 7.3 represents a schematic diagram and outlines the bioavailability pathways of phenolic compounds and different bioactive properties

## 7.6 ANTICANCER ACTIVITY OF PHENOLIC COMPOUNDS

### 7.6.1 Prostate Cancer

Prostate cancer is among the prevalent types of cancer in the male population of every country, including the United States. Flavonoids, which are randomly distributed in fruits, vegetables, and other plant products, have been reported to exhibit first-rate anticancer activity, especially in prostate cancer. This section summarizes the latest information on the impact of Phenols and their derivatives on PC, examining their actions, effectiveness, and possible uses. Up-regulation of apoptosis by phenolic compounds is one of the ways phenolic compounds control the growth of prostate cancer.

As an example, Mindiédiba et al. showed that the methanolic extract of *Striga hermonthica* causes apoptosis by activating caspase-3 in prostate cancer cells, specifically 22RV1 and C4-2 cells, thereby exerting antiproliferative effects (Mindiédiba et al., 2022). Consistent with this, Zhou et al. demonstrated that extracts from *Linum usitatissimum* (flaxseed) can trigger apoptosis of LNCaP prostate cancer cells through caspase-3 and that Bax pathways exist in the hydroxycinnamic acids, including caffeic acid, for the inhibition potential of prostate cancer cell invasion and metastasis (Zhou et al., 2020).

Experiments involving flavonoid and curcumin have shown the ability of these phenolic compounds to overcome the growth and proliferation of prostate cancer cells and prevent their metastasis. These compounds are effective because they define and modulate the androgen function and reduce PSA levels, which are important components of the progression of prostate cancer. The

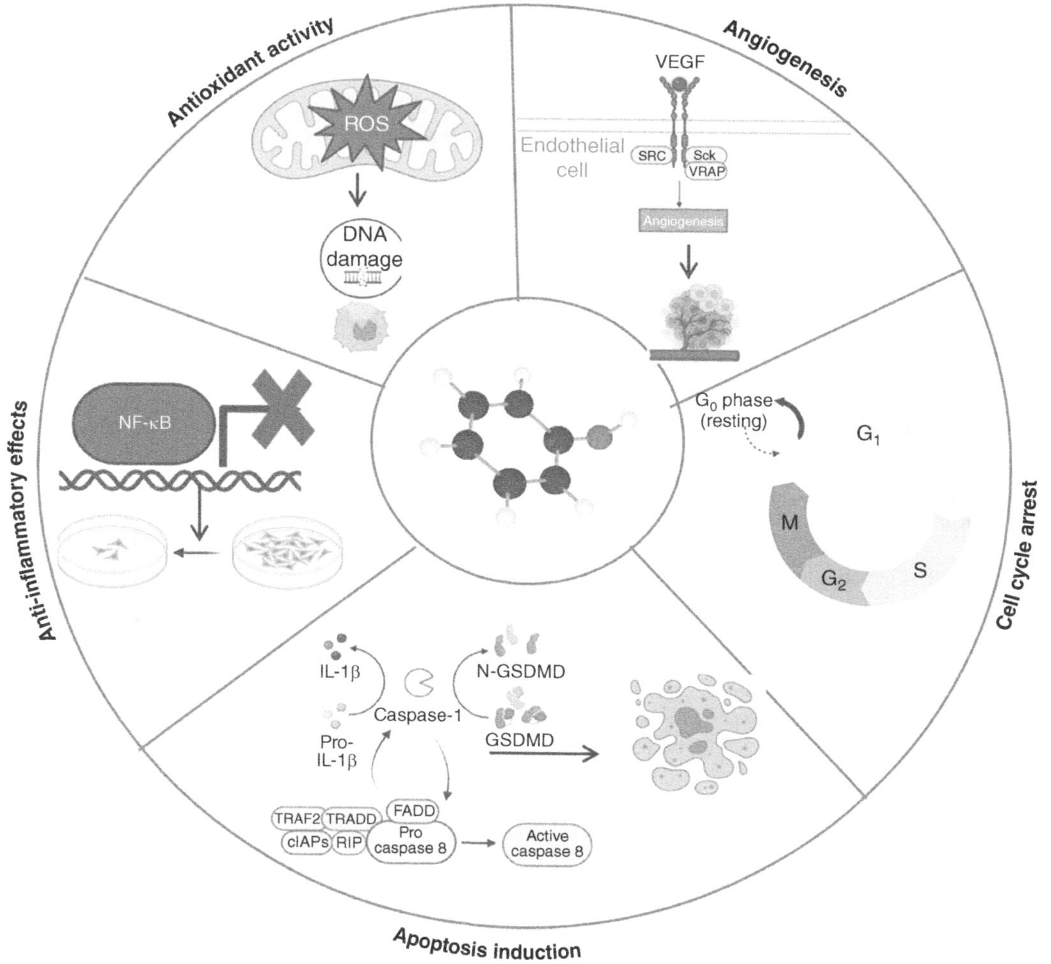


**FIGURE 7.3** Different bioactive properties of phenolic compounds.

findings have also provided evidence on how phenolic compounds inhibit inflammation in the prostate gland and oxidative stress, which are risk factors for cancer.

Furthermore, it has been established that certain phenolic compounds cause apoptosis (Figure 7.3) sensitization. Szliszka et al. have observed that the phenolic compound Artepillin C derived from Brazilian green propolis increased the pro-apoptotic effect of TRAIL on LNCaP prostate cancer cells. This is realized through the occurrence of TRAIL-R2 receptors and caspases that are thought to play a significant role in apoptosis (Szliszka et al., 2012). Furthermore, the ethanolic extract of Brazilian green propolis was reported to suppress prostate cancer cell TARP and promote TRAIL-induced apoptosis, which indicates that propolis and phenolic constituents might be good supporters in prostate cancer treatment (Szliszka et al., 2011).

In epidemiological investigations, dietary phenolic compounds also play a significant role in the prevention of prostate cancer. Russo et al. conducted a case–control study in Sicily and discovered that phenolic acids, specifically ferulic and caffeic acids, which are consumed in the diet, were inversely related to prostate carcinoma. It was found that these compounds decrease the proliferation rate and suppress oncogenes in prostate adenocarcinoma cells (Russo et al., 2017). The overall observations made in this study, therefore, suggest that phenolic compounds may reduce the risk of prostate cancer.

**TABLE 7.1**
**Bioavailability of Phenolic Compounds**

| Phenolic Compound | Source | Bioavailability Period | Factors Affecting Bioavailability | Improvement Strategies |
|---|---|---|---|---|
| **Resveratrol** | Grapes, red wine | 1–2 hours | Rapid metabolism in the liver, low plasma concentration | Nano formulations, liposomal delivery, combination with piperine |
| **Curcumin** | Turmeric | <1 hour | Poor solubility, rapid metabolism, low systemic absorption | Curcumin nanoparticles, phospholipid complexes, black pepper (piperine) |
| **Quercetin** | Onions, apples | 2–4 hours | Limited water solubility, fast degradation, poor intestinal absorption | Encapsulation in lipid-based carriers, glycosylation |
| **Epigallocatechin gallate (EGCG)** | Green tea | 1–3 hours | Instability in the digestive tract, low absorption, rapid degradation | Use of catechin derivatives, sustained-release formulations |
| **Gallic acid** | Berries, tea | 1–2 hours | High metabolism, fast clearance through urine | Co-delivery with other bioactive compounds |
| **Ferulic acid** | Rice bran, corn, oats | 1–2 hours | Poor absorption, rapid elimination | Structural modification, co-administration with fatty acids |
| **Anthocyanins** | Berries (blueberries, raspberries) | 2–6 hours | Instability in the gut, rapid breakdown, low permeability across cell membranes | Encapsulation in liposomes, polymeric nanoparticles |
| **Kaempferol** | Broccoli, Spinach | 4–8 hours | Low water solubility, high first-pass metabolism | Micellar systems, emulsions, phytosomes |

Additionally, the ability of phenolic compounds to act as adjuvant agents, thereby enhancing the effectiveness of traditional cancer treatments, has been considered. For instance, Eskra et al. observed that ellagic acid and anthocyanins, sourced from black raspberries, affect chemotherapy in castration-resistant prostate cancer cells, and these compounds improve the efficiency of taxane-based treatments. This has underscored the potential of using phenolic compounds as co-adjuvants to conventional cancer therapies.

Furthermore, phenolic compounds may also exert an impact on cancer signaling pathways controlling cancer cell growth and development. For example, Hsu et al. report that 6-shogaol, derived from ginger, has anticancer efficacy, particularly through the immunological elements of cancer-associated tumors (Hsu et al., 2015). This suggests that phenolic compounds may exert multiple functions in cancer biology, on both the tumor cell and its microenvironment.

Consequently, these compounds have been shown to possess profound anti-tumor activities in prostate cancer through the processes of apoptosis, alteration of signaling pathways, and enhancement of chemotherapy. The growing body of evidence supports the potential of these natural compounds in the prevention and treatment of prostate cancer, warranting further research into their clinical applications.

### 7.6.2 Colon Cancer

Colon cancer is a diet-dependent disease and hence falls easily within the sphere of diet manipulation. Some other compounds that belong to the phenolic category and are present in edible products such as berries, grapes, and green tea have demonstrated the ability to control the growth of colon cancer cells. Another way phenolic compounds reduce the incidence of cancer is by reducing the production of free radicals. For instance, Ademosun et al. showed that phenolic extracts from orange peels prevent the peroxyl radical-induced oxidation in colon cancer cells. This finding implied that the extracts protect against oxidative stress, a factor that promotes the growth of cancerous cells (Ademosun et al., 2016).

In a recent study, Špoljarić et al. discussed extracts of phenolic-rich grape pomace, which have been utilized in biotransformation, highlighting that such compounds contribute to antiproliferative effects, including the induction of apoptosis in colon cancer cells (Špoljarić et al., 2023). This is also evident as Chang et al. pointed out that dietary flavonoids can affect cell growth signaling pathways as well. Recommended approaches to reducing DNA damage can greatly decrease the chances of developing colorectal cancer (Chang et al., 2018).

However, individual phenolic compounds have been found to suppress HDACs, which are major players in gene expression related to cancer cell proliferation. Saenglee et al. revealed that the extract of phenolic compounds from peanut testa has an inherent potential for inhibiting the growth of colon cancer cells by restraining HDAC activity (Saenglee et al., 2016). This mechanism is vital as cancer-related genes can be reactivated by HDAC inhibition. Furthermore, cancer cells can be arrested in their cell cycle. Other compounds isolated from cloves are also beneficial and include eugenol, a phenolic that has been found to have worthy anticancer effects. A decreased survival rate, DNA fragmentation, and cell cycle arrest by eugenol were also observed by Jaganathan et al., who noted that eugenol has the potential to increase ROS activity in colon cancer cell lines. The above finding is consistent with work conducted by Fadilah et al., revealing that eugenol can influence COX-2 regulation, a gene linked to both inflammation and cancer (Fadilah et al., 2017).

Other issues affecting the cancer-preventive potentials of phenolic compounds include the bioavailability and metabolism of the compounds in the colon, as most colorectal cancers originate there. Research shows that phenolic compounds are subject to metabolic degradation by gut microbiota, an outcome that can increase the activity of these compounds. For instance, the differential metabolites arising from the olive oil phenolics exhibit cell cycle inhibition and apoptosis of colon cancer cells as described by Hazas et al. (2017). This highlights the need to incorporate foods rich in phenolic compounds into recommended anticancer diets.

Phenolic compounds target colon cancer using all three mechanisms, depending on their molecular targets, affecting antioxidant function, gene regulation, and promotion of apoptosis. The effectiveness depends on the specific pharmacological characteristics, including chemical composition, lipid solubility, and first-pass metabolism in the gastrointestinal tract. Further studies on these compounds may shed light on novel anticancer dietary approaches in humans.

### 7.6.3 Breast Cancer

Phenolic compounds have been investigated comprehensively in relation to breast cancer, which is one of the leading causes of cancer mortality in women. On this basis, one potential anticancer activity of phenolic compounds is proven to induce apoptotic death in breast cancer cells. In addition, several phenolic compounds, such as resveratrol, epigallocatechin gallate (EGCG), and curcumin, have been found to have inhibitory effects on breast cancer cell growth and pro-apoptotic effects. They can also inhibit the estrogen receptors that are essential in the growth of hormone-sensitive breast cancer cells. Moreover, phenols exhibit anti-inflammatory and antioxidant properties that can slow the development of breast cancer.

Various epidemiological investigations have pointed out that increased phenol contents in foods significantly reduce the risk of breast cancer. According to a study by Tayhan et al., specific polyphenol derivatives reduced cell viability in MCF-7 breast cancer cells, underscoring the potential of phenolic compounds as efficient therapeutic alternatives for cancer treatment without harming healthy breast cells (Tayhan et al., 2023). Along the same line, Ma et al. have described that sesamol, which is a phenolic compound for sesame oil, is capable of activating apoptosis alongside diminishing cellular proliferation in ER by affected cells, and given these findings, sesamol might be used in combination with targeted therapies (Ma et al., 2021).

This assertion is boosted by the works of Joo et al., who pointed out that 10-gingerol reduced the viability and motility of MDA-MB-231 breast cancer cells; evidence suggests that this was accomplished through the down-regulation of Akt and p38MAPK signaling molecules (Joo et al., 2015). Also, phenolic compounds were found to possess antioxidant activity; thus, the antioxidant activity of phenolic compounds is used to explain their anticancer effects. Abrahim et al. revealed that different types of phenolic compounds, such as EGCG and quercetin, exhibit anti-breast cancer cell growth by modulating the ROS level to induce apoptosis, but not to promote additional multiplication (Abrahim et al., 2012). This antioxidant mechanism is critical, as excessive oxidative stress is a major cause of cancer progression.

It is worth mentioning that the cytotoxicity of phenolic compounds has also been associated with their actions toward epigenetic regulation. For instance, Li et al. revealed how genistein, a phytoestrogen with recognized beneficial effects on breast cancer, can prime the hormonal therapy by reactivating ERα in the ER-negative breast cancer cells via epigenetic mechanisms (epigenetic changes) (Li et al., 2013). This suggests that dietary phenolic compounds act not only as direct antioxidants but also as positive modulators of the cancer signaling networks. Besides apoptosis and antioxidant effects, phenolic compounds can also suppress cancer cell migration and invasion, which are important steps in cancer metastasis.

Domínguez et al. stated that the ethyl acetate fraction from *Bursera copallifera* showed cytotoxicity and migration inhibition in breast cancer cells, which may be the result of specific phenolic compounds in the organization of cancer spread (Domínguez et al., 2018). This assertion is boosted by the works of Joo et al., who pointed out that 10-gingerol reduced the viability and motility of MDA-MB-231 breast cancer cells; evidence suggests that this was accomplished through the down-regulation of Akt and p38MAPK signaling molecules (Joo et al., 2015).

Also, phenolic compounds were found to possess antioxidant activity; thus, the antioxidant activity of phenolic compounds is used to explain their anticancer effects. Abrahim et al. revealed that different types of phenolic compounds like EGCG and quercetin exhibit anti-breast-cancerous cell growth through modulating ROS levels for inducing apoptosis (Abrahim et al., 2012). In addition, the study by Martino et al. identified that Annurca apple polyphenol extract killed MDA-MB-231 selectively through ROS and activation of the JNK signaling pathway, thus arresting the cell cycle and causing apoptosis (Martino et al., 2019). These phenolic compounds exhibit anticancer effects in breast carcinoma through the induction of apoptosis, modulation of oxidative stress, epigenetics, and inhibition of migration and invasion. Such multiple effects emphasize the promise of these compounds as therapeutic agents in breast cancer intervention.

### 7.6.4 Lung Cancer

Lung cancer, which is among the most lethal cancers worldwide, has also been an area of focus of phenolic-based approaches. The potential role of phenolic compounds in cancer prevention is based mostly on their abilities to suppress cancer cells, which are found in fruits, vegetables, and natural herbs, revealing multiple biological activities to inhibit cancer cells' proliferation, induce apoptosis, and control the signaling pathways that can facilitate tumor growth and metastasis.

Among all the phenolic compounds researched on lung cancer, curcumin, a spice derived from *Curcuma longa*, is one of them. The suppression of the adiponectin/NF-κB/MMP signaling by curcumin has the ability to attenuate the growth and metastasis of NSCLC cells, including A549 cells. More specifically, curcumin has been shown to prevent cancer cell migration and invasion abilities in these cancer cells in a dose-dependent manner, thus pointing to curcumin as a therapeutic agent for the treatment of lung cancer (Tsai et al., 2015). In support of the above, Townsend et al. also observed that dietary polyphenols such as curcumin could influence cell survival pathways and demonstrate anticarcinogenicity (Townsend et al., 2015).

Another phenolic compound is laricitrin, and this flavonoid was identified to reverse dendritic cell suppression due to lung cancer. Chang and his colleagues compared the effect of laricitrin with control and found that laricitrin suppresses the STAT3 signaling pathway, which is upregulated in several cancers for the immune checkpoint (Chang et al., 2016). This compound not only optimizes the occurrence of cancer cells but also stimulates the immunity to lung cancer and proves the biological function of cancer treatments.

The anticancer activities of phenolic compounds also depend on their antioxidative properties. Choi et al. pointed out that phenolic extracts of tomatoes present strong antioxidative and cancer cell-inhibiting activities; these reduce oxidative stress and prevent DNA damage in lung cancer cells (Choi et al., 2011). Phenolic compounds' ROS scavenging characteristics are significant because high ROS levels are related to cancer promotion and metastasis. Moreover, substances like CAPE have shown potent anticancer properties in primary lung cancer cells. According to Ulaşlı et al., CAPE can affect certain cellular signaling involved in carcinogenesis and induce apoptosis as well as suppress cell growth of A549 cells that are subjected to carcinogenic substances such as benzo(a)pyrene 202. This suggests that phenolic compounds can work cooperatively with traditional chemotherapy to increase its effectiveness while at the same time having a shielding influence on normal cells.

Moreover, the cytotoxic impact of some phenolic compounds has been evidenced in the literature. Klongkumnuankarn et al. also assessed the cytotoxic and antimigratory potential of phenolic ingredients in *Dendrobium brymerianum* and observed the reduction of migratory and invasive potential of lung cancer cells (Klongkumnuankarn et al., 2015). This indicates that certain phenolic compounds are able to affect the growth of cancer cells and their metastatic potential; in particular, phenolic compounds seem to have multidirectional action in the fight against lung cancer through promotion of apoptosis, enhancement of antioxidant defense, modulation of immune functions, and prevention of metastasis. The given compounds manifold biochemical effects confirm their opposite ability to serve as promoters in lung cancer development and treatment.

### 7.6.5 Bladder Cancer

Several studies have attributed the development of bladder cancer to chronically reinflamed and oxidative-induced damage, and therefore, phenolic compounds could be the ideal candidates for the prevention and treatment of bladder cancer. Several epidemiological investigations indicate that an increased intake of phenols can reduce the rate of bladder cancer. An important group of compounds, which are found in many plants, phenols possess anticancer properties that can suppress tumor growth, promote cancer cell death, and alter signal transduction processes that regulate cell division, migration, and invasive potential.

They include the flavonolignan, silibinin, that has been investigated in relation to its impact on bladder cancer. According to Philippou et al., silibinin possesses potential anticancer effects in in vitro and in vivo models of bladder cancer, especially because it can inhibit the recurrence of T1 G3 bladder tumors when used as intravesical treatment (Philippou et al., 2013). This has the potential of showing that silibinin could be used as a new therapeutic agent in bladder cancer, though the results from clinical investigations are still lacking.

Another major phenolic compound is the ECG, and is mostly found in green teas. Lee et al. pointed out that their study is significant as EGCG has direct cytotoxic effects on bladder cancer cells, similar to mitomycin C. The study pointed out that EGCG improves the effect of chemotherapy in bladder cancer by changing microRNA; in this case, miR-31-5p has been revealed to have an oncosuppressive role (Lee et al., 2019). One of the most important features of this extraordinary molecule, EGCG, is that it has a twofold effect: it kills cancer cells and improves the efficacy of known treatments.

Besides silibinin and EGCG, phenolic extract from the extra virgin olive oil also exhibits strong anticancer potential. Spagnuolo et al. also confirm that this extract causes apoptosis and alters several antiproliferative signaling pathways in human bladder cancer cell lines, especially when in different stages of tumor progression (Spagnuolo et al., 2022). It may be envisaged that the antioxidant nature of the olive oil phenols is crucial for the observed carcinogenicity effects, such as inhibition of proliferation and induction of apoptosis of the cancer cells.

Furthermore, Puranik et al. reviewed that in vitro antioxidant and anticancer properties of phenolic compounds containing extracts of *Simarouba glauca* leaves. The results that they found show that these extracts have marked deleterious effects on the T-24 bladder cancer cell line due to the rich chemical profile of the plant, including flavonoids and phenolic acids. This goes along with the potential of plant-derived phenolic compounds as therapeutic inhibitors in bladder carcinoma. In connection with the action of phenolic compounds, the significance of the PI3K/AKT/mTOR signaling pathway in bladder cancer has also been considered.

Li et al. revealed that gypenoside, a phenolic compound, affects bladder cancer cells by suppressing this abnormal pathway because it is involved in the development of cancer (Li et al., 2022). This supports our proposition that modulation of specific signaling pathways by phenolic compounds could offer better treatment benefits for bladder cancer. Therefore, the reported phenolics possess strong effects against bladder cancer through diverse mechanisms such as apoptosis, alteration of signaling, and synergism with other therapies. These pharmacophores' diverse biological activities support the idea of using these compounds for the prevention and treatment of bladder cancer.

### 7.6.6 Skin Cancer

Polyphenolics still acting as antioxidants in green tea, grape seed extract, and turmeric have proven to have an anticancerous effect on skin cancer. All of these compounds demonstrate multiple anticancer properties that include the ability to suppress tumor formation, promote cancer cell death, and prevent oxidative damage. The antioxidant activity is one of the primary ways that phenolic compounds are known to exhibit their anticancer properties. Phenolic compounds as described by Gado et al. help to reduce the effects of some factors, such as ultraviolet radiation, which is a major cause of skin cancer.

These compounds have been said to counter oxidative stress and DNA harm due to UV exposure, so as to decrease skin cancer incidences Gado et al. (2021). This protective influence is especially important for NMSC and melanoma since oxidative stress is considered one of the major factors in carcinoma development. It was also ascertained that phenolic compounds exhibit some antioxidant activity and can cause apoptosis in cancer cells.

A published work by Baloghová showed that different phenolic compounds derived from spices possess anticancer properties in treating skin cancer using apoptotic signaling (Baloghová, 2023). This is, however, important because the process of inducing apoptosis in malignant cells is one of the most effective strategies for combating cancers.

Furthermore, researchers note that specific phenolic compounds in food help to prevent the proliferation of skin cancer cells. For instance, Mehta and Paliwal showed that flavonoids and other phenolic compounds existing in the seeds of Annona squamosa affect cancer cells and have anticancer properties regarding skin cancer cell lines (Mehta & Paliwal, 2018). The compounds possess anticancer properties because they inhibit the cell cycle and induce cell death.

Phenolic compounds also exert their effects by the ability to intervene with signaling pathways connected to cancer development. Like any other study, Yumrutaş et al. also identified that CGA and caffeic acids trigger apoptosis in prostate cancer, and likely the same goes for skin cancer cells (Yumrutaş et al., 2018). As specific signaling pathways in cancer cells are deregulated, phenolic compounds can also improve the effectiveness of more traditional treatments and give an additive effect when administered with them.

Additionally, the role of phenolic compounds as chemopreventive agents has been underlined in several investigations. For instance, Elansary et al. observed that phenolic compounds of *Catalpa speciosa*, as well as other plants, perform antiproliferative characteristics against various cancer cell lines and hence could possess anticancer-preventing competence (Elansary et al., 2019). Of special preventive interest is the chemopreventive elemental action, which can significantly decrease the likelihood of the tumor's formation, especially in cervical and skin carcinomas. The phenolic compounds reveal several anticancer effects against skin cancer by providing antioxidant defense mechanisms, apoptotic effects, antiproliferative effects, and modulating cancer-associated molecular targets. The foregoing biological activities further point to the utility of these compounds as anti-skin cancer therapeutic agents.

### 7.6.7 Pancreatic Cancer

Pancreatic cancer is one of the most dangerous and rapidly progressive malignancies with a low five-year survival rate. These phenolic compounds have been reported for their in vitro antiproliferative effects on pancreatic cancer cells and their potential to induce apoptosis. Moreover, phenolic compounds can suppress angiogenesis, thereby affirming that tumor growth and metastasis will not occur. The three phenolic compounds that have gained attention in the research on pancreatic cancer include EGCG, a major component of green tea.

In a study, it was shown that the combined anticancer effect of EGCG and hyperthermia is synergistic against human pancreatic cancer PANC-1 cells. These treatments also increase the uptake of apoptotic cells by the mitochondria, which also shows that EGCG can significantly improve the effectiveness of hyperthermia therapy in the treatment of pancreatic cancer (Lu et al., 2019). Based on the observation of the ability of EGCG to induce apoptosis and inhibit cell proliferation, it can be routinely used to complement the treatment of pancreatic cancer.

CGA is yet another phenolic compound, which has the therapeutic possibility for pancreatic cancer treatment. Consequently, the study by Elansary et al. established that CGA exerted potent antiproliferative effects against numerous cancer cell lines, including those derived from the pancreas; the authors noted that this effect required further investigation. More specifically, the study showed that CGA is able to reposition cell cycle checkpoints, like other phenolic compounds, and induce apoptosis; these mechanisms explain the anticancer effect. This knowledge is very limiting and brings forth the need for CGA as a possible therapeutic remedy for pancreatic cancer.

Furthermore, the phenolic compounds from *Moringa oleifera* have been studied for antioxidant, anticancer potentials, and antimicrobial activities. Mumtaz et al. also found that there is an indication of massive cytotoxicity of the leaves' phenolic extracts of *Moringa* on pancreatic cancer cells. According to the study, the compounds under analysis prevented cell signaling that enhances the growth of cancer cells, thereby contributing to the suppression of tumor growth (Mumtaz et al., 2021). This finding highlights the impact of phenolics produced in the human body from consuming *Moringa* products in the treatment of pancreatic cancer. Another interesting effect involves the involvement of phenolic compounds in regulating the tumor microenvironment.

Nevertheless, the study by Kim et al. was confined to anti-malarial compounds, not phenolic compounds, that directly influence the tumor microenvironment of pancreatic carcinoma. Thus, more evidence is needed to substantiate the specificity of the respective claims with phenolic compounds and chemokine receptors (Kim et al., 2012). Besides, given these results, Ouyang et al. highlighted the importance of developing new approaches to treating pancreatic cancer, as evidenced by the low effectiveness of traditional treatments like gemcitabine. The inclusion of phenolic compounds in treatment regimens enhances the effectiveness of the treatments and reduces the barriers characteristic of pancreatic cancer (Ouyang et al., 2016,

p. 237). Moreover, there is evidence that phenolic compounds can lead to apoptosis via a variety of actions.

Some phenolic compounds have been shown to affect the levels of GADD45a and p53; thus, Jang et al. have shown that these molecules influence apoptosis of pancreatic cancer cells. This mechanism is especially important since p53 gene mutations are frequent in pancreatic cancer and cause resistance to most traditional treatments (Jang et al., 2021). Therefore, the phenolic compounds possess marked anticancer effects in pancreatic cancer by inducing apoptosis, affecting the cell cycle expression, and potentially changing the tumor physiochemistry. With all these described biological activities, the compounds can be implicated in therapeutic applications in the prevention and treatment of pancreatic cancer.

### 7.6.8 Leukemia

In research involving phenolic compounds, they have also scrutinized another type of cancer, which is leukemia, a group of cancers of the blood. They can also prevent the spreading of leukemia cells, disrupting the timing of the cycle, and reducing the biological growth of the malignant cells. Aside from their ability to attack leukemic cells, these compounds might also be able to enhance the effects of existing therapeutic agents, in which case the treatment could be in the form of combination therapy. One such compound is curcumin, which research has shown to be an effective anti-leukemia phytochemical. The authors have also pointed out that curcumin poses a potential risk to patients who might have further advances in the treatment of their illnesses, especially blood cancers, including leukemias. It is known that curcumin induces cell death and arrests the growth of multiple leukemia cell lines. Hence, it could be a plausible candidate for therapeutic use (Kelkel et al., 2010).

Furthermore, the research conducted by Fatima et al. revealed that apigenin, one of the phenolic compounds, promotes autophagy in leukemia cells, reinforcing the concept of cellular programmed death by phenolic compounds (Fatima et al., 2015). The anticancer effects of phenolic compounds are also attributed to their antioxidant properties. Casagrande et al. stated that the hydroethanolic extract from the leaves of *Jacaranda decurrens* exhibits antioxidant and cytotoxic properties against leukemia cells, with the possible aid of phenolic constituents in the extracts toward these activities (Casagrande et al., 2014) Similarly, Njuguna et al. found that the phenolic extracts of *Oxygonum sinuatum* exhibited high levels of antioxidant activity and antiproliferative activity against leukemia cells, suggesting their potential therapeutic use (Njuguna et al., 2018).

In addition, the investigation carried out by Gazwi and colleagues highlighted that compounds such as phenols, tannins, and glycosides could be purified through ethanol and water extraction processes, which may contribute to their effect in combating leukemia (Gazwi et al., 2023). Such solubility is a prerequisite for the formulation of successful therapies that rely on naturally occurring phenolics. In addition to curcumin and apigenin, other related phenolic compounds are helpful in the management of leukemia. For example, Hwang et al. reviewed the uses of several natural products, including phenolic compounds that promote apoptosis in cancerous leukemia and reduce the cells' ability to resist chemotherapy drugs (Hwang et al., 2019). This showed how phenolic compounds play a complex role in improving the efficacy of treatment modalities.

In addition to that, the antagonistic and additive interactions of phenolic compounds in the presence of other therapeutic agents are also investigated. A study by Trachtenberg et al. demonstrated that the combination of hydroxycinnamic acid derivatives and carnosic acid is selectively toxic to chronic lymphocytic leukemia cells (Trachtenberg et al., 2021). This provides insight into the possibility that phenolic compounds may serve as adjuvants to some anticancer agents explored and should be rationally used in combination therapies. Intrinsic anticancer activities of phenolic compounds in leukemia include, but are not limited to, inducing apoptosis, scavenging free radicals, and improving advances on existing treatment. Table 7.2 shows the anticancer mechanism of action of phenolic compounds.

**TABLE 7.2**
**Anticancer Mechanism of Action of Phenolic Compounds**

| Phenolic Compound | Cancer Type | Mechanism of Action |
|---|---|---|
| Curcumin | Colon cancer | Induces apoptosis, inhibits NF-κB and Wnt/β-catenin pathways, and reduces inflammation. |
| | Breast cancer | Inhibits cell proliferation, induces apoptosis, and downregulates ERα and HER2 expression. |
| | Pancreatic cancer | Suppresses tumor growth, induces apoptosis, and enhances chemosensitivity to gemcitabine. |
| Resveratrol | Colon cancer | Reduces cell proliferation, induces apoptosis, and inhibits angiogenesis via VEGF downregulation. |
| | Breast cancer | Inhibits estrogen receptor activity and induces apoptosis through mitochondrial pathways. |
| | Prostate cancer | Suppresses androgen receptor signaling and induces apoptosis; inhibits cell migration. |
| Quercetin | Lung cancer | Induces apoptosis, inhibits cell migration, and reduces inflammation by modulating NF-κB signaling. |
| | Bladder cancer | Suppresses tumor growth, induces apoptosis, and enhances the efficacy of chemotherapy. |
| | Prostate cancer | Inhibits the growth of prostate cancer cells through cell cycle arrest and induction of apoptosis. |
| EGCG | Breast cancer | It inhibits proliferation, induces apoptosis, and inhibits angiogenesis and metastasis. |
| | Lung cancer | Reduces cell growth, induces apoptosis, and modulates various signaling pathways (e.g., PI3K/Akt). |
| | Skin cancer | Induces apoptosis and inhibits tumor promotion through multiple signaling pathways. |
| Kaempferol | Bladder cancer | Induces apoptosis, inhibits cell proliferation, and suppresses metastasis. |
| | Skin cancer | Inhibits cell growth and induces apoptosis; modulates apoptosis-related proteins. |
| Anthocyanins | Colon cancer | Induces apoptosis and inhibits inflammation and cell proliferation. |
| | Lung cancer | Modulates antioxidant defines, induces apoptosis, and inhibits cell invasion. |
| Gallic acid | Prostate cancer | Induces apoptosis and inhibits proliferation by modulating signaling pathways. |
| | Leukemia | Induces apoptosis and cell cycle arrest; affects multiple signaling pathways involved in cell survival. |

## 7.7 CONCLUSION

Various phenolic compounds exhibit significant anticancer activity against a wide range of tumors, including leukemia, by inducing apoptosis, arresting the cell cycle, and regulating different signaling pathways. The presence of antioxidant properties in these compounds also contributes to cancer immunotherapy by effectively combating oxidative stress, which is a prevalent phenomenon during cancer. All these factors, along with numerous other biological activities, contribute to the recognition of these compounds as having potential in both the treatment and prevention of cancer. Further studies are needed to fully understand how phenolic compounds work at a molecular level against various types of cancer and how such treatments can be used in conjunction with conventional therapies. Moreover, studies on the bioavailability and metabolism of these compounds will be critical in ensuring their effective use in clinical treatment protocols.

## COMPETING INTERESTS

The authors declare that they have no competing interests.

## DECLARATION OF GENERATIVE AI AND AI-ASSISTED TECHNOLOGIES IN THE WRITING PROCESS

During the preparation of this work, the book chapter author(s) have not taken any assistance from AI, and the book chapter author(s) reviewed and edited the content as needed and take(s) full responsibility for the content of the publication.

## REFERENCES

Ademosun, A., Oboh, G., Passamonti, S., Tramer, F., Žiberna, L., Boligon, A., …, & Athayde, M. (2016). Phenolic composition of orange peels and modulation of redox status and matrix metalloproteinase activities in primary (caco-2) and metastatic (LoVo and LoVo/ADR) colon cancer cells. European Food Research and Technology, 242(11), 1949–1959. https://doi.org/10.1007/s00217-016-2694-0

Anantharaju, P., Prathima, C., Vimalambike, M., & Madhunapantula, S. (2016). An overview on the role of dietary phenolics for the treatment of cancers. Nutrition Journal, 15(1). https://doi.org/10.1186/s12937-016-0217-2

Bakrim, S., Omari, N., Bakri, Y., Lee, L., & Bouyahya, A. (2022). Dietary phenolic compounds as anticancer natural drugs: recent update on molecular mechanisms and clinical trials. Foods, 11(21), 3323. https://doi.org/10.3390/foods11213323

Baloghová, J. (2023). Spice-derived phenolic compounds: potential for skin cancer prevention and therapy. Molecules, 28(17), 6251. https://doi.org/10.3390/molecules28176251

Baptista, F. (2024). Exploring the role of phenolic compounds in chronic kidney disease: a systematic review. Molecules, 29(11), 2576. https://doi.org/10.3390/molecules29112576

Belhadi, F., Ouafi, S., & Bouguedoura, N. (2020). Phytochemical composition and pharmacological assessment of callus and parent plant of *Asteriscus graveolens* (Forssk.) Less. from Algerian Sahara. Tropical Journal of Pharmaceutical Research, 19(9), 1895–1901. https://doi.org/10.4314/tjpr.v19i9.14

Braakhuis, A., Campion, P., & Bishop, K. (2016). Reducing breast cancer recurrence: the role of dietary polyphenolics. Nutrients, 8(9), 547. https://doi.org/10.3390/nu8090547

Bray, F., Laversannem M., Sung, H., Ferlay, J., Siegel, RL., Soerjomataram, I., & Jemal, A. (2024). Global cancer statistics 2022: GLOBOCAN estimates of incidence and mortality worldwide for 36 cancers in 185 countries. CA: A Cancer Journal for Clinicians, 74(3), 229–263. https://doi.org/10.3322/caac.21834

Cañas, S., Rebollo-Hernanz, M., Braojos, C., Benítez, V., Ferreras-Charro, R., Dueñas, M., …, & Martín-Cabrejas, M. (2022). Understanding the gastrointestinal behavior of the coffee pulp phenolic compounds under simulated conditions. Antioxidants, 11(9), 1818. https://doi.org/10.3390/antiox11091818

Casagrande, J., Macorini, L., Antunes, K., Santos, U., Campos, J., Dias-Júnior, N., …, & Souza, K. (2014). Antioxidant and cytotoxic activity of hydroethanolic extract from jacaranda decurrens leaves. PLoS One, 9(11), e112748. https://doi.org/10.1371/journal.pone.0112748

Chang, H., Lin, L., Zhou, Y., Ye, F., & Zhao, G. (2018). Dietary flavonoids and the risk of colorectal cancer: an updated meta-analysis of epidemiological studies. Nutrients, 10(7), 950. https://doi.org/10.3390/nu10070950

Chang, W., Hung, J., Jian, S., Lin, Y., Wu, C., Hsu, Y., …, & Kuo, P. (2016). Laricitrin ameliorates lung cancer-mediated dendritic cell suppression by inhibiting signal transducer and activator of transcription 3. Oncotarget, 7(51), 85220–85234. https://doi.org/10.18632/oncotarget.13240

Chen, Y., Chang, H., Deng, J., Chen, J., Huang, S., Lin, I., …, & Huang, G. (2013). Hispolon from *Phellinus linteus* induces g0/g1 cell cycle arrest and apoptosis in nb4 human leukaemia cells. The American Journal of Chinese Medicine, 41(06), 1439–1457. https://doi.org/10.1142/s0192415x13500961

Choi, S., Kim, H., Kim, H., Lee, I., Kozukue, N., Levin, C., …, & Friedman, M. (2011). Free amino acid and phenolic contents and antioxidative and cancer cell-inhibiting activities of extracts of 11 greenhouse-grown tomato varieties and 13 tomato-based foods. Journal of Agricultural and Food Chemistry, 59(24), 12801–12814. https://doi.org/10.1021/jf202791j

Duan, J., Guo, H., & Fang, Y. (2021). The mechanisms of wine phenolic compounds for preclinical anticancer therapeutics. Food & Nutrition Research, 65. https://doi.org/10.29219/fnr.v65.6507

Dudonné, S., Dal-Pan, A., Dubé, P., Varin, T., & Desjardins, Y. (2016). Potentiation of the bioavailability of blueberry phenolic compounds by co-ingested grape phenolic compounds in mice, revealed by targeted metabolomic profiling in plasma and feces. Food & Function, 7(8), 3421–3430. https://doi.org/10.1039/c6fo00902f

Elansary, H., Szopa, A., Kubica, P., Al-Mana, F., Mahmoud, E., El-Abedin, T., …, & Ekiert, H. (2019). Phenolic compounds of *Catalpa speciosa*, *Taxus cuspidata*, and *Magnolia acuminata* have antioxidant and anticancer activity. Molecules, 24(3), 412. https://doi.org/10.3390/molecules24030412

Fadilah, F., Yanuar, A., Arsianti, A., Andrajati, R., & Paramita, R. (2017). In silico study, synthesis, and cytotoxic activity of esterification of eugenol and gallic acid against ht-29 cell line. Oriental Journal of Chemistry, 33(6), 3009–3014. https://doi.org/10.13005/ojc/330638

Falcão, S., Calhelha, R., Touzani, S., Lyoussi, B., Ferreira, I., & Vilas-Boas, M. (2019). In vitro interactions of Moroccan propolis phytochemical's on human tumor cell lines and anti-inflammatory properties. Biomolecules, 9(8), 315. https://doi.org/10.3390/biom9080315

Fantini, M., Benvenuto, M., Masuelli, L., Frajese, G., Tresoldi, I., Modesti, A., …, & Bei, R. (2015). In vitro and in vivo antitumoral effects of combinations of polyphenols, or polyphenols and anticancer drugs: perspectives on cancer treatment. International Journal of Molecular Sciences, 16(5), 9236–9282. https://doi.org/10.3390/ijms16059236

Fatima, H., Khan, K., Zia, M., Ur-Rehman, T., Mirza, B., & Haq, I. (2015). Extraction optimization of medicinally important metabolites from *Datura innoxia* Mill.: an in vitro biological and phytochemical investigation. BMC Complementary and Alternative Medicine, 15(1). https://doi.org/10.1186/s12906-015-0891-1

Gado, F., Digiacomo, M., Salsano, J., Macchia, M., & Manera, C. (2021). Phenolic compounds in prevention and treatment of skin cancers: a review. Current Medicinal Chemistry, 28(33), 6730–6752. https://doi.org/10.2174/0929867328666210324160324

Gazwi, H., Omar, M., & Mahmoud, M. (2023). Phytochemical analysis, antioxidant capacities, and in vitro biological activities of the extract of seed coat as by-products of pea. BMC Chemistry, 17(1). https://doi.org/10.1186/s13065-023-00911-8

Hazas, M., Piñol, C., Macià, A., & Motilva, M. (2017). Hydroxytyrosol and the colonic metabolites derived from virgin olive oil intake induce cell cycle arrest and apoptosis in colon cancer cells. Journal of Agricultural and Food Chemistry, 65(31), 6467–6476. https://doi.org/10.1021/acs.jafc.6b04933

Hsu, Y., Hung, J., Tsai, Y., Tsai, E., Huang, M., Hou, M., …, & Kuo, P. (2015). 6-shogaol, an active constituent of dietary ginger, impairs cancer development and lung metastasis by inhibiting the secretion of cc-chemokine ligand 2 (ccl2) in tumor-associated dendritic cells. Journal of Agricultural and Food Chemistry, 63(6), 1730–1738. https://doi.org/10.1021/jf504934m

Huang, G., Deng, J., Huang, S., & Hu, M. (2011). Hispolon induces apoptosis and cell cycle arrest of human hepatocellular carcinoma hep3b cells by modulating erk phosphorylation. Journal of Agricultural and Food Chemistry, 59(13), 7104–7113. https://doi.org/10.1021/jf201289e

Hwang, D., Kim, M., Park, H., Jeong, M., Jung, W., & Kim, B. (2019). Natural products and acute myeloid leukemia: a review highlighting mechanisms of action. Nutrients, 11(5), 1010. https://doi.org/10.3390/nu11051010

Jang, H., Yang, J., Hong, E., Jo, E., Lee, S., Lee, S., …, & Kang, H. (2021). Chelidonine induces apoptosis via gadd45a-p53 regulation in human pancreatic cancer cells. Integrative Cancer Therapies, 20. https://doi.org/10.1177/15347354211006191

Jin, M., Kim, I., Park, J., Dong, M., Na, C., & Yoo, H. (2015). Pharmacokinetic profile of eight phenolic compounds and their conjugated metabolites after oral administration of *Rhus verniciflua* extracts in rats. Journal of Agricultural and Food Chemistry, 63(22), 5410–5416. https://doi.org/10.1021/acs.jafc.5b01724

Kelkel, M., Jacob, C., Dicato, M., & Diederich, M. (2010). Potential of the dietary antioxidants resveratrol and curcumin in prevention and treatment of hematologic malignancies. Molecules, 15(10), 7035–7074. https://doi.org/10.3390/molecules15107035

Kim, J., Kim, N., Kim, S., Kim, Y., & Yeo, W. (2021). Immobilization of phenol-containing compounds via electrochemical activation of a urazole derivative. Bulletin of the Korean Chemical Society, 43(2), 236–240. https://doi.org/10.1002/bkcs.12458

Kim, J., Yip, M., Xiang-dong, S., Li, H., Hsin, L., LaBarge, S., …, & Vaidehi, N. (2012). Identification of antimalarial compounds as novel antagonists to chemokine receptor cxcr4 in pancreatic cancer cells. PLoS One, 7(2), e31004. https://doi.org/10.1371/journal.pone.0031004

Klongkumnuankarn, P., Busaranon, K., Chanvorachote, P., Sritularak, B., Jongbunprasert, V., & Likhitwitayawuid, K. (2015). Cytotoxic and antimigratory activities of phenolic compounds fromdendrobium brymerianum. Evidence-Based Complementary and Alternative Medicine, 2015, 1–9. https://doi.org/10.1155/2015/350410

Kwon, Y. (2018). Food-derived polyphenols inhibit the growth of ovarian cancer cells irrespective of their ability to induce antioxidant responses. Heliyon, 4(8), e00753. https://doi.org/10.1016/j.heliyon.2018.e00753

Lee, H., Chen, Y., Chang, W., Li, W., Ke, H., & Kuo, P. (2019). Effects of epigallocatechin gallate (EGCG) on urinary bladder urothelial carcinoma—next-generation sequencing and bioinformatics approaches. Medicina, 55(12), 768. https://doi.org/10.3390/medicina55120768

Li, X., Liu, H., Lv, C., Du, J., Lian, F., Zhang, S., …, & Zeng, Y. (2022). Gypenoside-induced apoptosis via the pi3k/akt/mtor signaling pathway in bladder cancer. Biomed Research International, 2022, 1–15. https://doi.org/10.1155/2022/9304552

Lima, V., Oliveira-Tintino, C., Santos, E., Pereira-de-Morais, L., Tintino, S., Freitas, T., …, & Coutinho, H. (2016). Antimicrobial and enhancement of the antibiotic activity by phenolic compounds: gallic acid, caffeic acid and pyrogallol. Microbial Pathogenesis, 99, 56–61. https://doi.org/10.1016/j.micpath.2016.08.004

Lu, C., Chen, W., Hsieh, C., Kuo, Y., & Chao, C. (2019). Thermal cycling-hyperthermia in combination with polyphenols, epigallocatechin gallate and chlorogenic acid, exerts synergistic anticancer effect against human pancreatic cancer panc-1 cells. PLoS One, 14(5), e0217676. https://doi.org/10.1371/journal.pone.0217676

Mehta, S., & Paliwal, S. (2018). Phytochemical analysis, liquid chromatography, and mass spectroscopy and in vitro anticancer activity of annona squamosa seeds linn. Asian Journal of Pharmaceutical and Clinical Research, 11(3), 101. https://doi.org/10.22159/ajpcr.2018.v11i3.22808

Mindiédiba, J., Tangbadioa, H., Aminata, P., Abdoulaye, S., Emmanuel, A., & Georges, A. (2022). Antitumoral effect of *Striga hermonthica* (Delile) Benth. Methanolic extract is mediated by alterations on procaspase-3 and cyclin b expression in prostate cancer cell lines. Journal of Pharmacognosy and Phytotherapy, 14(1), 1–7. https://doi.org/10.5897/jpp2022.0619

Muller, A., Sarker, S., Saleem, I., & Hutcheon, G. (2019). Delivery of natural phenolic compounds for the potential treatment of lung cancer. DARU Journal of Pharmaceutical Sciences, 27(1), 433–449. https://doi.org/10.1007/s40199-019-00267-2

Mumtaz, M., Kausar, F., Hassan, M., Javaid, S., & Arif, M. (2021). Anticancer activities of phenolic compounds from *Moringa oleifera* leaves: in vitro and in silico mechanistic study. Beni-Suef University Journal of Basic and Applied Sciences, 10(1). https://doi.org/10.1186/s43088-021-00101-2

Njuguna, D., Mbuthia, K., Mutuku, C., Jepkorir, M., Ndung'u, J., Mwangangi, R., …, & Mwitari, P. (2018). Phytochemical composition and in vitro anti-proliferative activity of *Oxygonum sinuatum* (Meisn.) dammer on selected cancerous cells. Journal of Complementary and Alternative Medical Research, 6(2), 1–9. https://doi.org/10.9734/jocamr/2018/44413

Oskoueian, E., Abdullah, N., Ahmad, S., Saad, W., Omar, A., & Ho, Y. (2011). Bioactive compounds and biological activities of *Jatropha curcas* L. kernel meal extract. International Journal of Molecular Sciences, 12(9), 5955–5970. https://doi.org/10.3390/ijms12095955

Ouyang, G., Liu, Z., Huang, S., Li, Q., Li, X., Miao, X., …, & Wen, Y. (2016). Gemcitabine plus cisplatin versus gemcitabine alone in the treatment of pancreatic cancer: a meta-analysis. World Journal of Surgical Oncology, 14(1). https://doi.org/10.1186/s12957-016-0813-9

Palafox-Carlos, H., Ayala-Zavala, J., & González-Aguilar, G. (2011). The role of dietary fiber in the bioaccessibility and bioavailability of fruit and vegetable antioxidants. Journal of Food Science, 76(1). https://doi.org/10.1111/j.1750-3841.2010.01957.x

Pandya, N., Prakash, T., Kotresha, D., Kamurthy, H., & Nadendla, R. (2013). Antitumor and antioxidant status of *Terminalia catappa* against ehrlich ascites carcinoma in swiss albino mice. Indian Journal of Pharmacology, 45(5), 464. https://doi.org/10.4103/0253-7613.117754

Philippou, Y., Hadjipavlou, M., Khan, S., & Rané, A. (2013). Complementary and alternative medicine (cam) in prostate and bladder cancer. BJU International, 112(8), 1073–1079. https://doi.org/10.1111/bju.12062

Puranik, S., Ghagane, S., Nerli, R., Jalalpure, S., & Hiremath, M. (2017). Evaluation of in vitro antioxidant and anticancer activity of *Simarouba glauca* leaf extracts on t-24 bladder cancer cell line. Pharmacognosy Journal, 9(6), 906–912. https://doi.org/10.5530/pj.2017.6.142

Quirós-Sauceda, A., Chen, C., Blumberg, J., Astiazarán-García, H., Wall-Medrano, A., & González-Aguilar, G. (2017). Processing 'ataulfo' mango into juice preserves the bioavailability and antioxidant capacity of its phenolic compounds. Nutrients, 9(10), 1082. https://doi.org/10.3390/nu9101082

Rein, M., Renouf, M., Cruz-Hernandez, C., Actis-Goretta, L., Thakkar, S., & Pinto, M. (2013). Bioavailability of bioactive food compounds: a challenging journey to bioefficacy. British Journal of Clinical Pharmacology, 75(3), 588–602. https://doi.org/10.1111/j.1365-2125.2012.04425.x

Ribas-Agustí, A., Martín-Belloso, O., Soliva-Fortuny, R., & Elez-Martínez, P. (2017). Food processing strategies to enhance phenolic compounds bioaccessibility and bioavailability in plant-based foods. Critical Reviews in Food Science and Nutrition, 58(15), 2531–2548. https://doi.org/10.1080/10408398.2017.1331200

Russo, G., Campisi, D., Mauro, M., Regis, F., Reale, G., Marranzano, M., …, & Morgia, G. (2017). Dietary consumption of phenolic acids and prostate cancer: a case-control study in sicily, southern italy. Molecules, 22(12), 2159. https://doi.org/10.3390/molecules22122159

Saenglee, S., Jogloy, S., Patanothai, A., & Senawong, T. (2016). Cytotoxic effects of peanut phenolic compounds possessing histone deacetylase inhibitory activity on human colon cancer cell lines. Turkish Journal of Biology, 40, 1258–1271. https://doi.org/10.3906/biy-1601-23

Sharifi-Rad, J., Seidel, V., Izabela, M., Monserrat-Mequida, M., Sureda, A., Ormazábal, V., …, & Cho, W. (2023). Phenolic compounds as nrf2 inhibitors: potential applications in cancer therapy. Cell Communication and Signaling, 21(1). https://doi.org/10.1186/s12964-023-01109-0

Spagnuolo, C., Moccia, S., Tedesco, I., Crescente, G., Volpe, M., Russo, M., …, & Russo, G. (2022). Phenolic extract from extra virgin olive oil induces different anti-proliferative pathways in human bladder cancer cell lines. Nutrients, 15(1), 182. https://doi.org/10.3390/nu15010182

Špoljarić, K., Šelo, G., Pešut, E., Martinović, J., Planinić, M., Tišma, M., …, & Bucić-Kojić, A. (2023). Antioxidant and antiproliferative potentials of phenolic-rich extracts from biotransformed grape pomace in colorectal cancer. BMC Complementary Medicine and Therapies, 23(1). https://doi.org/10.1186/s12906-023-03852-w

Szliszka, E., Zydowicz, G., Janoszka, B., Dobosz, C., Kowalczyk-Ziomek, G., & Król, W. (2011). Ethanolic extract of Brazilian green propolis sensitizes prostate cancer cells to trail-induced apoptosis. International Journal of Oncology, 38(4). https://doi.org/10.3892/ijo.2011.930

Szliszka, E., Zydowicz, G., Mizgała, E., & Król, W. (2012). Artepillin c (3,5-diprenyl-4-hydroxycinnamic acid) sensitizes lncap prostate cancer cells to trail-induced apoptosis. International Journal of Oncology, 41(3), 818–828. https://doi.org/10.3892/ijo.2012.1527

Trachtenberg, A., Sidoryk, K., Alreate, S., Muduli, S., Leś, A., Cybulski, M., …, & Danilenko, M. (2021). Structure-activity relationship of hydroxycinnamic acid derivatives for cooperating with carnosic acid and calcitriol in acute myeloid leukemia cells. Biomedicines, 9(11), 1517. DOI: 10.3390/biomedicines9111517

Tsai, J., Liu, P., Chen, Y., Chou, S., Cheng, Y., Hwang, J., …, & Chong, I. (2015). Curcumin inhibits non-small cell lung cancer cells metastasis through the adiponectin/NF-κB/MMPs signaling pathway. PLoS One, 10(12), e0144462. https://doi.org/10.1371/journal.pone.0144462

Vinson, J., Demkosky, C., Navarre, D., & Smyda, M. (2012). High-antioxidant potatoes: acute in vivo antioxidant source and hypotensive agent in humans after supplementation to hypertensive subjects. Journal of Agricultural and Food Chemistry, 60(27), 6749–6754. https://doi.org/10.1021/jf2045262

Xia, T., Yao, J., Zhang, J., Duan, W., Zhang, B., Xie, X., …, & Wang, M. (2018). Evaluation of nutritional compositions, bioactive compounds, and antioxidant activities of shanxi aged vinegars during the aging process. Journal of Food Science, 83(10), 2638–2644. https://doi.org/10.1111/1750-3841.14356

Xiang, Z., Wu, X., Liu, X., & Chen, G. (2020). Antioxidant phenolic compounds from Chinese white olive (*Canarium album* L.). Bangladesh Journal of Botany, 49(4), 1029–1035. https://doi.org/10.3329/bjb.v49i4.52535

Yumrutaş, Ö., Pehlivan, M., Güven, C., Bozgeyik, İ., Bozgeyik, E., Yumrutaş, P., …, & Üçkardeş, F. (2018). Investigation of cytotoxic effect of salvia pilifera extracts and synthetic chlorogenic and caffeic acids on du145 prostate cancer cells line. Kahramanmaraş Sütçü İmam Üniversitesi Tarım Ve Doğa Dergisi. https://doi.org/10.18016/ksudobil.302249

Zhou, Q., Pan, H., & Li, J. (2019). Molecular insights into potential contributions of natural polyphenols to lung cancer treatment. Cancers, 11(10), 1565. https://doi.org/10.3390/cancers11101565

Zhou, X., Huang, N., Chen, W., Tang, X., Mahdavi, B., Raoofi, A., …, & Atabati, H. (2020). Hplc phenolic profile and induction of apoptosis by *Linum usitatissimum* extract in lncap cells by caspase3 and bax pathways. AMB Express, 10(1). https://doi.org/10.1186/s13568-020-01138-9

# 8 Neuroprotective Effects of Phenolic Compounds in Neurodegenerative Disorders

*Walaa A. El-Kashak, Eman Y. Ahmed, Mai M. Elghonemy, Rehab F. Taher, Rania F. Ahmed, Ahmed Sabt, Rasha Z. Batran, Asmaa S. Abd Elkarim, Naglaa M. Ammar, Zeinab A. El-Gendy, Heba A. Hassan, Mahmoud I. Nassar, Ahmed H. Gaara, Tahia M. Kubacy, Abdulsalam M. Kassem, Sherif E. Emam, Mohamed F. Abdel-Hameed, Ahmed M. Nagy, Ahmed F. Essa, Yusuke Kasai, Hiroshi Imagawa, and Abdelsamed I. Elshamy**

## 8.1 INTRODUCTION

Alzheimer's disease (AD) is a progressive neurodegenerative disorder related to aging and leads to impaired cognitive function (Ibrahim et al. 2024). In 2012, the World Health Organization reported that more than 35.6 million people worldwide have dementia, and approximately 60%–70% of this population is diagnosed with AD. This disease was first described by Alois Alzheimer in 1906, based on the autopsy detection of accumulated amyloid plaques, vascular anomalies, and neurofibrillary tangles in a patient who suffered from severe cognitive impairments (Chen et al. 2021).

AD (named after Alois Alzheimer, the German neurologist) refers to a slowly progressive neurodegenerative disease that exhibits three sets of symptoms: (1) cognitive dysfunction, which involves memory loss and language difficulties, (2) non-cognitive symptoms, which include psychiatric symptoms, for example, delusions, hallucinations, and depression, (3) difficulties with carrying out the daily activities such as shopping, eating, and dressing (Burns et al. 1990; Crous-Bou et al. 2017; Feldman and Woodward 2005). Natural products have afforded a rich source of compounds that have found many applications in the fields of medicine, pharmacy, and biology (Husen 2022, 2023a,b, 2024a,b). A significant number of important commercialized drugs have been obtained from natural sources, by structural modification of naturally occurring compounds, or by the synthesis of new compounds following a natural compound as a model. The potential impact of naturally occurring and naturally derived plant compounds in the prevention of neurological disturbances and even amelioration of memory and cognitive function has been reported in many studies, where the majority of the FDA-approved medications for AD (tacrine, galantamine, and donepezil) have a heterocyclic ring system, a closer examination of their molecular structures shows how promising heterocyclic compounds can be as effective therapeutic choices. Furthermore, heterocyclic compounds are widely used in therapeutics and

* Corresponding author: elshamynrc@yahoo.com; ai.el-shamy@nrc.sci.eg


DOI: 10.1201/9781003527671-8

exhibit a diverse range of biological properties. Coumarin, chromone, and benzofuran derivatives are drawing much attention because of their natural origin and wide pharmacological applications (Patel et al. 2024). The multi-factorial nature of AD has been considered the main reason for the absence of a cure so far. Several research groups have developed multitarget candidate drugs aimed at interfering with multiple disease pathways, rather than targeting only one. Many multipotent compounds have been designed and studied by repositioning well-known AChEI classical drugs, such as tacrine or donepezil moieties, which have been hybridized with various functional groups (Leon et al. 2013; Nepovimova et al. 2014; Santos et al. 2016).

Numerous studies have been carried out to improve the potency and selectivity of some naturally occurring compounds by developing synthetic or semisynthetic derivatives as acetylcholinesterase inhibitors from these compounds. Tacrine, galantamine, metrifonate, rivastigmine, or donepezil are drugs that inhibit AChE, enhance cholinergic transmission, and increase the level of Ach. These drugs have been used to decrease the AD symptoms that are caused by cholinergic neuron degeneration and injured transmission. AChE inhibitors have many side effects, such as nausea, diarrhea, vomiting, abdominal pain, skin rash, and dyspepsia (Bittner et al. 2023; McGleenon et al. 1999). Physostigmine, which is isolated from *Physostigma venenosum*, had many disadvantages, so it was taken out of the drug market. Tacrine, marketed under the trade name Cognex in clinical trials, was found to induce hepatotoxicity (Kim et al. 2024; Watkins et al. 1994).

This chapter compiles a broad range of naturally occurring phenolic compounds and herbal remedies that are beneficial for neurodegenerative diseases, along with their potential mechanisms of action.

## 8.2 NATURAL COMPOUNDS SUCH AS ACETYLCHOLINESTERASE INHIBITORS

### 8.2.1 Shikimate-Derived Compounds

#### 8.2.1.1 Flavonoids and Their Derivatives

Acacetin-7-O-rutinosid (linarin), isolated from *Buddleja davidii* (Buddlejaceae), was tested for its AChE activity, together with other different phenolic compounds, to study their Structure-activity relationship (SAR); it was found that linear is the only active compound toward AChE. SAR showed that 7-O-sugar, 4′-O-Me, and the length of the glycosidic linkage were important structural features for the inhibition of AChE (Fan et al. 2008). Structures similar to linarin, such as tilianin (Figure 8.1), were synthesized by methylation of apigenin-7-O-glucoside.

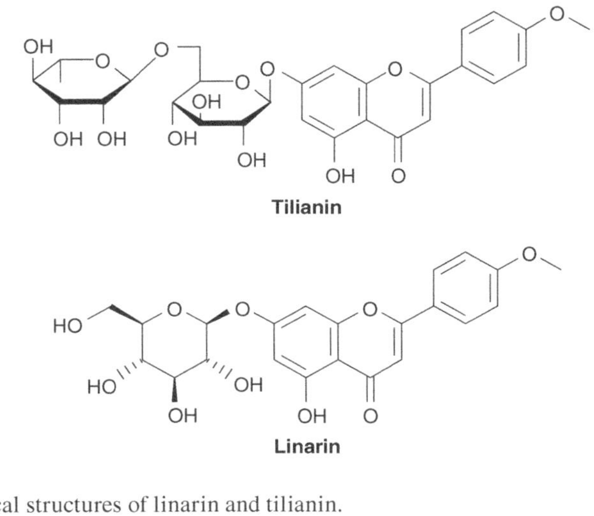


**FIGURE 8.1** Chemical structures of linarin and tilianin.

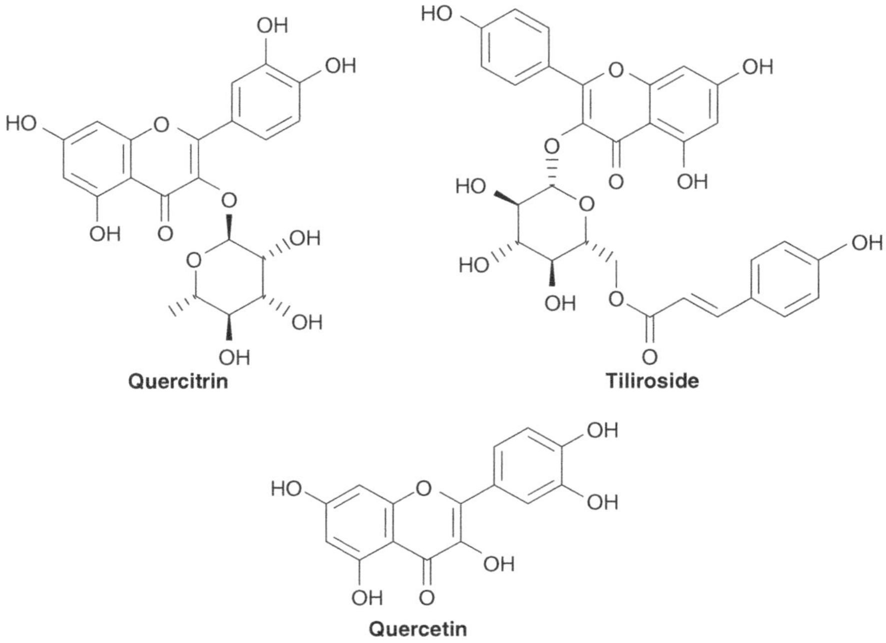


**FIGURE 8.2** Chemical structures of flavon-3-ols isolated from *Agrimonia Pilosa.*

*Agrimonia pilosa* (Rosaceae) is a source of flavon-3-ols that inhibit AChE, such as quercitrin, tiliroside, quercetin, and 3-methoxyquercetin (Figure 8.2). While tiliroside and quercetin were almost twice as active as alkaloid dehydroevodiamine, these compounds remained less active than berberine and tacrine (Jung and Park 2007; Khan et al. 2009).

The 7-neohesperidoside naringin reduces oxidative stress and cognitive decline while also lowering brain AChE activity in vivo (Chtourou et al. 2015; Kumar et al. 2010; Umukoro et al. 2018). Another example of flavonoids is icariin, a prenylated flavone isolated from *Epimedium* plants (Berberidaceae) that increases monoamine levels. Hence, it improves cognitive deterioration in mice, inhibits oxidative impairment, and also decreases the activity of AChE (He et al. 2010). In experimental mice, scopolamine-induced amnesia was counteracted by Glabridin, an isoflavone derived from the roots of the *Glycyrrhiza* species (Leguminosae) (Figure 8.3) (Cui et al. 2008). This action is associated with a reduction in brain AChE activity.

#### 8.2.1.2 Xanthones

Xanthones also represent inhibition activity toward AChE. Xanthones isolated from the leaves of *Gentiana campestris* (Gentianaceae) were reported to inhibit AChE. The inhibitory effect of bellidifolin, an AChE inhibitor as potent as galantamine, was linked to glucopyranosyl's absence and a methoxy group in the C-3 position (Urbain et al. 2004). Macluraxanthone isolated from Guttiferae plants showed a significant noncompetitive AChE inhibitory effect (Figure 8.4) (Khan et al. 2009; NAGORI et al. 2024; Vanessa and Mah 2021).

#### 8.2.1.3 Coumarins

Several coumarin derivatives inhibit AChE, such as scopoletin, marmesin, columbianetin, feronielloside, 4-methylumbelliferone, and bergapten (Figure 8.5) (Cavalli et al. 2008; Houghton et al. 2006; Howes and Houghton 2009). The inhibitory effect of simple coumarins as scopoletin and furanocoumarins as bergapten was found to show moderate to low activity, large substituents like the benzyloxy group in the coumarin's C-7 or at the furanocoumarin's C-5 and C-8 increased inhibitory

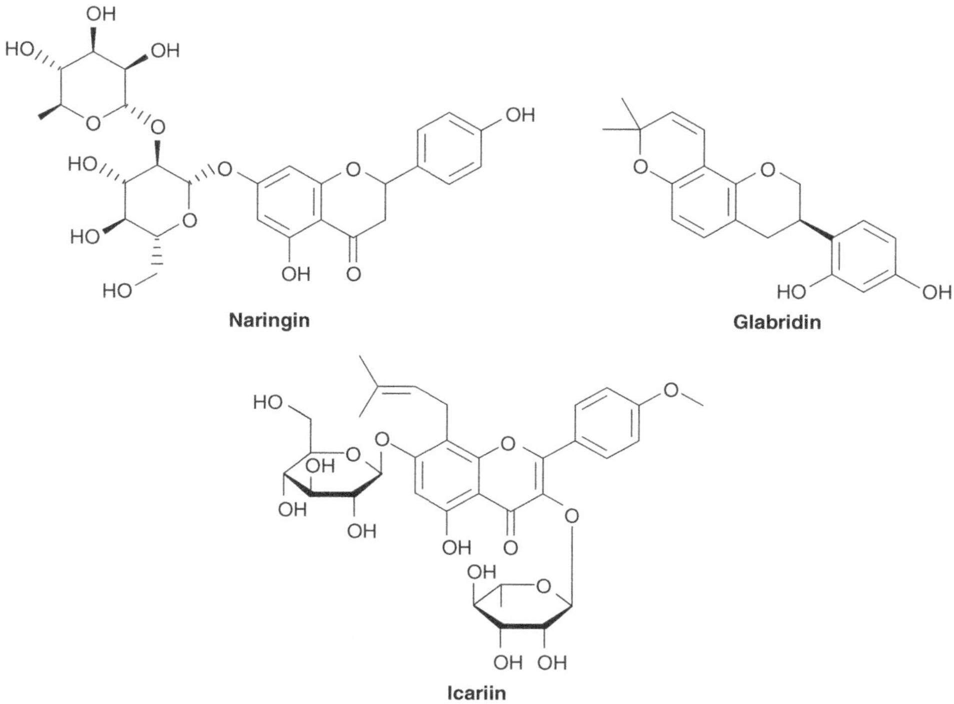


FIGURE 8.3 Chemical structures of naringin, icariin, and glabridin.

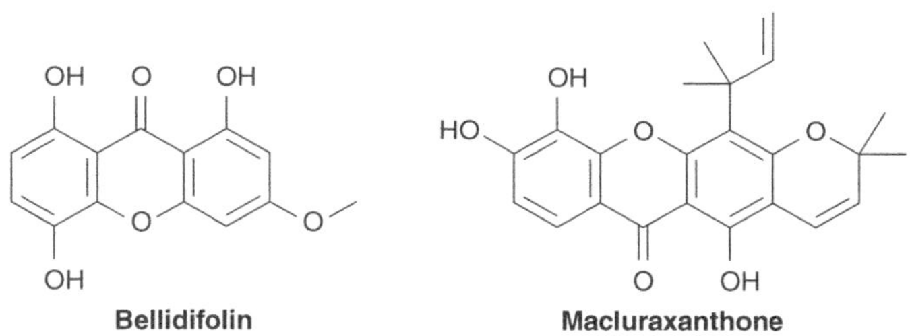


FIGURE 8.4 Structures of xanthones isolated from the leaves of *Gentiana campestris*.

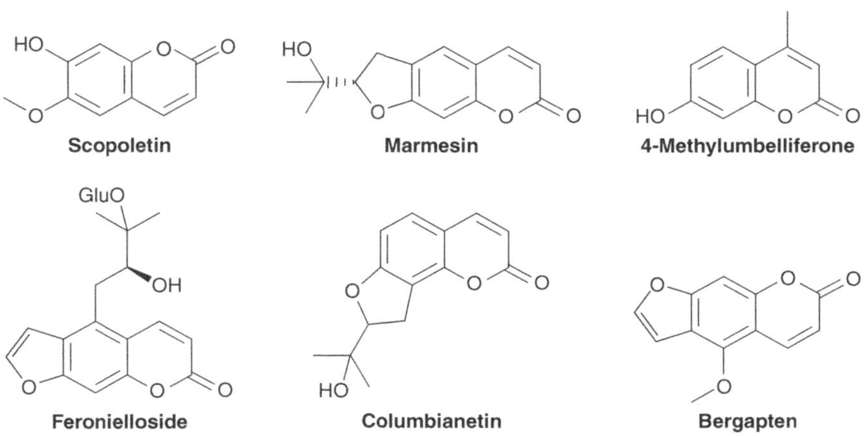


FIGURE 8.5 Structures of some naturally isolated coumarin derivatives inhibited AChE.

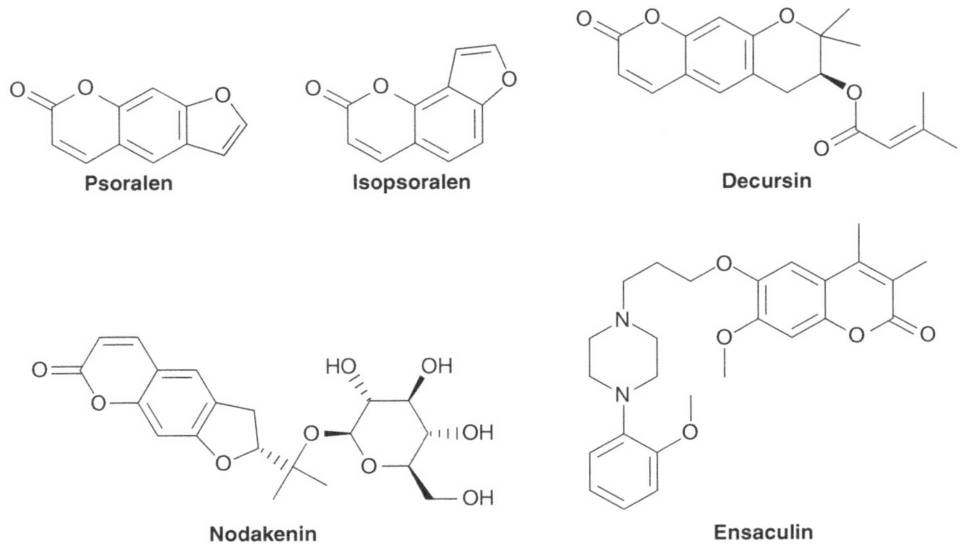


**FIGURE 8.6** Structures of coumarins isolated from *Psoralea species*.

potency in comparison to smaller substituents with methoxy and hydroxyl groups (Fallarero et al. 2008).

The effects of a small number of coumarins have been studied *in vivo*, as the two furanocoumarins, psoralen and isopsoralen, isolated from *Psoralea* species (family Leguminosae), decursin, and nodakenin, counteracted rodents' cognitive decline brought on by scopolamine, which is linked to AChE inhibition (Figure 8.6) (Howes and Houghton 2009). Ensaculin also inhibits AChE but modulates adrenergic, dopaminergic, and serotonergic functions (NMDA) *N*-methyl-D-aspartate receptor receptor antagonist. Thus, HCl salt (KA-672) is further subjected to clinical trials for AD (Teismann and Ferger 2000).

## 8.3 NATURAL COMPOUNDS AS SECRETASE INHIBITORS

### 8.3.1 Phenolic Compounds as Secretase Inhibitors

Phenolics are the largest class of natural products acting as BACE1 inhibitors (Choi et al. 2008a; Shimmyo et al. 2008a). The flavone luteolin (Figure 8.7) is considered one of the most potent naturally derived BACE1 inhibitors with an $IC_{50}$ value of 0.5 μM. On the other hand, the related flavonols myricetin, quercetin, kaempferol, and morin ($IC_{50}$ = 1.4–40 μM) showed moderate inhibitory potency toward BACE1 in an enzymatic assay. At the same time, the flavone apigenin lacked any inhibition effect, indicating that the C-3 hydroxy group is important to the inhibitory activity. This hypothesis was supported by docking studies, which demonstrated that the C-3 hydroxyl group forms a hydrogen bond with BACE1's Asp32, thereby maintaining the enzyme–inhibitor complex.

Lavandulyl flavanones, sophoraflavanone G, and leachianone A, extracted from *Sophora flavescens* (Fabaceae), are noncompetitive BACE1 inhibitors that reduce the extracellular cleavage product APPβ without affecting the intracellular full-length APP, which suggests the selective BACE1 inhibitory properties (Figure 8.8) (Hwang et al. 2008).

Furthermore, the isoflavone, neocorylin, extracted from *Psoralea corylifolia* seeds (Fabaceae), effectually suppressed BACE1 cleavage of APP with $IC_{50}$ 0.7 μM in the assay (Choi et al. 2008b; Narayanan et al. 2024). Active stilbenoids and phenylpropanoids, including resveratrol and cis-scripusin (Figure 8.9), isolated from *Smilax china* rhizomes were described as potent noncompetitive inhibitors with $IC_{50}$ values of 7.5 and 10 μM, respectively (Jeon et al. 2007).

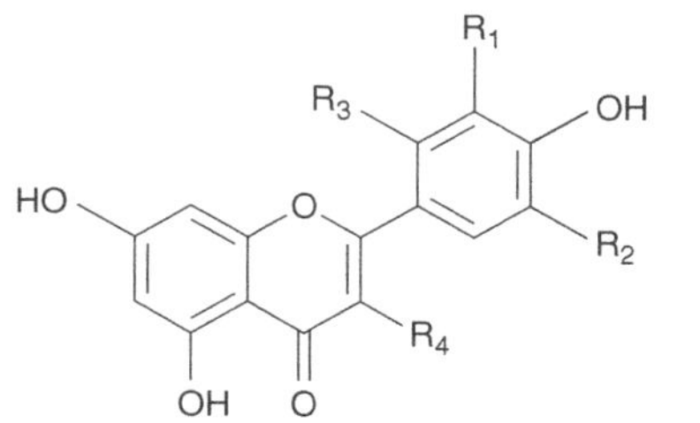

Myricetin; $R_1$ = OH, $R_2$ = OH, $R_3$ = H, $R_4$ = OH

Quercetin; $R_1$ = OH, $R_2$ = H, $R_3$ = H, $R_4$ = OH

Kaempferol; $R_1$ = H, $R_2$ = H, $R_3$ = H, $R_4$ = OH

Morin; $R_1$ = H, $R_2$ = H, $R_3$ = OH, $R_4$ = OH

Luteolin; $R_1$ = H, $R_2$ = OH, $R_3$ = H, $R_4$ = H

Apigenin; $R_1$ = H, $R_2$ = H, $R_3$ = H, $R_4$ = H

**FIGURE 8.7** Structures of some naturally active flavonoids as secretase inhibitors.

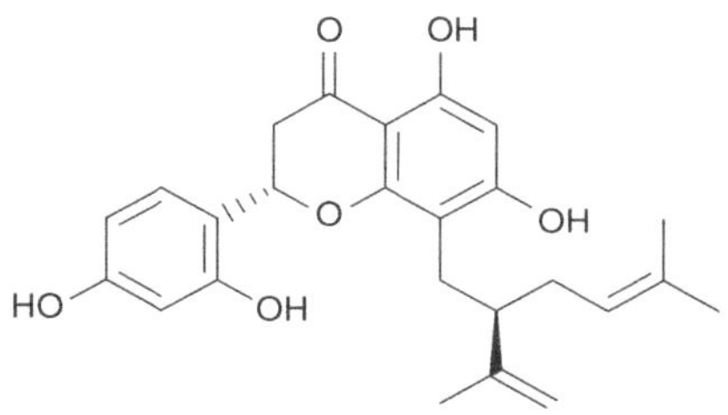

**Sophoraflavanone G**

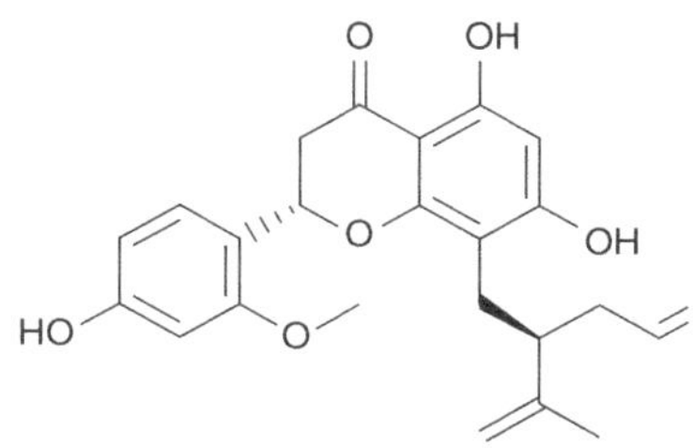

**Leachianone A**

**FIGURE 8.8** Structures of Lavandulyl flavanones.

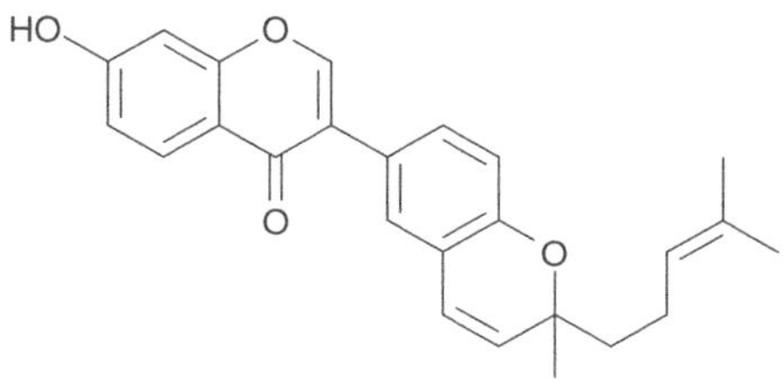

**Neocorylin**

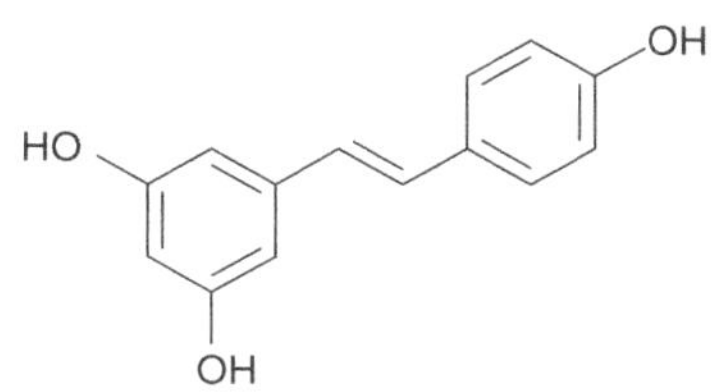

**Resveratrol**

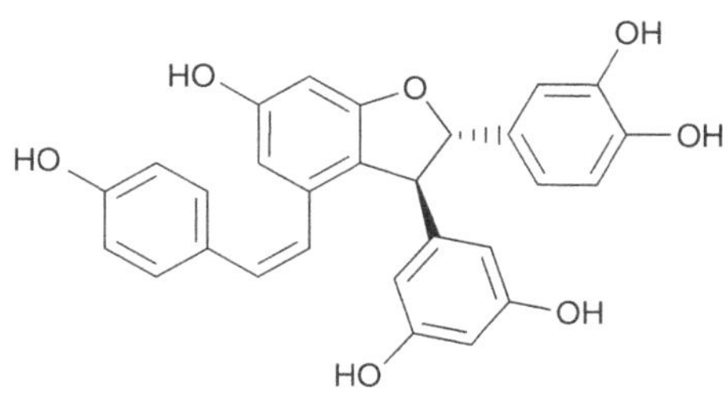

***cis*-Scripusin**

**FIGURE 8.9** Structures of neocorylin, resveratrol, and cis-scripusin isolated from *Psoralea corylifola and Smilax china.*

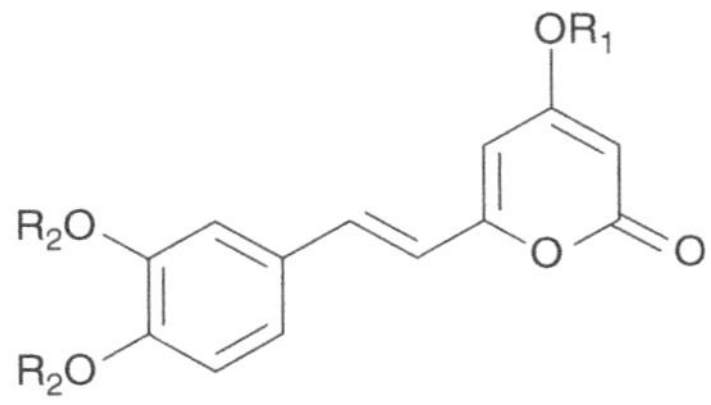


**Hispidin; R$_1$ = R$_2$ = H**

4-methoxy-6-(3′, 4′methylene-dioxystyryl)-2-pyrone; R$_1$ = Me, R$_2$ = Dioxolane

4-methoxy-6-(3′, 4′methylene-dihycroxystyryl)-2-pyrone; R$_1$ = Me, R$_2$ = H

**FIGURE 8.10** Structures of hispidin isolated from *Phellinus linteus* and its synthesized polyketide.

### 8.3.2 Polyketides as Secretase Inhibitors

Due to its potent secretase inhibition effect, the polyketide hispidin, isolated from *Phellinus linteus*, a fungal mycelium (Hymenochaetales), was identified as a lead compound for synthesizing other polyketide-derived candidates as secretase inhibitors; 4-methoxy-6-(3′,4′-methylene-dioxystyryl)-2-pyrone and 4-methoxy-6-(3′,4-methylene-dihycroxystyryl)-2-pyrone (Figure 8.10). The reduced potency of the synthesized analogs suggests that the catechol moiety is not critical for the inhibition properties (Park et al. 2004).

## 8.4 NATURAL COMPOUNDS AS ANTI-AGGREGATION AND CLEARANCE AGENTS

Most compounds acting on aggregation and clearance of β-amyloid inhibit fibrillogenesis and oligomerization. Generally, these compounds can promote the disruption of the weak bonds between the β-sheet fibril residues via interference with hydrogen-bonding or electrostatic interactions. These compounds are usually low molecular weight and have either lipophilic properties or hydrophilic characteristics due to polar substituents. Moreover, several compounds are non-toxic and possess multifunctional features, such as enzyme inhibitors, antioxidants, and anti-inflammatory properties, which could help overcome the diverse pathologies of AD (Gervais et al. 2007).

### 8.4.1 Polyketides Promoting Anti-Aggregation and Clearance

Rifampicin is a semisynthetic polyketide isolated from *Amycolatopsis rifamycinica* (Pseudonocardiaceae) and is mostly used to treat leprosy and tuberculosis. By attaching itself to the aggregates and preventing fibril adherence to the cell surface, it suppresses Aβ aggregation and averts amyloid fibril toxicity. However, due to its poor blood–brain barrier (BBB) permeability, rifampicin lacks an important prerequisite for an applicable neurological drug lead. Currently, a combination of rifampicin and the semisynthetic tetracycline antibiotic, doxycycline, has been advanced in Phase III clinical trials to assess its effect on AD biomarkers in the cerebral spinal fluid (Meng et al. 2008). Generally, it was found that tetracyclines interact with amyloid fibrils *via* H-bond formation through their polar substituent, in addition to hydrophobic interactions through aromatic moieties and thus prevent the formation of amyloid fibril and enhance the fibrils disassembly. Minocyclinethe, semisynthetic tetracycline compound, exerted significant neuroprotective effects both *in vitro* and *in vivo*. It inhibited Aβ fibril and microglial activation (Figure 8.11). In addition, it downregulates caspase-3 and reduces the production of hyperphosphorylated tau (Forloni et al. 2001; Liu et al. 2024).

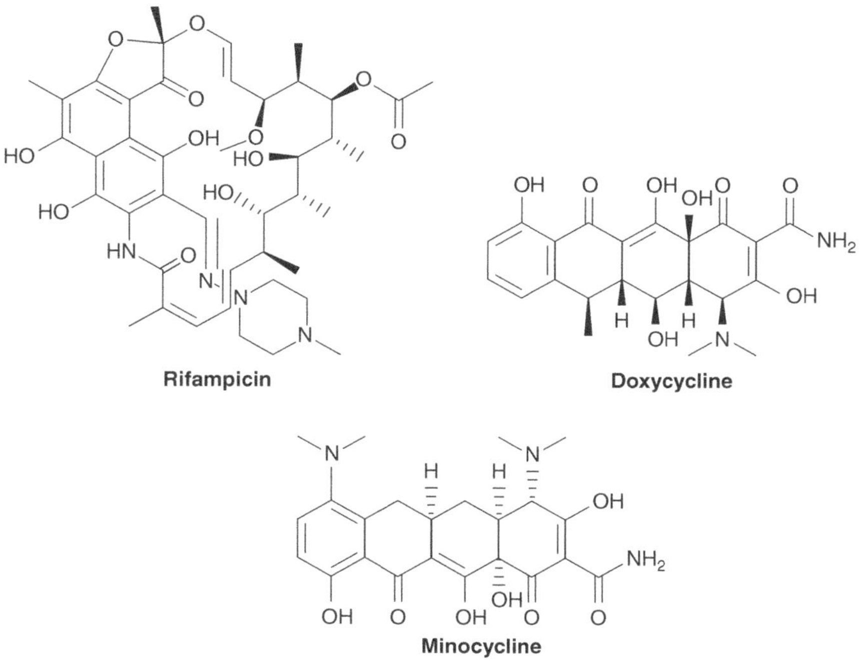


**FIGURE 8.11** Structures of rifampicin, doxycycline, and minocycline.

### 8.4.2 Shikimate and Sugar-Derived Compounds Promoting Anti-Aggregation and Clearance

Curcumin, is a well-known natural polyphenol compound in traditional medicine isolated from *Curcuma longa* belonging to the family *Zingiberaceae*. It exhibited a beneficial neuroprotective effect by blocking microglial activation. It also inhibits Aβ oligomers, binds the existing amyloid plaques, and reduces amyloid levels *in vivo*. Moreover, curcumin has efficient BBB permeability and exerts many other beneficial effects due to its anti-inflammatory and antioxidant properties. Presumably, these mechanisms are associated with the observed neuroprotective activity (Figure 8.12) (Gao et al. 2024; Yang et al. 2005). SAR studies revealed that the prerequisites for optimal activity involve two phenyl rings with polar substituents for H-bond formation. It is interesting to note that Longvida, a curcumin formulation, was assessed in Phase II clinical trials for AD (NCT01001637) due to several appealing features, including effective BBB permeability and low toxicity. Furthermore, several flavonoidal compounds were reported for their anti-aggregation properties. Myricetin showed potent activity attributed to the 3′,4′-dihydroxyl group on the B ring, which enhanced its selective, reversible binding to the neurotoxic oligomers. Myricetin also inhibits BACE-1, interferes with Aβ conformational changes, and exhibits antioxidant activity (Figure 8.12) (Shimmyo et al. 2008b).

Nordihydroguaiaretic acid isolated from *Larrea divaricata* (Zygophyllaceae) inhibited the formation of fibrogenic Aβ (fAβ) and enhanced its degradation. It showed higher potency than rifamycin. It also inhibited the aggregation of reactive oxygen species (ROS) and $Ca^{2+}$, which is related to its antioxidant effect (Goodman et al. 1994; Ono et al. 2002). Resveratrol, a polyphenolic compound commonly found in grapes, induces the degradation of Aβ aggregates and enhances the intracellular Aβ clearance. Recently, it has been reported that resveratrol disrupts $Aβ_{42}$ hydrogen bonding, blocking fibril formation, and destabilizing the preformed $fAβ_{42}$ *in vitro* (Figure 8.9) (Feng et al. 2009). Currently, resveratrol, in combination with glucose and malate, is being evaluated in

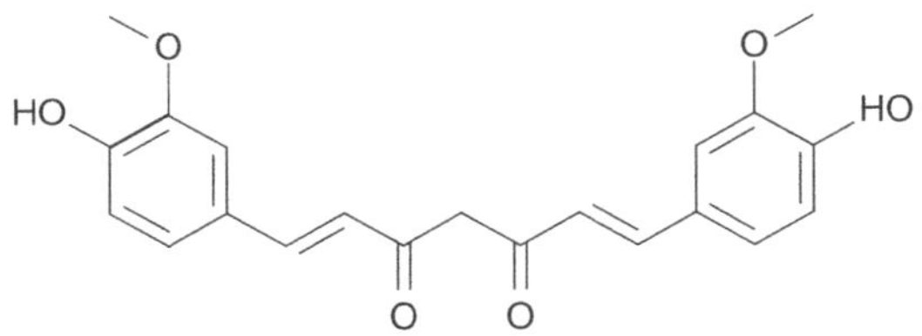

**Curcumin**

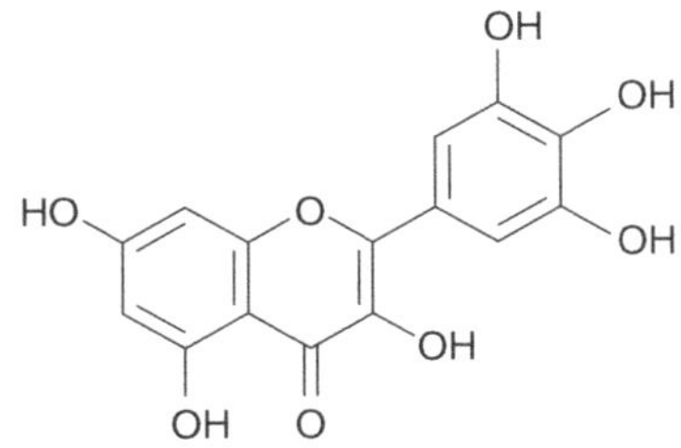

**Myricetin**

**FIGURE 8.12** Structures of curcumin, and myricetin.

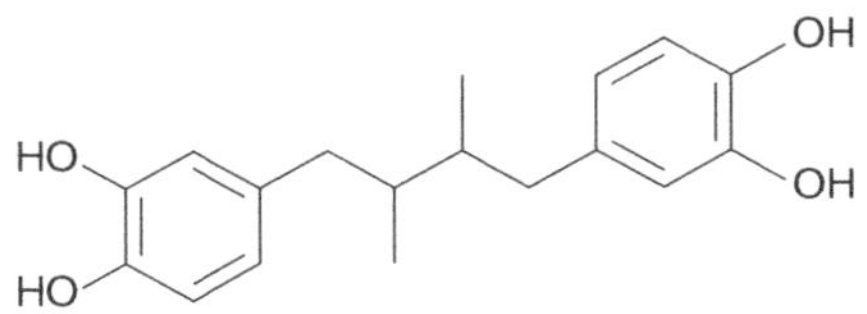

**Nordihydroguaiaretic acid**

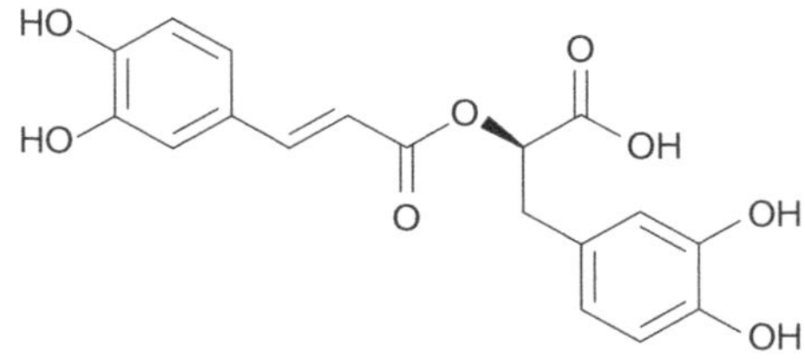

**Rosmarinic acid**

**FIGURE 8.13** Structures of nordihydroguaiaretic acid and rosmarinic acid.

Phase III clinical trials; # NCT00678431 (Zhu et al. 2018). It was also found that rosmarinic acid isolated from *Rosmarinus officinalis* (Rosemary) affects AD-associated pathways like DNA fragmentation, caspase-3 activation, lipid peroxidation, ROS formation, and tau protein hyperphosphorylation (Figure 8.13) (Yang et al. 2005).

### 8.4.3 Natural Compounds as Anti-Tau Phosphorylation

Scientists have sought dietary sources, including plant extracts and preparations, to control and relieve neurodegenerative disorders (Gu et al. 2010; Howes and Houghton 2009). Several natural anti-tau plant products are polyphenols, such as the linear diarylheptanoid curcumin from (*C. longa*) extract (Abdul-Rahman et al. 2024; Shytle et al. 2009). This extract increased the production of the anti-inflammatory cytokine IL-4 and was observed to act as an antioxidant, reducing tau and Aβ levels in experimental investigations (Shytle et al. 2012).

The inhibition of human tau aggregation was induced by cinnamon (*Cinnamonium zeylanicum*) extract due to the presence of cinnamaldehyde and procyanidin oligomers of the catechins/epicatechin classes (Peterson et al. 2009). Grape seed (*Vitis vinifera*) extracts were shown to possess similar procyanidins that inhibited tau neurotoxic aggregates (Pasinetti et al. 2010).

Paclitaxel, derived from the Pacific Yew, *Taxus brevifolia*, a well-known microtubule-stabilizing drug, has exhibited promising results in neurodegenerative tauopathy by mitigating tau pathology consequences in a mouse experimental model (Foa et al. 1994; Ma et al. 2024; Singh et al. 2024). Furthermore, oleocanthal from olive oil (*Olea europaea*) (Li et al. 2009), anthraquinone emodin from the root and rhizome of rhubarb (Pickhardt et al. 2005), and estrone, 17-β-estradiol, and β-sitosterol from young coconut juice (*Cocos nucifera*) (Radenahmad et al. 2011) were found to decrease the levels of tau significantly. Furthermore, ginsenoside Rg1, a steroid glycoside derived from *Panax ginseng*, decreased in vitro and in vivo tau hyperphosphorylation and neurotoxicity (Li et al. 2011; Malik et al. 2023; Manju and Bharadvaja 2024). Likewise, anti-inflammatory, anti-amyloidogenic, and anti-tangle effects were exhibited by aged garlic (*Allium sativum*) extract, which has S-allyl-cysteine and diallyl-disulfide (sulfur-containing constituents) (Figure 8.14) (Chauhan 2006).

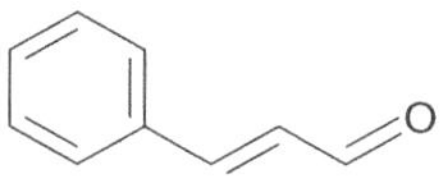

**Cinnamaldehyde**

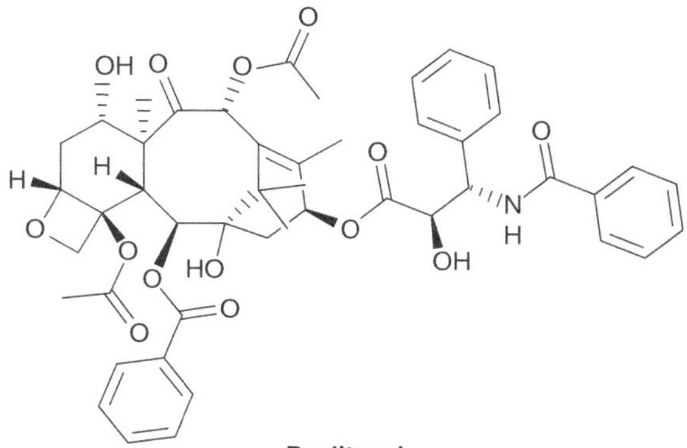

**Paclitaxel**

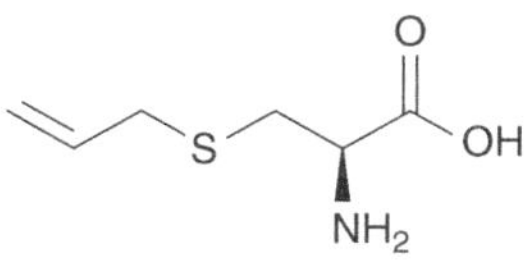

**S-Allyl-Cysteine**

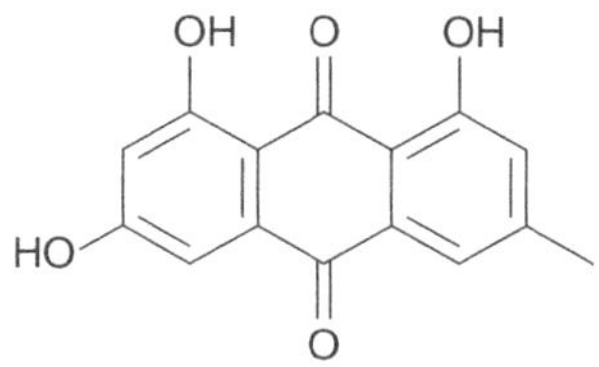

**Emodin**

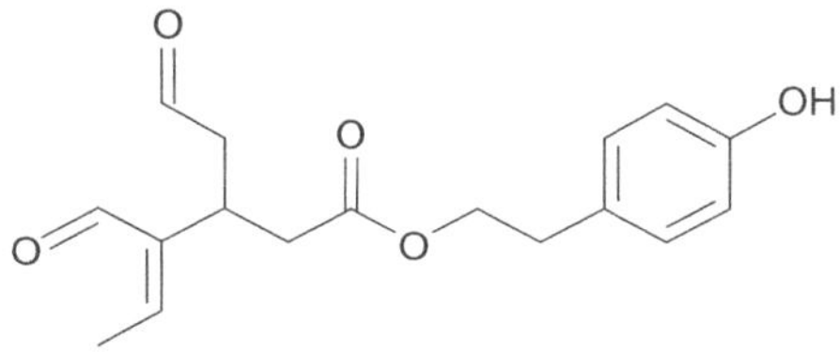

**Oleocanthal**

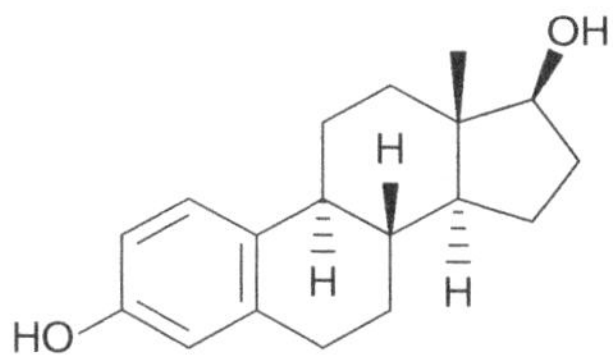

**17-β-estradiol**

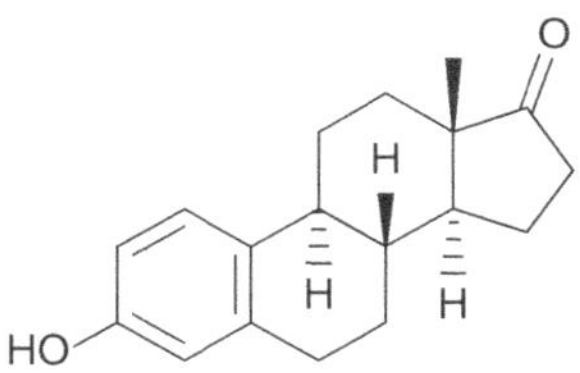

**Estrone**

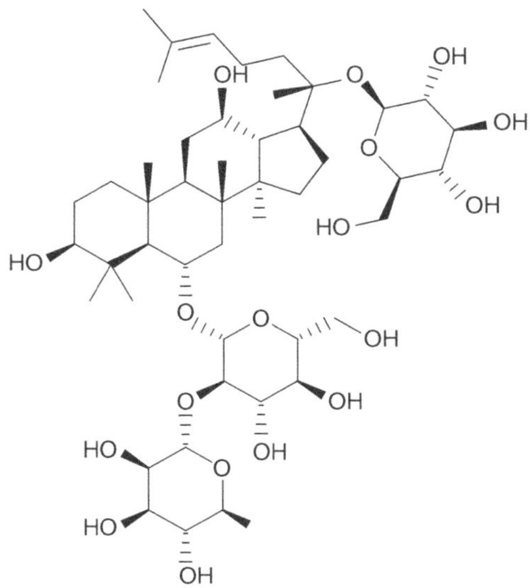

**Ginsenoside**

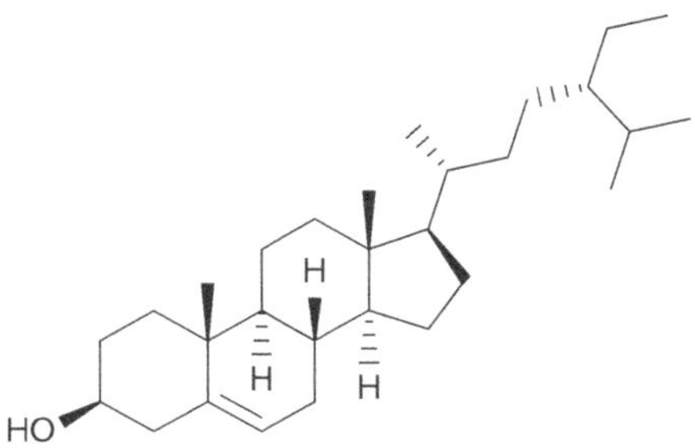

**β-Sitosterol**

**FIGURE 8.14** Structures of naturally phenolic compounds as anti-tau phosophyralation.

## 8.5 CONCLUSION AND FUTURE REMARKS

Polyphenols have a wide range of beneficial biological roles for human health and are harmless in nature. They can stop the activity of different biomarkers of inflammation, oxidative stress, autophagy, and neurotoxicity. In reality, they employ a range of methods linked to various signaling pathways, resulting in notable immunomodulatory and neuroprotective effects. Phenolic compounds have been shown to have a variety of neuroprotective effects on the brain, including promoting memory and cognitive function, inhibiting neuroinflammation, and shielding neurons from damage caused by neurotoxins. Furthermore, polyphenols and whole meals high in polyphenols can promote neurogenesis in humans and animals.

Clinical trials should be conducted sooner rather than later to assess the applicability of the proposed processes involving phenolic substances in human brain health, as well as the reported preclinical outcomes. Certain plant-derived polyphenols, such as curcumin, quercetin, caffeic acid, rosmarinic acid, and resveratrol from grapes and wine, are strong neuroprotectants that target distinct signaling pathways. It is advised that clinical trials be conducted to thoroughly assess the safety, efficacy, and potential negative effects of these substances.

## COMPETING INTERESTS

The authors declare that they have no competing interest.

## DECLARATION OF GENERATIVE AI AND AI-ASSISTED TECHNOLOGIES IN THE WRITING PROCESS

During the preparation of this work the book chapter author(s) have not taken any assistance from AI, and the book chapter author(s) reviewed and edited the content as needed and take(s) full responsibility for the content of the publication.

## REFERENCES

Abdul-Rahman T, Awuah WA, Mikhailova T, Kalmanovich J, Mehta A, Ng JC, Coghlan MA, Zivcevska M, Tedeschi AJ, de Oliveira EC, Kumar A, Cantu-Herrera E, Lyndin M, Sikora K, Alexiou A, Bilgrami AL, Al-Ghamdi KM, Perveen A, Papadakis M, Ashraf GM. (2024) Antioxidant, anti-inflammatory and epigenetic potential of curcumin in Alzheimer's disease. BioFactors (Oxford, England) 50 (4):693–708. doi:10.1002/biof.2039

Bittner N, Funk CS, Schmidt A, Bermpohl F, Brandl EJ, Algharably EE, Kreutz R, Riemer TG (2023) Psychiatric adverse events of acetylcholinesterase inhibitors in Alzheimer's disease and Parkinson's dementia: Systematic review and meta-analysis. Drugs & Aging 40 (11):953–964.

Burns A, Jacoby R, Levy R (1990) Psychiatric phenomena in Alzheimer's disease. I: Disorders of thought content. The British Journal of Psychiatry 157 (1):72–76.

Cavalli A, Bolognesi ML, Minarini A, Rosini M, Tumiatti V, Recanatini M, Melchiorre C (2008) Multi-target-directed ligands to combat neurodegenerative diseases. Journal of Medicinal Chemistry 51 (3):347–372.

Chauhan NB (2006) Effect of aged garlic extract on APP processing and tau phosphorylation in Alzheimer's transgenic model Tg2576. Journal of Ethnopharmacology 108 (3):385–394. doi:10.1016/j.jep.2006.05.030

Chen X, Drew J, Berney W, Lei W (2021) Neuroprotective natural products for Alzheimer's disease. Cells 10 (6):1309.

Choi S-H, Hur J-M, Yang E-J, Jun M, Park H-J, Lee K-B, Moon E, Song K-S (2008a) β-Secretase (BACE1) inhibitors from *Perilla frutescens* var. acuta. Archives of Pharmacal Research 31:183–187.

Choi YH, Yon GH, Hong KS, Yoo DS, Choi CW, Park WK, Kong JY, Kim YS, Ryu SY (2008b) In vitro BACE-1 inhibitory phenolic components from the seeds of *Psoralea corylifolia*. Planta Medica 74 (11):1405–1408. doi:10.1055/s-2008-1081301

Chtourou Y, Gargouri B, Kebieche M, Fetoui H (2015) Naringin abrogates cisplatin-induced cognitive deficits and cholinergic dysfunction through the down-regulation of AChE expression and iNOS signaling pathways in hippocampus of aged rats. Journal of Molecular Neuroscience 56:349–362.

Crous-Bou M, Minguillón C, Gramunt N, Molinuevo JL (2017) Alzheimer's disease prevention: From risk factors to early intervention. Alzheimer's Research & Therapy 9:1–9.

Cui Y-M, Ao M-Z, Li W, Yu L-J (2008) Effect of glabridin from *Glycyrrhiza glabra* on learning and memory in mice. Planta Medica 74 (04):377–380.

Fallarero A, Oinonen P, Gupta S, Blom P, Galkin A, Mohan CG, Vuorela PM (2008) Inhibition of acetylcholinesterase by coumarins: The case of coumarin 106. Pharmacological Research 58 (3–4):215–221.

Fan P, Hay A-E, Marston A, Hostettmann K (2008) Acetylcholinesterase-inhibitory activity of linarin from *Buddleja davidii*, structure-activity relationships of related flavonoids, and chemical investigation of *Buddleja nitida*. Pharmaceutical Biology 46 (9):596–601.

Feldman H, Woodward M (2005) The staging and assessment of moderate to severe Alzheimer disease. Neurology 65 (6_suppl_3):S10–S17.

Feng Y, Wang XP, Yang SG, Wang YJ, Zhang X, Du XT, Sun XX, Zhao M, Huang L, Liu RT (2009) Resveratrol inhibits beta-amyloid oligomeric cytotoxicity but does not prevent oligomer formation. Neurotoxicology 30 (6):986–995. doi:10.1016/j.neuro.2009.08.013

Foa R, Norton L, Seidman AD (1994) Taxol (paclitaxel): A novel anti-microtubule agent with remarkable anti-neoplastic activity. International Journal of Clinical & Laboratory Research 24 (1):6–14. doi:10.1007/bf02592403

Forloni G, Colombo L, Girola L, Tagliavini F, Salmona M (2001) Anti-amyloidogenic activity of tetracyclines: Studies in vitro. FEBS Letters 487 (3):404–407. doi:10.1016/S0014-5793(00)02380-2

Gao F, Liang W, Chen Q, Chen B, Liu Y, Liu Z, Xu X, Zhu R, Cheng L (2024) A curcumin-decorated nanozyme with ROS scavenging and anti-inflammatory properties for neuroprotection. Nanomaterials 14 (5):389.

Gervais F, Paquette J, Morissette C, Krzywkowski P, Yu M, Azzi M, Lacombe D, Kong X, Aman A, Laurin J, Szarek WA, Tremblay P (2007) Targeting soluble Aβ peptide with tramiprosate for the treatment of brain amyloidosis. Neurobiology of Aging 28 (4):537–547. doi:10.1016/j.neurobiolaging.2006.02.015

Goodman Y, Steiner MR, Steiner SM, Mattson MP (1994) Nordihydroguaiaretic acid protects hippocampal neurons against amyloid β-peptide toxicity, and attenuates free radical and calcium accumulation. Brain Research 654 (1):171–176. doi:10.1016/0006-8993(94)91586-5

Gu Y, Nieves JW, Stern Y, Luchsinger JA, Scarmeas N (2010) Food combination and Alzheimer disease risk: A protective diet. Archives of Neurology 67 (6):699–706. doi:10.1001/archneurol.2010.84

He X-L, Zhou W-Q, Bi M-G, Du G-H (2010) Neuroprotective effects of icariin on memory impairment and neurochemical deficits in senescence-accelerated mouse prone 8 (SAMP8) mice. Brain Research 1334:73–83.

Houghton PJ, Ren Y, Howes M-J (2006) Acetylcholinesterase inhibitors from plants and fungi. Natural Product Reports 23 (2):181–199.

Howes M-JR, Houghton PJ (2009) Acetylcholinesterase inhibitors of natural origin. International Journal of Biomedical and Pharmaceutical Sciences 3 (SI1):67–86.

Husen A (2022) Herbs, Shrubs and Trees of Potential Medicinal Benefits. CRC Press, Boca Raton, FL. doi:10.1201/9781003205067

Husen A (2023a) Exploring Poisonous Plants: Medicinal Values, Toxicity Response and Other Potential Uses. CRC Press, Boca Raton, FL. doi:10.1201/b23017

Husen A (2023b) Antidiabetic Medicinal Plants and Herbal Treatments. CRC Press, Boca Raton, FL. doi:10.1201/b23347

Husen A (2024a) Antimalarial Medicinal Plants. Taylor & Francis Group, LLC. ISBN: 9781032457154. doi:10.1201/9781003378396

Husen A (2024b) Medicinal Spices and Condiment Crops. Taylor & Francis Group. ISBN: 9781032469478. doi:10.1201/9781003387046

Hwang EM, Ryu YB, Kim HY, Kim D-G, Hong S-G, Lee JH, Curtis-Long MJ, Jeong SH, Park J-Y, Park KH (2008) BACE1 inhibitory effects of lavandulyl flavanones from *Sophora flavescens*. Bioorganic & Medicinal Chemistry 16 (14):6669–6674. doi:10.1016/j.bmc.2008.05.080

Ibrahim WW, Sayed RH, Abdelhameed MF, Omara EA, Nassar MI, Abdelkader NF, Farag MA, Elshamy AI, Afifi SM (2024) Neuroprotective potential of Erigeron bonariensis ethanolic extract against ovariectomized/D-galactose-induced memory impairments in female rats in relation to its metabolite fingerprint as revealed using UPLC/MS. Inflammopharmacology 32 (2):1091–1112.

Jeon SY, Kwon SH, Seong YH, Bae K, Hur JM, Lee YY, Suh DY, Song KS (2007) β-secretase (BACE1)-inhibiting stilbenoids from Smilax Rhizoma. Phytomedicine 14 (6):403–408. doi:10.1016/j.phymed.2006.09.003

Jung M, Park M (2007) Acetylcholinesterase inhibition by flavonoids from *Agrimonia pilosa*. Molecules 12 (9):2130–2139.

Khan MTH, Orhan I, Şenol F, Kartal M, Şener B, Dvorská M, Šmejkal K, Šlapetová T (2009) Cholinesterase inhibitory activities of some flavonoid derivatives and chosen xanthone and their molecular docking studies. Chemico-Biological Interactions 181 (3):383–389.

Kim B-J, Bak S-B, Bae S-J, Jin H-J, Park SM, Kim Y-R, Jung D-H, Song C-H, Kim Y-W, Kim S-C (2024) Protective effects of red ginseng against tacrine-induced hepatotoxicity: An integrated approach with network pharmacology and experimental validation. Drug Design, Development and Therapy 549–566.

Kumar A, Prakash A, Dogra S (2010) Naringin alleviates cognitive impairment, mitochondrial dysfunction and oxidative stress induced by D-galactose in mice. Food and Chemical Toxicology 48 (2):626–632.

Leon R, Garcia AG, Marco-Contelles J (2013) Recent advances in the multitarget-directed ligands approach for the treatment of Alzheimer's disease. Medicinal Research Reviews 33 (1):139–189.

Li L, Liu J, Yan X, Qin K, Shi M, Lin T, Zhu Y, Kang T, Zhao G (2011) Protective effects of ginsenoside Rd against okadaic acid-induced neurotoxicity in vivo and in vitro. Journal of Ethnopharmacology 138 (1):135–141. doi:10.1016/j.jep.2011.08.068

Li W, Sperry JB, Crowe A, Trojanowski JQ, Smith AB, 3rd, Lee VM (2009) Inhibition of tau fibrillization by oleocanthal via reaction with the amino groups of tau. Journal of Neurochemistry 110 (4):1339–1351. doi:10.1111/j.1471-4159.2009.06224.x

Liu S, Xu L, Shen Y, Wang L, Lai X, Hu H (2024) Qingxin Kaiqiao Fang decreases Tau hyperphosphorylation in Alzheimer's disease via the PI3K/Akt/GSK3β pathway in vitro and in vivo. Journal of Ethnopharmacology 318:117031. doi:10.1016/j.jep.2023.117031

Ma R, Mu Q, Xi Y, Liu G, Liu C (2024) Nanotechnology for tau pathology in Alzheimer's disease. Materials Today Bio 27:101145. doi:10.1016/j.mtbio.2024.101145

Malik J, Mandal SC, Choudhary S, Parihar S, Rahamathulla M (2023) Herbal Medicines for Management of Alzheimer's Disease. In: Dhara AK, Mandal SC (eds) Role of Herbal Medicines: Management of Lifestyle Diseases. Springer Nature Singapore, Singapore, pp 231–250. doi:10.1007/978-981-99-7703-1_12

Manju, Bharadvaja N (2024) Exploring the potential therapeutic approach using ginsenosides for the management of neurodegenerative disorders. Molecular Biotechnology 66 (7):1520–1536. doi:10.1007/s12033-023-00783-2

McGleenon B, Dynan K, Passmore A (1999) Acetylcholinesterase inhibitors in Alzheimer's disease. British Journal of Clinical Pharmacology 48 (4):471.

Meng F, Marek P, Potter KJ, Verchere CB, Raleigh DP (2008) Rifampicin does not prevent amyloid fibril formation by human islet amyloid polypeptide but does inhibit fibril thioflavin-T interactions: Implications for mechanistic studies of β-cell death. Biochemistry 47 (22):6016–6024. doi:10.1021/bi702518m

Nagori K, Pradhan M, Nakhate KT, Thakur A, Badwaik HR, Sharma MK, Dubey A (2024) Insights into Computational Repurposing of Drugs for Alzheimer's Disease. Computational Biology in Drug Discovery and Repurposing: 337. DOI:10.1201/9781003455424-16

Narayanan AP, Jayan J, Sudevan ST, Dhyani A, Zachariah SM, Mathew B (2024) Flavonoid and chalcone scaffolds as inhibitors of BACE1: Recent updates. Combinatorial Chemistry & High Throughput Screening 27 (9):1243–1256. doi:10.2174/1386207326666230731092409

Nepovimova E, Uliassi E, Korabecny J, Pena-Altamira LE, Samez S, Pesaresi A, Garcia GE, Bartolini M, Andrisano V, Bergamini C (2014) Multitarget drug design strategy: quinone–tacrine hybrids designed to block amyloid-β aggregation and to exert anticholinesterase and antioxidant effects. Journal of Medicinal Chemistry 57 (20):8576–8589.

Ono K, Hasegawa K, Yoshiike Y, Takashima A, Yamada M, Naiki H (2002) Nordihydroguaiaretic acid potently breaks down preformed Alzheimer's beta-amyloid fibrils in vitro. Journal of Neurochemistry 81 (3):434–440. doi:10.1046/j.1471-4159.2002.00904.x

Park IH, Jeon SY, Lee HJ, Kim SI, Song KS (2004) A beta-secretase (BACE1) inhibitor hispidin from the mycelial cultures of *Phellinus linteus*. Planta Medica 70 (2):143–146. doi:10.1055/s-2004-815491

Pasinetti GM, Ksiezak-Reding H, Santa-Maria I, Wang J, Ho L (2010) Development of a grape seed polyphenolic extract with anti-oligomeric activity as a novel treatment in progressive supranuclear palsy and other tauopathies. Journal of Neurochemistry 114 (6):1557–1568. doi:10.1111/j.1471-4159.2010.06875.x

Patel P, Shakya R, Asati V, Kurmi BD, Verma SK, Gupta GD, Rajak H (2024) Furan and benzofuran derivatives as privileged scaffolds as anticancer agents: SAR and docking studies (2010 to till date). Journal of Molecular Structure 1299:137098.

Peterson DW, George RC, Scaramozzino F, LaPointe NE, Anderson RA, Graves DJ, Lew J (2009) Cinnamon extract inhibits tau aggregation associated with Alzheimer's disease in vitro. Journal of Alzheimer's Disease: JAD 17 (3):585–597. doi:10.3233/jad-2009-1083

Pickhardt M, Gazova Z, von Bergen M, Khlistunova I, Wang Y, Hascher A, Mandelkow EM, Biernat J, Mandelkow E (2005) Anthraquinones inhibit tau aggregation and dissolve Alzheimer's paired helical filaments in vitro and in cells. The Journal of Biological Chemistry 280 (5):3628–3635. doi:10.1074/jbc.M410984200

Radenahmad N, Saleh F, Sawangjaroen K, Vongvatcharanon U, Subhadhirasakul P, Rundorn W, Withyachumnarnkul B, Connor JR (2011) Young coconut juice, a potential therapeutic agent that could significantly reduce some pathologies associated with Alzheimer's disease: Novel findings. The British Journal of Nutrition 105 (5):738–746. doi:10.1017/s0007114510004241

Santos MA, Chand K, Chaves S (2016) Recent progress in multifunctional metal chelators as potential drugs for Alzheimer's disease. Coordination Chemistry Reviews 327:287–303.

Shimmyo Y, Kihara T, Akaike A, Niidome T, Sugimoto H (2008a) Flavonols and flavones as BACE-1 inhibitors: Structure–activity relationship in cell-free, cell-based and in silico studies reveal novel pharmacophore features. Biochimica et Biophysica Acta (BBA) – General Subjects 1780 (5):819–825. doi:10.1016/j.bbagen.2008.01.017

Shimmyo Y, Kihara T, Akaike A, Niidome T, Sugimoto H (2008b) Multifunction of myricetin on A beta: Neuroprotection via A conformational change of A beta and reduction of A beta via the interference of secretases. Journal of Neuroscience Research 86 (2):368–377. doi:10.1002/jnr.21476

Shytle RD, Bickford PC, Rezai-zadeh K, Hou L, Zeng J, Tan J, Sanberg PR, Sanberg CD, Roschek B, Jr., Fink RC, Alberte RS (2009) Optimized turmeric extracts have potent anti-amyloidogenic effects. Current Alzheimer Research 6 (6):564–571. doi:10.2174/156720509790147115

Shytle RD, Tan J, Bickford PC, Rezai-Zadeh K, Hou L, Zeng J, Sanberg PR, Sanberg CD, Alberte RS, Fink RC, Roschek B (2012) Optimized turmeric extract reduces β-amyloid and phosphorylated Tau protein burden in Alzheimer's transgenic mice. Current Alzheimer Research 9 (4):500–506. doi:10.2174/156720512800492459

Singh H, Das A, Khan MM, Pourmotabbed T (2024) New insights into the therapeutic approaches for the treatment of tauopathies. Neural Regeneration Research 19 (5):1020–1026. doi:10.4103/1673-5374.385288

Teismann P, Ferger B (2000) Effects of ensaculin on dopamine metabolite levels and K+-induced glutamate release. European Journal of Pharmacology 398 (2):247–250.

Umukoro S, Kalejaye HA, Ben-Azu B, Ajayi AM (2018) Naringenin attenuates behavioral derangements induced by social defeat stress in mice via inhibition of acetylcholinesterase activity, oxidative stress and release of pro-inflammatory cytokines. Biomedicine & Pharmacotherapy 105:714–723.

Urbain A, Marston A, Queiroz EF, Ndjoko K, Hostettmann K (2004) Xanthones from *Gentiana campestris* as new acetylcholinesterase inhibitors. Planta Medica 70 (10):1011–1014.

Vanessa VV, Mah SH (2021) Xanthone: Potential acetylcholinesterase inhibitor for Alzheimer's disease treatment. Mini Reviews in Medicinal Chemistry 21 (17):2507–2529.

Watkins PB, Zimmerman HJ, Knapp MJ, Gracon SI, Lewis KW (1994) Hepatotoxic effects of tacrine administration in patients with Alzheimer's disease. JAMA 271 (13):992–998.

Yang F, Lim GP, Begum AN, Ubeda OJ, Simmons MR, Ambegaokar SS, Chen PP, Kayed R, Glabe CG, Frautschy SA, Cole GM (2005) Curcumin inhibits formation of amyloid β oligomers and fibrils, binds plaques, and reduces amyloid in vivo. Journal of Biological Chemistry 280 (7):5892–5901. doi:10.1074/jbc.M404751200

Zhu CW, Grossman H, Neugroschl J, Parker S, Burden A, Luo X, Sano M (2018) A randomized, double-blind, placebo-controlled trial of resveratrol with glucose and malate (RGM) to slow the progression of Alzheimer's disease: A pilot study. Alzheimer's & Dementia (New York, N Y) 4:609–616. doi:10.1016/j.trci.2018.09.009

# 9 Diabetes Management Using Phenolic Compounds

*Deepa Mundekkad*, Anjali R. Mallya, and William C. Cho**

## 9.1 INTRODUCTION

Diabetes mellitus is a metabolic disorder that is chronic and prevalent in the whole world. It has emerged as a global health crisis and a challenge to public health over the last few decades. The number of people affected by diabetes is found to have increased dramatically in recent years. Although it affects individuals of all ages, its prevalence is higher in older adults. The increased occurrence of diabetes in the elderly population could be endorsed to a combination of factors such as physiological changes associated with aging (resulting in reduced insulin sensitivity, beta cell function, and increased fat mass) (Bellary et al., 2021), lifestyle factors (sedentary life due to age-related movement issues, obesity, poor diet) (Bellettiere et al., 2021), polypharmacy (the use of multiple medications for treating various ailments) (Artasensi et al., 2020), reduced physical activity, and chronic disease conditions (comorbidities like hypertension and heart diseases) (Arumugam et al., 2020). This has become a cause for concern due to the various complications that seriously affect the functioning of other organs if left untreated. Conditions like cardiovascular diseases (including heart attacks and strokes), nephropathy or kidney damage, retinopathy or even eye damage that can lead to blindness, neuropathy resulting in severe nerve damage, foot ulcers leading to amputation, and an increased risk of various types of infections are some of the main complications that influence the progression of diabetes into a severe and potentially fatal disease.

Diabetes is characterized by elevated blood glucose levels, a condition termed hyperglycemia, that occurs due to the body's inability to produce or effectively use insulin, a hormone that regulates blood sugar. The worldwide surge in diabetes cases is undeniably linked to significant shifts in lifestyle patterns (Jwala and Kumari, 2024). The globalized world has witnessed a dramatic transformation in dietary habits, physical activity levels, and overall living conditions, all of which have contributed to the diabetes epidemic. Furthermore, the transition from traditional, plant-based diets to diets high in processed foods, saturated fats, and added sugars has been a major contributor. Modern lifestyles characterized by sedentary occupations and reduced physical activity have also increased obesity rates, a primary risk factor for type 2 diabetes (Gonzalez Ramirez and Bolaños Muñoz, 2023). Additionally, the fast-paced nature of modern life has led to increased stress and sleep disturbances, both of which can impact insulin resistance.

Diabetes can be categorized into three different types. Table 9.1 depicts the different types and characteristics of diabetes. The most common type of diabetes is type 2 diabetes, often associated with obesity and stemming from a sedentary lifestyle and Type 1 diabetes manifests when the pancreas fails to produce insulin that is adequate for the body, and this gradually results in resistance that the body develops to insulin. Despite advancements in medical treatment, the management of diabetes remains a significant challenge. The rising prevalence of diabetes has prompted an urgent search for alternative and complementary therapeutic strategies. However, despite the high prevalence of diabetes, awareness about the disease has led to improved access to prevention along with better treatment strategies. In recent years, there has been growing interest in the potential benefits of natural compounds for preventing and managing chronic diseases (Husen, 2022, 2023a,b,

* Corresponding authors: nascprojectdirector@nehrucolleges.com; deepamundekkad@gmail.com


DOI: 10.1201/9781003527671-9

**TABLE 9.1**
**Types of Diabetes and Specific Characteristics of Each Type**

| Type of Diabetes | Characteristics |
|---|---|
| Type 1 diabetes | • Autoimmune condition<br>• The insulin-producing cells are destroyed by the body's immune system destroys<br>• Typically developed during childhood or adolescence |
| Type 2 diabetes | • The most standard type<br>• Often results from obesity and a sedentary lifestyle<br>• The body typically becomes resistant to insulin<br>• The pancreas gradually fails to produce adequate insulin |
| Gestational diabetes | • Develop during pregnancy in women<br>• May have never experienced diabetes previously<br>• Typically resolves after childbirth<br>• The likelihood of developing type 2 diabetes in the future is high |

2024a,b; Mutha et al., 2021; Rahman et al., 2021; Sun and Shahrajabian, 2023; Vodovotz et al., 2020). Among these, polyphenols are crucial among the heterogeneous groups of naturally occurring chemical compounds abundantly found in plants. These natural compounds have attracted particular attention due to their diverse biological activities. This chapter delves into the therapeutic potential of phenolic compounds in diabetes management, exploring their mechanisms of action, sources, and prospective applications in functional foods and nutraceuticals.

## 9.2 THE PERPETUATING CYCLE OF OXIDATIVE STRESS AND PHENOLIC COMPOUNDS

The metabolic activities of the body result in the generation of a set of molecules called free radicals that contain highly reactive unpaired electrons. The imbalance between the production of these free radicals and the speed with which the body neutralizes them results in the overproduction of such free radicals. This can lead to severe damage to cells, proteins, and DNA. Generally, in this condition, the body sets off oxidative stress, a response initiated by the body's antioxidant defenses. The antioxidant system, at times, is overwhelmed by the imbalance, contributing to various diseases. This overwhelming response could be controlled by using antioxidant compounds like curcumin (Mundekkad and Cho, 2023). Oxidative stress is known to contribute to the pathogenesis of diabetes. Hyperglycemia is a condition reported with elevated blood glucose levels accompanied by increased formation of advanced glycation end products (AGEs) (Papachristoforou et al., 2020) resulting in the activation of the polyol pathway leading to reactive oxygen species (ROS) generation. However, this condition is very complex and bidirectional. Oxidative stress can damage pancreatic beta cells, impairing insulin production (Eguchi et al., 2021). Additionally, ROS can induce insulin resistance, further exacerbating hyperglycemia. Other factors responsible for oxidative stress in diabetes are inflammatory responses and the generation of uncoupled nitric oxide synthases that generate superoxide instead of nitric oxide. The interplay between these factors and oxidative stress creates a vicious cycle (Figure 9.1) that leads to the pathogenesis of diabetes and its related complications.

Antioxidants are molecules that help to neutralize these harmful free radicals, thus preventing oxidative damage to cells. Primarily, they act as a defense system of the cell, protecting cells from the detrimental effects of oxidative stress. The body produces antioxidants, such as glutathione and superoxide dismutase (Gusti et al., 2021). However, there are antioxidants that are exogenously supplied and taken as supplements. These are obtained from sources like fruits, vegetables, nuts, seeds, and whole grains. By consuming antioxidant-rich foods or taking appropriate supplements,

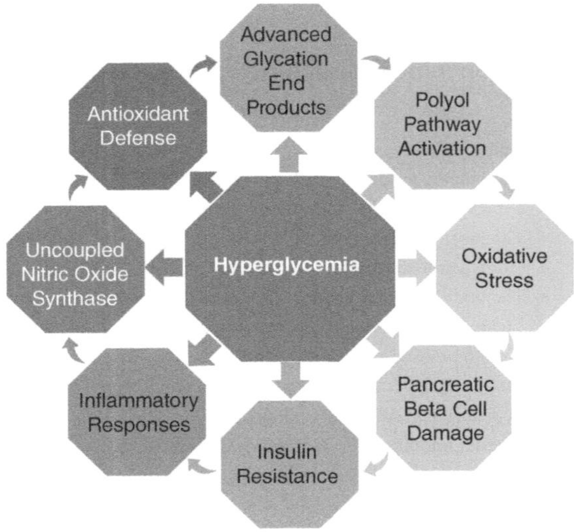


**FIGURE 9.1 The key elements to consider in the management of diabetes** - the complex interplay between hyperglycemia, oxidative stress, and the development of diabetic complications highlights the vicious cycle that can occur, where the components contribute to the progression of diabetes and its complications.

individuals can bolster their body's defense against oxidative stress and mitigate the potential risk of developing chronic diseases.

Phenolic compounds, a diverse group of phytochemicals, have garnered significant attention for their potent antioxidant properties. These natural substances are synthesized by plants as a defense mechanism against oxidative stress. Phenolic compounds can donate a hydrogen atom to stabilize free radicals, forming a less reactive phenoxyl radical (Nagarajan et al., 2020). They can also transfer an electron to a free radical, neutralizing its activity by the process of Single Electron Transfer (SET). Another mechanism of action of phenolic compounds is the sequential proton loss electron transfer (SPLET), which is a combination of proton donation and electron transfer, often observed in aqueous environments. Metal ions like iron and copper catalyze Fenton and Haber-Weiss reactions, resulting in the formation of harmful free radicals (Tvrdá and Benko, 2020). Phenolic compounds can chelate these metal ions and prevent this catalytic reaction. They can also inhibit enzymes (like lipoxygenase and cyclooxygenase) that are involved in oxidative stress. Thus, by modulating the redox signaling, phenolic compounds can impact cellular responses to oxidative stress.

## 9.3 PHENOLIC COMPOUNDS AND DIABETES COMPLICATIONS

Metabolism-induced oxidative stress plays a decisive role in the progression of diabetic complications. The excessive production of ROS in diabetes inevitably leads to structural and functional impairment to the make-up of many cellular components, proteins, RNA, lipids, and DNA (Bhatti et al., 2022). Further, complications in other organs also happen due to untreated diabetes. Continuous stress may result in other predicaments like diabetic retinopathy, diabetic nephropathy, diabetic neuropathy, and diabetic foot (Figure 9.2). Many of these debilitating conditions are curable. However, the intricate relationship between oxidative stress and diabetic complications can become serious if left untreated, leading to severe consequences.

### 9.3.1 Diabetic Retinopathy

Adults with diabetes reach a stage where elevated blood sugar levels impair the blood vessels in the retina inciting blindness. This situation is called diabetic retinopathy. There are mainly two types of diabetic

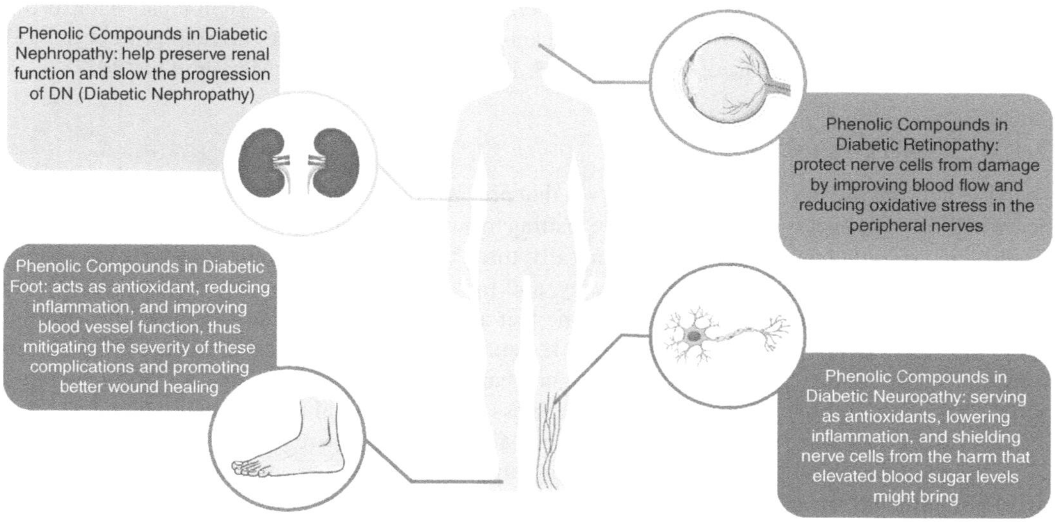


**FIGURE 9.2** Various diabetic complications developed because of oxidative stress and the role of phenolic compounds in mitigating these effects.

retinopathy – non-proliferative diabetic retinopathy (NPDR) and proliferative diabetic retinopathy (PDR) (Sardarinia et al., 2022). NPDR is an early stage of blindness where tiny bulges in the blood vessels (microaneurysms) may lead to bleeding or hemorrhages in the retina. It is also characterized by areas of inflammation, called cotton wool spots. PDR, on the other hand, is an advanced stage of blindness in diabetic patients where new blood vessel growth called neovascularization is observed. Leakage of blood and fluid into the eye occurs, which may lead to the potential detachment of the retina from where it is attached at the back of the eye. Early symptoms are not so visible, but blurred vision, floating spots, dark spots or flashes of light during advanced stages, and vision loss are common. Oxidative stress is a major contributing factor to the pathogenesis of diabetic retinopathy (Kang and Yang, 2020), which induces damage to retinal cells, leading to inflammation, blood vessel leakage, and ultimately vision loss. This condition occurs when elevated doses of blood sugar damage the blood vessels in the retina, typically as in diabetic retinopathy. Over time, this damage can lead to vision loss and even blindness. Early detection and treatment are essential to prevent severe vision problems. Diabetes duration, i.e., longer duration, combined with an inadequate regulation of blood sugar level, increases the risk. High blood pressure and pregnancy can increase the risk and worsen the existing retinopathy, whereas family history is also known to have an important role in the advancement of diabetic retinopathy.

### 9.3.2 Diabetic Nephropathy

Diabetic nephropathy is a kidney disease that develops in people with diabetes, where the kidneys' filtering units are damaged, leading to a build-up of waste products in the blood and, eventually, kidney failure. Over time, the high blood sugar level damages the blood vessels in the kidneys, leading to chronic inflammation in the kidneys (Akhtar et al., 2020). Some of the diabetic people may be more susceptible to diabetic nephropathy due to genetic factors as well. The prognostic signs of diabetic nephropathy include swelling in the feet and ankles, often accompanied by fatigue and a loss of appetite, followed by nausea, shortness of breath, high blood pressure, and trouble sleeping among other symptoms. Apart from the normal causes mentioned, oxidative stress is also found to promote inflammation, endothelial dysfunction, and mesangial cell proliferation, culminating in diabetic nephropathy. These chronic conditions can cause kidney failure, requiring dialysis or a kidney transplant in cases of end-stage kidney disease. Regular blood pressure control and monitoring of kidney function are crucial for preventing diabetic nephropathy. ACE – Angiotensin-Converting Enzyme inhibitors or angiotensin receptor blockers are medications that can help protect the kidneys (Banerjee et al., 2022; Jacobsen et al., 2003; Kao and Huang, 2021; Merker et al., 2021).

### 9.3.3 Diabetic Neuropathy

Diabetic neuropathy is a damage to the nerves that prevails in people with diabetes. It can affect any nerve in the body, resulting in nerve damage leading to numbness, pain, weakness, and tingling in the hands, feet, and other parts of the body. Generally, three types of diabetic neuropathy are identified – peripheral neuropathy, autonomic neuropathy, and focal neuropathy. Peripheral neuropathy is the most common type, affecting the nerves in the feet and legs with numbness and tingling, burning pain, and weakness of the muscles. Autonomic neuropathy usually affects the nerves that control some of the involuntary functions like digestion, heart rate, and blood pressure, resulting in constipation, diarrhea, erectile dysfunction, and dizziness. Normally focal neuropathy is found to affect a single nerve, resulting in sudden weakness or pain in a specific area, such as the face or hand. Conventionally, high blood sugar levels, chronic inflammation, and genetic factors are reasons for developing diabetic neuropathy. Oxidative stress-induced damage to nerve cells is yet another factor known to have an effect on the development of diabetic neuropathy. In severe cases, diabetic neuropathy can cause ulcers and infections, which may require amputation (Anandhanarayanan et al., 2022; Quiroz-Aldave et al., 2023; Sasaki et al., 2020).

### 9.3.4 Diabetic Foot

In diabetic patients, neuropathy and ischemia can lead to a complication called diabetic foot, which can manifest in foot ulcers, infections, and, later, can even result in amputation if left untreated. Diabetic foot ulcers are difficult to heal as they are often followed by infections, leading to various other complicated situations like cellulitis, osteomyelitis, and gangrene. Severe gangrene in diabetic foot infections will further lead to amputations that will significantly impact the order of the life of the patients. The financial burden of diabetic foot is substantial, including costs associated with healthcare, lost productivity, and reduced quality of life (Edmonds et al., 2021). The major cause of diabetic foot, other than diabetes, is high blood sugar levels, where the nerves and blood vessels get damaged, leading to poor blood flow to the feet. This will result in an increased risk of ulceration and infection. Generally, it is observed that conditions like bunions, calluses, and corns also lead to foot ulcers resulting in nerve damage and numbness in the feet (Reardon et al., 2020).

Phenolic compounds, the natural antioxidants found in most plants and also in almost all plant-based foods, have shown high promise in mitigating oxidative stress and, consequently, reducing the risk of these diabetic complications. These compounds possess potent antioxidant properties, enabling them to neutralize harmful ROS and protect cells from damage, thus helping mitigate the detrimental effects of high blood sugar levels. They can directly scavenge ROS, preventing their interaction with cellular components and reducing oxidative damage. By neutralizing harmful free radicals and reducing inflammation, phenolic compounds contribute to improved insulin sensitivity, glucose uptake, and overall cardiovascular health. Incidentally, they can also stimulate the body's endogenous antioxidant defense systems, indirectly enhancing its ability to combat oxidative stress. Studies have highlighted the beneficial effects of specific phenolic compounds, such as anthocyanins, flavonoids, and resveratrol, in preventing complexities like diabetic retinopathy, nephropathy, neuropathy, and cardiovascular disease. Incorporating a diet that is rich in plant-based foods containing phenolic compounds can be a valuable strategy for diabetes patients to manage their condition and reduce the risk of associated complications. Further, many of the phenolic compounds have anti-inflammatory activities, which can assist in reducing inflammation associated with diabetic complications. They can also interact with various signaling pathways involved in the development of diabetic complications, thereby mitigating the toxic effect of ROS molecules.

Research on the benefits of phenolic compounds in treating diabetic complications is ongoing (Hassan et al., 2021; Serina and Castilho, 2022; Sharma et al., 2022); the available evidence suggests that these natural compounds may offer promising therapeutic potential for complications related to diabetes. One of the major complications of diabetes is the risk of heart disease and stroke – phenolic compounds are found to mitigate the chances of having cardiovascular disease by directly scavenging the harmful ROS that are implicated in the development of atherosclerosis (plaque build-up in arteries). They are also known to improve the blood lipid profile by lowering the levels of low-density lipoprotein (LDL) cholesterol and increasing levels of high-density lipoprotein cholesterol (HDLC). Improved glycemic control offered by phenolic compounds may help to improve blood sugar level and bring it to normal rates (Sarkar et al., 2022).

## 9.4 PHENOLIC COMPOUNDS AND BLOOD GLUCOSE CONTROL

There is increasing evidence that suggests that phenolic compounds play a major role in maintaining healthy blood glucose levels. This is because phenolic compounds are functionally very diverse with antioxidant activity and antimicrobial properties, with an ability to affect signaling pathways. Recent studies have shown that phenolic compounds can regulate blood glucose levels (Fernandes et al., 2022) and thus help in preventing blood glucose level imbalances which can potentially lead to serious conditions such as diabetes and hypoglycemia, if left untreated. The different categories of phenolic compounds present in fruits, vegetables, and even beverages like tea and coffee, can have promising effects on glucose metabolism. These compounds are found to enhance the insulin sensitivity of the body and thus modulate the absorption of glucose by the body. These compounds can also protect the body against oxidative assault that may result in insulin resistance and thereby type 2 diabetes. Many of these antidiabetic compounds (like flavonoids, and anthocyanins), are exemplary molecules used to manage diabetes from time immemorial. The yellow-colored polyphenolic compound called curcumin present in turmeric, obtained from the rhizomes of *Curcumin longa*, is known to have immense power to control many of the complications associated with diabetes (Mundekkad and Cho, 2023).

### 9.4.1 Impact of Phenolic Compounds on Insulin Resistance, Glucose Uptake, and Utilization

Insulin is a hormone that is crucial for maintaining blood sugar levels in the body. Insulin allows the cells to take in glucose that is needed for energy. Without proper functioning of insulin, the blood sugar levels tend to become too high, resulting in diabetes. When the cells in the body become less responsive to this hormone, resistance build-up and this will lead to an uprise in blood glucose level, leading eventually to type 2 diabetes. Insulin resistance is developed as a result of impaired insulin signaling, which, in turn, is the result of oxidative stress (Khalid et al., 2021). Phenolic compounds can obliterate insulin resistance by improving insulin sensitivity through the neutralization of the reactive oxygen/nitrogen species. Inflammation is another condition that favors insulin resistance (Li et al., 2022) as chronic inflammation is recognized as a key factor in the progression of insulin resistance. Phenolic compounds can reduce inflammation by inhibiting pro-inflammatory mediators like cytokines and, thus, mitigate insulin resistance (Pannucci et al., 2023a). Activating insulin signaling pathways that influence the metabolic fate of insulin is another strategy that the phenolic compounds rely on for managing insulin resistance. For example, phenolic compounds can activate the insulin receptor substrate (IRS) proteins, which are the key signaling molecules (Su et al., 2023), mediating the effects of insulin by stimulating the phosphorylation of these proteins. The activation of IRS proteins enhances insulin's ability to stimulate glucose uptake and help reduce blood sugar levels. Modulation of phosphatidylinositol 3-kinase (PI3K) activity is another mode of control that leads to increased transport of glucose in cells. Protein tyrosine phosphatases or PTPs are enzymes that counteract the effects of insulin by dephosphorylation of IRS proteins. Phenolic compounds are found to inhibit PTPs, thereby conserving the insulin signaling. Another major regulator is glucose transporter type 4 (GLUT4) – a key protein in glucose trafficking. Phenolic compounds can enhance the translocation of this protein from storage vesicles inside the cells and take them to the cell membrane, thus increasing glucose uptake (Bu Kang and Huang Chiang, 2022).

### 9.4.2 Potential for Glucose-Lowering Effects

The polyphenolic compounds, due to their exceptional biological activities like antioxidant, anti-inflammatory, and antidiabetic properties, are known to lower blood sugar levels through diverse mechanisms. Some of them, in particular, can inhibit α-glucosidase, which is an enzyme responsible for converting carbohydrates into glucose that subsequently will reduce the rate of glucose absorption from the intestine, thus lowering blood glucose levels. Lowering blood sugar levels can lead to the body being forced to utilize insulin much more effectively, thus enhancing insulin sensitivity. This is mainly achieved by either raising the levels of expression of insulin receptors or by modifying the signaling pathways associated with insulin activity. The antioxidant activity of polyphenolics protects the cells from oxidative stress, which is a major factor frequently implicated in

**TABLE 9.2**
**Pre-Clinical Studies on Phenolics**

| | Findings | References |
|---|---|---|
| **Flavonoids from blueberries** | Blueberry extract feeding in diabetic rats led to decreased body weight gain, lowered fasting glucose, and improved insulin response. The polyphenols were shown to activate signaling pathways related to glucose metabolism. | Basu et al. (2021) |
| **Resveratrol and metabolic syndrome** | This study showed that resveratrol, reduced blood glucose levels, and improved insulin sensitivity in obese mice. The mechanism involved activation of AMPK (AMP-activated protein kinase) and modulation of gluconeogenesis in the liver. | Barber et al. (2022) |
| **Curcumin and diabetes** | This study demonstrated that curcumin, a phenolic compound found in turmeric, enhanced insulin sensitivity and reduced fasting blood glucose levels in diabetic rats. Curcumin also decreased markers of oxidative stress and inflammation. | Vafaeipour et al. (2022) |
| **Chlorogenic acid** | In diabetic rats, chlorogenic acid, commonly found in coffee, decreased postprandial glucose levels and improved glucose tolerance by inhibiting intestinal glucose absorption and enhancing insulin sensitivity. | Singh et al. (2023) |
| **Quercetin and insulin resistance** | In a rat model of insulin resistance, quercetin addition led to considerable improvement in fasting blood glucose and insulin levels, attributed to enhanced GLUT4 translocation and improved glucose uptake in skeletal muscle. | Oyedemi et al. (2020) |
| **Green tea polyphenols** | An investigation into the effects of catechins (primarily EGCG) from green tea on diabetic mice showed reduced blood glucose and improved glucose tolerance. It was linked to enhanced β-cell function and reduced oxidative stress. | Wan et al. (2024) |
| **Grapeseed extract** | This study found that grapeseed extracts improved glucose tolerance and reduced metabolic disorders in diabetic mice via modulation of glucose transporters and enhancement of insulin signaling pathways. | Martín and Ramos (2021) |
| **Cinnamon extract** | In a type 2 diabetes rat model, cinnamon polyphenols improved glucose uptake and insulin sensitivity while lowering blood glucose levels. The effects were attributed to enhanced insulin receptor activity. | Cheng et al. (2012) |
| **Olive leaf extract** | The administration of olive leaf extract led to appreciable reduction in blood glucose and normalized lipid profiles in diabetic rats. The active phenolic components were linked to improved antioxidant capacity and reduced insulin resistance. | Berköz et al. (2023) |
| **Lycopene from tomatoes** | Lycopene supplementation produced notable reduction in blood glucose levels and improvements in the antioxidant status of diabetic rats. The study suggested that lycopene enhances the action of insulin. | Leh and Lee (2022) |

the advancement of type 2 diabetes. Although pre-clinical and clinical studies have confirmed the role of polyphenolic compounds in lowering glucose levels, many of these studies could not fully elucidate the mode of action of these compounds. There are a lot of options available to study the glucose-lowering effects of polyphenolic compounds and examine their efficacy and safety to apply as therapeutic agents for diabetes management (Blahova et al., 2021).

### 9.4.3 Pre-Clinical/Clinical Studies on Phenolic Compounds and Blood Glucose Control

Pre-clinical studies directed at the effect of phenolic compounds on blood glucose control and its effect on diabetes are potentially found to have promising results. In vitro studies carried out using cell culture models emphasized that phenolic compounds, tested in various cell culture models, including hepatocytes, myocytes, and adipocytes, were found to impact glucose uptake, insulin signaling, and glucose metabolism (Bu Kang and Huang Chiang, 2022). In clinical studies, the glucose metabolism and insulin sensitivity were effectively augmented upon treatment with phenolic compounds, and the studies demonstrated that in individuals with type 2 diabetes, the patients/participants experienced noticeable improvements in fasting blood sugar, HbA1c levels, and insulin sensitivity (Tables 9.2 and 9.3).

**TABLE 9.3**
**Clinical Studies on Phenolics**

| | Findings | References |
|---|---|---|
| **Pomegranate juice and glycemic control** | Pomegranate juice consumption was associated with noticeable reduction in fasting glucose and improved insulin sensitivity in patients with type 2 diabetes. | Banihani et al. (2014) |
| **Cinnamon and blood glucose** | This study showed that supplementation with cinnamon reduced the levels of fasting blood glucose and enhanced lipid profiles in individuals with type 2 diabetes. | Wu et al. (2022) |
| **Blueberry supplementation** | A clinical trial involving daily blueberry supplementation showed a sizable reduction in fasting glucose and insulin levels in individuals with metabolic syndrome. | Stull et al. (2015) |
| **Cocoa and insulin sensitivity** | This randomized controlled trial found that consuming dark chocolate led to improvements in insulin sensitivity and significant reductions in fasting glucose levels in obese participants. | Jafarirad et al. (2018) |
| **Lycopene and type 2 diabetes** | Higher dietary intake of lycopene was associated with a reduced risk of developing type 2 diabetes in a longitudinal study of middle-aged participants, indicating that its antioxidant properties may support glucose metabolism. | Gao et al. (2019) |
| **Green tea and glycemic response** | A meta-analysis of trials involving clinical examination indicated that green tea consumption improved glucose metabolism and insulin sensitivity, particularly in overweight individuals. | Liu et al. (2013) |
| **Olive leaf extract and glycemic control** | Participants in a Mediterranean diet that was fortified with olive oil and olive leaf extract showed low levels in fasting blood glucose with high insulin sensitivity. | De Bock et al. (2013) |
| **Quercetin and glycemic control** | A randomized, double-blind placebo-controlled trial demonstrated that quercetin supplementation improved fasting blood glucose and HbA1c levels in patients with type 2 diabetes. | Li et al. (2024) |
| **Resveratrol in humans** | In a study with type 2 diabetes patients, resveratrol supplementation led to decreased fasting glucose and improved insulin sensitivity. | Zare Javid et al. (2017) |
| **Grape polyphenols** | A study involving daily supplementation of grape polyphenols demonstrated improvements in glycemic control markers, particularly postprandial glucose levels. | Choleva et al. (2022) |

## 9.5 DIETARY SOURCES OF PHENOLIC COMPOUNDS FOR DIABETES MANAGEMENT

Effective management and prevention strategies are essential to combat diabetes mellitus, which is characterized by insulin resistance, hyperglycemia, and associated complications. Recent global health concerns over this highlight the role of dietary patterns in diabetes management, especially the consumption of foods rich in bioactive compounds, such as phenolic compounds that are found abundantly in fruits, vegetables, whole grains, legumes, nuts, and beverages like tea and red wine (Xavier et al., 2024). Phenolic compounds are well-known for their antioxidant, anti-inflammatory, and antidiabetic properties, contributing to overall metabolic health. Phenolic compounds can modulate glucose metabolism (Shahwan et al., 2022), thereby enhancing insulin sensitivity, and can also reduce oxidative stress, which will aid in managing and preventing diabetes. The various sources of phenolic compounds and their effect are depicted in Figure 9.3.

### 9.5.1 Dietary Patterns and Diabetes Prevention/Management

Worldwide, there is an emerging trend in adopting a diet that emphasizes the use of whole, plant-based foods. This pattern is gaining momentum for its potential benefits in managing and preventing diabetes. Different types of diets that have emerged over the years, like the Mediterranean diet, plant-based diets, and the DASH (Dietary Approaches to Stop Hypertension) diet are being followed by people all over because they are rich in phenolic compounds and have been associated with a lower risk of type 2 diabetes.

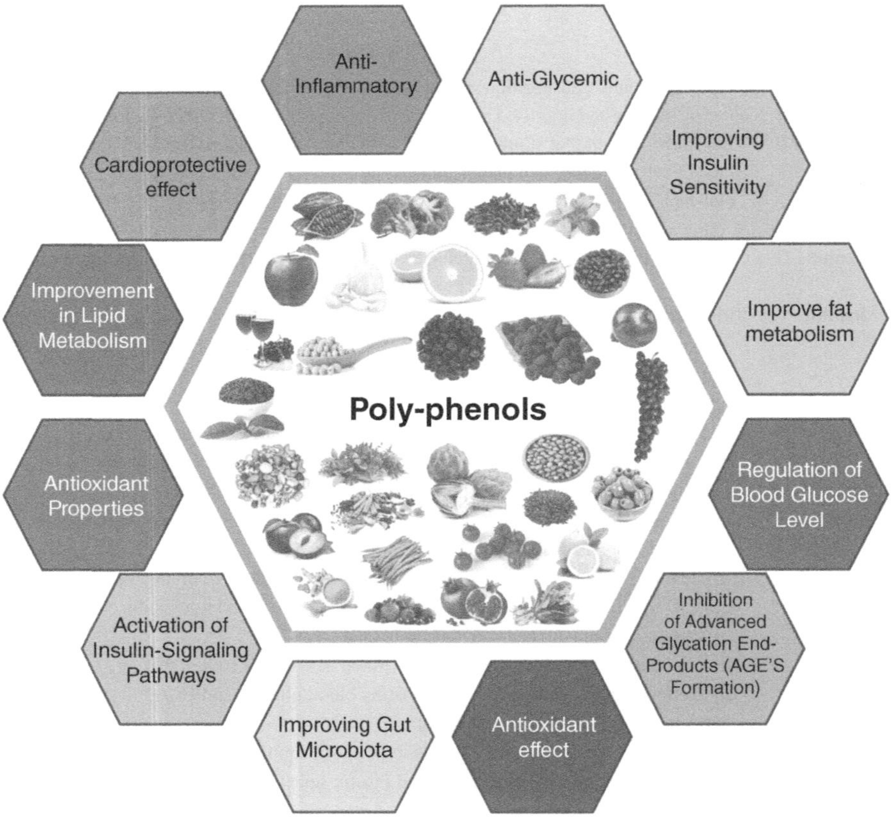


**FIGURE 9.3** Various dietary sources of phenolic compounds (fruits, vegetables, whole grains, etc.) and their effect on diabetes.

**TABLE 9.4**
**Various Types of Diets and Their Benefits**

| Type of Diet | Items Included in the Diet | Benefits |
|---|---|---|
| **Mediterranean diet** | Fruits, vegetables, nuts, legumes, whole grains, olive oil, and moderate intake of red wine | Improved glycemic control and reduced inflammation |
| **Plant-based diets** | Whole plant foods, including fruits, vegetables, whole grains, and legumes | Lower incidences of type 2 diabetes, improved fasting glucose levels and insulin sensitivity |
| **DASH** | Fruits, vegetables, low-fat dairy products, whole grains, and nuts | Beneficial effects on blood pressure and blood glucose levels |

While the Mediterranean diet is characterized by high consumption of fruits, vegetables, nuts, legumes, whole grains, olive oil, and moderate intake of red wine (Guasch-Ferré and Willett, 2021), the plant-based diet focuses on whole plant foods, including fruits, vegetables, whole grains, and legumes. The DASH diet emphasizes the use of fruits, vegetables, low-fat dairy products, whole grains, nuts, and other compounds known for their high antioxidant content (Singh et al., 2022). All these types provide various compounds that are rich in phenolics like flavonoids and phenolic acids. The various types of diets, the items that can be included in each of these diets, and the benefits of including these phenolic compounds can be seen in Table 9.4. Table 9.5 includes the key dietary sources of phenolic compounds and their major benefits.

**TABLE 9.5**
**Key Dietary Sources of Phenolic Compounds and Their Benefits**

| Dietary Source | Category | Examples | Benefits |
|---|---|---|---|
| **Fruits** | Berries | Blueberries, strawberries, and blackberries | Rich in anthocyanins, which have been associated with improved insulin sensitivity and glycemic control |
| | Apples | | Contain quercetin and chlorogenic acid, which may improve blood sugar levels |
| | Citrus fruits | | High in flavanones such as hesperidin; some studies suggest a link to reduced diabetes risk |
| **Vegetables** | Leafy greens | spinach, kale | Rich in phenolic acids and secondary metabolites that promote health |
| | Red and purple vegetable | Carrots, beetroot, red cabbage, and eggplant | Contain proanthocyanidins and other flavonoids with potential antidiabetic effects |
| **Whole Grains** | Oats | | Provide avenanthramides, phenolic compounds that may help lower blood sugar |
| | Barley and quinoa | | Rich in fiber and polyphenols that contribute to better glycemic control |
| | Beans and lentils | | Excellent sources of flavonoids and other polyphenols, which can help in managing blood glucose levels |
| **Nuts and Seeds** | Walnuts and almonds | | Contain various phenolic compounds and healthy fats that help reduce inflammation and may enhance insulin sensitivity |
| **Beverages** | Green tea and black tea | | High in catechins, particularly epigallocatechin gallate (EGCG), which has been shown to improve glucose metabolism |
| | Coffee | | Contains chlorogenic acid and other polyphenols associated with a lower risk of type 2 diabetes |
| | Dark chocolate and cocoa | | Rich in flavonoids, particularly epicatechin which has been linked to improved insulin sensitivity and endothelial function |

### 9.5.2 Role of Phenolic Compounds in Diabetes Prevention

The management and prevention of diabetes can be made substantial through the integration of phenolic-rich foods into dietary patterns and by following a holistic approach toward dietary practices that emphasizes the use of whole, plant-based foods. This is not only a viable strategy for improving metabolic health and glycemic control but also this emerges as a tool to maintain a more sustainable and health-promoting lifestyle. The all-round benefits that the phenolics offer could be attributed to their antioxidant, anti-inflammatory, and regulatory metabolic activities.

#### 9.5.2.1 Antioxidant Effects of Phenolic Compounds and Diabetes Prevention

ROS are produced as a result of metabolic stress, and the fate of cells is determined by the levels of ROS in the body (Mundekkad and Cho, 2022). Supplementing the body with antioxidants like curcumin will enhance the innate immunity of the body (Mundekkad and Cho, 2023). When the antioxidant system is compromised, chronic oxidative stress ensues, followed by the development of insulin resistance and, subsequently, type 2 diabetes. The variety of phenolic compounds is known to possess strong antioxidant tenacity that can neutralize various types of ROS, consequently protecting the pancreatic β-cells and other tissues from oxidative damage. This method mainly depends on the mitigation of ROS. Another approach by the phenolic compounds is the promotion or enhancement of the activity of the endogenous antioxidant enzymes (like catalase or superoxide dismutase) that are intrinsically produced by the immune system in the body. This promotion of antioxidant enzyme production will further reduce oxidative stress and its detrimental effects on the pancreatic cells.

#### 9.5.2.2 Anti-Inflammatory Properties of Phenolic Compounds and Diabetes Prevention

Persistent low-grade inflammation, observed in many diabetic patients, is a significant factor contributing to the pathogenesis of insulin resistance and type 2 diabetes mellitus. Though subtle, the chronic inflammatory state is responsible for a progressive activation of the immune system (Megha et al., 2021) and is often triggered by lifestyle choices like poor dietary choices and a sedentary lifestyle resulting in obesity. At this point, the adipose tissue releases pro-inflammatory mediators like cytokines, leading to systemic inflammation. These mediators are known to dysregulate various metabolic processes and interfere with signaling pathways, especially the insulin signaling pathway, which can result in minimal uptake of glucose, leading to insulin resistance (Abdalla, 2024). This leaves a conducive environment for the development of type 2 diabetes. In addition to adipose tissues, the inflammatory state of the cells can affect other organs like the liver, which is associated with glucose metabolism. Phenolic compounds exert their anti-inflammatory activity by inhibiting pro-inflammatory cytokines and mediators (such as TNF-α and IL-6) and modulating signaling pathways (such as NF-κB). They can also enhance the function of adipose tissue and thus reduce inflammation. Considering the connection between inflammation, adipose tissue, and obesity, if these factors are brought under control, diabetes can also be controlled as obesity is known as a major risk factor for developing type 2 diabetes.

#### 9.5.2.3 Modification of Insulin Signaling Pathways by Phenolic Compounds and Diabetes Prevention

Primarily, two pathways govern the potential of phenolics to improve insulin sensitivity mechanisms – enhanced insulin action where they enhance the translocation of glucose transporters, like GLUT4, to the cell membrane, thereby increasing glucose uptake by muscle and fat cells. Enhanced insulin action is guided by the reduction of oxidative stress, anti-inflammatory effects, and improved glucose uptake, that was already discussed (Abdalla, 2024). However, the impact of phenolic compounds on insulin signaling pathways plays a crucial role in altering these pathways, as is evidenced by their interaction with various compounds and molecules. Phenolic compounds activate a crucial regulator of cellular energy homeostasis, AMP-activated protein kinase (AMPK), thereby enhancing insulin sensitivity and promoting glucose uptake in the peripheral tissues. The phosphoinositide 3-kinase (PI3K)/Akt is another target in an insulin signaling pathway that is vital for insulin action, the modification of which will improve insulin sensitivity and glucose metabolism. Gene expression modulation by phenolic compounds results in the up- or down-regulation of crucial genes involved in glucose metabolism and insulin signaling (Pannucci et al., 2023b), resulting in the increased/decreased expression of proteins that facilitate insulin action and, in turn, improved metabolic health. In addition, many phenolic compounds are known to influence the secretion of incretin hormones, like glucagon-like peptide-1 (GLP-1), which helps in enhanced insulin secretion in response to meal intake while inhibiting the release of glucagon.

#### 9.5.2.4 Regulation of Glucose Metabolism by Phenolic Compounds and Diabetes Prevention

Phenolic compounds are known to influence various metabolic pathways, especially the ones that are involved in carbohydrate digestion and hepatic glucose production. By inhibiting some of the key carbohydrate-digesting enzymes, such as α-glucosidase and α-amylase, they can regulate glucose metabolism by slowing the absorption of glucose from the gut and lowering the postprandial blood glucose levels (Li et al., 2023). Moreover, glycemic control using phenolic compounds can reduce the risk of developing type 2 diabetes. Evidence from pre-clinical and clinical studies collaborates with this philosophy. These studies imply that phenolic compounds can stabilize blood glucose levels by controlling carbohydrate digestion and absorption. Additionally, they can affect glycogenolysis and gluconeogenesis by influencing glucose homeostasis controlled by the liver by moderating the activity of some of the key regulatory enzymes (like phosphoenolpyruvate carboxykinase (PEPCK) and glucose-6-phosphatase) that are actively involved in the processes of gluconeogenesis and glycogenolysis (Legouis et al., 2022).

#### 9.5.2.5 Modulation of Gut Microbiota by Phenolic Compounds and Diabetes Prevention

Dysbiosis is a phenomenon where there is an imbalance in gut microbiota known to influence metabolic processes and immune functions implicated in various metabolic disorders, including insulin resistance (Kasprzak-Drozd et al., 2021) and type 2 diabetes. Phenolic compounds can significantly dominate these microbial communities by influencing the prebiotic effects and metabolite production (Rodríguez-Daza et al., 2021). Phenolic compounds provide substrates for these beneficial gut bacteria and promote their growth, thereby maintaining a healthy gut environment. This is important because conditions like dysbiosis can lead to a leaky gut where endotoxins can enter the bloodstream and trigger inflammation, and this can ultimately contribute to insulin resistance. Therefore, by restoring the gut microbiota, phenolic compounds bring about a balance and potentially improve metabolic health, reducing the risk of developing insulin resistance and type 2 diabetes. Further, the fermentation of phenolic compounds and dietary fibers leads to the production of short-chain fatty acids (SCFAs) that are well-known for their anti-inflammatory properties. People with high insulin resistance and type 2 diabetes will benefit from the intake of these SCFAs as they help to regulate systemic inflammation (Zheng et al., 2021). SCFAs such as acetate, propionate, and butyrate can help to enhance insulin sensitivity and improve glucose metabolism. Butyrate is also known to influence expression levels of some receptors (like GPR43) and modulation of signaling pathways that enhance insulin sensitivity. Elevated levels of SCFA also promote the increased production of hormones like peptide YY (PYY) and GLP-1. These two hormones are involved in appetite regulation and glycemic control (Alhabeeb et al., 2021).

#### 9.5.2.6 Impact on Weight Management by Phenolic Compounds and Diabetes Prevention

Controlling obesity through weight management has a prominent role in the prevention of diabetes (ElSayed et al., 2023). Promoting satiety by maintaining a balanced energy intake and enhanced fat metabolism is a modest strategy to control obesity. Effective intake of phenolic-rich foods will help to have an increased feeling of fullness. It will help in properly utilizing the fat storage mechanisms and help mitigate the risk of developing obesity-related disorders, including type 2 diabetes. Some phenolic compounds are known to influence the energy the body expends to digest, absorb, and metabolize nutrients – a term called the thermic effect of food (TEF) (Al-Ibraheem et al., 2024). Phenolic compounds and fiber-rich foods require more energy to process, resulting in a higher overall calorie expenditure. They also influence the signaling pathways related to lipid metabolism, influencing the body to use fat for its energy requirements. For example, when the enzymes involved in fat oxidation are activated, it results in the increased breakdown of stored fats into fatty acids, which can later be used as a source of energy. This will reduce visceral fat and is critical for improving insulin sensitivity and overall metabolic health (Kolb et al., 2021). The phenolic compounds can also modulate the secretion of certain signaling molecules produced by adipose tissue, called adipokines, that can regulate metabolism and insulin sensitivity.

## 9.6 PHENOLIC COMPOUNDS IN NUTRACEUTICALS AND FUNCTIONAL FOODS

Nutraceuticals and functional foods have a wide range of health benefits, and phenolic compounds are a vital component of their production and development. The dietary intake of antioxidants and other beneficial compounds can be enhanced by incorporating these bioactive compounds into food and supplement products, which will enhance overall health and help consumers in disease prevention. The development of effective and safe products in the field of nutraceuticals is crucial and the health-promoting properties of phenolic compounds will be a boost to this industry. The advantage is that these phenolic compounds can be marketed as supplements and can be included in the food as isolated compounds, extracts, or even concentrates. Phenolic compounds contribute to various health benefits like antioxidant and anti-inflammatory activities, which underpin their incorporation into nutraceuticals and functional foods. The incorporation of phenolic compounds into functional food is known to manage metabolism to improve insulin sensitivity, regulate glucose metabolism, and facilitate fat metabolism, making them suitable substitutes in diabetes management and obesity prevention (Fernandes et al., 2022). Cardiovascular and gut health can also be improved through regular consumption of foods rich in phenolic compounds.

The general considerations that are influencing the development of functional foods and nutraceuticals enriched with phenolic compounds are explained (Figure 9.4). These deliberations are important as it is observed that the demand for functional foods is growing exponentially with

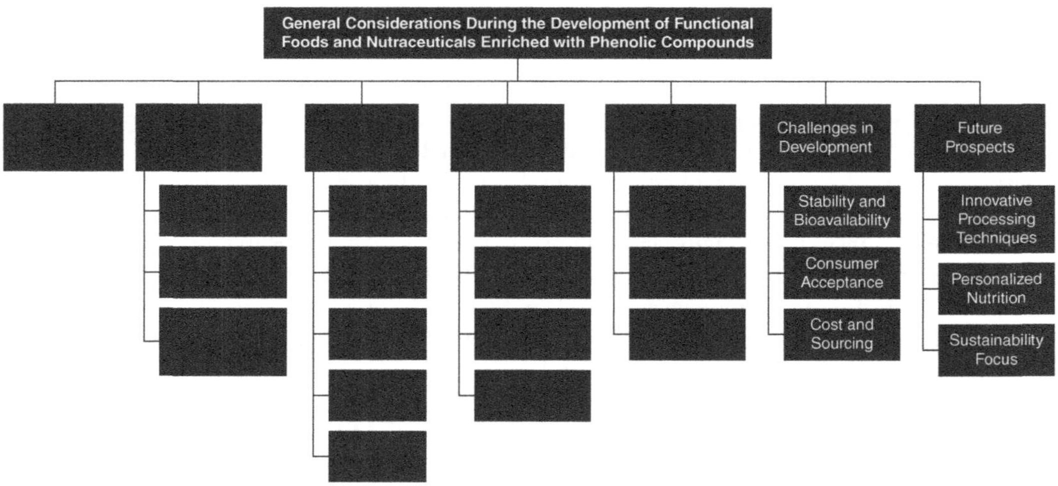


**FIGURE 9.4** General considerations influencing the development of functional foods and nutraceuticals enriched with phenolic compounds. This figure highlights the key factors impacting the formulation and development of functional foods and nutraceuticals enriched with phenolic compounds.

increased consumer awareness about the health benefits associated with phenolic compounds. This drives the next point – clean eating – where there is a shift in health trends toward preventive healthcare and wellness, which, in turn, encourages the incorporation of phenolic-rich foods into diets. The role of ongoing scientific research in identifying new phenolic compounds cannot be discounted at this point – in fact, the innovations in extraction and formulation technologies are significant factors that can enhance product efficacy and stability (Khadhraoui et al., 2021). However, there should be an understanding of the country's regulatory frameworks regarding health claims on food products, as these can influence the development of any pharmaceutical or nutraceutical compound. Compliance with safety standards and health regulations is crucial.

## 9.7 CHALLENGES AND FUTURE PERSPECTIVES

Significant developments have occurred over the years in diabetes management, yet controlling diabetes remains a complex challenge due to the multifaceted nature of the disease and its diverse risk factors (Skou et al., 2022). Dietary interventions, especially those incorporating bioactive compounds such as phenolics, are among the promising strategies to enhance glycemic control and reduce complications. These naturally occurring substances, abundantly found in fruits, vegetables, and beverages like tea and coffee, exhibit antioxidant and anti-inflammatory properties that could strengthen metabolic health and overall well-being. However, despite the potential benefits of this integrated approach, the incorporation of phenolic compounds into diabetes management faces several challenges, including variability in bioavailability, lack of standardized formulations, and insufficient clinical evidence, among others. As researchers continue to explore these compounds and the various ways to blend and infuse them into the dietary plans, it is crucial that we address these challenges early and devise strategies for their incorporation into everyday nutritional practices. The obstacles and limitations currently encountered in leveraging phenolic compounds for diabetes management and the future perspectives that may guide effective utilization in clinical settings are discussed below.

The potential effects of phenolic compounds on diabetes management are highlighted by their role in improving insulin sensitivity, reducing oxidative stress, and modulating inflammatory responses. Recent research in this domain has significantly advanced our understanding of their positive influence on glucose metabolism and lipid profiles, contributing to better glycemic control. Some clinical studies (Table 9.3) with various phenolic compounds have shown promising results, indicating their ability to lower blood sugar levels and enhance overall metabolic health. Furthermore, selected in vitro and animal studies have provided insights into the mechanisms through which these compounds exert their effects, such as improving pancreatic function and enhancing glucose uptake in muscle cells. Although these findings are promising, further research is essential to establish optimal dosages, identify specific compounds with the most beneficial effects, and translate these results into practical dietary recommendations for individuals with diabetes. There are several challenges and limitations that researchers face in establishing the role of phenolic compounds in diabetes management, as detailed in Figure 9.5.

## 9.8 CONCLUSION

Phenolic compounds are bioactive compounds that are abundantly found in fruits, vegetables, grains, and beverages like tea and coffee. They have garnered significant attention for their potential role in managing diabetes, as they exhibit various health-promoting properties, including antioxidant, anti-inflammatory, and insulin-sensitizing effects. Worldwide research indicates that phenolic compounds may improve glycemic control, enhance insulin sensitivity, and reduce the risk of diabetes-related complications. While the preliminary findings are promising, there remains a critical need for further research to fully understand their mechanisms of action, optimal dosages, and the most effective food sources for their use as a therapeutic agent.

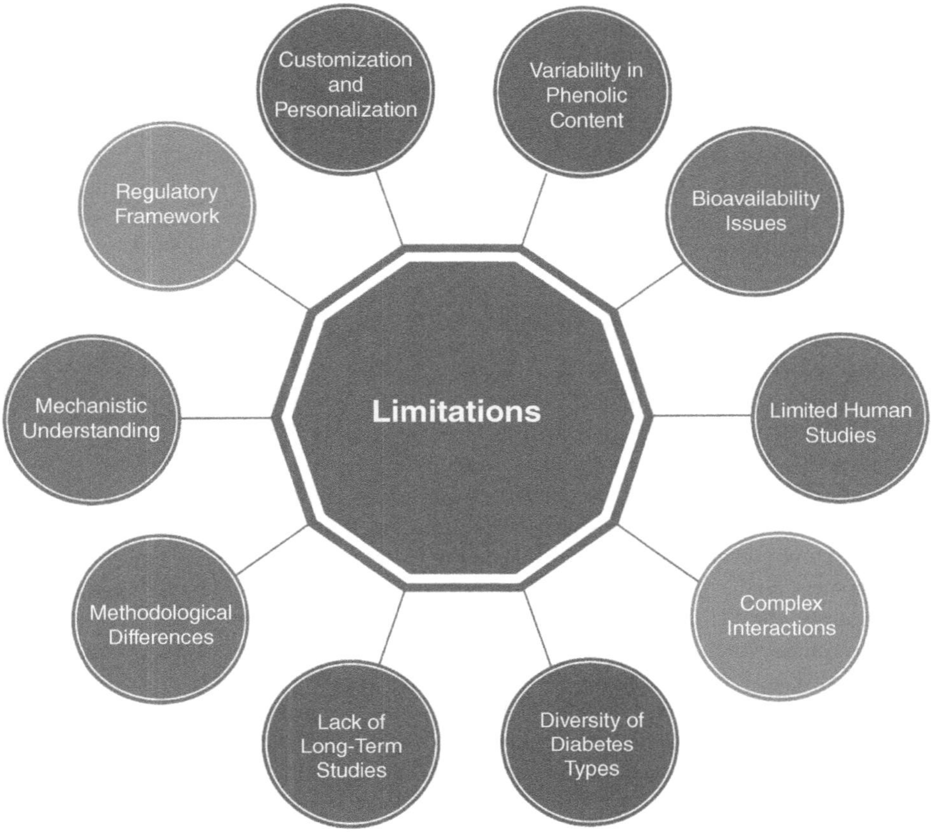


**FIGURE 9.5** This figure outlines the various challenges and limitations encountered by researchers in determining the efficacy of phenolic compounds in diabetes management. These limitations influenced by factors such as absorption, distribution, metabolism, and excretion, pose a significant challenge in assessing their true impact.

However, despite the potential benefits of phenolic compounds, there are challenges associated with translating these research findings into practical dietary recommendations. It is essential to conduct well-designed clinical trials that assess the effects of specific phenolic-rich foods and their impact on various stages of diabetes management. Moreover, individual variability in response to these compounds necessitates personalized approaches in dietary planning. Such research can lead to evidence-based guidelines, enabling healthcare professionals to recommend phenolic-rich foods more effectively as part of a comprehensive diabetes management strategy, prompting more research and, in turn, products that are easily available to the common man for diabetes management.

Incorporating phenolic-rich foods into a balanced diet can significantly contribute to overall health and well-being for individuals with diabetes. Foods such as berries, nuts, dark chocolate, and whole grains not only provide essential nutrients but also enhance the dietary portfolio with beneficial phytochemicals. Encouraging the consumption of these foods, alongside other healthy practices like regular physical activity and weight management, may help to optimize diabetes control and improve the quality of life. Ultimately, a holistic approach that includes the integration of phenolic compounds into daily nutrition can pave the way for more effective diabetes management, underscoring the need for ongoing research and public health initiatives that promote healthy-eating habits. An integrated approach where medical practitioners, policymakers, and nutritionists work together and come up with a cohesive document and work plan may help in reducing the burden of treatment failure for diabetes.

## REFERENCES


Abdalla, M. M. I. (2024). Insulin resistance as the molecular link between diabetes and Alzheimer's disease. *World Journal of Diabetes* **15**, 1430.

Akhtar, M., Taha, N. M., Nauman, A., Mujeeb, I. B., and Al-Nabet, A. D. M. (2020). Diabetic kidney disease: past and present. *Advances in Anatomic Pathology* **27**, 87–97.

Alhabeeb, H., AlFaiz, A., Kutbi, E., AlShahrani, D., Alsuhail, A., AlRajhi, S., Alotaibi, N., Alotaibi, K., AlAmri, S., and Alghamdi, S. (2021). Gut hormones in health and obesity: the upcoming role of short chain fatty acids. *Nutrients* **13**, 481.

Al-Ibraheem, A. M. T., Hameed, A. T. A. Z., Marsool, M. D. M., Jain, H., Prajjwal, P., Khazmi, I., Nazzal, R. S., Al-Najati, H. M. H., Al-Zuhairi, B. H. Y. K., and Razzaq, M. (2024). Exercise induced cytokines, diet, and inflammation and their role in adipose tissue metabolism. *Health Science Reports* **7**, e70034.

Anandhanarayanan, A., Teh, K., Goonoo, M., Tesfaye, S., & Selvarajah, D. (2022). Diabetic neuropathies. In L.J. De Groot, G. Chrousos, K. Dungan, et al. (Eds.), Endotext. South Dartmouth (MA): MDText.com, Inc.

Artasensi, A., Pedretti, A., Vistoli, G., and Fumagalli, L. (2020). Type 2 diabetes mellitus: a review of multi-target drugs. *Molecules* **25**, 1987.

Arumugam, V. A., Thangavelu, S., Fathah, Z., Ravindran, P., Sanjeev, A. M. A., Babu, S., Meyyazhagan, A., Yatoo, M. I., Sharun, K., and Tiwari, R. (2020). COVID-19 and the world with co-morbidities of heart disease, hypertension and diabetes. *Journal of Pure and Applied Microbiology* **14**, 1623–1638.

Banerjee, D., Winocour, P., Chowdhury, T., De, P., Wahba, M., Montero, R., Fogarty, D., Frankel, A., Karalliedde, J., and Mark, P. B. (2022). Management of hypertension and renin-angiotensin-aldosterone system blockade in adults with diabetic kidney disease: Association of British Clinical Diabetologists and the Renal Association UK guideline update 2021. *BMC Nephrology* **23**, 1–31.

Banihani, S., Makahleh, S., El-Akawi, Z., Al-Fashtaki, R., Khabour, O., Gharibeh, M., Saadah, N., Al-Hashimi, F., and Al-Khasieb, N. (2014). Fresh pomegranate juice ameliorates insulin resistance, enhances β-cell function, and decreases fasting serum glucose in type 2 diabetic patients. *Nutrition Research* **34**, 862–867.

Barber, T. M., Kabisch, S., Randeva, H. S., Pfeiffer, A. F., and Weickert, M. O. (2022). Implications of resveratrol in obesity and insulin resistance: a state-of-the-art review. *Nutrients* **14**, 2870.

Basu, A., Crew, J., Ebersole, J. L., Kinney, J. W., Salazar, A. M., Planinic, P., and Alexander, J. M. (2021). Dietary blueberry and soluble fiber improve serum antioxidant and adipokine biomarkers and lipid peroxidation in pregnant women with obesity and at risk for gestational diabetes. *Antioxidants* **10**, 1318.

Bellary, S., Kyrou, I., Brown, J. E., and Bailey, C. J. (2021). Type 2 diabetes mellitus in older adults: clinical considerations and management. *Nature Reviews Endocrinology* **17**, 534–548.

Bellettiere, J., LaMonte, M. J., Healy, G. N., Liles, S., Evenson, K. R., Di, C., Kerr, J., Lee, I.-M., Rillamas-Sun, E., and Buchner, D. (2021). Sedentary behavior and diabetes risk among women over the age of 65 years: the OPACH study. *Diabetes Care* **44**, 563–570.

Berköz, M., Kahraman, T., Shamsulddin, Z. N., and Krośniak, M. (2023). Antioxidant and anti-inflammatory effect of olive leaf extract treatment in diabetic rat brain. *Journal of Basic and Clinical Physiology and Pharmacology* **34**, 187–196.

Bhatti, J. S., Sehrawat, A., Mishra, J., Sidhu, I. S., Navik, U., Khullar, N., Kumar, S., Bhatti, G. K., and Reddy, P. H. (2022). Oxidative stress in the pathophysiology of type 2 diabetes and related complications: Current therapeutics strategies and future perspectives. *Free Radical Biology and Medicine* **184**, 114–134.

Blahova, J., Martiniakova, M., Babikova, M., Kovacova, V., Mondockova, V., and Omelka, R. (2021). Pharmaceutical drugs and natural therapeutic products for the treatment of type 2 diabetes mellitus. *Pharmaceuticals* **14**, 806.

Bu Kang, B., and Huang Chiang, B. (2022). A novel phenolic formulation for treating hepatic and peripheral insulin resistance by regulating GLUT4-mediated glucose uptake. *Journal of Traditional and Complementary Medicine* **12**, 195–205.

Cheng, D. M., Kuhn, P., Poulev, A., Rojo, L. E., Lila, M. A., and Raskin, I. (2012). In vivo and in vitro antidiabetic effects of aqueous cinnamon extract and cinnamon polyphenol-enhanced food matrix. *Food Chemistry* **135**, 2994–3002.

Choleva, M., Matalliotaki, E., Antoniou, S., Asimomyti, E., Drouka, A., Stefani, M., Yannakoulia, M., and Fragopoulou, E. (2022). Postprandial metabolic and oxidative stress responses to grape pomace extract in healthy Normal and overweight/obese women: a randomized, double-blind, Placebo-Controlled Crossover Study. *Nutrients* **15**, 156.

De Bock, M., Derraik, J. G., Brennan, C. M., Biggs, J. B., Morgan, P. E., Hodgkinson, S. C., Hofman, P. L., and Cutfield, W. S. (2013). Olive (*Olea europaea* L.) leaf polyphenols improve insulin sensitivity in middle-aged overweight men: a randomized, placebo-controlled, crossover trial. *PLoS One* **8**, e57622.

Edmonds, M., Manu, C., and Vas, P. (2021). The current burden of diabetic foot disease. *Journal of Clinical Orthopaedics and Trauma* **17**, 88–93.

Eguchi, N., Vaziri, N. D., Dafoe, D. C., and Ichii, H. (2021). The role of oxidative stress in pancreatic β cell dysfunction in diabetes. *International Journal of Molecular Sciences* **22**, 1509.

ElSayed, N. A., Aleppo, G., Aroda, V. R., Bannuru, R. R., Brown, F. M., Bruemmer, D., Collins, B. S., Hilliard, M. E., Isaacs, D., and Johnson, E. L. (2023). 8: standards of care in diabetes—2023. *Diabetes Care* 46, S128–S139.

Fernandes, I., Oliveira, J., Pinho, A., and Carvalho, E. (2022). The role of nutraceutical containing polyphenols in diabetes prevention. *Metabolites* **12**, 184.

Gao, Q., Zhong, C., Zhou, X., Chen, R., Xiong, T., Hong, M., Li, Q., Kong, M., Han, W., and Sun, G. (2019). The association between intake of dietary lycopene and other carotenoids and gestational diabetes mellitus risk during mid-trimester: a cross-sectional study. *British Journal of Nutrition* **121**, 1405–1412.

Gonzalez Ramirez, G., and Bolaños Muñoz, L. (2023). Relationship of sedentary lifestyle with obesity and comorbidities. *In Physical Activity and bariatric surgery*, pp. 3–16. Springer.

Guasch-Ferré, M., and Willett, W. (2021). The Mediterranean diet and health: A comprehensive overview. *Journal of Internal Medicine* **290**, 549–566.

Gusti, A. M., Qusti, S. Y., Alshammari, E. M., Toraih, E. A., and Fawzy, M. S. (2021). Antioxidants-related superoxide dismutase (SOD), catalase (CAT), glutathione peroxidase (GPX), glutathione-S-transferase (GST), and nitric oxide synthase (NOS) gene variants analysis in an obese population: a preliminary case-control study. *Antioxidants* **10**, 595.

Hassan, M. A., Xu, T., Tian, Y., Zhong, Y., Ali, F. A. Z., Yang, X., and Lu, B. (2021). Health benefits and phenolic compounds of *Moringa oleifera* leaves: a comprehensive review. *Phytomedicine* **93**, 153771.

Husen, A. (2022). Herbs, Shrubs and Trees of Potential Medicinal Benefits. CRC Press, 6000 Broken Sound Parkway NW, Suite 300, Boca Raton, FL 33487–2742. https://doi.org/10.1201/9781003205067

Husen, A. (2023a). Exploring Poisonous Plants: Medicinal Values, Toxicity Response and Other Potential Uses. CRC Press, 6000 Broken Sound Parkway NW, Suite 300, Boca Raton, FL 33487–2742. https://doi.org/10.1201/b23017

Husen, A. (2023b). Antidiabetic Medicinal Plants and Herbal Treatments. CRC Press, 6000 Broken Sound Parkway NW, Suite 300, Boca Raton, FL 33487–2742. https://doi.org/10.1201/b23347

Husen, A. (2024a). Antimalarial Medicinal Plants. Taylor & Francis Group, LLC, 6000 Broken Sound Pkwy. NW ISBN: 9781032457154. https://doi.org/10.1201/9781003378396

Husen, A. (2024b). Medicinal Spices and Condiment Crops. Taylor & Francis Group, LLC, 6000 Broken Sound Pkwy. NW ISBN: 9781032469478. https://doi.org/10.1201/9781003387046

Jacobsen, P., Andersen, S., Jensen, B. R., and Parving, H.-H. (2003). Additive effect of ACE inhibition and angiotensin II receptor blockade in type I diabetic patients with diabetic nephropathy. *Journal of the American Society of Nephrology* **14**, 992–999.

Jafarirad, S., Ayoobi, N., Karandish, M., Jalali, M.-T., Haghighizadeh, M. H., and Jahanshahi, A. (2018). Dark chocolate effect on serum adiponectin, biochemical and inflammatory parameters in diabetic patients: a randomized clinical trial. *International Journal of Preventive Medicine* **9**.

Jwala, J., and Kumari, B. (2024). Exploring the Link between Dietary Choices and Diabetes Mellitus: An In-Depth Analysis.

Kang, Q., and Yang, C. (2020). Oxidative stress and diabetic retinopathy: molecular mechanisms, pathogenetic role and therapeutic implications. *Redox Biology* **37**, 101799.

Kao, T.-W., and Huang, C.-C. (2021). Blood pressure management and renal protection: revisiting hypertensive nephropathy. *Journal of the Chinese Medical Association* **84**, 911–916.

Kasprzak-Drozd, K., Oniszczuk, T., Stasiak, M., and Oniszczuk, A. (2021). Beneficial effects of phenolic compounds on gut microbiota and metabolic syndrome. *International Journal of Molecular Sciences* **22**, 3715.

Khadhraoui, B., Ummat, V., Tiwari, B., Fabiano-Tixier, A., and Chemat, F. (2021). Review of ultrasound combinations with hybrid and innovative techniques for extraction and processing of food and natural products. *Ultrasonics Sonochemistry* **76**, 105625.

Khalid, M., Alkaabi, J., Khan, M. A., and Adem, A. (2021). Insulin signal transduction perturbations in insulin resistance. *International Journal of Molecular Sciences* **22**, 8590.

Kolb, H., Kempf, K., Röhling, M., Lenzen-Schulte, M., Schloot, N. C., and Martin, S. (2021). Ketone bodies: from enemy to friend and guardian angel. *BMC Medicine* **19**, 1–15.

Legouis, D., Faivre, A., Cippà, P. E., and de Seigneux, S. (2022). Renal gluconeogenesis: an underestimated role of the kidney in systemic glucose metabolism. *Nephrology Dialysis Transplantation* **37**, 1417–1425.

Leh, H. E., and Lee, L. K. (2022). Lycopene: a potent antioxidant for the amelioration of type II diabetes mellitus. *Molecules* **27**, 2335.

Li, F., Zeng, K., & Ming, J. (2023). Lowering glycemic levels via gastrointestinal tract factors: the roles of dietary fiber, polyphenols, and their combination. Critical Reviews in Food Science and Nutrition, 63(1), 1–37. DOI: 10.1080/10408398.2023.2278169

Li, H., Meng, Y., He, S., Tan, X., Zhang, Y., Zhang, X., Wang, L., and Zheng, W. (2022). Macrophages, chronic inflammation, and insulin resistance. *Cells* **11**, 3001.

Li, N., Cui, C., Xu, J., Mi, M., Wang, J., and Qin, Y. (2024). Quercetin intervention reduced hepatic fat deposition in patients with nonalcoholic fatty liver disease: a randomized, double-blind, placebo-controlled crossover clinical trial. *The American Journal of Clinical Nutrition* **120**, 507–517.

Liu, K., Zhou, R., Wang, B., Chen, K., Shi, L.-Y., Zhu, J.-D., and Mi, M.-T. (2013). Effect of green tea on glucose control and insulin sensitivity: a meta-analysis of 17 randomized controlled trials. *The American Journal of Clinical Nutrition* **98**, 340–348.

Martín, M. Á., and Ramos, S. (2021). Dietary flavonoids and insulin signaling in diabetes and obesity. *Cells* **10**, 1474.

Megha, K., Joseph, X., Akhil, V., and Mohanan, P. (2021). Cascade of immune mechanism and consequences of inflammatory disorders. *Phytomedicine* **91**, 153712.

Merker, L., Bautsch, B.-W., Ebert, T., Guthoff, M., and Isermann, B. (2021). Nephropathy in diabetes. *Experimental and Clinical Endocrinology & Diabetes* **129**, S60–S63.

Mundekkad, D., and Cho, W. C. (2022). Mitophagy induced by metal nanoparticles for cancer treatment. *Pharmaceutics* **14**, 2275.

Mundekkad, D., and Cho, W. C. (2023). Applications of curcumin and its nanoforms in the treatment of cancer. *Pharmaceutics* **15**, 2223.

Mutha, R. E., Tatiya, A. U., and Surana, S. J. (2021). Flavonoids as natural phenolic compounds and their role in therapeutics: an overview. *Future Journal of Pharmaceutical Sciences* **7**, 1–13.

Nagarajan, S., Nagarajan, R., Kumar, J., Salemme, A., Togna, A. R., Saso, L., and Bruno, F. (2020). Antioxidant activity of synthetic polymers of phenolic compounds. *Polymers* **12**, 1646.

Oyedemi, S. O., Nwaogu, G., Chukwuma, C. I., Adeyemi, O. T., Matsabisa, M. G., Swain, S. S., and Aiyegoro, O. A. (2020). Quercetin modulates hyperglycemia by improving the pancreatic antioxidant status and enzymes activities linked with glucose metabolism in type 2 diabetes model of rats: In silico studies of molecular interaction of quercetin with hexokinase and catalase. *Journal of Food Biochemistry* **44**, e13127.

Pannucci, E., Spagnuolo, L., De Gara, L., Santi, L., and Dugo, L. (2023a). Phenolic compounds as preventive and therapeutic agents in diabetes-related oxidative stress, inflammation, advanced glycation end-products production and insulin sensitivity. *Discovery Medicine* **35**, 715–732.

Pannucci, E., Spagnuolo, L., De Gara, L., Santi, L., and Dugo, L. (2023b). Phenolic compounds as preventive and therapeutic agents in diabetes-related oxidative stress, inflammation, advanced glycation end-products production and insulin sensitivity. *Discovery Medicine* **35**.

Papachristoforou, E., Lambadiari, V., Maratou, E., and Makrilakis, K. (2020). Association of glycemic indices (hyperglycemia, glucose variability, and hypoglycemia) with oxidative stress and diabetic complications. *Journal of Diabetes Research* **2020**, 7489795.

Quiroz-Aldave, J., Durand-Vásquez, M., Gamarra-Osorio, E., Suárez-Rojas, J., Roseboom, P. J., Alcalá-Mendoza, R., Coronado-Arroyo, J., Zavaleta-Gutiérrez, F., Concepción-Urteaga, L., and Concepción-Zavaleta, M. (2023). Diabetic neuropathy: past, present, and future. *Caspian Journal of Internal Medicine* **14**, 153.

Rahman, M. M., Rahaman, M. S., Islam, M. R., Rahman, F., Mithi, F. M., Alqahtani, T., Almikhlafi, M. A., Alghamdi, S. Q., Alruwaili, A. S., and Hossain, M. S. (2021). Role of phenolic compounds in human disease: current knowledge and future prospects. *Molecules* **27**, 233.

Reardon, R., Simring, D., Kim, B., Mortensen, J., Williams, D., and Leslie, A. (2020). The diabetic foot ulcer. *Australian Journal of General Practice* **49**, 250–255.

Rodríguez-Daza, M. C., Pulido-Mateos, E. C., Lupien-Meilleur, J., Guyonnet, D., Desjardins, Y., and Roy, D. (2021). Polyphenol-mediated gut microbiota modulation: toward prebiotics and further. *Frontiers in Nutrition* **8**, 689456.

Sardarinia, M., Asgari, S., Hizomi Arani, R., Eskandari, F., Azizi, F., Khalili, D., and Hadaegh, F. (2022). Incidence and risk factors of severe non-proliferative/proliferative diabetic retinopathy: more than a decade follow up in the Tehran Lipids and Glucose Study. *Journal of Diabetes Investigation* **13**, 317–327.

Sarkar, D., Christopher, A., and Shetty, K. (2022). Phenolic bioactives from plant-based foods for glycemic control. *Frontiers in Endocrinology* **12**, 727503.

Sasaki, H., Kawamura, N., Dyck, P. J., Dyck, P. J. B., Kihara, M., and Low, P. A. (2020). Spectrum of diabetic neuropathies. *Diabetology International* **11**, 87–96.

Serina, J. J. C., and Castilho, P. C. M. F. (2022). Using polyphenols as a relevant therapy to diabetes and its complications, a review. *Critical Reviews in Food Science and Nutrition* **62**, 8355–8387.

Shahwan, M., Alhumaydhi, F., Ashraf, G. M., Hasan, P. M., and Shamsi, A. (2022). Role of polyphenols in combating type 2 Diabetes and insulin resistance. *International Journal of Biological Macromolecules* **206**, 567–579.

Sharma, P., Hajam, Y. A., Kumar, R., and Rai, S. (2022). Complementary and alternative medicine for the treatment of diabetes and associated complications: a review on therapeutic role of polyphenols. *Phytomedicine Plus* **2**, 100188.

Singh, A. K., Singla, R. K., and Pandey, A. K. (2023). Chlorogenic acid: a dietary phenolic acid with promising pharmacotherapeutic potential. *Current Medicinal Chemistry* **30**, 3905–3926.

Singh, R. B., Nabavizadeh, F., Fedacko, J., Pella, D., Vanova, N., Jakabcin, P., Fatima, G., Horuichi, R., Takahashi, T., and Mojto, V. (2022). Dietary approaches to stop hypertension via indo-mediterranean foods, may be superior to DASH diet intervention. *Nutrients* **15**, 46.

Skou, S. T., Mair, F. S., Fortin, M., Guthrie, B., Nunes, B. P., Miranda, J. J., Boyd, C. M., Pati, S., Mtenga, S., and Smith, S. M. (2022). Multimorbidity. *Nature Reviews Disease Primers* **8**, 48.

Stull, A. J., Cash, K. C., Champagne, C. M., Gupta, A. K., Boston, R., Beyl, R. A., Johnson, W. D., and Cefalu, W. T. (2015). Blueberries improve endothelial function, but not blood pressure, in adults with metabolic syndrome: a randomized, double-blind, placebo-controlled clinical trial. *Nutrients* **7**, 4107–4123.

Su, J., Tang, L., Luo, Y., Xu, J., & Ouyang, S. (2023). Research progress on drugs for diabetes based on insulin receptor/insulin receptor substrate. Biochemical Pharmacology, 212, 115830. https://doi.org/10.1016/j.bcp.2023.115830

Sun, W., and Shahrajabian, M. H. (2023). Therapeutic potential of phenolic compounds in medicinal plants—natural health products for human health. *Molecules* **28**, 1845.

Tvrdá, E., and Benko, F. (2020). Free radicals: what they are and what they do. *In Pathology*, pp. 3–13. Elsevier.

Vafaeipour, Z., Razavi, B. M., and Hosseinzadeh, H. (2022). Effects of turmeric (*Curcuma longa*) and its constituent (curcumin) on the metabolic syndrome: an updated review. *Journal of Integrative Medicine* **20**, 193–203.

Vodovotz, Y., Barnard, N., Hu, F. B., Jakicic, J., Lianov, L., Loveland, D., Buysse, D., Szigethy, E., Finkel, T., and Sowa, G. (2020). Prioritized research for the prevention, treatment, and reversal of chronic disease: recommendations from the lifestyle medicine research summit. *Frontiers in Medicine* **7**, 585744.

Wan, C., Ouyang, J., Li, M., Rengasamy, K. R., and Liu, Z. (2024). Effects of green tea polyphenol extract and epigallocatechin-3-O-gallate on diabetes mellitus and diabetic complications: recent advances. *Critical Reviews in Food Science and Nutrition* **64**, 5719–5747.

Wu, T., Huang, W., He, M., and Yue, R. (2022). Effects of cinnamon supplementation on lipid profiles among patients with metabolic syndrome and related disorders: a systematic review and meta-analysis. *Complementary Therapies in Clinical Practice* **49**, 101625.

Xavier, J. R., Sameer, B., Gupta, D., Mehta, S., & Chauhan, O. P. (2024). Bioactive compounds of foods: phytochemicals and peptides. Food and Humanity, 2, 100354. DOI:10.1016/j.foohum.2024.100354

Zare Javid, A., Hormoznejad, R., Yousefimanesh, H. A., Zakerkish, M., Haghighihneshn, M. H., Dehghan, P., and Ravanbakhsh, M. (2017). The impact of resveratrol supplementation on blood glucose, insulin, insulin resistance, triglyceride, and periodontal markers in type 2 diabetic patients with chronic periodontitis. *Phytotherapy Research* **31**, 108–114.

Zheng, S., Zhang, H., Liu, R., Huang, C.-L., Li, H., Deng, Z.-Y., and Tsao, R. (2021). Do short chain fatty acids and phenolic metabolites of the gut have synergistic anti-inflammatory effects?–New insights from a TNF-α-induced Caco-2 cell model. *Food Research International* **139**, 109833.

# 10 Anti-Obesity Activity of Phenolic Compounds

*Sumitra Nain, Arvind Kumar Patel, Omsatyam, Shankul Kumar, Vinod Kumar Singh, and Vedant Kumar Prajapati[*]*

## 10.1 INTRODUCTION

### 10.1.1 Overview of Obesity

Obesity is a major global health concern caused by excessive body fat buildup, which can lead to negative health outcomes. It is influenced by a complex interplay of genetic, environmental, and lifestyle factors, and its prevalence is increasing worldwide. The World Health Organization (WHO) has emphasized the importance of addressing obesity due to its connection with various chronic diseases and its impact on public health systems. Despite various measures, no country has successfully halted the rising trend of obesity, which continues to impose a significant burden on global healthcare systems (Kebekova et al., 2024; Seidell, 2024). Obesity has been associated with an increased risk of cardiovascular disease, type 2 diabetes, hypertension, fatty liver disease, obstructive sleep apnea, and certain malignancies. It causes approximately 300,000 fatalities each year and generates significant healthcare expenses. The illness also has an impact on mental health, which complicates management and intervention options (Ahmed et al., 2024). Obesity is caused by a combination of genetic factors, environmental influences, and lifestyle choices, including sedentary behavior and poor dietary habits. Global trade liberalization, urbanization, and economic expansion have exacerbated the obesity epidemic by fostering obesogenic conditions (Lan and Sulaiman, 2024).

### 10.1.2 Global Prevalence of Obesity

Obesity is a major public health concern around the world, affecting millions of people from all demographics. According to recent estimates, over 2 billion people and 380 million children and adolescents are overweight or obese, with rising rates observed in low- and middle-income nations (Seidell, 2024). In 2016, an estimated 650 million adults worldwide were classed as obese, accounting for roughly 13% of the adult population. In North America, the obesity rate among adults was 42.4% in 2017–2018 (Park and Blanck, 2024). Obesity prevalence among children and adolescents is reported at 8.5%, with considerable geographical variations ranging from 0.4% in Vanuatu to 28.4% in Puerto Rico. Obesity rates have increased 1.5-fold between 2000 and 2023, which is an alarming trend (Zhang et al., 2024). The global incidence of maternal obesity is anticipated to be 20.9%, rising to 23.3% by 2030. Overweight and obesity in pregnancy affect roughly half of all pregnancies worldwide, emphasizing the need for enhanced maternal health services (Kent et al., 2024). Despite these worrying statistics, specific regions, particularly in Asia, report lower prevalence rates, which may be understated due to incorrect body mass index (BMI) standards for local people (Seidell, 2024).

[*] Corresponding author: vedant.mpharm2012@gmail.com

DOI: 10.1201/9781003527671-10

### 10.1.3 Complication of Obesity

Obesity is a serious health risk that has been related to several complications impacting various physiological systems. One of the most serious consequences is cardiovascular disease, which includes hypertension, coronary artery disease, and heart failure. Excess adipose tissue adds to insulin resistance, which raises the risk of type 2 diabetes. Obese people frequently experience respiratory problems. Obstructive sleep apnea and obesity hypoventilation syndrome develop as a result of fat accumulation around the neck and chest, which causes breathing difficulty when sleeping. Musculoskeletal disorders, especially osteoarthritis, are widespread because of the increased mechanical pressure on weight-bearing joints, resulting in discomfort and limited movement (Chandrasekaran and Weiskirchen, 2024; Lima et al., 2023). Obesity has been linked to several malignancies, including colorectal, breast, and endometrial cancers. The specific processes are unclear; however, they may include chronic inflammation and hormone abnormalities (Burke, 2010; Camilleri et al., 2017).

Obesity has been related to psychological impacts such as depression and anxiety, which are often the result of societal stigma and a lower quality of life. Obesity can also cause non-alcoholic fatty liver disease (NAFLD), which is characterized by fat accumulation in the liver and may proceed to cirrhosis or liver failure. Obesity also exacerbates disorders like hypertension and dyslipidemia, raising the risk of stroke. Given these significant problems, addressing obesity through lifestyle changes, medical interventions, and public health efforts is crucial for improving individual health and reducing healthcare burdens (Figure 10.1) (Hittepole, 2024; Luzi, 2021; Prendergast et al., 2022).

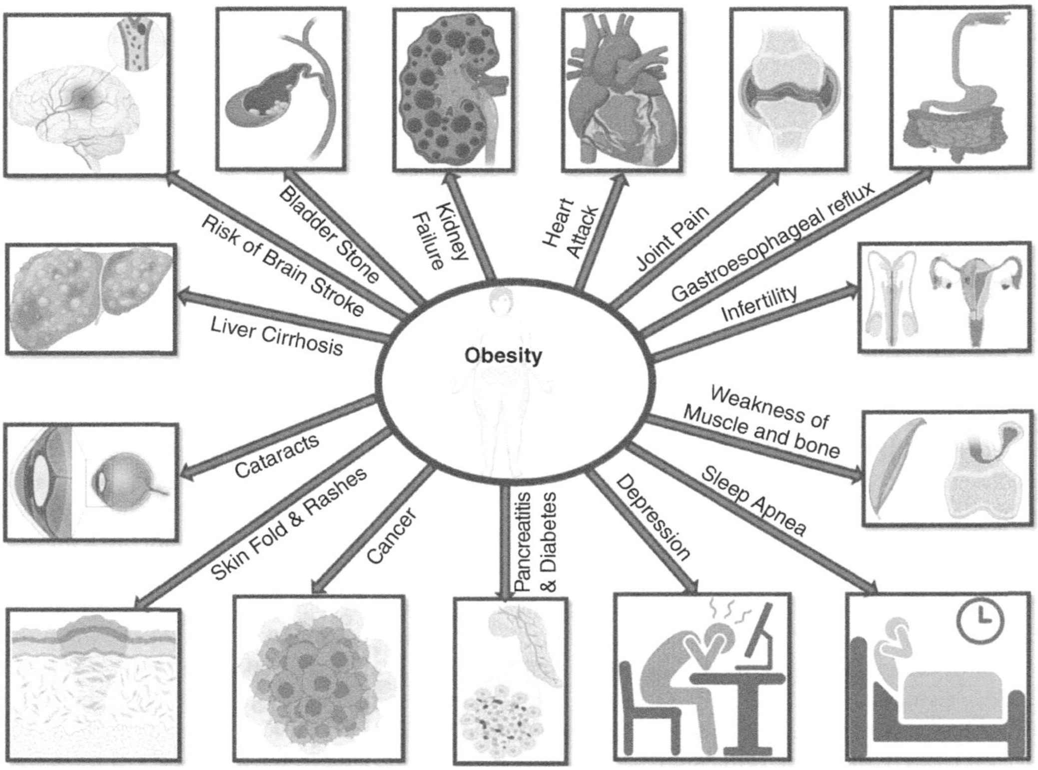


**FIGURE 10.1** Obesity is a multifactorial disorder linked to diverse systemic complications, including cardiovascular disease, stroke risk, liver cirrhosis, renal failure, musculoskeletal disorders, gastrointestinal reflux, sleep apnea, depression, infertility, dermatological conditions, ocular complications, and increased cancer susceptibility.

### 10.1.4 Intervention and Policy for Obesity Control

Several nations have introduced legislation to enhance food quality and limit added sugar consumption, but these measures have not adequately addressed the obesity problem. Current guidelines focus on diagnostic and therapeutic methods that primary care practitioners can use to treat obesity effectively. While obesity is a significant public health concern, it is essential to understand the broader socioeconomic and cultural factors that contribute to its prevalence. Addressing obesity requires comprehensive and diversified methods that include policy changes, public health initiatives, and individual lifestyle improvements to effectively reduce this global epidemic (Pandey et al., 2022; Peri and Eisenberg, 2024).

### 10.1.5 Role of Diet Management and Bioactive Compounds in Obesity Control

Diet and bioactive chemicals play an increasingly important role in reducing obesity and related metabolic diseases. Bioactive substances, notably polyphenols, and other phytochemicals have demonstrated potential in regulating energy metabolism, appetite, and inflammation, thereby contributing to effective weight management measures (Konstantinidi and Koutelidakis, 2019).

#### 10.1.5.1 Dietary Patterns and Energy Balance

A well-balanced diet, particularly the sustainable Mediterranean diet, emphasizes whole foods and can help with weight management by encouraging a good energy balance (Falakacılar and Yücecan, 2024). Reducing energy-dense foods and employing techniques such as calorie restriction can help individuals lose weight (Rodrigues et al., 2022). A well-balanced diet rich in whole grains, lean meats, fruits, and vegetables helps to control calorie consumption while increasing satiety. The Mediterranean and Dietary Approaches to Stop Hypertension (DASH) diets, which emphasize nutrient-dense meals, aid in weight loss by limiting the consumption of energy-dense, processed foods. Portion control and careful eating help to prevent excessive calorie consumption. Physical activity supplemented dietary efforts by increasing energy expenditure. Long-term obesity control requires sustainable lifestyle improvements, rather than restricted diets (Hall et al., 2012).

#### 10.1.5.2 Bioactive Compounds and Their Mechanisms

Bioactive substances, such as catechins, capsaicinoids, and polyphenols, can boost thermogenesis, promote fat oxidation, and regulate appetite (Bawa et al., 2017). These chemicals also affect lipid metabolism and inflammatory pathways, both of which are crucial in the treatment of obesity-related conditions (Pelczyńska et al., 2023).

#### 10.1.5.3 Functional Foods and Health Benefits

Specific functional foods, such as green tea, berries, and nuts, include bioactive chemicals that may aid in weight loss through a variety of processes, including increased satiety and reduced fat absorption (Konstantinidi and Koutelidakis, 2019). The inclusion of these foods in daily meals can provide a comprehensive strategy for preventing and managing obesity (Bawa et al., 2017). While the effects of bioactive chemicals in weight management appear intriguing, more research is required. Understanding the correct quantities and combinations of these substances will be critical to establishing effective dietary interventions (Breda and de Kok, 2018).

### 10.1.6 Introduction to Phenolic Compounds and Their Health Benefits

Phenolic chemicals, generally derived from plants, are known for their numerous health benefits, which are largely due to their antioxidant, anti-inflammatory, and anti-carcinogenic qualities. These chemicals, which include flavonoids and phenolic acids, are abundant in fruits, vegetables, nuts, and beverages such as tea and wine (Nemzer et al., 2025; Rayess et al., 2024). Their health-promoting properties are linked to a reduction in oxidative stress, which is crucial in preventing chronic diseases such as cardiovascular disease, diabetes, and certain malignancies (Kyazimova and Kornyakova, 2024). Phenolic chemicals improve immunological function and brain health, demonstrating their diverse impact on nutrition. Furthermore, the anti-obesity effects of phenolic compounds are gaining popularity due to their diverse mechanisms that address numerous areas of obesity treatment. These chemicals can regulate adipogenesis, increase thermogenesis, and improve metabolic profiles, making them interesting candidates for obesity intervention techniques (Özkaya, 2021). The bioavailability of these chemicals is regulated by their chemical structure and the food matrix in which they are ingested, emphasizing the relevance of dietary context in maximizing their health effects (Alaya et al., 2024). While the health advantages of phenolic compounds are well-documented, further research is needed to fully understand their modes of action and maximize their bioavailability in human diets (Sattar et al., 2024).

## 10.2 PHENOLIC COMPOUNDS HAVE ANTI-OBESITY ACTIVITY

According to their chemical structure, a wide range of organic molecules known as phenolic compounds are distinguished by their incorporation of one or more hydroxyl (–OH) groups joined to an aromatic benzene ring (Hättenschwiler and Vitousek, 2000). The following is a definition of phenolic compounds based on their functions (Table 10.1):

*Antioxidants*: By rescuing free radicals and lowering oxidative stress, phenolic substances function as potent antioxidants. Examples include phenolic acids (ferulic acid) and flavonoids (catechins, quercetin) (Aceituno-Medina et al., 2015; Tessema et al., 2023).

*Antimicrobial agents*: Certain phenolic compounds help protect plants and preserve food by preventing the emergence of bacteria, fungi, and viruses. Phenolic acids (gallic acid) and tannins are two examples (Lobiuc et al., 2023).

*Anti-inflammatory agents*: By blocking inflammatory enzymes and pathways, several phenolics aid in the reduction of inflammation. Examples include resveratrol (found in grapes) and curcumin (found in turmeric) (Gupta et al., 2013).

*Contributors to color and pigment*: Phenolic chemicals, particularly flavonoids and anthocyanins, give plants their pigmentation, which influences the colors of their fruits, flowers, and leaves. Red cabbage and berries are two examples of anthocyanins (Mattioli et al., 2020).

*Plant growth regulators*: Certain phenolics have an impact on the germination of seeds and the establishment of roots, among other aspects of plant growth and development. Salicylic acid is an example of a plant hormone (Jovanović et al., 2024).

*Agents of bitterness and astringency*: Phenolic chemicals give several foods and drinks their distinct flavors, such as the astringency of tea and wine and the bitterness of coffee. Catechins and tannins are two examples (Lesschaeve and Noble, 2005).

*UV protection*: Phenolic chemicals in plants absorb UV rays, shielding cells from harm brought on by prolonged exposure to sunshine. Flavonoids in plant epidermal tissues are one example (Table 10.1) (Ferreyra et al., 2021).

## 10.3 MECHANISMS OF ANTI-OBESITY ACTIVITY OF PHENOLIC COMPOUNDS

Phenolic compounds, naturally occurring in plants, have been extensively studied for their potential to combat obesity. These compounds exert their anti-obesity effects through multiple mechanisms, targeting various physiological processes. Recent studies have employed network pharmacology to elucidate the anti-obesity mechanisms of polyphenols. For example, phenolics from *Artemisia argyi* have been shown to target multiple pathways, including PI3K-Akt and cAMP signaling, to combat obesity. These findings highlight the complexity of polyphenol interactions at the molecular level (Liu et al., 2024). Below is a detailed exploration of the key mechanisms supported by evidence from recent studies.

**TABLE 10.1**
**Major Classes of Phenolic Compounds**

| Compounds | Examples | Sources | Structure | Ref. |
|---|---|---|---|---|
| Phenolic acids and simple phenols | Derivatives of benzoic acid, such as gallic acid | *Emblica officinalis*, *Terminalia chebula*, *Mangifera indica*, *Camellia sinensis* | 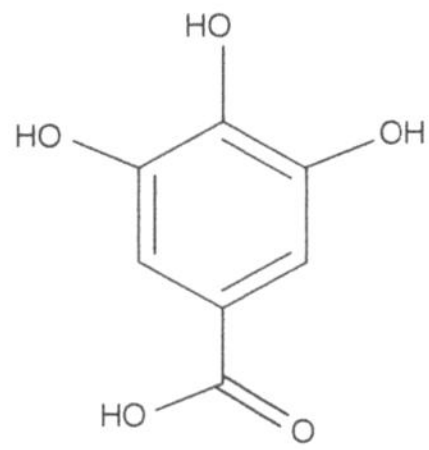  | Deka et al. (2022) |
| | Derivatives of cinnamic acid | *Cinnamomum cassia* | 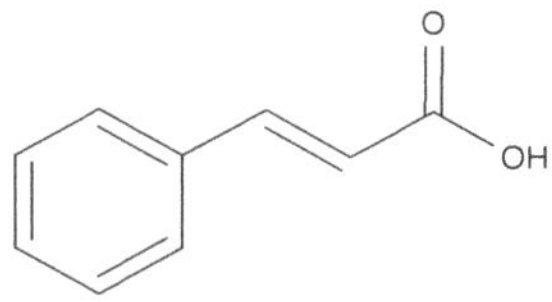  | |
| Flavonoids | Flavanols quercetin | Apple, grapes, onions | 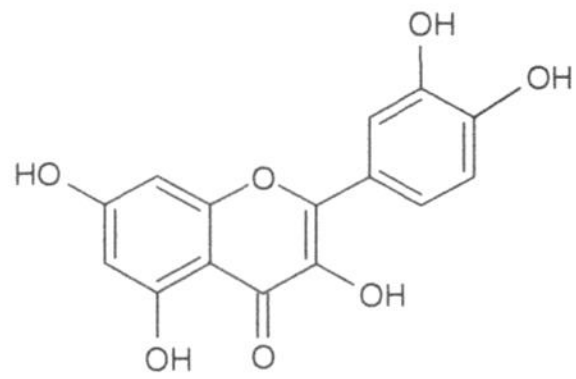  | Aceituno-Medina et al. (2015) |
| | Flavones apigenin | Parsley, chamomile, celery, and oregano | 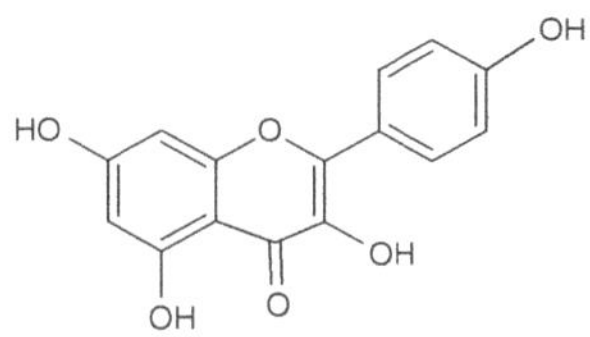  | Mushtaq et al. (2023) |
| | Flavanols catechins | Green tea, black tea, cocoa, and certain fruits like apples, berries, and grapes | 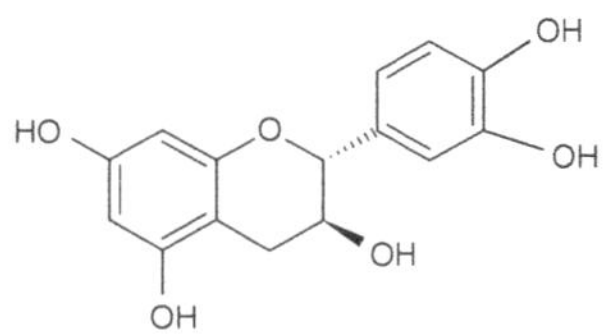  | Salvamani et al. (2014) |
| | Flavanones, such as naringenin | Cherries, tomatoes, and a few herbs, such as Greek oregano | 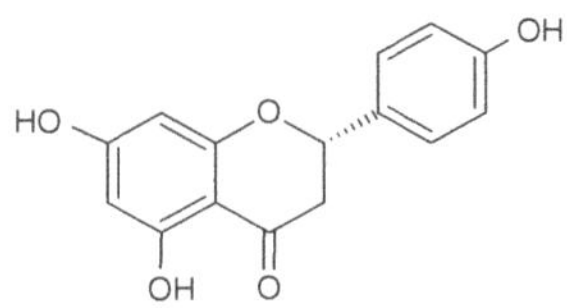  | Manchope et al. (2022) |
| | Isoflavones, such as daidzein | Alfalfa, red clover, soybean, and some legumes | 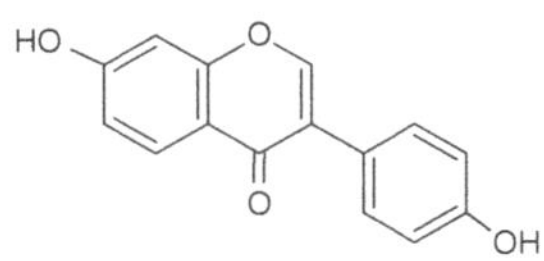  | Huang et al. (2023) |
| | Anthocyanins, which give plants their red, purple, and blue hues, include cyanidin | Red-colored berries like grapes, blackberry, cherry, cranberry | 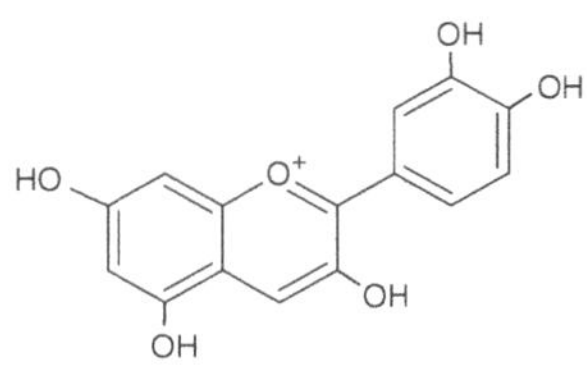  | Singh et al. (2022) |

*(Continued)*

**TABLE 10.1** *(Continued)*
**Major Classes of Phenolic Compounds**

| Compounds | Examples | Sources | Structure | Ref. |
|---|---|---|---|---|
| Stilbenes | Monomeric stilbene Resveratrol | Grapes, red wine, and peanuts | 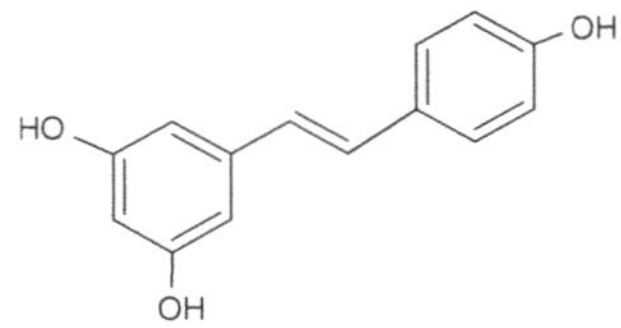 | Wang et al. (2019) |
| | Pinosylvin | Pinewood | 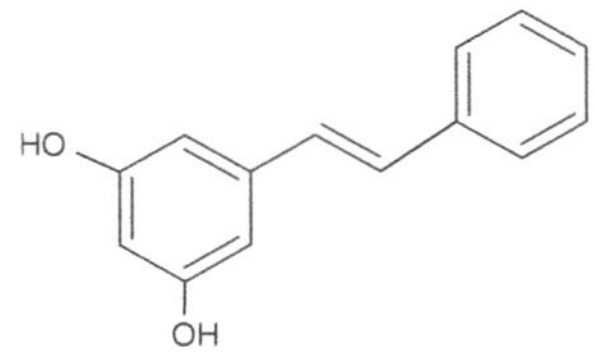 | Chakraborty et al. (2021) |
| | Oligomeric and polymeric stilbenes Viniferins | Grapes and wines | 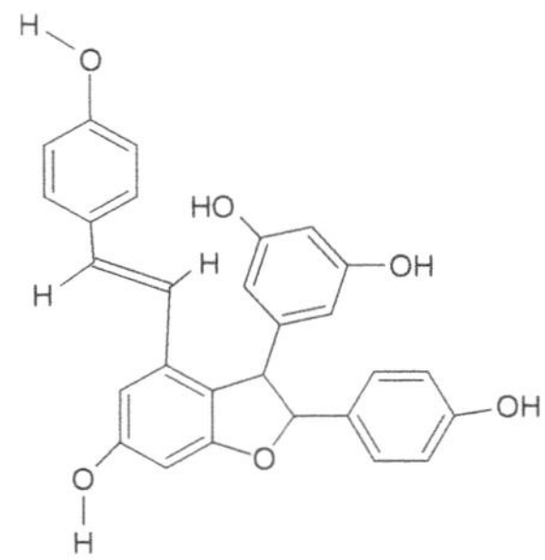 | Fuloria et al. (2022) |
| Lignans | Secoisolariciresinol (SECO) | *Linum usitatissimum*, or flaxseeds, the seeds of sesame (*Sesamum indicum*), whole grains (barley, oats, rye, and wheat), vegetables (garlic, asparagus, carrots, and broccoli) | 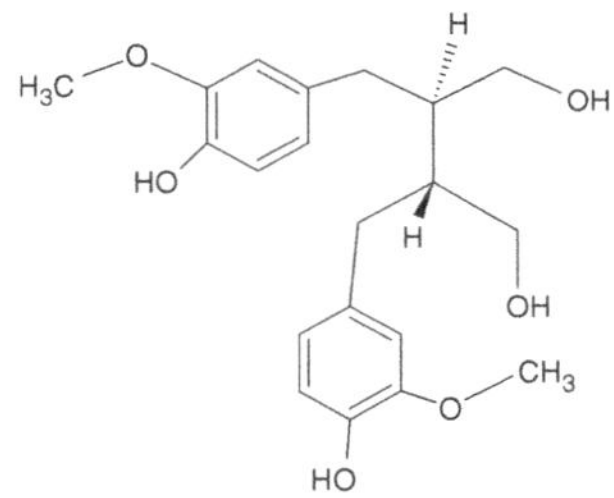 | Tavčar Benković et al. (2017) |

## 10.3.1 Regulation of Lipid Metabolism

Phenolic compounds influence lipid metabolism by inhibiting lipogenesis (fat synthesis) and promoting lipolysis (fat breakdown). For instance, rose polyphenols have been shown to downregulate genes such as Scd1, Hmgcr, and Acat1, which are involved in lipid synthesis, while upregulating genes like adipose triglyceride lipase (ATGL) and hormone-sensitive lipase (HSL), which promote fat breakdown (Ye et al., 2023). Similarly, quercetin and epigallocatechin gallate (EGCG) inhibit key enzymes like pancreatic lipase, reducing fat absorption in the intestine (Carrasco-Pozo et al., 2019).

### 10.3.1.1 Inhibition of Adipocyte Differentiation and Proliferation

Adipogenesis, the process of fat cell formation, is a critical target for anti-obesity interventions. Phenolic compounds such as curcumin, ellagic acid, and resveratrol inhibit the differentiation of pre-adipocytes into mature adipocytes by downregulating key transcription factors like PPAR-γ and C/EBP-α (He et al., 2024). For example, phenolic extracts from *Hibiscus sabdariffa* have been shown to reduce lipid accumulation in human-adipose-derived stem cells by downregulating PPAR-γ and aP2, two markers of adipogenesis (Singh et al., 2023).

#### 10.3.1.2 Promotion of Thermogenesis and Energy Expenditure

Brown adipose tissue (BAT) plays a pivotal role in energy expenditure through thermogenesis, a process mediated by uncoupling protein 1 (UCP1). Certain polyphenols, such as resveratrol and pterostilbene, activate AMPK-PGC-1α signaling, leading to increased UCP1 expression and enhanced thermogenesis in BAT (Wen et al., 2024). This mechanism not only increases energy expenditure but also promotes the browning of white adipose tissue (WAT), converting it into beige adipose tissue, which resembles BAT in function (Abra and Assis, 2020).

#### 10.3.1.3 Modulation of Gut Microbiota

The gut microbiota plays a crucial role in energy homeostasis and metabolic health. Polyphenols, such as those from pomegranate peel, tea, and blueberries, modulate the composition of gut microbiota by increasing the abundance of beneficial bacteria (e.g., *Lactobacillus*, *Bacteroides*) and reducing pathogenic bacteria (e.g., *Blautia*). This modulation enhances the production of short-chain fatty acids (SCFAs), which improve intestinal barrier function, reduce inflammation, and regulate lipid metabolism (Shi et al., 2023; Tian et al., 2024).

#### 10.3.1.4 Suppression of Inflammation and Oxidative Stress

Chronic inflammation and oxidative stress are key contributors to obesity and its comorbidities. Polyphenols exhibit potent anti-inflammatory and antioxidant properties, suppressing the production of pro-inflammatory cytokines (e.g., TNF-α, IL-6) and inhibiting the activation of inflammatory pathways like TLR4 signaling (Suo et al., 2023). For example, tea polyphenols have been shown to reduce inflammation by inhibiting histone deacetylases (HDACs) and enhancing intestinal tight junction expression, thereby preserving gut barrier integrity (Tian et al., 2024).

#### 10.3.1.5 Regulation of Appetite and Food Intake

Certain polyphenols, such as EGCG, have been shown to suppress appetite and reduce food intake by modulating neurohormonal pathways involved in energy balance. These compounds influence the expression of hormones like leptin and ghrelin, which regulate hunger and satiety. Additionally, polyphenols like curcumin and quercetin have been reported to inhibit digestive enzymes, such as pancreatic lipase, further reducing fat absorption and calorie intake (Carrasco-Pozo et al., 2019; He et al., 2024).

#### 10.3.1.6 Targeting mTOR Signaling

The mechanistic target of the rapamycin (mTOR) signaling pathway is a key regulator of cell growth, metabolism, and energy homeostasis. Polyphenols like curcumin and resveratrol have been shown to modulate mTOR signaling, thereby inhibiting adipogenesis and promoting lipolysis. This mechanism highlights the potential of polyphenols to target obesity at the molecular level (Cao et al., 2022).

#### 10.3.1.7 Improvement of Insulin Sensitivity and Glucose Metabolism

Obesity is often associated with insulin resistance and impaired glucose metabolism. Polyphenols, such as those from green tea and blueberries, improve insulin sensitivity by enhancing glucose uptake in skeletal muscle and reducing hepatic glucose production (Jiao et al., 2019). For instance, EGCG has been shown to activate adenosine monophosphate-activated protein kinase (AMPK), a key regulator of glucose and lipid metabolism, thereby improving insulin sensitivity (Carrasco-Pozo et al., 2019).

#### 10.3.1.8 Synergistic Effects of Polyphenol-Rich Diets

The Mediterranean diet, rich in polyphenols from olive oil, fruits, and vegetables, has been associated with a lower risk of obesity. The combination of polyphenols from different sources may exert synergistic effects, enhancing their anti-obesity potential. For example, the combination of pomegranate polyphenols and inulin has been shown to reduce body weight and improve lipid metabolism in high-fat diet (HFD)-induced obesity (Shi et al., 2023).

#### 10.3.1.9 Modulation of Adipokine Production

Phenolic compounds have been demonstrated to significantly impact the production of adipokines, particularly adiponectin, which is essential for enhancing fatty acid oxidation and insulin sensitivity. Various studies suggest that these compounds can modulate adipokine secretion, thereby potentially alleviating obesity-related complications. An increase in adiponectin has been observed with the use of grape polyphenol concentrate 'Enoant,' which significantly elevates adiponectin levels while reducing pro-inflammatory adipokines such as visfatin and resistin in models of metabolic syndrome (Derdemezis et al., 2011). Specific polyphenols, including quercetin and resveratrol, have shown the capacity to reduce visfatin secretion, which may indirectly promote higher adiponectin levels (Derdemezis et al., 2011). Phenolic compounds exhibit anti-inflammatory effects that can mitigate adipose tissue inflammation, a critical factor in adipokine dysregulation (Kirichenko et al., 2022). These compounds also stimulate fatty acid β-oxidation and inhibit adipocyte differentiation, contributing to improved metabolic profiles (Bullón-Vela et al., 2018). While the evidence supports the beneficial role of phenolic compounds in enhancing adiponectin levels and improving metabolic health, further research is necessary to fully elucidate the mechanisms and long-term effects of these compounds on adipokine regulation and obesity management.

#### 10.3.1.10 Downregulation of Lipogenic Gene Expression

The modulation of gene expression involved in lipogenesis by phenolic compounds represents a significant area of research, particularly concerning the transcription factor sterol regulatory element-binding protein-1c (SREBP-1c). These compounds have been demonstrated to decrease the expression of SREBP-1c, which plays a crucial role in regulating lipogenic genes in the liver. This response can be attributed to various mechanisms, including the direct inhibition of SREBP-1c and the modulation of other transcription factors.

*Role of SREBP-1c in Lipogenesis*: SREBP-1c is essential for lipid synthesis, controlling genes responsible for cholesterol and fatty acid production (Guasch-Ferré and Willett, 2021). Its expression is induced by insulin, linking it to metabolic processes in the liver.

The impact of phenolic compounds, such as hesperidin and kaempferol, has been shown to inhibit the expression of SREBP-1c and other adipogenic genes, leading to reduced triglyceride accumulation in adipocytes (Gómez-Zorita et al., 2017). Polyphenols also enhance fatty acid β-oxidation, further counteracting lipogenesis (Rodriguez-Ramiro et al., 2016).

*Broader implications*: While phenolic compounds exhibit beneficial effects on lipogenesis, it is essential to consider that excessive modulation of SREBP-1c could disrupt normal lipid metabolism, potentially leading to adverse effects in certain contexts, such as in individuals with metabolic disorders.

## 10.4 EVIDENCE-BASED SCIENTIFIC STUDY OF PHENOLIC FOR ANTI-OBESITY ACTIVITY

### 10.4.1 *In-vitro* Studies of Phenolic Compounds for Weight Management

Numerous in vitro studies have examined the effects of phenolic compounds on lipid accumulation and adipocyte development, suggesting that they may possess anti-adipogenic properties. A summary of the noteworthy findings is provided in Table 10.2.

The results of in vitro research show that phenolic chemicals have a major effect on lipid accumulation and adipocyte development. By suppressing key adipogenic genes and reducing triglyceride

**TABLE 10.2**
**Findings of *In-Vitro* Studies of Phenolic Compounds for Weight Management**

| S. No. | Studies | Findings | Ref. |
|---|---|---|---|
| 1 | Phenolic compounds inhibit 3T3-L1 adipogenesis depending on the stage of differentiation and their binding affinity to PPARγ | Resveratrol and quercetin prevented lipid buildup during differentiating. Myricetin and apigenin worked well in the early and late phases, whereas hesperidin worked best in the middle. The early phases of differentiation were the main ones affected by phenolic acids. Downregulation of genes such as Scd1 and Lpl was linked to these effects, which were also linked to the compounds' affinity for binding to PPARγ. | Aranaz et al. (2019) |
| 2 | Phenolic compounds apigenin, hesperidin, and kaempferol reduce in vitro lipid accumulation in human adipocytes | All phases of adipogenesis were blocked by hesperidin, whereas early-stage genes were impacted by kaempferol. All three substances decreased the buildup of triglycerides in mature adipocytes, in part by promoting lipolysis and, in the case of hesperidin and apigenin, by inhibiting lipogenesis. | Gómez-Zorita et al. (2017) |
| 3 | Phenolic compounds from coffee by-products modulate adipogenesis-related inflammation, mitochondrial dysfunction, and insulin resistance in adipocytes, via insulin/PI3K/AKT signaling pathways | Adipocyte dysfunction brought on by inflammation was lessened by phenolic extracts, which also encouraged adipocyte browning, hindered adipogenesis, and reduced macrophage-adipocyte interactions. | Rebollo-Hernanz et al. (2019) |
| 4 | Polyphenols from Indian cereal grains inhibit 3T3-L1 adipogenesis through modulating early and late phase adipogenic markers | The polyphenols found in pearl millet, proso millet, small millet, and kodo millet showed promise in preventing obesity by preventing lipid buildup during adipocyte differentiation. | Huhtamaki and Gamesman (2022) |

accumulation, quercetin, resveratrol, apigenin, hesperidin, and kaempferol exhibit stage-specific inhibitory effects on adipogenesis. Furthermore, polyphenols from Indian cereal grains efficiently inhibit lipid formation, while phenolics from coffee by-products encourage adipocyte browning and mitigate inflammation-induced adipocyte dysfunction. These findings suggest that phenolic compounds, which target adipocyte growth and lipid metabolism, hold considerable promise as anti-obesity medications. To confirm these results in vivo and investigate potential clinical uses, further investigation is required.

### 10.4.2 Animal Studies of Phenolic Compounds for Weight Management

Widely included in plant-based diets, phenolic chemicals have been thoroughly investigated for their possible impacts on animal models of obesity, weight gain, and fat metabolism. According to research, these bioactive substances affect energy expenditure, lipid metabolism, and adipogenesis, which helps people control their weight and prevent fat from accumulating. Decrease in Body Weight and Fat Mass: Using rodent models (mice and rats), several studies have shown that dietary supplementation with phenolic compounds, including quercetin, curcumin, resveratrol, and catechins (from green tea), significantly reduces body weight, fat mass, and adipocyte size. In models of obesity brought on by a HFD, these consequences are more noticeable (Table 10.3).

**TABLE 10.3**
**Findings of Animal Studies of Phenolic Compounds for Weight Management**

| S. No. | Studies | Finding | Ref. |
|---|---|---|---|
| 1 | Phenolic compounds rutin and o-coumaric acid ameliorate obesity induced by high-fat diet in rats | In this study, Wistar rats given a high-fat diet (HFD) were used to test the anti-obesity benefits of rutin and o-coumaric acid. Adipose tissue weight, liver weight, and body weight were all considerably decreased by supplementing with these substances. Improvements were also observed in leptin, insulin, and serum lipid profiles | Hsu et al. (2009) |
| 2 | Quercetin decreases high-fat diet induced body weight gain and accumulation of hepatic and circulating lipids in mice | In mice fed a high-fat diet, this study showed that quercetin supplementation significantly reduced body weight increases and lipid buildup in the liver and circulation. | Hoek-Van Den Hil et al. (2014) |
| 3 | Monoterpene phenolic compound thymol prevents high-fat diet induced obesity in murine model | Serum lipid levels, visceral fat-pad weights, and total weight growth were all significantly reduced in Wistar rats given a high-fat diet after receiving thymol. Additionally, the research noted increased antioxidant capacity and increases in insulin and leptin sensitivity. | Haque et al. (2014) |
| 4 | Effects of polyphenolic natural products on the lipid profiles of rats fed high fat diets | This study examined how lipid profiles in rats given a high-fat diet were affected by quercetin, morin, and tannic acid. Liver fat content, triglyceride levels, low-density lipoprotein cholesterol, and plasma total cholesterol were all markedly decreased by tannic acid and morin. While quercetin had no effect on other lipid measures, it did raise high-density lipoprotein cholesterol. | Yugarani et al. (1992) |
| 5 | Ferulic acid lowers body weight and visceral fat accumulation via modulation of enzymatic, hormonal, and inflammatory changes in a mouse model of high-fat diet-induced obesity | In mice fed a high-fat diet, ferulic acid treatment dramatically decreased visceral fat buildup and body weight growth. Additionally, the medication reduced insulin and blood glucose levels, altered inflammatory markers, and enhanced plasma lipid profiles. | Melo et al. (2017) |

According to the combined findings of these research, several phenolic compounds can reduce weight gain and fat deposition in animal models, mainly via enhancing insulin sensitivity, lipid metabolism, and inflammatory responses.

### 10.4.3 Clinical Trials of Phenolic Compounds for Weight Management

Human clinical trials on phenolic compounds indicate potential benefits in weight management, fat reduction, and metabolic regulation. Studies on green tea extract (GTE), resveratrol, curcumin, and coffee polyphenols show varying efficacy, with dosages ranging from 50 mg to 2,400 mg daily. Further large-scale trials are needed to confirm their therapeutic potential (Table 10.4).

Although their effectiveness varies, the results of human clinical research indicate that phenolic substances can affect fat metabolism and weight control. Due to its ability to regulate appetite and boost metabolism, GTE showed notable weight reduction, a decreased waist circumference, and better lipid profiles. Combining polyphenol supplements with calorie restriction and exercise did not yield significant overall weight reduction advantages, although it was effective for postmenopausal women. Phenolic-rich coffee enhances body composition without changing total weight. Although encouraging, further research is needed to validate the optimal doses and long-term benefits of phenolic compounds in the treatment of obesity.

**TABLE 10.4**
**Outcomes of Clinical Trials of Phenolic Compounds for Weight Management**

| S. No. | Studies | Dosage | Outcomes | Ref. |
|---|---|---|---|---|
| 1 | Anti-obesity effects of polyphenol intake: current status and future possibilities | High-dose GTE over 12 weeks. | Women who are obese can lose a significant amount of weight, reduce their waist circumference, and lower their LDL and total cholesterol levels without experiencing any negative consequences. Ghrelin secretion suppression and elevated adiponectin levels might be part of the process. | Boccellino and D'Angelo (2020) |
| 2 | The effects of polyphenol supplementation in addition to calorie restricted diets and/or physical activity on body composition parameters: a systematic review of randomized trials | Varied polyphenol supplements. | Nothing substantial to improve weight and fat reduction above calorie restriction and exercise alone. But in postmenopausal women, isoflavone and soy products, when paired with lifestyle modifications, demonstrated further anti-obesity benefits. | Llaha and Zamora-Ros (2020) |
| 3 | Consumption of a coffee rich in phenolic compounds may improve the body composition of people with overweight or obesity: preliminary insights from a randomized, controlled and blind crossover study | Three cups daily for 12 weeks | Despite individuals' imbalanced diets, there were improvements in the ratios of body fat to skeletal muscle mass without appreciable changes in body weight. | Fernández-Cardero et al. (2024) |

## 10.5 PHENOLIC COMPOUNDS IN FUNCTIONAL FOODS AND SUPPLEMENTS

Phenolic compounds represent a class of metabolites that are synthesized through the secondary metabolic pathways of plants. The categorization of plant phenolics is anchored in the term 'phenol,' which refers to a chemical structure characterized by a phenyl ring that possesses one or more hydroxyl groups. Various classes of phenolic compounds have been systematically classified based on their fundamental structural framework (Tsao, 2010). The spectrum of phenolic compounds encompasses a range of simple to relatively low molecular weight entities, including those with a singular aromatic ring, as well as more intricate and sizable tannins, along with their derivative polyphenols. Polyphenols include a diverse array of compounds such as flavonoids, phenolic acids, tannins, lignans, and coumarins, which are naturally occurring in a variety of plant-derived foods, including fruits, vegetables, cereals, roots, and foliage (Rana et al., 2022). The ingestion of phenolic compounds, whether through functional foods or dietary supplements, is associated with a multitude of health advantages, primarily attributable to their properties as antioxidants, anti-inflammatories, and anticancer agents. These compounds, which are prevalent in fruits, vegetables, and various beverages, play a vital role in reducing oxidative stress and promoting overall health (Bocsan et al., 2022; Dixit et al., 2025).

### 10.5.1 Properties of Phenolic Compounds

#### 10.5.1.1 Antioxidant Properties

Phenolic compounds demonstrate pronounced antioxidant properties, which facilitate the neutralization of reactive oxygen species (ROS) and mitigate oxidative stress, a known contributor to a multitude of diseases. They augment the body's antioxidant defense mechanisms by modulating enzymatic activity and safeguarding cellular structures from deleterious effects (Bocsan et al., 2022).

#### 10.5.1.2 Anti-Inflammatory Effects

These compounds possess the capacity to significantly attenuate inflammation, which is associated with chronic diseases, including cardiovascular disease and diabetes. Through the regulation of inflammatory pathways, phenolic compounds may assist in the management of conditions such as arthritis and other inflammatory ailments (Yahfoufi et al., 2018).

#### 10.5.1.3 Cardiovascular and Cancer Protection

Regular intake of foods abundant in phenolic compounds is correlated with diminished risks of cardiovascular diseases and select cancers (Liao and Meng, 2023). Their capacity to enhance endothelial function and lower blood pressure plays a crucial role in promoting cardiovascular health (Hodgson and Croft, 2010).

### 10.5.2 Phenolic-Rich Dietary Interventions

Phenolic-rich dietary interventions encompass the ingestion of alimentary products and supplements characterized by elevated concentrations of polyphenolic bioactive compounds. These interventions are increasingly acknowledged for their potential efficacy in the management of obesity and its concomitant metabolic disorders. Polyphenols, which are abundantly present in fruits, vegetables, tea, coffee, cocoa, and whole grains, serve a pivotal role in the modulation of lipid metabolism, the attenuation of oxidative stress, and the alteration of gut microbiota composition. By impeding lipogenesis (the synthesis of fat) and facilitating lipolysis (the breakdown of fat), polyphenols contribute to the diminution of adipose tissue accumulation. Furthermore, they may enhance insulin sensitivity and support cardiovascular health, thereby presenting a holistic approach to metabolic well-being (Bocsan et al., 2022).

In addition, polyphenols possess the ability to modulate the gut microbiome by promoting the proliferation of beneficial bacterial strains, which subsequently affects metabolic processes and energy balance. This modulation of gut microbiota represents an emerging mechanism through which phenolic-rich diets aid in weight management and enhance overall health (Wang et al., 2022).

Dietary interventions encompass the integration of polyphenol-rich alimentary products such as berries, apples, dark chocolate, green tea, and moderate consumption of red wine. The supplementation with standardized extracts (e.g., resveratrol, catechins, and quercetin) constitutes a domain of ongoing research to deliver consistent dosing for therapeutic applications. Nevertheless, although the advantages of polyphenols appear promising, it is essential to acknowledge that their efficacy may be influenced by variables such as bioavailability, individual health status, and overall dietary patterns (Luca et al., 2020; Rodriguez-Mateos et al., 2014).

### 10.5.3 Commercially Available Supplements and Their Effectiveness

Commercially available dietary supplements have experienced a notable surge in popularity due to their purported health benefits, particularly in domains such as cardiovascular health, weight regulation, and inflammation management. Omega-3 fatty acids, derived from fish oil, rank among the most extensively researched supplements, with substantial empirical evidence demonstrating their efficacy in reducing triglyceride concentrations and mitigating the risk of cardiovascular disease (Manson et al., 2019). Curcumin, the principal bioactive compound found in turmeric, is renowned for its anti-inflammatory and antioxidant properties; however, its clinical efficacy is frequently hindered by suboptimal bioavailability, a challenge that is being addressed through various advanced formulations (Gupta et al., 2013). GTE, which is abundant in catechins, has been correlated with enhanced metabolic rates and moderate weight reduction, in addition to exhibiting cardioprotective properties (Hodgson and Croft, 2010; Rana et al., 2022). Resveratrol, a polyphenolic compound found in red grapes, represents another supplement that has shown potential in promoting cardiovascular health and anti-aging effects; however, findings from human studies remain inconclusive and require further investigation (Baur and Sinclair, 2006).

While these dietary supplements present potential advantages, their efficacy is contingent upon various factors, including dosage, formulation, and individual health status. In general, further extensive longitudinal clinical trials are imperative to substantiate these effects and formulate standardized guidelines for their application in health enhancement and disease prevention.

### 10.5.4 Safety and Recommended Dosages of Common Supplements

Safety and proper dosing are critical when using dietary supplements. For example, omega-3 fatty acids (EPA and DHA) are generally recommended at a combined dose of approximately 1,000 mg daily for general cardiovascular support, although higher doses (2–4 g/day) may be advised under medical supervision for high triglyceride levels (Manson et al., 2019). Curcumin, the active ingredient in turmeric, is typically taken at 500–2,000 mg/day. However, its poor bioavailability is often overcome by formulations containing piperine, which can enhance absorption significantly (Gupta et al., 2013). For GTE, a daily dosage of 250–500 mg is common, but caution is warranted as high doses may potentially lead to liver toxicity in sensitive individuals (Hodgson and Croft, 2010). Resveratrol dosages typically range from 150 to 500 mg daily, although its optimal therapeutic range remains under investigation (Baur and Sinclair, 2006).

## 10.6 CHALLENGES IN BIOAVAILABILITY AND METABOLISM OF PHENOLIC COMPOUNDS

### 10.6.1 Challenges in the Bioavailability of Phenolic Compounds

Many phenolic compounds, including flavonoids, exhibit low water solubility and are prone to degradation under physiological conditions, which reduces their absorption in the gastrointestinal tract (Teng et al., 2023). These characteristics contribute to the poor bioavailability of phenolic compounds. Liver enzymes and intestinal flora extensively metabolize phenolic substances, producing conjugated metabolites that may have lower biological activity. Additionally, the dietary matrix can influence the bioavailability of phenolic compounds, altering their release and absorption through interactions with other nutrients and components (Tomas et al., 2024).

### 10.6.2 Challenges in the Metabolism of Phenolic Compounds

Individual differences in phenolic chemical metabolism can be attributed to several variables, including the critical role of the gut microbiota in the biotransformation of phenolic compounds and the disparate metabolic consequences resulting from distinct microbial communities (Favari et al., 2024). Inter-individual variability can be further exacerbated by genetic variations that affect the activity of enzymes involved in phenolic metabolism. Additionally, the metabolism and bioavailability of phenolic chemicals are influenced by age, sex, ethnicity, and health state (Souza et al., 2015).

Numerous processes, such as conjugation (e.g., glucuronidation, sulfation), methylation, and deglycosylation, are involved in the metabolism of phenolic substances. Numerous metabolites, some of which may have unique biological functions, can be produced by these processes. Nevertheless, it isn't easy to forecast and improve the bioavailability of phenolic compounds due to the intricacy of these routes (Thai et al., 2014).

## 10.7 FUTURE PERSPECTIVES ON ENHANCING BIOAVAILABILITY AND METABOLISM

### 10.7.1 Biotechnological Approaches

By transforming them into more bioactive metabolites, steered fermentation processes can enhance the release of phenolic compounds from plant matrices and increase their bioavailability (Thai et al., 2014). Additionally, phenolic compounds can be modified by enzyme-catalyzed processes to increase their stability and bioactivity. For example, methylation of flavones has been demonstrated to improve intestinal absorption and metabolic stability (Walle, 2007).

### 10.7.2 Nanotechnology and Encapsulation

Phenolic chemicals can be better delivered to target tissues and shielded from degradation by being encapsulated in nanocarriers, including liposomes, nanoparticles, and microemulsions (El Monfalouti and Kartah, 2024). The solubility, stability, and controlled release of phenolic compounds can be enhanced by loading them into lipid-based nanocarriers, thereby increasing their bioavailability (Esfanjani et al., 2018).

### 10.7.3 Precision Nutrition and Personalized Approaches

Metabotypes can be identified with the aid of developments in metabolomics and gut microbiomics, enabling customized nutrition plans tailored to each person's unique phenolic chemical metabolism (Quesada-Vázquez et al., 2024). Phenolic compounds can be delivered to their intended locations of action without being prematurely digested by developing tailored delivery systems, such as those that use certain coatings or carriers (Ohara and Ohyama, 2014).

### 10.7.4 Sustainable Extraction and Formulation

Novel methods of extraction, such as ultrasound-assisted extraction and natural deep eutectic solvents, can enhance the yield and purity of phenolic compounds, thereby increasing their suitability for medicinal applications. The bioefficacy and shelf life of phenolic compounds can be enhanced by employing stability and bioefficacy techniques, such as co-crystallization and amorphous solid dispersion (Tomas et al., 2024).

### 10.7.5 Synergistic Effects and Combination Therapies

Through synergistic interactions, mixing or combining certain phenolic compounds with other bioactive molecules can enhance their biological effects and improve bioavailability. Prebiotic and probiotic formulations that improve their bioconversion into bioactive metabolites can be developed as a result of the interactions between phenolic chemicals and gut microbiota (Table 10.5) (Alaya et al., 2024).

Because obesity is impacted by behavioral, environmental, and hereditary variables, it is difficult to treat with a single therapy strategy. Given the complex nature of obesity, combined therapy approaches that incorporate phenolic substances with lifestyle modifications and other therapies might be required. Although phenolic compounds present a promising natural substitute for synthetic medications, there are certain obstacles to overcome before using them to treat obesity. To overcome these challenges and fully utilize phenolic chemicals in the fight against obesity, a more thorough study is required, including clinical trials and technical developments (Ghosh et al., 2021; Monisha et al., 2024).

**TABLE 10.5**
**Strategies to Improve Bioavailability of Phenolic Compounds**

| Strategy | Description | Ref. |
|---|---|---|
| Fermentation | Enhance release and bioconversion of phenolic compounds into bioactive metabolite | Tomas et al. (2024), Thai et al. (2014) |
| Nano-encapsulation | Protect phenolic compounds from degradation and improves delivery to target tissues | Esfanjani et al. (2018), El Monfalouti and Kartah (2024) |
| Enzymatic modification | Improves stability and bioactivity through methylation or other enzymatic reactions | Contente et al. (2021), Walle (2007) |
| Specific nutrition | Personalized approaches based on gut microbiota and metabotyping | Duynhoven et al. (2011), Quesada-Vázquez et al. (2024) |
| Innovative extraction techniques | To enhance yield and purity of phenolic compounds | Kandemir et al. (2022), Tomas et al. (2024) |

## 10.8 CONCLUSION AND FUTURE PROSPECTS

We conclude this study as plant-based diets are rich in phenolic compounds, which have shown promise in preventing obesity through several mechanisms. Targeting lipid metabolism, adipogenesis, thermogenesis, gut microbiota, inflammation, and insulin sensitivity, phenolic substances have a complex influence on preventing obesity. Their promise as safer and more natural alternatives to pharmaceutical anti-obesity medications is highlighted by their ability to influence these processes. To maximize their therapeutic use and address issues with absorption and clinical efficacy, further study is necessary. By altering the expression of genes linked to fat storage, substances such as apigenin and kaempferol have been shown in vitro to suppress adipogenesis and lower triglyceride accumulation in human adipocytes. These results are corroborated by research on animals. For example, resveratrol has been shown to activate the AMPK and SIRT1 pathways, thereby improving energy metabolism and reducing obesity in animal models. However, there are difficulties in converting these advantages into applications for people. Due to individual variables, including age, gender, and heredity, as well as variations in doses, bioavailability, and research designs, clinical trials have yielded conflicting outcomes. Furthermore, our understanding of the effectiveness of phenolic compounds is complicated by their intricate relationships with other dietary components and by the gut microbiota's metabolism of them. To create regular dosage schedules, evaluate any adverse effects, and validate the therapeutic potential of phenolic compounds in the management of obesity, future research should focus on well-planned, extended clinical trials. To fully utilize the benefits of these chemicals in the fight against obesity, it will be essential to address aspects that alter bioavailability and individual responses.

## COMPETING INTERESTS

The authors declare that they have no competing interests.

## DECLARATION OF GENERATIVE AI AND AI-ASSISTED TECHNOLOGIES IN THE WRITING PROCESS

During the preparation of this work, the book chapter author(s) have not taken any assistance from AI, and the book chapter author(s) reviewed and edited the content as needed and take(s) full responsibility for the content of the publication.

## REFERENCES


Abra CE, Assis JR (2020). Polyphenolic compounds in the loss of body fat. Scientific Electronic Archives, 13(8), 76–93.

Aceituno-Medina, Marysol, Sandra Mendoza, Beatriz A Rodríguez, José María Lagaron, and Amparo López-Rubio. 2015. "Improved Antioxidant Capacity of Quercetin and Ferulic Acid during In-Vitro Digestion through Encapsulation within Food-Grade Electrospun Fibers." *Journal of Functional Foods* 12: 332–41. https://doi.org/10.1016/j.jff.2014.11.028.

Ahmed, Zahoor, Naseem Aslam Channa, Shazia Parveen Solangi, Saud Farooque, Zeba Hussain Soomro, Naeem H. Quraishi, and Zulfiqar Ali Dahri. 2024. "Journal of Population Therapeutics & Clinical Pharmacology." *SSRN Electronic Journal* 31 (1): 451–60. https://doi.org/10.2139/ssrn.4919734.

Alaya, Ikbel Ben, Gilberto Alves, João Lopes, and Luís R Silva. 2024. "Use of Encapsulated Polyphenolic Compounds in Health Promotion and Disease Prevention: Challenges and Opportunities." *Macromol* 4 (4): 805–42.

Aranaz, Paula, David Navarro-Herrera, María Zabala, Itziar Miguéliz, Ana Romo-Hualde, Miguel López-Yoldi, J Alfredo Martínez, José Luis Vizmanos, Fermín I Milagro, and Carlos Javier González-Navarro. 2019. "Phenolic Compounds Inhibit 3T3-L1 Adipogenesis Depending on the Stage of Differentiation and Their Binding Affinity to PPARγ." *Molecules* 24 (6): 1045.

Baur, Joseph A, and David A Sinclair. 2006. "Therapeutic Potential of Resveratrol: The in Vivo Evidence." *Nature Reviews. Drug Discovery* 5 (6): 493–506. https://doi.org/10.1038/nrd2060.

Bawa, Sa'eed Halilu, Anna Harton, Joanna Myszkowska-Ryciak, Danuta Gajewska, and Marquitta Webb. 2017. "The Role of Selected Bioactive Compounds in Teas, Spices, Cocoa and Coffee in Body Weight Control." *Polish Journal of Applied Sciences* 1 (2): 56–66.

Boccellino, Mariarosaria, and Stefania D'Angelo. 2020. "Anti-Obesity Effects of Polyphenol Intake: Current Status and Future Possibilities." *International Journal of Molecular Sciences* 21 (16): 5642.

Bocsan, Ioana C, Dan C Măgureanu, Raluca M Pop, Antonia M Levai, Ştefan O Macovei, Ioana M Pătraşca, and Anca D Buzoianu. 2022. "Antioxidant and Anti-Inflammatory Actions of Polyphenols from Red and White Grape Pomace in Ischemic Heart Diseases." *Biomedicines*. https://doi.org/10.3390/biomedicines10102337.

S. G. van Breda, T. M. de Kok, Mol. Nutr. Food Res. 2018, 62, 1700597. https://doi.org/10.1002/mnfr.201700597

Bullón-Vela, M Vanessa, Itziar Abete, J Alfredo Martínez, and M Angeles Zulet. 2018. "Chapter 6 – Obesity and Nonalcoholic Fatty Liver Disease: Role of Oxidative Stress." In, edited by Amelia Marti del Moral and Concepción María B T – *Obesity Aguilera García*, 111–33. Academic Press. https://doi.org/10.1016/B978-0-12-812504-5.00006-4.

Burke, Carol A. 2010. "Colonic Complications of Obesity." *Gastroenterology Clinics of North America* 39 (1): 47–55.

Camilleri, Michael, Harmeet Malhi, and Andres Acosta. 2017. "Gastrointestinal Complications of Obesity." *Gastroenterology* 152 (7): 1656–70. https://doi.org/10.1053/j.gastro.2016.12.052.

Cao, Yunyun, Shuai Han, Han Lu, Yi Luo, Tianyi Guo, Qi Wu, and Feijun Luo. 2022. "Targeting MTOR Signaling by Dietary Polyphenols in Obesity Prevention." *Nutrients* 14 (23): 5171.

Carrasco-Pozo, Catalina, María Jose Cires, and Martin Gotteland. 2019. "Quercetin and Epigallocatechin Gallate in the Prevention and Treatment of Obesity: From Molecular to Clinical Studies." *Journal of Medicinal Food* 22 (8): 753–70.

Chakraborty, Debolina, Kriti Gupta, and Sagarika Biswas. 2021. "A Mechanistic Insight of Phytoestrogens Used for Rheumatoid Arthritis: An Evidence-Based Review." *Biomedicine & Pharmacotherapy* 133: 111039. https://doi.org/10.1016/j.biopha.2020.111039.

Chandrasekaran, Preethi, and Ralf Weiskirchen. 2024. "The Signaling Pathways in Obesity-related Complications." *Journal of Cell Communication and Signaling* 18 (2): e12039.

Contente, Martina Letizia, Francesca Annunziata, Pietro Cannazza, Silvia Donzella, Cecilia Pinna, Diego Romano, Lucia Tamborini, Francisco Geraldo Barbosa, Francesco Molinari, and Andrea Pinto. 2021. "Biocatalytic Approaches for an Efficient and Sustainable Preparation of Polyphenols and Their Derivatives." *Journal of Agricultural and Food Chemistry* 69 (46): 13669–81.

Deka, Himangshu, Ananta Choudhury, and Biplab Kumar Dey. 2022. "An Overview on Plant Derived Phenolic Compounds and Their Role in Treatment and Management of Diabetes." *Journal of Pharmacopuncture* 25 (3): 199.

Derdemezis, Christos S, Dimitrios N Kiortsis, Vasilis Tsimihodimos, Maria P Petraki, Patra Vezyraki, Moses S Elisaf, and Alexandros D Tselepis. 2011. "Effect of Plant Polyphenols on Adipokine Secretion from Human SGBS Adipocytes." *Biochemistry Research International* 2011 (1): 285618.

Dixit, V, K K Chaubey, D Dayal, P B Chole, M BT, R K Bachheti, P C Gupta, A Bachheti, and L A Worku. 2025. "Natural Products in Aging: Cellular Mechanisms and Emerging Therapeutics for Age-Related Disorders." *Journal of Aging Research* 2025 (1): 6868732.

Duynhoven, John van, Elaine E Vaughan, Doris M Jacobs, Robèr A. Kemperman, Ewoud J J van Velzen, Gabriele Gross, Laure C Roger, Sam Possemiers, Age K Smilde, and Joël Doré. 2011. "Metabolic Fate of Polyphenols in the Human Superorganism." *Proceedings of the National Academy of Sciences* 108 (supplement_1): 4531–38.

El Monfalouti, Hanae, and Badr Eddine Kartah. 2024. "Enhancing Polyphenol Bioavailability through Nanotechnology: Current Trends and Challenges."

Esfanjani, Afshin Faridi, Elham Assadpour, and Seid Mahdi Jafari. 2018. "Improving the Bioavailability of Phenolic Compounds by Loading Them within Lipid-Based Nanocarriers." *Trends in Food Science & Technology* 76: 56–66.

Falakacılar, Çağla Pınarlı, and Sevinç Yücecan. 2024. "The Impact of Sustainability Courses: Are They Effective in Improving Diet Quality and Anthropometric Indices?." *Nutrients* 16 (11): 1700.

Favari, Claudia, José Fernando Rinaldi de Alvarenga, Lorena Sánchez-Martínez, Nicole Tosi, Cristiana Mignogna, Eleonora Cremonini, Claudine Manach, Letizia Bresciani, Daniele Del Rio, and Pedro Mena. 2024. "Factors Driving the Inter-Individual Variability in the Metabolism and Bioavailability of (Poly) Phenolic Metabolites: A Systematic Review of Human Studies." *Redox Biology*, 103095.

Fernández-Cardero, Álvaro, José Luis Sierra-Cinos, Laura Bravo, and Beatriz Sarriá. 2024. "Consumption of a Coffee Rich in Phenolic Compounds May Improve the Body Composition of People with Overweight or Obesity: Preliminary Insights from a Randomized, Controlled and Blind Crossover Study." *Nutrients* 16 (17): 2848.

Ferreyra, María Lorena Falcone, Paloma Serra, and Paula Casati. 2021. "Recent Advances on the Roles of Flavonoids as Plant Protective Molecules after UV and High Light Exposure." *Physiologia Plantarum* 173 (3): 736–49.

Fuloria, Shivkanya, Mahendran Sekar, Farrah S Khattulanuar, Siew H Gan, Nur N Rani, Subban Ravi Vetriselvan Subramaniyan, et al. 2022. "Chemistry, Biosynthesis and Pharmacology of Viniferin: Potential Resveratrol-Derived Molecules for New Drug Discovery, Development and Therapy." *Molecules*. https://doi.org/10.3390/molecules27165072.

Ghosh, Shampa, Srividya Manchala, Manchala Raghunath, Gaurav Sharma, Abhishek K Singh, and Jitendra K Sinha. 2021. "Role of Phytomolecules in the Treatment of Obesity: Targets, Mechanisms and Limitations." *Current Topics in Medicinal Chemistry* 21 (10): 863–77.

Gómez-Zorita, Saioa, Arrate Lasa, Naiara Abendaño, Alfredo Fernández-Quintela, Andrea Mosqueda-Solís, Maria Pilar Garcia-Sobreviela, Jose M Arbonés-Mainar, and Maria P Portillo. 2017. "Phenolic Compounds Apigenin, Hesperidin and Kaempferol Reduce in Vitro Lipid Accumulation in Human Adipocytes." *Journal of Translational Medicine* 15 (1): 237. https://doi.org/10.1186/s12967-017-1343-0.

Guasch-Ferré, Marta, and Walter C Willett. 2021. "The Mediterranean Diet and Health: A Comprehensive Overview." *Journal of Internal Medicine* 290 (3): 549–66.

Gupta, Subash C, Sridevi Patchva, and Bharat B Aggarwal. 2013. "Therapeutic Roles of Curcumin: Lessons Learned from Clinical Trials." *The AAPS Journal* 15: 195–218.

Hall, Kevin D, Steven B Heymsfield, Joseph W Kemnitz, Samuel Klein, Dale A Schoeller, and John R Speakman. 2012. "Energy Balance and Its Components: Implications for Body Weight Regulation." *The American Journal of Clinical Nutrition* 95 (4): 989–94. https://doi.org/10.3945/ajcn.112.036350.

Haque, Mohammad Rafiul, S H Ansari, A K Najmi, and Mohammed Afroz Ahmad. 2014. "Monoterpene Phenolic Compound Thymol Prevents High Fat Diet Induced Obesity in Murine Model." *Toxicology Mechanisms and Methods* 24 (2): 116–23.

Hättenschwiler, S, and P M Vitousek. 2000. "The Role of Polyphenols in Terrestrial Ecosystem Nutrient Cycling." *Trends in Ecology & Evolution* 15 (6): 238–43.

He, Lan, Zhan Su, and Shuangshuang Wang. 2024. "The Anti-Obesity Effects of Polyphenols: A Comprehensive Review of Molecular Mechanisms and Signal Pathways in Regulating Adipocytes." *Frontiers in Nutrition* 11: 1393575.

Hithamani, Gavirangappa, and Ponesakki Ganesan. 2022. "Polyphenols from Indian Cereal Grains Inhibit 3T3-L1 Adipogenesis through Modulating Early and Late Phase Adipogenic Markers." *Food Bioscience* 50: 102075.

Hittepole, Bridget. 2024. "Obesity: Complications, Complexities and Care Considerations." *International Journal of Disease Reversal and Prevention* 6 (2): 12.

Hodgson, Jonathan M, and Kevin D Croft. 2010. "Tea Flavonoids and Cardiovascular Health." *Molecular Aspects of Medicine* 31 (6): 495–502. https://doi.org/10.1016/j.mam.2010.09.004.

Hoek-Van Den Hil, E F, Evert M van Schothorst, Inge van der Stelt, Hans J Swarts, Dini Venema, Manuela Sailer, Jacques J Vervoort, Peter C Hollman, Ivonne M Rietjens, and Jaap Keijer. 2014. "Quercetin Decreases High-Fat Diet Induced Body Weight Gain and Accumulation of Hepatic and Circulating Lipids in Mice." *Genes & Nutrition* 9(5): 418.

Hsu, Chin-Lin, Chi-Hao Wu, Shih-Li Huang, and Gow-Chin Yen. 2009. "Phenolic Compounds Rutin and O-Coumaric Acid Ameliorate Obesity Induced by High-Fat Diet in Rats." *Journal of Agricultural and Food Chemistry* 57 (2): 425–31.

Huang, Shuo, Baowen Qi, Ling Yang, Xue Wang, Jing Huang, Ya Zhao, Yonghe Hu, and Wenjing Xiao. 2023. "Phytoestrogens, Novel Dietary Supplements for Breast Cancer." *Biomedicine & Pharmacotherapy* 160: 114341. https://doi.org/10.1016/j.biopha.2023.114341.

Jiao, Xinyao, Yuehua Wang, Yang Lin, Yuxi Lang, Enhui Li, Xiuyan Zhang, Qi Zhang, Ying Feng, Xianjun Meng, and Bin Li. 2019. "Blueberry Polyphenols Extract as a Potential Prebiotic with Anti-Obesity Effects on C57BL/6 J Mice by Modulating the Gut Microbiota." *The Journal of Nutritional Biochemistry* 64: 88–100.

Jovanović, Ivana, Nicole Frantová, Jhonny E Alba-Mejía, Lenka Porčová, Vratislav Psota, Jana Asszonyi, Radim Cerkal, and Tomáš Středa. 2024. "Role of Total Polyphenol Content in Seed Germination Characteristics of Spring Barley Varieties amidst Climate Change." *Scientific Reports* 14 (1): 23818. https://doi.org/10.1038/s41598-024-74795-6.

Kandemir, Kevser, Merve Tomas, David Julian McClements, and Esra Capanoglu. 2022. "Recent Advances on the Improvement of Quercetin Bioavailability." *Trends in Food Science & Technology* 119: 192–200.

Kebekova, Aizat, Sakhabat Mashirapova, Rahul Kumar, Gulam Ahmed Raza Quadri, Ali Kamran, and Ranjan Kumar Yadav. 2024. "Obesity." *InterConf* 41 (185): 390–98. https://doi.org/10.51582/interconf.19-20.01.2024.049.

Kent, Lisa, Meabh McGirr, and Kelly-Ann Eastwood. 2024. "Global Trends in Prevalence of Maternal Overweight and Obesity: A Systematic Review and Meta-Analysis of Routinely Collected Data Retrospective Cohorts." *International Journal of Population Data Science* 9 (2): 6.

Kirichenko, Tatiana V, Yuliya V Markina, Anastasia I Bogatyreva, Taisiya V Tolstik, Yurgita R Varaeva, and Antonina V Starodubova. 2022. "The Role of Adipokines in Inflammatory Mechanisms of Obesity." *International Journal of Molecular Sciences* 23 (23). https://doi.org/10.3390/ijms232314982.

Konstantinidi, Melina, and Antonios E Koutelidakis. 2019. "Functional Foods and Bioactive Compounds: A Review of Its Possible Role on Weight Management and Obesity's Metabolic Consequences." *Medicines* 6 (3). https://doi.org/10.3390/medicines6030094.

Kyazimova, N. D., and V. V. Kornyakova. 2024. "Influence of Polyphenolic Compounds on Human Health and the Course of a Number of Diseases." *Scientific Bulletin of the Omsk State Medical University* 4 (1): 87–91. https://doi.org/10.61634/2782-3024-2024-13-87-91.

Lan, Xiaoyue, and Norhasmah Sulaiman. 2024. "Public Policies on Obesity: A Literature Review of Global Challenges and Response Strategies." *Cureus* 16 (6).

Lesschaeve, Isabelle, and Ann C Noble. 2005. "Polyphenols: Factors Influencing Their Sensory Properties and Their Effects on Food and Beverage Preferences[2]." *The American Journal of Clinical Nutrition* 81 (1): 330S–335S. https://doi.org/10.1093/ajcn/81.1.330S.

Liao, Yunshu, and Qinghua Meng. 2023. "Protection against Cancer Therapy-Induced Cardiovascular Injury by Planed-Derived Polyphenols and Nanomaterials." *Environmental Research* 238: 116896. https://doi.org/10.1016/j.envres.2023.116896.

Lam, B. C., A. Y. Lim, S. L. Chan, M. P. Yum, N. S. Koh, and E. A. Finkelstein. 2023. "The Impact of Obesity: A Narrative Review." *Singapore Medical Journal* 64 (3):163–71.

Liu, Yongxiang, Jinhai Luo, and Baojun Xu. 2024. "Elucidation of Anti-Obesity Mechanisms of Phenolics in *Artemisiae argyi* Folium (Aiye) by Integrating LC–MS, Network Pharmacology, and Molecular Docking." *Life* 14 (6): 656.

Llaha, Fjorida, and Raul Zamora-Ros. 2020. "The Effects of Polyphenol Supplementation in Addition to Calorie Restricted Diets and/or Physical Activity on Body Composition Parameters: A Systematic Review of Randomized Trials." *Frontiers in Nutrition* 7: 84.

Lobiuc, Andrei, Naomi-Eunicia Pavăl, Ionel I Mangalagiu, Roxana Gheorghiţă, Gabriel-Ciprian Teliban, Dorina Amăriucăi-Mantu, and Vasile Stoleru. 2023. "Future Antimicrobials: Natural and Functionalized Phenolics." *Molecules (Basel, Switzerland)* 28 (3). https://doi.org/10.3390/molecules28031114.

Luca, Simon Vlad, Irina Macovei, Alexandra Bujor, Anca Miron, Krystyna Skalicka-Woźniak, Ana Clara Aprotosoaie, and Adriana Trifan. 2020. "Bioactivity of Dietary Polyphenols: The Role of Metabolites." *Critical Reviews in Food Science and Nutrition* 60 (4): 626–59. https://doi.org/10.1080/10408398.2018.1546669.

Luzi, L (Editor). 2021. *Thyroid, Obesity and Metabolism: Exploring Links Between Thyroid Function, Obesity, Metabolism and Lifestyle*. Springer Nature.

Manchope MF, Ferraz CR, Borghi SM, Rasquel-Oliveira FS, Franciosi A, Bagatim-Souza J, Dionisio AM, Casagrande R, Verri Jr WA. Therapeutic role of naringenin to alleviate inflammatory pain. InTreatments, mechanisms, and adverse reactions of anesthetics and analgesics 2022 Jan 1 (pp. 443–455). Academic Press.

Manson, JoAnn E, Nancy R Cook, I-Min Lee, William Christen, Shari S Bassuk, Samia Mora, Heike Gibson, et al. 2019. "Marine N-3 Fatty Acids and Prevention of Cardiovascular Disease and Cancer." *The New England Journal of Medicine* 380 (1): 23–32. https://doi.org/10.1056/NEJMoa1811403.

Mattioli, Roberto, Antonio Francioso, Luciana Mosca, and Paula Silva. 2020. "Anthocyanins: A Comprehensive Review of Their Chemical Properties and Health Effects on Cardiovascular and Neurodegenerative Diseases." *Molecules (Basel, Switzerland)* 25 (17). https://doi.org/10.3390/molecules25173809.

Melo, T S De, P R Lima, KMMB Carvalho, T M Fontenele, F R N Solon, A R Tomé, T L G De Lemos, S G da Cruz Fonseca, F A Santos, and Vietla Satyanarayana Rao. 2017. "Ferulic Acid Lowers Body Weight and Visceral Fat Accumulation via Modulation of Enzymatic, Hormonal and Inflammatory Changes in a Mouse Model of High-Fat Diet-Induced Obesity." *Brazilian Journal of Medical and Biological Research* 50 (1): e5630.

Monisha, D, D N Swathi, and R Deveswaran. 2024. "Harnessing the Power of Nutraceuticals." *A New Approach to Obesity Treatment* 36 (12): 140–60.

Mushtaq, Zarina, Nabeelah Bibi Sadeer, Muzzamal Hussain, Mahwish, Suliman A Alsagaby, Muhammad Imran, Tamseela Mumtaz, Maryam Umar, Ambreen Tauseef, and Waleed Al Abdulmonem. 2023. "Therapeutical Properties of Apigenin: A Review on the Experimental Evidence and Basic Mechanisms." *International Journal of Food Properties* 26 (1): 1914–39.

Nemzer, Boris V, Fadwa Al-Taher, Diganta Kalita, Alexander Y Yashin, and Yakov I Yashin. 2025. "Health-Improving Effects of Polyphenols on the Human Intestinal Microbiota: A Review." *International Journal of Molecular Sciences* 26 (3). https://doi.org/10.3390/ijms26031335.

Ohara, Masaru, and Yoshihiko Ohyama. 2014. "Delivery and Application of Dietary Polyphenols to Target Organs, Tissues and Intracellular Organelles." *Current Drug Metabolism* 15 (1): 37–47.

Özkaya, Şebnem Özgen. 2021. "Yaşam Kalitesi ve Fonksiyonel Besinler." *Fenerbahçe Üniversitesi Sağlık Bilimleri Dergisi* 1 (1): 62–68.

Pandey, Aparna, Krishnan Viswanathan, Prabhu Manickam Natrajan, and Jayadevan Sreedharan. 2022. "Obesity: Overview of a Universal Health Conundrum." *Journal of Clinical and Diagnostic Research* 16 (2). https://doi.org/10.7860/jcdr/2022/51930.15973.

Park, Sohyun, and Heidi M Blanck. 2024. "Age, Sex, Ethnicity, and Other Sources of Variation in the Prevalence of Obesity." In *Handbook of Obesity, Two-Volume Set*, Vol 1-53. CRC Press.

Pelczyńska, Marta, Małgorzata Moszak, Agnieszka Wesołek, and Paweł Bogdański. 2023. "The Preventive Mechanisms of Bioactive Food Compounds against Obesity-Induced Inflammation." *Antioxidants (Basel, Switzerland)* 12 (6). https://doi.org/10.3390/antiox12061232.

Peri, Katya, and Mark Eisenberg. 2024. "Review on the Update in Obesity Management : Epidemiology," 1–10. https://doi.org/10.1136/bmjph-2023-000247.

Prendergast, Heather, Carissa Tyo, Christopher Colbert, Morgan Kelley, and Ruth Pobee. 2022. "Medical Complications of Obesity: Heightened Importance in a COVID Era." *International Journal of Emergency Medicine* 15 (1): 29. https://doi.org/10.1186/s12245-022-00431-7.

Quesada-Vázquez, Sergio, Itziar Eseberri, Francisco Les, Patricia Pérez-Matute, María Herranz-López, Claude Atgié, Marta Lopez-Yus, Paula Aranaz, José A Oteo, and Xavier Escoté. 2024. "Polyphenols and Metabolism: From Present Knowledge to Future Challenges." *Journal of Physiology and Biochemistry* 80 (3): 603–25.

Rana, Ananya, Mrinal Samtiya, Tejpal Dhewa, Vijendra Mishra, and Rotimi E Aluko. 2022. "Health Benefits of Polyphenols: A Concise Review." *Journal of Food Biochemistry* 46 (10): e14264. https://doi.org/10.1111/jfbc.14264.

Rayess, Youssef El, Nancy Nehme, Samar Azzi-Achkouty, and Sofi G Julien. 2024. "Wine Phenolic Compounds: Chemistry, Functionality and Health Benefits." *Antioxidants* 13 (11): 1312.

Rebollo-Hernanz, Miguel, Qiaozhi Zhang, Yolanda Aguilera, Maria A Martín-Cabrejas, and Elvira Gonzalez de Mejia. 2019. "Phenolic Compounds from Coffee By-Products Modulate Adipogenesis-Related Inflammation, Mitochondrial Dysfunction, and Insulin Resistance in Adipocytes, via Insulin/PI3K/AKT Signaling Pathways." *Food and Chemical Toxicology* 132: 110672.

Rodrigues, Marina, Caroline Bertoncini-Silva, Anderson Gregorio Joaquim, Cássia Dias Machado, Leandra Náira Zambelli Ramalho, Daniela Carlos, Priscila Giacomo Fassini, and Vivian Marques Miguel Suen. 2022. "Beneficial Effects of Eugenol Supplementation on Gut Microbiota and Hepatic Steatosis in High-Fat-Fed Mice." *Food & Function* 13 (6): 3381–90. https://doi.org/10.1039/d1fo03619j.

Rodriguez-Mateos, Ana, David Vauzour, Christian G Krueger, Dhanansayan Shanmuganayagam, Jess Reed, Luca Calani, Pedro Mena, Daniele Del Rio, and Alan Crozier. 2014. "Bioavailability, Bioactivity and Impact on Health of Dietary Flavonoids and Related Compounds: An Update." *Archives of Toxicology* 88 (10): 1803–53. https://doi.org/10.1007/s00204-014-1330-7.

Rodriguez-Ramiro, I, D Vauzour, and A M Minihane. 2016. "Polyphenols and Non-Alcoholic Fatty Liver Disease: Impact and Mechanisms." *Proceedings of the Nutrition Society* 75 (1): 47–60.

Salvamani, Shamala, Baskaran Gunasekaran, Noor Azmi Shaharuddin, Siti Aqlima Ahmad, and Mohd Yunus Shukor. 2014. "Antiartherosclerotic Effects of Plant Flavonoids." *BioMed Research International* 2014: 480258. https://doi.org/10.1155/2014/480258.

Sattar, Sarah, Noor Ali Salman, Farah Ali Dawood, and Ali A Al-fahham. 2024. "The Chemical Structure and Clinical Significance of Phenolic Compounds." https://doi.org/10.58806/ijhmr.2024.v3i10n06.

Seidell, Jacob C. 2024. "Worldwide and Regional Prevalence of Obesity." In *Handbook of Obesity, Two-Volume Set*, Vol 1-48. CRC Press.

Shi, Haidan, Xuezhi Li, Chen Hou, Li Chen, Yuhuan Zhang, and Jianke Li. 2023. "Effects of Pomegranate Peel Polyphenols Combined with Inulin on Gut Microbiota and Serum Metabolites of High-Fat-Induced Obesity Rats." *Journal of Agricultural and Food Chemistry* 71 (14): 5733–44.

Singh, Durgeshwer, Khushboo Kumari, and Salim Ahmed. 2022. "Natural Herbal Products for Cancer Therapy." In *Understanding Cancer* (pp. 257–68). Academic Press. https://doi.org/10.1016/B978-0-323-99883-3.00010-X.

Singh, Manisha, Thilini Thrimawithana, Ravi Shukla, Charles Stephen Brennan, and Benu Adhikari. 2023. "Impact of Phenolic Extracts and Potassium Hydroxycitrate of *Hibiscus sabdariffa* on Adipogenesis: A Cellular Study." *International Journal of Food Science and Technology* 58 (3): 1204–18.

Souza, Jacqueline Elis De, Livia Marques Casanova, and Sônia Soares Costa. 2015. "Bioavailability of Phenolic Compounds: A Major Challenge for Drug Development?" *Revista Fitos* 9 (1): 55–67.

Suo, Hao, Mohammad Rezaul Islam Shishir, Qi Wang, Mingfu Wang, Feng Chen, and Ka Wing Cheng. 2023. "Red Wine High-Molecular-Weight Polyphenolic Complex Ameliorates High-Fat Diet-Induced Metabolic Dysregulation and Perturbation in Gut Microbiota in Mice." *Journal of Agricultural and Food Chemistry* 71 (18): 6882–93.

Tavčar Benković, Eva, Dušan Žigon, Vladimir Mihailović, Tanja Petelinc, Polona Jamnik, and Samo Kreft. 2017. "Identification, in Vitro and in Vivo Antioxidant Activity, and Gastrointestinal Stability of Lignans from Silver Fir (Abies Alba) Wood Extract." *Journal of Wood Chemistry and Technology* 37 (6): 467–77.

Teng, Hui, Yimei Zheng, Hui Cao, Qun Huang, Jianbo Xiao, and Lei Chen. 2023. "Enhancement of Bioavailability and Bioactivity of Diet-Derived Flavonoids by Application of Nanotechnology: A Review." *Critical Reviews in Food Science and Nutrition* 63 (3): 378–93.

Tessema, F. B., Y. H. Gonfa, T. B. Asfaw, M. G. Tadesse, and R. K. Bachheti. 2023. "Antioxidant Activity of Flavonoids and Phenolic Acids from *Dodonaea angustifolia* Flower: HPLC Profile and PASS Prediction." *Journal of Chemistry* 2023 (1): 8315711.

Thai, Huynh Nguyen, John van Camp, Guy Smagghe, and Katleen Raes. 2014. "Improved Release and Metabolism of Flavonoids by Steered Fermentation Processes: a review International journal of molecular sciences, 15(11), 19369–19388."

Tian, Baoming, Pinjiao Huang, Yizhu Pan, Hong Gu, Kai Yang, Zhengxun Wei, and Xiangchun Zhang. 2024. "Tea Polyphenols Reduced Obesity by Modulating Gut Microbiota-SCFAs-Barrier and Inflammation in High-Fat Diet-Induced Mice." *Molecular Nutrition & Food Research* 68 (24): 2400685.

Tomas, Merve, Yuxi Wen, Wei Liao, Lizhu Zhang, Chao Zhao, David Julian McClements, Elifsu Nemli, Mustafa Bener, Resat Apak, and Esra Capanoglu. 2024. "Recent Progress in Promoting the Bioavailability of Polyphenols in Plant-Based Foods." *Critical Reviews in Food Science and Nutrition*: 1–22.

Tsao, Rong. 2010. "Chemistry and Biochemistry of Dietary Polyphenols." *Nutrients* 2 (12): 1231–46. https://doi.org/10.3390/nu2121231.

Walle, Thomas. 2007. "Methylation of Dietary Flavones Greatly Improves Their Hepatic Metabolic Stability and Intestinal Absorption." *Molecular Pharmaceutics* 4 (6): 826–32.

Wang, Xiaofei, Yue Qi, and Hao Zheng. 2022. "Dietary Polyphenol, Gut Microbiota, and Health Benefits." *Antioxidants (Basel, Switzerland)* 11 (6). https://doi.org/10.3390/antiox11061212.

Wang, Yuhan, Yujing Li, Chuting He, Bo Gou, and Moshi Song. 2019. "Mitochondrial Regulation of Cardiac Aging." *Biochimica et Biophysica Acta (BBA) – Molecular Basis of Disease* 1865 (7): 1853–64. https://doi.org/10.1016/j.bbadis.2018.12.008.

Wen, Xueping, Yufei Song, Mei Zhang, Yiping Kang, Dandan Chen, Hui Ma, Fajun Nan, Yanan Duan, and Jingya Li. 2024. "Polyphenol Compound 18a Modulates UCP1-Dependent Thermogenesis to Counteract Obesity." *Biomolecules* 14 (6): 618.

Yahfoufi, Nour, Nawal Alsadi, Majed Jambi, and Chantal Matar. 2018. "The Immunomodulatory and Anti-Inflammatory Role of Polyphenols." *Nutrients* 10 (11). https://doi.org/10.3390/nu10111618.

Ye, Yuyang, Yuya Kawaguchi, Asahi Takeuchi, Ni Zhang, Ryosuke Mori, Maihemuti Mijiti, Arata Banno, Toshitaka Okada, Naoto Hiramatsu, and Satoshi Nagaoka. 2023. "Rose Polyphenols Exert Antiobesity Effect in High-Fat–Induced Obese Mice by Regulating Lipogenic Gene Expression." *Nutrition Research* 119: 76–89.

Yugarani, T, B K H Tan, M Teh, and N P Das. 1992. "Effects of Polyphenolic Natural Products on the Lipid Profiles of Rats Fed High Fat Diets." *Lipids* 27: 181–86.

Zhang, Xinyue, Jiaye Liu, Yinyun Ni, Cheng Yi, Yiqiao Fang, Qingyang Ning, Bingbing Shen, Kaixiang Zhang, Yang Liu, and Lin Yang. 2024. "Global Prevalence of Overweight and Obesity in Children and Adolescents: A Systematic Review and Meta-Analysis." *JAMA Pediatrics* 178 (8): 800–13.

# 11 Role of Phenolic Compounds in Skin Health and Dermatological Applications

*Tilahun Belayneh Asfaw*, Fekade Beshah Tessema, and Rakesh Kumar Bachheti*

## 11.1 INTRODUCTION

Phenolic compounds are a class of organic compounds that contain one or more phenol groups attached to the benzene ring. These phenolic compounds consist of a hydroxyl group (–OH) attached to an aromatic ring. These substances are very small in amount and naturally occurring antioxidants that can be found in a wide range of plant-based foods and medicines, including tea, coffee, spices, fruits, and vegetables (Beya et al., 2021; Husen, 2022, 2023a,b, 2024a,b). The importance of phenolic compounds on human health, such as their anti-inflammatory, anti-carcinogenic, antioxidant, and antibacterial qualities, has been extensively researched.

Phenolic compounds are a diverse group of compounds that can be found in various biological systems, including plants, animals, and microorganisms (Carlsen et al., 2010). The main classes of phenolic compounds are: (1) **flavonoids** are types of phenolic compounds that are particularly active against different microorganisms. For instance, quercetin, rutin, kaempferol, catechin, and isorhapontigenin are among several flavonoids. (2) **Phenolic acids** are types of phenolic compounds that are found mostly in plants. Gallic acid, chlorogenic acid, and ellagic acid are a few of the most well-known phenolic acids. (3) **Lignans** are types of phenolic compounds that are found in plants. Enterolactone and enterodiol are the two most well-known examples. (4) **Stilbenes** are a type of phenolic compound. Resveratrol and pterostilbene are the best examples of stilbenes. (5) **Anthocyanins** are other important biologically active compounds found in colorful plants and flowers. The red, purple, and blue colors of many fruits and vegetables are caused by groupings of flavonoids called anthocyanins. Delphinidin-3-glucoside and cyanidin-3-glucoside are two examples. These groups of phenolic compounds are found in plants and are known for their antimicrobial, anti-parasite, anti-viral, and antioxidant properties (Belayneh Asfaw et al., 2024; Pebam et al., 2022; Tessema et al., 2023).

Based on their molecular structure, phenolic compounds can be classified into several types (Vuolo, Lima, & Maróstica Junior, 2019). (1) **Monophenols** are phenolic compounds with a single hydroxyl (–OH) group attached to the phenyl ring are known as monophenols. (2) **Diphenols** are phenolic substances with two hydroxyl (–OH) groups attached to the phenyl ring are known as diphenols. (3) **Triphenols** are phenolic compounds with three hydroxyl (–OH) groups attached to the phenyl ring are known as triphenols. (4) **Polyphenols** are phenolic substances with more than three hydroxyl (–OH) groups attached to the phenyl ring are known as polyphenols. Some typical chemical constituents of phenolic compounds are categorized according to the functional groups attached to them as follows: (1) the most prevalent functional groups in phenolic compounds are hydroxyl groups (**–OH**), which are active for most microbial, viral, and pathogenic microbes. (2) Carboxyl groups (**–COOH**) are other useful chemical components for the prevention and

* Corresponding author: tibebbelay1621@gmail.com


DOI: 10.1201/9781003527671-11

treatment of different ailments. Gallic acid and ellagic acid are two examples of phenolic acids that contain these groups. (3) Aldehyde groups (**CHO**) are also some of the most active groups of phenolic compounds for human ailments. Resveratrol and other stilbenes contain these groups. (4) Ketone groups (**C=O**) are a unique phenolic compounds that are the most effective components of plants used to treat different infectious diseases, such as cancer and some other infectious diseases. Quercetin is one of the flavonoids that contain these groups. (5) Ester groups (**COO–**) are also active ingredients of plants for the treatment of several diseases. Chlorogenic acid and other phenolic acids contain these groups. In general, most of the diversified phenolic compounds are due to the various ways in which these chemical components can be combined biosynthetically to form different structures and molecules with unique properties and functions (Anulika et al., 2016; Marchiosi et al., 2020; Pietta, 2000; Vuolo, Lima, & Maróstica Junior, 2019).

Because of their anti-inflammatory and antioxidant qualities, polyphenols can be utilized in medications. Polyphenols may be used in dermatological treatment for a variety of skin disorders (Aires, 2022). Aires' report claims that polyphenols enhance solar photoprotection of the epidermis in dermatological applications and function as antioxidants and anti-inflammatory agents in skin health. According to experimental research (Abeyrathne et al., 2022; Ketnawa et al., 2022), polyphenols support the epidermis's solar photoprotection and have anti-aging, anti-inflammatory, and antioxidant qualities.

Because of their anti-aging, anti-inflammatory, and antioxidant properties, phenolic compounds such as polyphenols are essential for skin health and dermatological uses (Butkeviciute et al., 2022). These substances, which are found in large quantities in plants, have been connected to promoting the epidermis that is used to sun photoprotection (Aires, 2022; Gugleva et al., 2021). However, because of their intrinsic properties, such as strong reactivity and low solubility, phenolic compounds provide difficulties when used in skincare formulations in pharmaceuticals and nutraceuticals (Lee, Park, & Shin, 2022). Phenolic compounds have been demonstrated in studies to postpone the aging process of the skin by controlling age-related signaling pathways in dermal fibroblasts, increasing suppleness, decreasing wrinkle formation, and promoting wound healing (Butkeviciute et al., 2022). Additionally, studies have shown that phenolic compounds have excellent penetration into the epidermis and dermis and exhibit antioxidant activity in skin layers, indicating their potential therapeutic effectiveness in dermatology (de Lima Cherubim et al., 2020).

Phenolic extracts are a sustainable and successful way to address dermatological issues and promote skin health in skincare products. According to dermatology, the pharmacological effects of phenolic compounds on oxidation processes, inflammation, vascular pathology, immunological response, precancerous and oncological lesions or formations, and microbial proliferation are the most significant advantages of these compounds for human health. Because the nature of phenolic compounds is designed to fit the phytochemical needs of plants and not the biopharmaceutical requirements for a specific route of delivery (dermal or other), their utilization in cutaneous formulations sets challenges to drug development (Gugleva et al., 2021). Additionally, these compounds play a vital role in dermatological and skin therapy applications through the antioxidant activity of phenolic compounds, highlighting the action mechanisms and structure-activity relationship (de Lima Cherubim et al., 2020). Several clinical studies have shown that the other main application of phenolic compounds is in the protection of skin cancer therapy, which is due to exposure of skin to solar ultraviolet (UV) radiation, causing adverse effects, such as inflammation, oxidative stress, and DNA damage (Gado et al., 2021). Melanoma and non-melanoma skin cancers (NMSCs) are the other skin disorders, the most lethal and commonest forms of skin cancers, that affect the largest population in the world.

The *in vivo* and *in vitro* studies showed that there is a strong positive correlation between the number of phenolic compounds that penetrated the epidermis and the antioxidant activity detected in the epidermis extract, and plant-based phenolic compounds demonstrated antioxidant activity and showed great permeability properties through the skin (Butkeviciute et al., 2022). All skin and dermal problems are treated with medicinal plants that were used by traditional healers and new lead compounds by researchers as a starting point for attaining cost-effective therapies (Chamcheu et al., 2022).

Phenolic compounds, which are natural antioxidants found in plants, have been shown to play a significant role in skin health and have various dermatological applications. The following six methods form the basis of how those phenolic chemicals work to safeguard the health of the skin (Anulika et al., 2016; Marchiosi et al., 2020; Pietta, 2000; Vuolo, Lima, & Maróstica Junior, 2019).

1. *Antioxidant Properties:* Strong antioxidant qualities of phenolic compounds aid in shielding the skin from oxidative stress and free radical damage. Wrinkles, other skin disorders, and early aging can be avoided in this way.
2. *Anti-Inflammatory Properties:* Because of their anti-inflammatory qualities, phenolic compounds can help lessen skin inflammation and the symptoms of a number of skin disorders, including psoriasis, eczema, and acne.
3. *Anti-Microbial Properties:* Because of their antibacterial qualities, certain phenolic compounds can help shield the skin from dangerous bacteria and fungus, lowering the risk of infections and enhancing skin health in general.
4. *Skin Brightening Properties:* By brightening and even out skin tone, certain phenolic compounds—including those in plant extracts like licorice root—can lessen the visibility of dark patches and hyperpigmentation.
5. *Wound Healing Properties:* By boosting cell proliferation, decreasing inflammation, and increasing collagen synthesis, phenolic substances have been shown to aid in wound healing. They are therefore useful in the treatment of small cuts, scratches, and other skin wounds.
6. *UV Protection:* It has been demonstrated that certain phenolic chemicals provide UV protection, hence lowering the risk of skin cancer and preventing sun damage.

## 11.2 DERMATOLOGY

The diagnosis, management, and prevention of illnesses and conditions affecting the skin, hair, nails, and mucous membranes are the main goals of the medical specialty known as dermatology (Ferreira et al., 2023). According to numerous research papers, dermatology is an essential medical specialty that is important in the diagnosis and treatment of a broad range of skin conditions, such as skin cancer, acne, eczema, psoriasis, dermatitis, hair loss, nail issues, and skin-related STDs. A wide range of problems affecting the skin, hair, and nails are referred to as dermatological conditions, and for the best care, a variety of therapeutic modalities are required. Apart from traditional medications and radiation therapy, the use of alternative remedies, especially those using plants high in phenolic compounds, has become more popular. Prevalent skin diseases such as dermatitis, eczema, and infectious skin illnesses, dermatological services in hospital settings offer crucial support, particularly using the synergetic effects of plant extracts (Ferreira et al., 2023). In medical practice, dermatology covers nearly 2000 disorders, which is a popular presentation since it includes the examination of skin, hair, and nails (Bickers et al., 2006). Dermatologists use a range of therapeutic approaches, including radiation therapy, medication, and natural remedies using plants that contain phenolic components, including flavonoids, terpenes, alkaloids, phenolic acids, anthocyanins, and other essential oils.

## 11.3 SKIN HEALTH

The term "skin health" describes the general health and integrity of the skin, which acts as an essential barrier of defense against a variety of environmental stresses and infections (Hassan, 2024; Woodby et al., 2020). The largest organ in the body, the skin is essential for temperature regulation, immune defense, feeling, and the synthesis of vitamin D (Mostafa & Hegazy, 2013).

In addition to extrinsic factors like exposure to UV light, pollution, and toxins, which can cause oxidative stress, DNA damage, and premature aging, intrinsic factors like genetics and aging can also have an impact on skin health (Chen et al., 2021). Practices like good hygiene, skincare regimens, shielding the skin from damaging UV rays, and using antioxidants like vitamins C and E to fight oxidative damage and promote skin resilience are all part of maintaining healthy skin (Liao et al., 2022). Since skin conditions can have a substantial impact on a person's physical, emotional, and social well-being, putting skin health first is crucial for overall well-being and quality of life.

Humans are protected by their skin from harmful substances, pathogens, UV light, chemicals, and physical, mechanical, and thermal harm. The skin's other vital functions include immunological surveillance, vitamin D synthesis, detecting stimuli, and controlling a number of the body's vital physiological functions. However, the way the skin behaves and functions is greatly altered by aging, illnesses, and environmental factors. Topical treatment with medical cosmetology, active chemicals, and other physical measures of photoprotection are typically used as treatment and preventative techniques for a variety of skin problems, particularly photo-aging.

## 11.4 THE ROLE OF PHENOLIC COMPOUNDS IN SKIN HEALTH

The treatment and prevention strategies for various skin diseases, including photo-aging, usually include topical treatment with medical cosmetology, active ingredients, and other physical means of photoprotection such as clothing, window films, glass, and sunglasses. In recent times, increasing attention has been given to the role of nutrition in skin health, with certain dietary components rich with phenolic compounds to alleviate photo-aging, pathogens, UV light, chemicals, mechanical, thermal, and physical injuries, as well as hazardous substances and their symptoms. Diet, nutrition, and herbal medicines are adequate alternative approaches to increase skin health consciousness and alleviate and prevent both endogenous and exogenous aging symptoms of the skin. Therefore, the consumption of fruits, vegetables, nuts, and legumes, which contain high antioxidants and polyphenolic, can have a low glycemic index diet, probiotics, and phytoestrogens that should be adopted to enhance skin health (Ahmed & Mikail, 2024).

Phenolic compounds have demonstrated potent activities against skin diseases due to their antioxidant properties, which help counteract the negative effects of factors like solar UV radiation and airborne particulate matter (PM) that contribute to skin disorders (Abeyrathne et al., 2022; Ketnawa et al., 2022). These compounds exhibit anti-inflammatory effects, induce apoptosis, and reduce DNA damage, thus playing a crucial role in preventing and treating skin cancers, including melanoma and NMSC (Chen et al., 2021; Mani et al., 2021). Furthermore, phenolic compounds have been shown to protect the skin from premature aging brought on by environmental pollutants like PM by increasing cellular antioxidant capacity, lowering reactive oxygen species (ROS) levels, and mitigating oxidative damage to proteins, lipids, and nucleic acids (Martic, Jansen-Dürr, & Cavinato, 2022). Phenolic compounds' varied bioactive characteristics emphasize their great potential in skincare and dermatology research and make them attractive options for creating treatments against a range of skin conditions.

Plants are rich in phenolic chemicals, which have many health advantages, especially for skin and dermatological conditions. These secondary metabolites exhibit antioxidant, anti-inflammatory, antimicrobial, and anti-aging properties (Singh & Yadav, 2022). They can alleviate symptoms and inhibit the development of various skin disorders, including wrinkles, acne, and even potentially life-threatening conditions like cancer (Damayanti et al., 2023; Xu et al., 2021). Table 11.1 presents a comprehensive overview of the roles that phenolic compounds play in maintaining skin health, detailing their sources and specific mechanisms of action. It is noteworthy that nearly all categories of these phenolic compounds exhibit comparable effects and modes of action against agents that contribute to skin diseases. This suggests a shared biochemical pathway through which they exert their protective effects, highlighting the potential of phenolic compounds as beneficial constituents in skincare and dermatological treatments.

**TABLE 11.1**
**Phenolic Compounds, Their Roles in Skin Health, Sources, and Their Mechanism of Actions**

| Phenolic Compound | Sources | Key Benefits for Skin Health | Mechanism of Action | References |
|---|---|---|---|---|
| Resveratrol and its derivatives | Grapes, red wine, berries | Antioxidant, anti-inflammatory, improves skin elasticity, protects against UV damage | Scavenge free radicals, reduce oxidative stress, inhibit inflammation | Marko and Pawliczak (2023), Ratz-Łyko and Arct (2019), Woodby et al. (2020) |
| Curcumin | Turmeric | Anti-inflammatory, antioxidant, improves skin elasticity | Inhibits inflammatory cytokines, reduces oxidative stress, enhances collagen production | Gonzalez and Sethi (2012), Gopinath and Karthikeyan (2018) |
| Quercetin | Apples, onions, berries, tea, some fruits | Antioxidant, anti-inflammatory, protects against UV damage | Neutralizes free radicals, reduces inflammation, stabilizes mast cells | Asfaw et al. (2023), Vicentini et al. (2011) |
| Epigallocatechin gallate (EGCG) | Green tea | Antioxidant, anti-inflammatory, reduces acne, protects against UV damage | Protects skin from oxidative damage, delaying signs of aging and reducing the risk of skin cancers, and inflammation associated with conditions like acne, eczema, and psoriasis, skin cancer and supports treatment of existing skin cancers. | Damayanti et al. (2023), Xu et al. (2021) |
| Ellagic acid | Pomegranates, strawberries | Antioxidant, anti-inflammatory, protects against UV damage | Neutralizes free radicals, reduces inflammation, inhibits collagenase | Kunchana et al. (2021), Tavares et al. (2023) |
| Catechins | Green tea, cocoa | Antioxidant, anti-inflammatory, improves skin texture | Scavenges free radicals reduce inflammation, improve blood flow | Kapoor et al. (2021), Megow et al. (2017) |
| Luteolin | Celery, peppers, parsley | Anti-inflammatory, antioxidant, supports skin healing | Reduces oxidative stress, inhibits inflammatory pathways | Gendrisch et al. (2021) |
| Apigenin | Chamomile, parsley | Anti-inflammatory, antioxidant, promotes wound healing | Reduces oxidative stress, enhances skin repair processes | Bhardawaj et al. (2023), Majma Sanaye et al. (2022) |
| Anthocyanins | Blueberries, cherries | Antioxidant, anti-inflammatory, improves skin hydration, anticancer | Scavenges free radicals reduce inflammation, improve skin moisture, and slow down or stop cancer cell proliferation, contributing to the control of tumor growth | Correia et al. (2021), Diaconeasa et al. (2020) |
| Tannins | Tea, red wine, oak barrels | Astringent, antioxidant, antimicrobial | Tightens skin neutralizes free radicals, inhibits microbial growth | Sun and Shahrajabian (2023) |

Phenolic compounds act on multiple molecular pathways, inhibiting pro-inflammatory mediators and enzymes associated with various diseases (Dhananjaya et al., 2016). Their therapeutic potential extends to treating hypertension, metabolic disorders, neurodegenerative diseases, and inflammatory conditions. However, the utilization of phenolic compounds in dermal formulations faces challenges due to their physicochemical properties. To overcome these limitations and enhance their bioavailability and therapeutic potency, lipid-based nanotechnologies, such as nanoemulsions, liposomes, and solid lipid nanoparticles, have been employed as delivery systems for these phytochemicals (Gugleva et al., 2021).

Because they guard against a variety of skin conditions and enhance general well-being, phenolic compounds are essential for healthy skin. These substances, which are found in large quantities in fruits, vegetables, and medicinal plants, have a strong antioxidant capacity and help fight off oxidative stress, inflammation, and UV-induced DNA damage, hence preventing melanoma and NMSC (Karağaç & Ceylan, 2023; Silva et al., 2022). Furthermore, polyphenols, a type of phenolic compound, demonstrate multiple biological benefits such as antioxidant, anti-inflammatory, and antimicrobial properties, making them promising candidates for treating skin diseases and hair loss with minimal adverse effects (Sun et al., 2022; Villegas-Aguilar et al., 2022). Overall, the diverse bioactive properties of phenolic compounds highlight their significance in preserving skin health and preventing age-related dermatological issues. Chronic wounds or skin exposure to UV irradiation disrupts fibroblast function by generating ROS, which may damage cell components and modify various signaling pathways. The antioxidant activity of plant-derived phenolic compounds may reverse the resulting imbalance (Merecz-Sadowska et al., 2021).

The phenolic compounds in plant-based foods and extracts have been extensively studied in dermatology due to their multifaceted properties, which contribute to their potential to promote skin health and mitigate various skin-related disorders (Maya-Cano, Arango-Varela, & Santa-Gonzalez, 2021). The antioxidant properties of phenolic compounds, particularly flavonoids, phenolic acids, and lignans, are crucial in neutralizing free radicals and preventing oxidative stress (Zeb, 2020). According to the oxidative stress theory, this process helps maintain skin integrity by reducing the accumulation of ROS and preventing premature aging, wrinkles, and other skin conditions associated with oxidative stress (Rinnerthaler et al., 2015). The antioxidant activity of phenolic compounds also scavenges ROS, thereby protecting skin cells from damage and promoting overall health. This is supported by the "free radical theory" concept, which suggests that oxidative stress is a primary cause of aging and age-related diseases (Rinnerthaler et al., 2015).

The anti-inflammatory properties of phenolic compounds, such as polyphenols and hydroxybenzoic acids, have been demonstrated to alleviate symptoms of skin conditions like acne, eczema, and psoriasis (Rahman et al., 2021). By inhibiting inflammatory pathways, these compounds can calm irritated skin, reduce inflammation, and promote healing (Fernandes et al., 2023). This is particularly relevant for acne-prone or sensitive skin, where inflammation can exacerbate symptoms. For example. The study reviewed by Fernandes et al. (2023) indicated that the antioxidant properties of polyphenols in the skin are well-explored. Indeed, a range of molecules such as antioxidant vitamins (particularly vitamins C and E), curcumin, coenzyme Q, lipoic acid, melatonin, resveratrol, and other polyphenols are safe through the diet or external application in the skin. As stated from the in vitro and in vivo studies, the most frequently explored effects of the immunomodulatory capacity of polyphenols are related to cancer, viruses, allergic diseases, and intestinal mucosal immune responses. For instance, polyphenols extracted from medicinal plants have been shown to have the ability to attenuate the NF-κB pathway and increase the NRF2 activity from the plant *Uvaria alba* extracts. These compounds have been shown to promote the regeneration of skin tissue, enhance collagen production, and reduce inflammation, thereby accelerating the healing process of wounds, cuts, and other skin injuries (Działo et al., 2016).

Throughout history, traditional medicinal plants have played a significant role in the development of formulations aimed at treating various skin ailments, including wounds, burns, and cuts (Vivo et al., 2021). Research indicates that contemporary Western pharmacopeias include only about 1%–3% of pharmacological agents specifically approved for dermal wound management, with approximately one-third of these treatments incorporating traditional herbal remedies having high amounts of phenolic compounds (Maver et al., 2015; Sun & Shahrajabian, 2023). In contrast to synthetic alternatives, which often come with high costs and potential side effects, traditional plants, recognized for their therapeutic properties in folklore, may offer a more effective and economically feasible option for wound care when subjected to scientific evaluation (Alemu, Misganaw, & Mengistu, 2020). Consequently, exploring the potential of these plants not only aligns with historical usage but also paves the way for integrating more accessible and holistic treatments into modern medicine, even if there are some challenges like culture and beliefs in the societies. With the World Health Organization's formal promotion of traditional medicine in poor nations, there is already an opportunity to include traditional medicine in a country's healthcare system (Chi, 1994).

### 11.4.1 Role of Phenolic Compounds on the Treatment of Skin Diseases

A broad spectrum of illnesses affecting the skin, hair, nails, and mucous membranes is referred to as skin disorders. The intensity and symptoms of these illnesses might differ, spanning from minor infections and irritations to malignant growths and chronic inflammatory disorders. Skin cancer, rosacea, warts, eczema, psoriasis, dermatitis, fungal infections, and acne are examples of common skin illnesses. The individual ailment and its underlying causes determine the course of treatment for skin diseases. To manage and control skin illnesses, dermatologists may suggest oral drugs, phototherapy, laser therapy, cryotherapy, surgical procedures, and lifestyle changes in addition to topical medications (such as corticosteroids, antibiotics, and antifungals). Natural substances like phenolic compounds have drawn interest recently because of their possible therapeutic benefits for managing diseases and maintaining healthy skin. One type of bioactive substance is phenolic compounds. A class of bioactive chemicals called phenolic compounds is present in plants, fruits, vegetables, and other natural sources. These compounds' anti-inflammatory, antibacterial, antioxidant, and anti-cancer properties can help your skin stay healthier. The potential applications of phenolic compounds in wound healing, managing a range of skin disorders, and protecting the skin from environmental stressors have all been studied.

The positive effects of phenolic compounds on skin health and disease management have been documented in a number of studies (Kaparekar & Anandasadagopan, 2022; Karağaç & Ceylan, 2023; Sun & Shahrajabian, 2023). For instance, studies have demonstrated that the polyphenols in green tea, resveratrol in red wine, and curcumin in turmeric have anti-inflammatory and antioxidant qualities that may help reduce the symptoms of inflammatory skin diseases like psoriasis and eczema (Contardi et al., 2021). Furthermore, the ability of phenolic substances such as gallic acid and ellagic acid to prevent the growth and proliferation of skin cancer cells has been studied (Anantharaju et al., 2016).

### 11.4.2 Antioxidant Properties

Numerous plant sources include phenolic compounds, which are rich in antioxidants and are essential for protecting the body from oxidative stress and the diseases it causes (Asfaw et al., 2022). These bioactive compounds, including flavonoids and phenolic acids, have the capacity to neutralize free radicals, thereby mitigating cellular damage resulting from lipid peroxidation (Asfaw et al., 2022; Rinnerthaler et al., 2015). The regular consumption of foods rich in antioxidants, such as wine, has been associated with protective effects against a range of health conditions, including cancer, atherosclerosis, and cardiovascular diseases (Stojiljković, Arsić, & Tadić, 2016). Moreover, the efficacy of phenolic compounds extends beyond health benefits; they also enhance the shelf life of food products by inhibiting oxidative degradation. In light of increasing concerns regarding the safety of synthetic antioxidants, there has been a burgeoning interest in harnessing natural antioxidants derived from plant sources for applications in both the food and pharmaceutical industries (Asfaw et al., 2022).

#### 11.4.2.1 Antioxidant Inhibition Mechanism of Action on Skin Disease Healing

##### *11.4.2.1.1 Protection Against Oxidative Stress*

Oxidative stress is a pathological state in which reactive oxygen/nitrogen species (ROS/RNS) overwhelm the antioxidative defenses of an organism, leading to oxidative modification of biological macromolecules like lipid, protein, DNA, tissue injury, and accelerated cellular diseases. Antioxidants are molecules that can neutralize or inhibit the harmful effects of free radicals and ROS in the body. Free radicals are unstable molecules that can damage cells, DNA, and proteins, leading to oxidative stress and inflammation. Antioxidants work by donating electrons to stabilize these free radicals, thereby reducing oxidative stress and protecting cells from damage. The high antioxidant capacity of natural sources as compared with synthetic ones has an interesting potential to prevent or minimize the oxidative stress that causes many chronic diseases (Becker et al., 2019).

##### *11.4.2.1.2 Neutralization of Free Radicals*

An antioxidant is a substance that prevents oxidation of biological molecules like proteins and DNA at low concentrations through the mechanism shown in Figure 11.1. Bioactive compounds are the natural antioxidants in plants and their fruits, having a significant role in human health due to their scavenging of free radicals, and antioxidant enzyme cofactors, protecting dietary antioxidants and food colorants. In recent years, there has been a surge of interest in the potential application of antioxidants in medical treatment and the development of skin care products containing high amounts of naturally obtained phenolic compounds. The antioxidant mechanism is that a single electron or electrons are transferred from the antioxidant molecule (i.e., phenolics—electron donor, AH or ArOH) to the oxidant (free radicals, Berker et al., 2007) as shown in Figure 11.1.

Where AH or ArOH are the phenolic compounds as shown in the following figure (Figure 11.2).

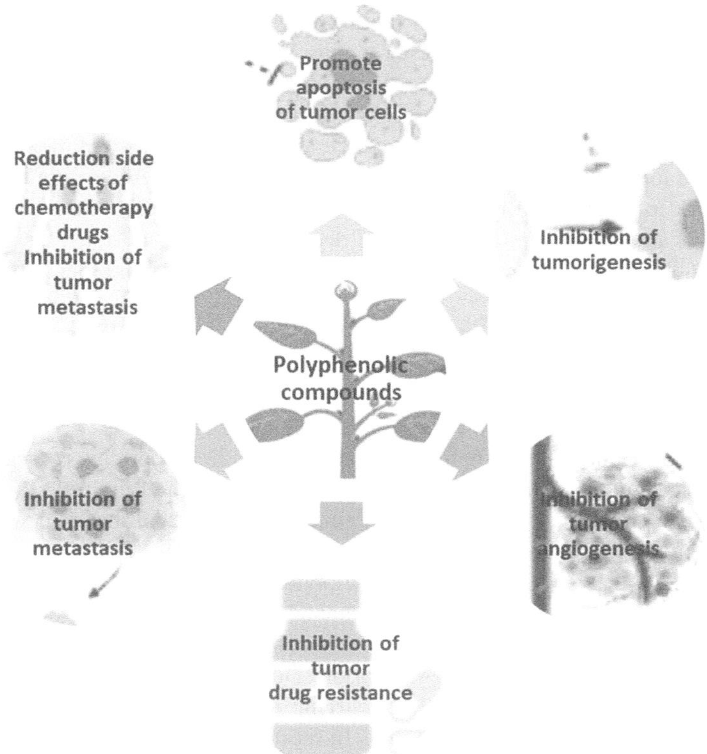


**FIGURE 11.1** Reaction steps for the antioxidant mechanisms.

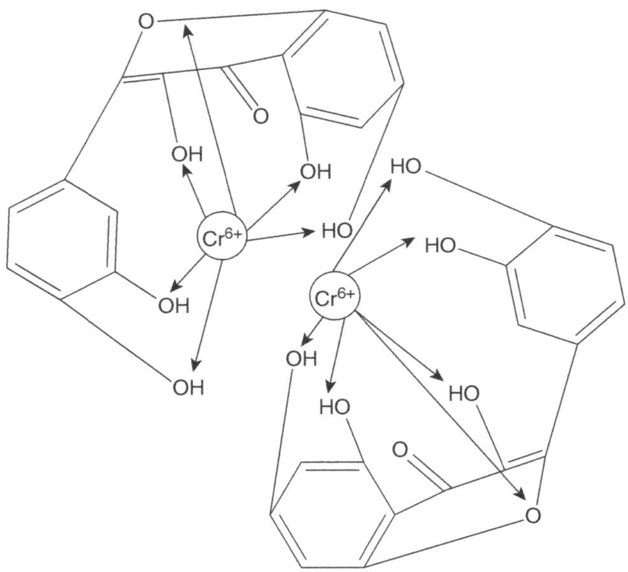


**FIGURE 11.2** Phenolic compounds: (A) A specific compound Quercetin from phenolic compounds and (B) the general structure of Flavonoids.

Antioxidants are essential for preserving skin health and advancing the treatment of a variety of skin disorders in the context of treating skin diseases. Skin conditions such as acne, eczema, psoriasis, dermatitis, and premature aging have been linked to oxidative stress in their early stages of development and progression based on the above mechanisms of action (Becker et al., 2019). Antioxidants help shield skin cells, preserve skin barrier function, lessen inflammation, and encourage skin repair and regeneration by scavenging free radicals and lowering oxidative damage. Antioxidants can also boost the efficacy of other skincare products in treating skin conditions and enhancing skin health. Examples of these include anti-inflammatory compounds and ingredients that promote wound healing (Liao et al., 2022; Vivo et al., 2021). According to R. H. Liu (2003), antioxidants work in concert with other active ingredients to increase the overall effectiveness of skincare products and treatments for a range of dermatological conditions. However, Butkeviciute et al. (2022) found that antioxidants also have a synergistic effect with other active ingredients.

### 11.4.2.2 Applications of Antioxidants in Dermatology Treatments

Because of its ability to combat free radicals and protect the skin from oxidative stress, antioxidants are important in dermatology treatments. Free radicals are unstable molecules that can damage skin cells, accelerate aging, and worsen a variety of skin conditions, such as wrinkles, hyperpigmentation, and skin cancer. Antioxidants help combat these detrimental effects by scavenging free radicals, reducing inflammation, and promoting skin regeneration and repair. Antioxidants are widely used in dermatology for a variety of reasons, such as wound healing, skin whitening, reducing aging, and preventing skin cancer.

When combined with skincare products, nutritional supplements, and dermatological treatments, antioxidants can generally help prevent environmental damage, enhance skin health, and boost the effectiveness of other skincare ingredients. Phenolic compounds, which are abundant in plants, hold a lot of promise for tissue repair and regeneration. Due to their antibacterial, anti-inflammatory, and antioxidant properties, these naturally occurring metabolites can be used to treat a range of skin disorders and aid in the healing of wounds. In tissue engineering, polyphenols are increasingly being employed to manage the tissue microenvironment, reduce oxidative stress, and enhance scaffold biocompatibility (Gao et al., 2021). Their unique chemical structure, which contains catechol or pyrogallol moieties, enables strong non-covalent and covalent bonds. This makes it possible to develop an extensive range of scaffolds, such as films, hydrogels, and nanofibers. These polyphenol-based scaffolds have shown promise in wound healing, bone regeneration, and electroactive tissue creation. The bioactive properties of phenolic compounds included in traditional medicines further emphasize their potential for tissue regeneration and repair medicines (Gao et al., 2021; Zhang et al., 2021). All things considered, phenolic compounds offer a versatile and promising approach to regenerative medicine and tissue engineering.

### 11.4.3 Anti-Inflammatory Effects

Because of their potential therapeutic uses in inflammatory chronic diseases, phenolic compounds' anti-inflammatory properties have attracted a lot of research. These substances, which are found in many different plants, have a wide range of biological functions, most notably the ability to regulate inflammatory pathways. By blocking pathways like NF-kB and mitogen-activated protein kinase (MAPK), phenolic substances can lower the production of pro-inflammatory cytokines and mediators, including nitric oxide (NO) and interleukins (Kauffmann & Castro, 2023; Kim et al., 2023). Antioxidant properties found in many phenolic compounds aid in reducing oxidative stress, which is a factor in chronic inflammation (Kim et al., 2023; Scarpa et al., 2024). Common plant sources of polyphenolic chemicals include curcumin and resveratrol, and different plant phenolic acids are regularly researched for their anti-inflammatory qualities (Mendonça et al., 2024).

Phenolic compounds' structural variety affects their modes of action and efficacy, making them ideal candidates for drug development with promising therapeutic potential. For diseases like diabetes and osteoporosis, where inflammation is a major factor, phenolic chemicals are being investigated as supplements to traditional treatments (Kim et al., 2023; Scarpa et al., 2024). Although phenolic compounds' anti-inflammatory qualities are encouraging, more study is required to completely understand their mechanisms and maximize their medicinal uses. In the field, it is still difficult to comprehend the intricacies of their interactions within biological systems.

#### 11.4.3.1 Reduction of Inflammation and Redness

One important area of research that highlights the potential medicinal applications of phenolic compounds is their ability to reduce inflammation and redness. Numerous investigations have shown that phenolic compounds have anti-inflammatory qualities via a variety of channels, including the alteration of inflammatory mediators. By lowering cytokines like COX-2 and iNOS in keratinocytes, phenolic substances like those in *Lonicera japonica* and Citri reticulatae pericarpium polyphenolic extract (LCPE) dramatically reduce inflammation (Jeong et al., 2023). Additionally, pomegranate phenolics, such as punicalagin and ellagic acid, protect skin cells from oxidative stress, which is often linked to inflammation (Yue et al., 2012). The use of nanocarrier technologies enhances the stability and bioavailability of phenolic compounds, improving their anti-inflammatory effects (Rakotondrabe et al., 2023).

Sources of phenolic compounds, such as plant extracts, are commonly employed for the reduction of inflammation and redness. Common beans and grape skins are rich in phenolic compounds, with studies showing that extracts from these sources can effectively reduce inflammation (Baron et al., 2021; Pendota et al., 2018). Various phenolic classes, including flavonoids and phenolic acids, have been identified as having significant anti-inflammatory effects (Jeong et al., 2023). While the anti-inflammatory properties of phenolic compounds are well-documented, it is essential to consider that their efficacy can vary based on extraction methods and the specific compounds involved. Further research is needed to optimize their use in clinical applications.

#### 11.4.3.2 Inhibition of Inflammatory Mediators

Resveratrol, curcumin, and polyphenols are antioxidants that can help calm and soothe irritated skin, lowering the redness, sensitivity, and irritation that come with rosacea, acne, and eczema. Polyphenols exhibit significant anti-inflammatory and antioxidant properties beneficial for skin health. These compounds, including flavonoids, lignans, and stilbenes, demonstrate anti-inflammatory activity through various cellular mechanisms both *in vitro* and *in vivo* (Mir & Masoodi, 2020). Polyphenols protect the skin from UV radiation-induced damage by inhibiting inflammation, oxidative stress, and DNA damage. They also combat skin aging by neutralizing free radicals, suppressing pro-inflammatory cytokines, and stimulating collagen synthesis. While their efficacy has been widely recognized in laboratory studies, clinical proof for preventing and treating major skin inflammatory conditions is still needed for most polyphenols. Future research should focus on identifying specific polyphenols that target particular inflammatory pathways in the skin, potentially leading to the development of more effective anti-inflammatory agents for skin health (Mir & Masoodi, 2020).

Many phenolic compounds possess anti-inflammatory properties by modulating key inflammatory mediators such as cytokines, chemokines, and enzymes like cyclooxygenase (COX) and lipoxygenase (LOX). Through inhibition of pro-inflammatory pathways, phenolic compounds can attenuate inflammation associated with conditions like acne, eczema, and rosacea.

### 11.4.4 Photoprotective Benefits

The photoprotective benefits of phenolic compounds are increasingly recognized in dermatology, particularly for their role in mitigating the harmful effects of UV radiation. These compounds, derived from various plant sources, exhibit antioxidant properties that help protect skin cells from oxidative stress and damage caused by UV exposure. Phenolic compounds scavenge ROS, reducing oxidative stress and preventing cellular damage (Fernandes et al., 2021). They help in repairing UV-induced DNA damage, thereby reducing the risk of skin cancer (Sharma et al., 2024). These compounds modulate inflammatory responses triggered by UV exposure, which can lead to skin aging and other dermatological issues (Ayad et al., 2024).

Plant extracts are the most common sustainable sources of phenolic compounds for the applications of photoprotective skin damage or prevention. Flavonoids and other phenolic compounds from plants are effective in absorbing UV radiation and providing a natural barrier against skin damage (Ghazi, 2022). Compounds like phlorotannins from marine algae have shown significant photoprotective and anti-aging benefits (Kalasariya & Pereira, 2022). Incorporating phenolic compounds into sunscreens enhances their protective efficacy against UV radiation, offering a safer alternative to synthetic filters (Ghazi, 2022). While the benefits of phenolic compounds in photoprotection are well-documented, it is essential to consider potential limitations, such as variability in efficacy based on extraction methods and concentrations used in formulations. Further research is needed to optimize their use in skincare products.

#### 11.4.4.1 Prevention of UV-Induced Damage

Exposure to UV solar radiation is the main cause of the high incidence of skin cancer. Plant phenolic compounds have anti-inflammatory, immunomodulatory, antioxidant, and DNA-repair properties and can be employed to prevent various skin diseases resulting from exposure to UV radiation. In vitro and in vivo research has evidenced the protective effects of phenolic compounds on biochemical processes that are induced or mediated by UV radiation and photoaging (Fernandes et al., 2021).

#### 11.4.4.2 Enhancement of Endogenous Antioxidant Defense

The enhancement of endogenous antioxidant defense by phenolic compounds derived from plants is a significant area of research, highlighting their role in combating oxidative stress. These compounds, produced as secondary metabolites, not only protect plants from pathogens but also contribute to human health by mitigating oxidative damage. Flavonoids and hydroxycinnamic acids are examples of phenolic chemicals that scavenge free radicals, hence lowering oxidative stress in both people and plants (Rammohan et al., 2023; Tariq & Ahmed, 2024). Depending on their quantity, several phenolics have both prooxidant and antioxidant properties, which affect how well they work overall to regulate oxidative stress (Nowak et al., 2022).

Because of their antioxidant properties, dietary consumption of foods high in phenolics has been associated with lower risks of chronic diseases, such as cancer and cardiovascular problems (Rammohan et al., 2023). High concentrations of phenolic chemicals, which have been demonstrated to have strong antioxidant activity and support general health, are found in a variety of herbs and plants (Foss, Przybyłowicz, & Sawicki, 2022). Although phenolic compounds are advantageous, it is important to keep in mind that their effects can differ depending on dietary circumstances and individual health situations. This implies that further investigation is required to ascertain the optimal levels of intake and the ways in which they interact with other nutritional elements.

### 11.4.5 Anti-Aging Effects

The anti-aging effects of phenolic compounds derived from plants are increasingly recognized due to their diverse biological activities, particularly their antioxidant and anti-inflammatory properties. These compounds, including flavonoids and polyphenols, have shown promise in mitigating the effects of aging on various body systems, especially the skin. The mechanisms of plant-derived phenolic compounds are antioxidant activity, anti-inflammatory effects, and collagen synthesis stimulation. Plant-derived phenolic compounds' potential applications are in making nutraceuticals and cosmeceutical products.

Phenolic compounds neutralize free radicals, reducing oxidative stress that contributes to aging (Dwevedi & Srivastava, 2024). They suppress pro-inflammatory cytokines, alleviating inflammation that accelerates skin aging (Abeyrathne et al., 2022; Dwevedi & Srivastava, 2024). Certain polyphenols promote collagen production, enhancing skin elasticity and reducing wrinkles (Dwevedi & Srivastava, 2024). The potential of flavonoids in dietary supplements to prevent aging-related illnesses is being investigated. Phenolics obtained from plants are used in skincare products to enhance the appearance and health of the skin (Gomez-Molina et al., 2024). Even though phenolic compounds have several potential advantages, more study is required to completely comprehend their workings and maximize their application in anti-aging treatments. Standardized procedures are necessary because some research indicates that the effectiveness of these substances can differ depending on the extraction techniques and storage conditions (Gomez-Molina et al., 2024).

Antioxidants like vitamins C and E, coenzyme Q10, and green tea extract can help reduce the signs of aging by protecting the skin from UV-induced damage, stimulating collagen production, and improving skin elasticity and firmness. Polyphenols, abundant in plants, exhibit potent anti-aging effects on skin through multiple mechanisms. They protect against UV radiation, stimulate collagen synthesis, and provide antioxidant and anti-inflammatory benefits. Structurally diverse polyphenols, including stilbenoids, flavonoids, and chalcones, effectively scavenge ROS and coordinate broader antioxidant responses. In aged human dermal fibroblasts, polyphenols regulate age-related signaling pathways, improving skin elasticity and firmness (Yar Khan et al., 2012). Their anti-aging mechanisms include preventing cellular senescence, targeting microRNA, influencing NO bioavailability, and promoting mitochondrial function (Luo et al., 2021). Dietary intervention studies support the health-promoting effects of polyphenol supplementation, particularly in the elderly population. Future research should focus on exploring the benefits of combining multiple polyphenols or polyphenol-rich foods for more physiologically relevant applications in preventing aging-related disorders (Luo et al., 2021).

#### 11.4.5.1 Reduction of Wrinkles and Fine Lines

The reduction of wrinkles and fine lines through phenolic compounds derived from plants has garnered significant attention in recent research. These compounds exhibit various biological properties that contribute to skin health, particularly in combating signs of aging. Phenolic compounds scavenge free radicals, reducing oxidative stress that contributes to skin aging (Kumar & Goel, 2019). They enhance skin cell longevity through pathways like MPK-1/ERK and insulin/IGF-1 signaling, which are crucial for maintaining skin homeostasis (Działo et al., 2016). Certain phenolic extracts, such as those from propolis, have been shown to modulate inflammation and tissue remodeling, leading to improved skin texture. The phenolic compounds are also important in the applications of anti-wrinkle efficacy activities, such as phenolic acids in plant extracts (Radić et al., 2024).

A study demonstrated that creams containing phenolic acids from propolis resulted in significant wrinkle reduction, with improvements of 25% and 34% for different concentrations over 28 days (Radić et al., 2024). The use of plant-derived phenolics offers a safer alternative to traditional anti-aging ingredients, aligning with consumer preferences for natural products (Gomez-Molina et al., 2024; Radić et al., 2024). While the efficacy of phenolic compounds in reducing wrinkles is promising, it is essential to consider the variability in their effectiveness based on extraction methods and storage conditions. This variability can influence the stability and activity of these compounds, potentially affecting their overall impact on skin health (Gomez-Molina et al., 2024).

#### 11.4.5.2 Improvement of Skin Elasticity

The improvement of skin elasticity through plant-derived phenolic compounds is supported by various studies highlighting their biological properties and mechanisms of action. These compounds, known for their antioxidant and anti-inflammatory effects, play a crucial role in enhancing skin health and combating aging. Antioxidant Properties: Phenolic compounds reduce oxidative stress, a key factor in skin aging, by neutralizing ROS (Lee, Park, & Shin, 2022; Nobile et al., 2021). They promote the synthesis of collagen and elastin in dermal fibroblasts, essential for maintaining skin elasticity and firmness (Lee, Park, & Shin, 2022). Inflammation Reduction: By mitigating inflammation, these compounds help prevent skin damage and promote healing (Gomez-Molina et al., 2024).

A study on a polyphenol-enriched supplement demonstrated significant improvements in skin elasticity and moisture levels among participants exposed to urban pollution (Nobile et al., 2021). Research indicates that topical application of phenolic compounds can alleviate symptoms of skin diseases and enhance overall skin appearance (Gomez-Molina et al., 2024). While the benefits of phenolic compounds are well-documented, it is essential to consider that individual responses may vary based on genetic factors and environmental influences, suggesting a need for personalized approaches in skincare.

#### 11.4.5.3 Stimulation of Collagen Production

Plant-based phenolic compounds promote the complex process of collagen synthesis via many mechanisms. These compounds exhibit potential as therapeutic agents for connective tissue disorders and skin health due to their capacity to promote collagen formation, modify fibroblast activity, and impact the structural integrity of collagen fibers.

##### *Mechanisms of Action*

*Gene Expression Regulation*: The expression of the collagen gene can be directly or indirectly impacted by plant phenolics through pathways such as TGF-beta1, which is necessary for collagen metabolism (Bondarenko et al., 2023).

*Fibrillogenesis Acceleration:* By stabilizing intermediate structures through hydrogen bonding, simple phenols like pyrogallol and resorcinol speed up the lag phase of fibril production and promote collagen fibrillogenesis (Kim, Gaidin, & Tarahovsky, 2018).

*Antioxidant Properties:* As antioxidants, phenolic chemicals shield fibroblasts from oxidative stress, which would otherwise hinder the synthesis of collagen (Merecz-Sadowska et al., 2021).

##### *Specific Compounds and Effects*

*Equol:* This polyphenolic chemical promotes cell regeneration and counteracts the effects of photoaging by dramatically increasing the expression of collagen and elastin in fibroblasts (Lephart, 2021).

*Flavonoid Glycosides:* Apigenin glycosides are among the compounds that promote collagen type I synthesis in fibroblasts, possibly via increasing prolidase activity and inhibiting matrix metalloproteinases (Galicka & Nazaruk, 2007).

Although there is a lot of promise in the stimulation of collagen formation by plant-derived phenolic compounds, it is important to take into account the fact that individual responses vary and that further research is necessary to completely comprehend the mechanisms and maximize their therapeutic uses.

## 11.5 DERMATOLOGICAL APPLICATIONS OF PHENOLIC COMPOUNDS

Phenolic compounds' antibacterial, wound-healing, antioxidant, and anti-inflammatory properties have sparked interest in their potential therapeutic use in dermatology (Ahmed & Mikail, 2024; Działo et al., 2016; Ferreira et al., 2023; Lee, Park, & Kwon, 2013). These ingredients offer a safe, all-natural way to improve skin health and cure a variety of skin conditions. The antioxidant properties of phenolic compounds help protect the skin from oxidative stress, UV radiation, and environmental pollutants, all of which can accelerate the aging process, the formation of wrinkles, and the onset of skin cancer. By scavenging free radicals and reducing oxidative damage, phenolic compounds can help maintain the integrity of the skin and prevent premature aging. Phenolic compounds are helpful in lowering the symptoms of inflammatory skin conditions such as psoriasis, acne, and eczema because of their anti-inflammatory properties (Watzl, 2008). These substances can help skin recover and mend by reducing pain, edema, and redness. Phenolic chemicals can also help treat skin problems caused by bacteria, fungus, and other organisms because of their antibacterial qualities. By preventing the growth of harmful germs on the skin, phenolic compounds can help prevent and treat skin infections. Moreover, the wound-healing properties of phenolic compounds accelerate the healing of burns, cuts, and wounds by encouraging skin regeneration, collagen synthesis, and tissue repair (Działo et al., 2016; Vivo et al., 2021). These compounds may promote good skin repair and reduce scarring.

To maximize their positive skin-benefitting effects, phenolic substances from plant sources, such as olive oil (Melguizo-rodríguez et al., 2021), green tea (Barbosa, 2007), and grape seeds (Ghafoor, Al-Juhaimi, & Choi, 2011), are often added to skincare products and treatments in dermatological applications. Phenolic compounds may be utilized in dermatology to treat a range of skin conditions, protect skin health, and promote skin renewal due to their anti-inflammatory, anti-microbial, antioxidant, and wound-healing properties.

### 11.5.1 Acne

The application of plant-derived phenolic compounds in treating acne vulgaris is gaining attention due to their multifaceted mechanisms of action. These compounds, such as resveratrol, quercetin, and epigallocatechin-3-gallate (EGCG), exhibit significant anti-inflammatory, antibacterial, and antioxidant properties, making them promising alternatives to conventional treatments.

#### 11.5.1.1 Mechanisms of Action

*Anti-inflammatory Effects:* Reduced inflammation in acne lesions is the result of phenolic compounds' reduction of oxidative stress indicators and pro-inflammatory cytokines (such as TNF-$\alpha$ and IL-1$\beta$) (dos Santos et al., 2019; Soleymani et al., 2020).

*Antibacterial Activity:* They reduce inflammation and the bacterial burden by preventing *Propionibacterium acnes*, the main culprit behind acne, from growing (dos Santos et al., 2019; Saviuc et al., 2017).

*Sebum Production Regulation:* A key factor in the pathophysiology of acne is the modulation of sebum production by certain phenolic chemicals (Ramsis et al., 2024).

#### 11.5.1.2 Clinical Applications

*Cosmeceuticals:* Products with EGCG and resveratrol are becoming popular topical therapies with few adverse effects (Hopkins, 2019).

*Formulations:* Increased antibacterial and anti-inflammatory properties have been demonstrated by multi-agent formulations that combine phenolic components with other natural extracts (Saviuc et al., 2017).

Phenolic chemicals show promise in treating acne, but further clinical research is required to fully determine their safety and effectiveness. On the other hand, despite their adverse consequences, the use of synthetic medicines is nevertheless common because of their proven efficacy.

### 11.5.2 Antibacterial and Anti-Inflammatory Properties

The use of phenolic compounds obtained from plants to treat acne is becoming more popular because of their antibacterial and anti-inflammatory qualities. These substances, which are present in a variety of plants, can successfully address the bacterial growth and inflammation that are the root causes of acne. The following sections detail their mechanisms and applications.

#### 11.5.2.1 Antibacterial Properties

Phenolic compounds exhibit significant antibacterial activity against acne-related bacteria, particularly *Cutibacterium acnes* and *Staphylococcus aureus* (Poomanee et al., 2022; Soleymani et al., 2020). Extracts from plants like *Bouea macrophylla* and *Quercus acutissima* have shown superior antibacterial effects compared to standard polyphenols (Kim et al., 2022; Poomanee et al., 2022).

#### 11.5.2.2 Anti-Inflammatory Mechanisms

These compounds reduce inflammation by inhibiting pro-inflammatory cytokines (e.g., IL-1β, TNF-α) and NO production (Lee, Park, & Shin, 2022; Soleymani et al., 2020). The antioxidant properties of phenolic compounds help mitigate oxidative stress, further alleviating inflammation associated with acne (Poomanee et al., 2022; Soleymani et al., 2020).

#### 11.5.2.3 Clinical Relevance

The growing interest in herbal treatments stems from the side effects of conventional therapies and antibiotic resistance (Abozeid et al., 2022). While promising, further clinical studies are necessary to validate the efficacy of these natural compounds in acne treatment (Soleymani et al., 2020).

Despite the potential benefits of plant-derived phenolic compounds, the variability in individual responses and the need for standardized formulations pose challenges in their clinical application. Further research is essential to optimize their use in acne management.

### 11.5.3 Reduction of Acne Lesions and Inflammation

Phenolic compounds extracted from plants have attracted attention because of their potential health advantages, particularly in relation to skin health, such as reducing inflammation and acne lesions. These compounds can be used to treat acne, a condition that frequently becomes worse by inflammation and the proliferation of certain bacteria such as "Propionibacterium acnes." Their antibacterial, anti-inflammatory, and antioxidant properties set them apart. These preventative or therapeutic measures are based on a variety of mechanisms, including antibacterial, anti-inflammatory, antioxidant, sebaceous gland modulation, and inhibition of keratinocyte proliferation effects.

Acne flares are linked to oxidative stress, which phenolic substances might mitigate by scavenging free radicals. Green tea catechins, for example, have been demonstrated to have potent antioxidant properties, lowering inflammation brought on by oxidative damage (Truong & Jeong, 2022). According to Truong and Jeong (2022), green tea polyphenols' antioxidant properties help to reduce skin irritation. Many phenolic compounds, such as curcumin found in turmeric, have demonstrated the ability to inhibit pro-inflammatory cytokines (like TNF-α and IL-6) and enzymes (such as COX-2). This action can significantly reduce inflammation associated with acne lesions (Zong et al., 2012). As indicated by Zong et al. (2012), the anti-inflammatory properties of curcumin and its potential application in inflammatory skin conditions showed great promising results. Certain plant-derived phenolics exhibit antimicrobial properties against "P. acnes." For example, extracts of green tea have been shown to inhibit the growth of this bacterium, thereby reducing the incidence of acne lesions (Alkufeidy et al., 2024). Some compounds, such as resveratrol, can modulate sebum production by influencing the activity of sebaceous glands. Overproduction of sebum is a key factor in the development of acne (Saad et al., 2020). Additionally, phenolic chemicals can affect the development and proliferation of keratinocytes, which are essential in acne because they help produce comedones, or plugged pores. These substances can aid in preventing acne by restoring normal keratinocyte activity (Działo et al., 2016).

Phenolic compounds are an interesting therapy option for acne lesions and inflammation. The benefits of the many acne treatment options can be increased by their numerous modes of action, which include antibacterial, anti-inflammatory, and antioxidant qualities. Further clinical study is needed to identify the most effective formulations and administration methods.

### 11.5.4 Eczema

Eczema, also known as atopic dermatitis, is a chronic inflammatory skin condition characterized by dry, itchy, and inflamed skin. The exact cause of eczema is not fully understood, but it is believed to be a combination of genetic and environmental factors. Some phenolic compounds have been investigated for their potential therapeutic effects on various skin conditions, including eczema. The potential relationship or treatment methods between eczema and phenolic compounds are mainly through anti-inflammatory effects, antioxidant activity, immune modulation, skin barrier function, and antimicrobial activity.

The exact mechanisms by which phenolic compounds exert their effects on eczema are still being elucidated, but various pathways are involved, such as the inhibition of inflammatory signaling pathways. Phenolic compounds can inhibit the activation of signaling pathways that contribute to inflammation, such as the NF-κB pathway (Gendrisch et al., 2021; Rahman et al., 2021). The other mode of action of the phenolic compounds on eczema is through the modulation of cytokine production. Phenolic compounds can suppress the production of inflammatory cytokines like TNF-α, IL-1β, and IL-6 (Wambe et al., 2019). Another mechanism is through the antioxidant defense pathway. Phenolic compounds neutralize ROS, which are known to contribute to skin inflammation and damage.

#### 11.5.4.1 Anti-Inflammatory and Anti-Itching Effects

Strong anti-inflammatory effects are exhibited by numerous phenolic substances. This is because of their capacity to suppress the synthesis of inflammatory mediators, such as prostaglandins and cytokines, which are essential components of the inflammatory cascade of eczema. Additionally, phenolic chemicals are powerful antioxidants that enhance anti-inflammatory properties. The severity of eczema is known to be exacerbated by oxidative stress, which can be prevented by scavenging free radicals. Some phenolic compounds can modulate the immune system, potentially reducing the hyper-reactive immune response that drives eczema.

#### 11.5.4.2 Improvement of Skin Barrier Function

Certain phenolic compounds can strengthen the skin barrier function. This is crucial in eczema, as a compromised skin barrier allows allergens and irritants to penetrate the skin, further exacerbating the condition. Some phenolic compounds have antimicrobial properties, which could be beneficial in managing eczema, as bacterial infections can complicate the condition, which helps to improve skin barrier function.

Some phenolic compounds can upregulate the expression of tight junction proteins, strengthening the skin barrier and reducing allergen penetration. For instance, phenolic compounds studied for their potential efficacy in eczema are resveratrol—a polyphenol found in grapes, curcumin—found in berries, and peanuts, quercetin—a polyphenol found in turmeric, EGCG—a flavonoid found in apples, onions, and berries, a catechin—found in green tea, and ferulic acid—a hydroxycinnamic acid found in rice bran and wheat germ (Arts & Hollman, 1998).

While promising, research on the use of phenolic compounds for eczema treatment is still in its early stages. Further clinical trials are needed to determine their efficacy, safety, and optimal dosage for eczema treatment. Some phenolic compounds can interact with medications or cause adverse effects. Because of their anti-inflammatory, antioxidant, immune-modulating, and skin-protective qualities, phenolic compounds therefore have the potential to be used as an additional or different strategy for the treatment of eczema. To guarantee their safe and efficient usage and to completely comprehend their role in the treatment of eczema, more research is needed.

#### 11.5.4.3 Inhibitory Effects on Melanin Synthesis

The process by which melanocytes create melanin pigments, known as melanin synthesis, is largely responsible for the color of our skin, hair, and eyes. Conditions like hyperpigmentation, which can be unsightly, can result from excessive melanin production. Therefore, the development of skin-lightening agents has shifted its focus to identifying ways to limit melanin formation. One promising approach to the treatment and prevention of skin disorders is the inhibition of melanin formation by phenolic compounds obtained from plants. These compounds, including flavonoids and phenolic acids, have demonstrated significant anti-pigmentary effects through various mechanisms, such as inhibiting tyrosinase activity and modulating melanogenesis-related signaling pathways. Phenolic compounds can inhibit the activity of tyrosinase, a key enzyme in melanin production, thereby reducing melanin synthesis (Ullah et al., 2023). These compounds influence multiple signaling pathways involved in melanogenesis, including the cyclic adenosine monophosphate (AMP) and Wnt pathways, which are crucial for regulating pigmentation (Ullah et al., 2023). Plant phenolics possess strong antioxidant capabilities, which help mitigate oxidative stress that can lead to skin aging and pigmentation disorders (Nisa et al., 2024).

Additionally, phenolic compounds have been found to modulate other cellular signaling pathways involved in melanin synthesis, such as the MAPK pathway and the cyclic AMP (cAMP) pathway. By targeting these pathways, phenolic compounds can regulate the expression of genes involved in melanin synthesis and exert their inhibitory effects. Certain spice-derived phenolic compounds have shown potential in preventing skin cancers by combating UV-induced damage and inhibiting cancer cell proliferation (Baloghová et al., 2023). Flavonoids have been linked to beneficial effects in treating chronic skin conditions such as vitiligo and psoriasis, highlighting their therapeutic versatility (Čižmárová et al., 2023). In conclusion, phenolic compounds' ability to prevent the manufacture of melanin presents exciting opportunities for the beauty and dermatology industries. Comprehending the complex mechanisms of action underlying these effects will facilitate the creation of innovative treatments and products aimed at hyperpigmentation and associated disorders.

### 11.5.5 Wound Healing

Hemostasis, inflammation, proliferation, and remodeling are the four separate processes that define wound healing under normal circumstances. When the skin is damaged, the skin tissue starts the process of forming a clot to seal up the region from the outside world and prevent further bleeding (Butkeviciute et al., 2022).

Antioxidants such as vitamin E, zinc, and selenium can promote skin repair and regeneration by accelerating wound healing, reducing scar formation, and protecting the skin barrier. Polyphenols have emerged as promising agents for wound healing, demonstrating multiple beneficial effects across different phases of the process. Tea polyphenols reduce inflammation, promote angiogenesis, and improve wound morphology (Zhao et al., 2023). By increasing cell proliferation, decreasing apoptosis, and lowering immunological reactions, resveratrol speeds up the healing process. Numerous Mediterranean medicinal plants have demonstrated wound-healing qualities, as have their phenolic components, such as flavonoids and tannins. Concerns over antibiotic resistance are addressed by polyphenols, which provide an alternative to antibiotics in the treatment of chronic wounds. To overcome this, polymer-based systems have been developed to improve polyphenol stability and control release kinetics at wound sites (Guimarães et al., 2021). These findings highlight the potential of polyphenols as natural wound healing agents and suggest avenues for developing more effective treatments.

#### 11.5.5.1 Promotion of Collagen Synthesis and Angiogenesis

The promotion of collagen synthesis and angiogenesis, along with the acceleration of wound closure, can be significantly influenced by plant-derived phenolic compounds. These compounds exhibit various bioactive properties that enhance the wound-healing process through multiple mechanisms.

#### 11.5.5.2 Collagen Synthesis Enhancement

*Polyphenolic Compounds*: Compounds from *Bacopa procumbens* have been shown to increase collagen deposition and organization, leading to improved scar quality (Martínez-Cuazitl et al., 2022). Resveratrol promotes collagen synthesis and dermal repair by upregulating genes associated with these processes (Hu, Zhang, & Chen, 2024). Injectable hydrogels containing plant-derived polysaccharides have demonstrated enhanced angiogenesis, contributing to faster wound healing (Liu et al., 2024). Phenolic extracts from wild blueberries promote endothelial cell migration, a critical step in angiogenesis (Adekeye et al., 2022).

#### 11.5.5.3 Acceleration of Wound Closure

*In vivo* studies showed that *B. procumbens* and resveratrol treatments have shown significant reductions in wound closure time, with effects observable within 48 hours (Liu et al., 2023b; Martínez-Cuazitl et al., 2022). The activation of TGF-β and related pathways has been implicated in the accelerated healing effects of these compounds (Martínez-Cuazitl et al., 2022). While the benefits of plant-derived phenolic compounds in wound healing are well-documented, it is essential to consider potential variability in individual responses and the need for further clinical studies to establish standardized treatment protocols.

## 11.6 TYPES OF TREATMENT METHODS FOR SKIN DISEASES

*Pharmacological Treatments:* Pharmaceutical interventions are crucial to the treatment of dermatological conditions because they provide customized medications that target-specific pathogenic pathways underlying skin conditions. Both topical and systemic medications, such as immunomodulators, retinoids, corticosteroids, and antibacterial medications, are commonly advised, depending on the nature and severity of the condition. Additionally, novel biological medications that target cytokines and immunological pathways have shown promise as therapies for inflammatory skin disorders such as psoriasis, atopic dermatitis, and hidradenitis suppurativa (Ferreira et al., 2023). These pharmaceutical therapies aim to improve the quality of life for those who suffer from dermatological problems by lowering symptoms, decreasing inflammation, and promoting skin healing.

*Radiation Therapy:* Radiation therapy, often used in conjunction with surgery or chemotherapy, has demonstrated efficacy in the treatment of various dermatological conditions, including cutaneous lymphomas, basal cell carcinomas, and keloids. Low-dose superficial radiation therapy (SDRT) is particularly effective for NMSCs, offering excellent cosmetic outcomes and high cure rates (Bhupalam et al., 2024). By delivering targeted ionizing radiation to affected areas, radiation therapy destroys cancer cells while minimizing damage to surrounding healthy tissues. Furthermore, adjuvant radiation therapy helps to prevent recurrence and metastasis, thereby improving long-term outcomes for patients with aggressive or high-risk skin cancers.

*Natural Treatment Methods with Phenolic Compounds:* In recent years, there has been growing interest in harnessing the therapeutic potential of phenolic compounds derived from plants for the management of dermatological conditions. Phenolic compounds, including phenolic acids, flavonoids, anthocyanins, alkaloids, terpenes, and essential oils, exhibit a wide range of biological activities, including antioxidant, anti-inflammatory, antimicrobial, and wound healing properties (de Lima Cherubim et al., 2020). Plant-derived phenolic compounds exert their therapeutic effects through various mechanisms, such as scavenging free radicals, modulating inflammatory pathways, inhibiting microbial growth, and promoting tissue repair

## 11.7 SOURCES OF PHENOLIC COMPOUNDS FOR DERMATOLOGICAL APPLICATIONS

### 11.7.1 Natural Sources

The dermatological applications of natural sources of phenolic compounds have attracted significant attention due to their anti-aging, anti-inflammatory, and antioxidant properties. These compounds, which are present in a wide variety of plants, may help treat skin disorders and enhance cosmetic formulations. These polyphenol-rich herbs promote healthy skin, particularly in psoriasis and atopic dermatitis sufferers (Di Salvo et al., 2023).

Wastewater from olive oil mills is another natural source of phenolic compounds that can be used to treat skin issues. This byproduct contains a significant amount of phenolic compounds, such as caffeic and ferulic acids, which have been demonstrated to be useful in treating inflammatory skin disorders and speeding up wound healing (Carrara et al., 2021). Numerous fruits, vegetables, and herbs are rich in polyphenols, which are integral to the Mediterranean diet and have been linked to a lower incidence of chronic skin disorders (Di Salvo et al., 2023).

Phenolic compounds scavenge free radicals, reducing oxidative stress and skin aging (Nisa et al., 2024). They help mitigate inflammation, making them suitable for treating conditions like psoriasis and chronic urticaria (Di Salvo et al., 2023). Polyphenols support skin protection against UV radiation, enhancing the efficacy of sunscreens (Aires, 2022; de Lima Cherubim et al., 2020). While the benefits of phenolic compounds in dermatology are well-documented, it is essential to consider potential limitations, such as variability in individual responses and the need for further research to optimize their use in clinical settings.

Numerous plant sources rich in phenolic compounds have been investigated for their potential dermatological applications. Table 11.2 highlights a variety of plants that are rich in phenolic compounds, which are highly beneficial for skin health and have promising applications in dermatology. Notably, *Aloe vera*, known for its soothing and hydrating properties, has been widely utilized in skincare products. Green tea is celebrated for its potent antioxidant properties, which help protect the skin from oxidative stress and premature aging. Similarly, pomegranate is recognized for its anti-inflammatory and healing effects, making it an important ingredient in the treatment of various skin conditions. Additionally, numerous other plants with high phenolic content are employed in the management of dermatological issues, underscoring the valuable role of plant-based compounds in promoting skin health. Examples include *A. vera*, which contains phenolic acids and flavonoids with anti-inflammatory and wound-healing properties (Maan et al., 2018), green tea extract rich in catechins with antioxidant and photoprotective effects (Bae et al., 2020), and chamomile extract containing apigenin (Tables 11.1 and 11.2) with anti-inflammatory and soothing properties (Arora et al., 2019). These natural treatments offer complementary or alternative options to conventional therapies, with the advantage of fewer side effects and a lower risk of resistance development.

The combination of pharmacological, radiation, and natural treatment techniques used in dermatology demonstrates the need for a multimodal approach to skin conditions. While medicines and radiation therapy remain the cornerstones of treatment, natural therapeutic techniques that employ phenolic compounds derived from plants offer possible supplemental solutions with favorable safety profiles and comprehensive advantages. Further research on the efficacy, safety, and mechanisms of action of phenolic compounds derived from plants is required to optimize their usage in dermatological treatment.

**TABLE 11.2**
**Plant Families, Genera, and Species That Are Known for High Amounts of Phenolic Compounds Beneficial for Skin Health**

| Plant Family | Genus | Species | Phenolic Compounds | Key Benefits for Skin Health | References |
|---|---|---|---|---|---|
| Theaceae | *Camellia* | Camellia sinensis | Epigallocatechin gallate (EGCG) | Antioxidant, anti-inflammatory, reduces acne, protects against UV damage | Damayanti et al. (2023) |
| Lythraceae | Punica | Punica granatum | Ellagic acid, punicalagins | Antioxidant reduces skin aging, inhibits UV-induced skin damage, enhances skin hydration | Kunchana et al. (2021), Tavares et al. (2023) |
| Zingiberaceae | Curcuma | Curcuma longa | Curcumin | Anti-inflammatory, antioxidant, enhances wound healing, improves skin texture | Gonzalez and Sethi (2012), Gopinath and Karthikeyan (2018) |
| Ericaceae | *Vaccinium* | Vaccinium corymbosum | Quercetin, anthocyanins | Antioxidant, anti-inflammatory, improves skin elasticity, reduces hyperpigmentation | Ivarsson et al. (2023) |
| Asteraceae | Matricaria | Matricaria chamomilla | Apigenin, luteolin | Anti-inflammatory, antioxidant, reduces skin irritation, enhances skin repair | Abedi et al. (2021) |
| Apiaceae | Apium | Apium graveolens | Luteolin, apigenin | Anti-inflammatory, antioxidant, improves skin barrier function, reduces skin redness | Prakoso et al. (2020) |
| Apiaceae | Petroselinum | Petroselinum crispum | Apigenin, luteolin | Antioxidant, anti-inflammatory, improves skin tone, reduces hyperpigmentation | Bahramsoltani et al. (2024) |

*Note:* The table includes the specific phenolic compounds found in each and their associated benefits

#### 11.7.1.1 Fruits and Vegetables

Phenolic compounds are a wide group of compounds originating from the secondary metabolism of plants found in different natural sources such as fruits, vegetables, tea, wine, and honey. Fruits are among the main natural sources of phenolic compounds. These compounds exert important antioxidant properties primarily associated with the presence of hydroxyl groups in their molecular structure. Fruits are the main dietary sources of phenolic compounds such as phenolic acids, flavonoids, stilbenes, and lignans (Table 11.3). Some of the phenolic compounds most commonly reported in the literature and their main fruit sources, as well as their chemical structure, are shown in Table 11.1 and Figure 11.1, respectively.

#### 11.7.1.2 Herbs and Spices

Phenolic compounds derived from herbs and spices are gaining attention for their potential dermatological applications, particularly in skin cancer prevention and cosmetic formulations. These compounds exhibit significant antioxidant, anti-inflammatory, and antimicrobial properties, making them valuable in skincare products. The following sections outline the primary sources

**TABLE 11.3**
**Phenolic Compounds and Their Main Fruit Sources (Lima et al., 2019)**

| Phenolic Compound | Fruit Source |
|---|---|
| Catechol | Blackberry (*Rubus fruticosus* L.) and blueberry (*Vaccinium corymbosum* L.) |
| Coumaric acid | Blackberry, blueberry, orange (*Citrus x sinensis* (L.) Osbeck), and mango (*Mangifera indica* L.) |
| Chlorogenic acid | Blackberry, blueberry, apple (*Malus domestica* Borkh.), and cherry (*Prunus avium* L.) |
| Ellagic acid | Pomegranate (*Punica granatum* L.), blackberry, blueberry, and mango |
| (-) Epicathechin | Blackberry, blueberry, pawpaw (*Asimina triloba* (L.) Dunal), and mango |
| Gallic acid | Grape (*Vitis vinifera* L.), blackberry, blueberry, jambolan (*Syzygium cumini* L.), Cornelian cherry (*Cornus mas* L.), and orange |
| Luteolin | Mango |
| Myricetin | Grape |
| Naringenin | Jambolan, cashew apple (*Anacardium occidentale* L.), guava (*Psidium guajava* L.) and banana (*Musa* spp.) |
| Protocatechuic acid | Orange |
| Punicalagin | Grape |
| Quercetin | Pomegranate |
| Quercetin-3-galactside | Grape, jambolan, blackberry, blueberry, guava, mango, apple, and orange |
| Rutin | Blackberry and blueberry |
| Sinapic acid | Apple, orange, and banana |
| Tannic acid | Jambolan, orange, and banana |
| Vanillic acid | Grape |

and benefits of phenolic compounds from herbs and spices. Notable sources include oregano, clove, and pomegranate peel, which contain high levels of phenolic compounds such as gallic acid and eugenol (Ulanowska & Olas, 2021). Phenolic compounds in herbs and spices include simple phenols, hydroxybenzoic acids, and flavonoids, each contributing to their antioxidant properties (Deng et al., 2021).

As shown in Table 11.4, several spices or herbs are sources of many phenolic compounds that can treat skin disease problems via medicine, nutraceuticals, or cosmetic applications. Spice-derived phenolic compounds have shown potential in preventing skin cancers by combating oxidative stress caused by UV radiation (Baloghová et al., 2023). The antioxidant properties of phenolic compounds support their use in cosmetics, enhancing skin protection and anti-aging

**TABLE 11.4**
**The Sources of Phenolic Compounds from Herbs and Spices That Can Be Applied in Dermatological Contexts**

| Herb/Spice | Phenolic Compounds | Dermatological Benefits | References |
|---|---|---|---|
| Turmeric | Curcumin | Anti-inflammatory | Hewlings and Kalman (2017) |
| Rosemary | Rosmarinic acid | Antiseptic, moisturization | Pérez-Sánchez et al. (2014) |
| Oregano | Carnosic acid | Antimicrobial | Sakkas and Papadopoulou (2017) |
| Lavender | Linalool | Skin calming, soothing | Cavanagh and Wilkinson (2005) |
| Sage | Carnosol | Anti-inflammatory | Iacopetta et al. (2023) |
| Cinnamon | Cinnamaldehyde | Antioxidant | Deng et al. (2021) |
| Clove | Eugenol | Antimicrobial | Ulanowska and Olas (2021) |
| Basil | Rosmarinic Acid | Anti-inflammatory | Yeşiloğlu and Şit (2012) |
| Thyme | Thymol | Antimicrobial | Salehi et al. (2018) |

**TABLE 11.5**
**Phenolic Compounds Found in Tea and Coffee, Their Sources, and Their Potential Dermatological Applications**

| Source | Phenolic Compounds | Potential Dermatological Applications | References |
|---|---|---|---|
| Green tea | Catechins, epigallocatechin gallate, theaflavins | Antioxidant, anti-inflammatory, skin rejuvenation | Xu et al. (2021) |
| Black tea | Catechins, theaflavins, thearubigins | Antioxidant, anti-inflammatory, skin protection | Aneja et al. (2004) |
| Oolong tea | Polyphenols, flavonoids | Skin hydration, wound healing | Hasegawa et al. (2013) |
| Coffee | Chlorogenic acid, caffeic acid, ferulic acid | Anti-inflammatory, scar reduction | Fukagawa et al. (2017) |

effects (de Lima Cherubim et al., 2020). Although the advantages of phenolic compounds are widely known, it is important to take into account the possibility of toxicity and the requirement for comprehensive safety assessments in dermatological applications.

#### 11.7.1.3 Tea and Coffee

Tea and coffee are rich in phenolic chemicals, which are recognized for their antibacterial, anti-inflammatory, and antioxidant qualities. They are advantageous for dermatological applications because of their qualities, especially in skincare compositions. Table 11.5 summarizes the most prevalent and potent phenolic chemicals in tea and coffee.

### 11.7.2 Synthetic Sources

A thorough understanding of the underlying dermatological problems and their pathophysiology is necessary for the manufacture of chemical compounds used to treat skin illnesses. Skin conditions fall into one of four categories: inflammatory (like psoriasis or eczema), infectious (like bacterial or fungal infections), autoimmune (like lupus), or neoplastic (like skin cancer). Targeted therapy involves selecting drugs that change certain biological processes, like cytokine receptors or enzymes, in order to effectively treat the illness and alleviate its symptoms. For instance, chemical reactions like acylation and alkylation can be utilized to create tiny molecules, and natural substances are helpful starting points for the synthesis of pharmaceuticals. Peptides and biologics, produced by solid-phase synthesis and recombinant DNA technologies, respectively, provide additional treatment alternatives by precisely altering immune responses or biological processes. These compounds function by inhibiting inflammatory pathways, preventing the growth of microorganisms, and regulating keratinocyte proliferation. Understanding these complexities is necessary to develop effective treatments while considering the potential drawbacks that could arise from concentrating on specific channels (Arora et al., 2019; Guo et al., 2023; Luong, Thanh, & Tran, 2020).

#### 11.7.2.1 Laboratory-Synthesized Phenolic Compounds

Laboratory-synthesized phenolic compounds have emerged as promising therapeutic agents for various skin diseases, including aging, cancer, and inflammatory conditions. These compounds display a range of beneficial biological activities, including antioxidant, anti-inflammatory, and enzyme inhibition properties. For example, hydroxy-3-phenylcoumarins significantly inhibit enzymes like tyrosinase, elastase, and collagenase, which are vital to the skin aging process. One such compound, 3-(4′-Bromophenyl)-5,7-dihydroxycoumarin, has demonstrated remarkable inhibitory effects on tyrosinase with an IC50 value of 1.05 μM, and it also exerts photoprotective effects without cytotoxicity to skin cells (Pintus et al., 2022). Additionally, phenolic compounds are recognized for their anticancer properties, including the ability to induce photoprotection, apoptosis, and autophagy in skin tumors, with studies suggesting that plant extracts rich in phenols may offer superior anticancer effects compared to their isolated counterparts (Bakrim et al., 2022). Furthermore, hydroxycinnamic acids and their derivatives are noted for their anti-inflammatory and antibacterial effects, proving effective in treating conditions such as dermatitis and promoting wound healing (Sajadimajd et al., 2020). While the therapeutic potential of these compounds is promising, further clinical trials are essential to confirm their efficacy and safety in human applications. A multifaceted treatment approach may be necessary to address the complexities of skin diseases fully.

## 11.8 THE MECHANISM OF ACTION OF PHENOLIC COMPOUNDS IN SKIN HEALTH

It has long been known that many phenolic compounds in nature have a wide variety of biological effects, such as anti-inflammatory, anti-cancer, antibacterial, and antioxidant qualities. In recent years, there has been increased interest in their possible application in dermatology, namely in the treatment of different skin disorders. Based on their structures, this concept paper investigates how phenolic chemicals work to cure skin conditions and offers support for their potential as a treatment.

Phenolic compounds exert their beneficial effects on skin health through various mechanisms of action, including antioxidant, anti-inflammatory, antimicrobial, and wound healing properties. These mechanisms work together to protect the skin from damage, reduce inflammation, combat harmful microorganisms, and promote the healing of wounds and injuries. The mechanism of the phenolic compounds to treat different skin ailments is described in detail in the following six mechanisms, with their mechanism of action.

### 11.8.1 Polyphenolics in Skin Disease Modulation

Since ancient times, natural products have been used as anti-inflammatory, antioxidant, analgesic, and anti-tumoral drugs. Currently, natural products are still important to support the development of new drugs for the most diverse diseases. Different from other conventional synthetic molecules, naturally derived molecules are characterized by a vast scaffold diversity and structural complexity. They can be classified according to their chemical structures into flavonoids, such as flavones, flavonols, neoflavonoids, chalcones, isoflavones, anthocyanidins, and proanthocyanidins and nonflavonoids, such as phenolic acids, stilbenoids, and phenolic amides. In the skin, phenolic compounds have proved to be promising molecules in the prevention of skin disorders and in the reduction of the healing time [A systematic review of natural products for skin applications: Targeting inflammation, wound healing, and photo-aging]. Phenolic compounds, particularly polyphenols, have been shown to exhibit significant interactions with skin disease-causing agents and participate in the healing process. Polyphenols, ubiquitous plant compounds, exhibit diverse health benefits by interacting with cellular signaling pathways rather than through direct antioxidant effects (Vauzour et al., 2010). They modulate key pathways like MAPK, PI3K/Akt, and NF-κB, influencing cellular processes involved in cancer, cardiovascular disease, and neurodegeneration (Liu et al., 2023a; Vauzour et al., 2010). In skin health, polyphenols combat oxidative stress-induced cellular senescence, inflammation, and cancer by targeting ROS-mediated signaling pathways and protecting biological macromolecules (Liu et al., 2023a). Particularly for skin cancer, polyphenols such as EGCG, resveratrol, and anthocyanins influence angiogenesis, apoptosis, and metastasis via a number of pathways, including mTOR/PI3K/Akt and EGFR/MAPK (Sajadimajd et al., 2020). Activating AMPK and blocking TLRs while blocking pro-inflammatory transcription factors, including AP-1 and NF-κB, are some of its anti-inflammatory qualities (Chu, 2014). These mechanisms underpin polyphenols' potential for treating and preventing illness. For instance, inflammatory proteins such as COX-2 and NF-κB can be bound by polyphenols, which inhibit their action and reduce inflammation (Guan et al., 2022). This mechanism is illustrated in the following figure (Figure 11.3).

Additionally, polyphenols can also chelate metal ions, such as copper and iron, which are involved in the formation of ROS that contribute to oxidative stress and skin damage. By sequestering these

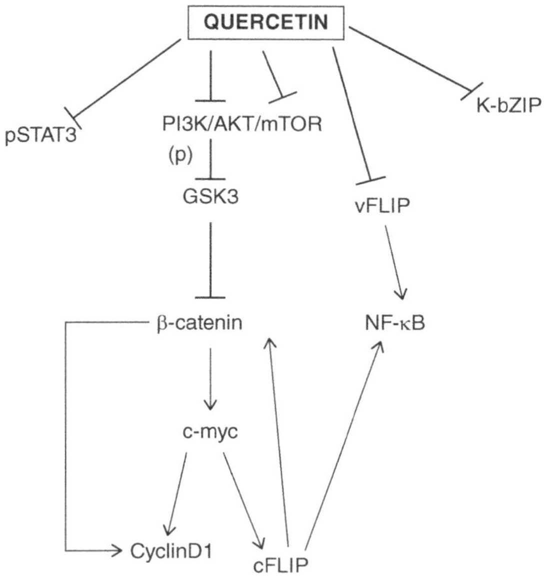


**FIGURE 11.3** Image of Polyphenol binding to NF-κB, inhibiting mechanism.

metal ions (Figure 11.4), polyphenols can reduce ROS formation and promote antioxidant defenses. Polyphenols, known for their antioxidant properties, can also act as prooxidants in the presence of metal ions like copper, leading to the formation of ROS (Iwasaki et al., 2011). This prooxidant activity is proposed as a mechanism for their anticancer and chemopreventive effects (Yar Khan et al., 2012). Polyphenols can mobilize intracellular copper ions, which are elevated in cancer cells, generating ROS and inducing cell death preferentially in cancer cells (Khan et al., 2014). This mechanism has been observed with various polyphenols, including luteolin, apigenin, epigallocatechin-3-gallate, and resveratrol (Khan et al., 2014). The ability of polyphenols to chelate metal ions, particularly copper and iron, is also relevant in Alzheimer's disease, where metal dyshomeostasis contributes to disease progression (Lakey-Beitia et al., 2021). The dual nature of polyphenols as antioxidants and prooxidants, coupled with their metal chelation properties, makes them promising candidates for various therapeutic applications, including cancer prevention and neurodegenerative disease treatment.

By stimulating the enzymes involved in wound closure, collagen formation, and skin hydration, polyphenols can promote several aspects of the healing process. As illustrated in Figure 11.5, quercetin, for instance, has been demonstrated to promote collagen synthesis via triggering the PI3K/Akt signaling pathway.

All things considered, the way that phenolic compounds interact with substances that cause skin diseases and the healing process is intricate and multidimensional, including a number of routes and molecular targets. To completely understand these processes and find possible therapeutic uses for polyphenols in the management of skin conditions, further investigation is required.

$$ROO^{\bullet} + AH/ArOH \longrightarrow ROOH + A^{\bullet}/ArO^{\bullet}$$

Or

$$ROO^{\bullet} + AH/ArOH \longrightarrow ROO^{-} + AH^{\bullet+}/ArOH^{\bullet+}$$

$$AH^{\bullet+}/ArOH^{\bullet+} + H_2O \rightleftharpoons A^{\bullet}/ArO^{\bullet} + H_3O^{+}$$

$$ROO^{-} + H_3O^{+} \rightleftharpoons ROOH + H_2O$$

**FIGURE 11.4** Polyphenols with metal ion chelate, such as quercetin and iron.

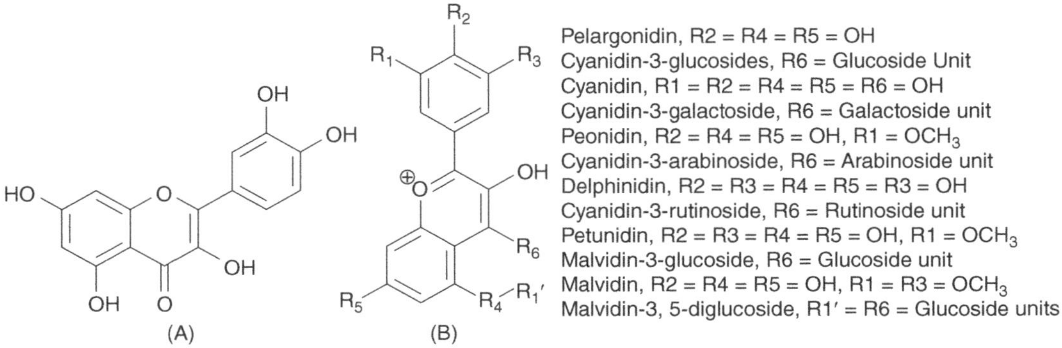


**FIGURE 11.5** Image of Quercetin activating PI3K/Akt signaling pathway.

### 11.8.2 Phenolic Compounds in Skin Disease Healing

Naturally occurring plant metabolites called phenolic compounds have great promise for both preventing and treating a range of skin conditions. These substances have antioxidant qualities that prevent oxidative stress and damage from UV radiation (Gado et al., 2021). From simple problems like wrinkles and acne to more serious disorders like skin cancer, they help reduce symptoms and prevent the development of skin conditions. Phenolics work in a variety of ways, such as regulating gene expression, modifying enzyme function, and exerting an antioxidant effect. They have been shown to be effective in treating bacterial infections, burns, wounds, and aged skin (Gomez-Molina et al., 2024). Interestingly, phenolic compounds can decrease DNA methylation, carcinogenesis, and tumor proliferation in skin malignancies while inducing photoprotection, apoptosis, and autophagy (Gado et al., 2021). Because of their low toxicity and natural nature, phenolic compounds offer a viable approach to managing skin health.

## 11.9 CHALLENGES AND FUTURE DIRECTIONS

Although phenolic compounds have shown promise in treating skin conditions, there are a number of issues that need to be resolved before they may be used in other contexts. These substances can be used to treat a variety of skin conditions since they have antibacterial, anti-inflammatory, and antioxidant qualities. However, problems like low stability and bioavailability make them less effective. The main obstacles and potential paths for phenolic compounds in the treatment of skin diseases are described in the sections that follow.

### 11.9.1 Challenges in Application

*Bioavailability:* Many phenolic compounds have low absorption rates, limiting their therapeutic potential (Gado et al., 2021).

*Stability:* Phenolic compounds can degrade under certain conditions, affecting their efficacy in formulations (Gado et al., 2021).

*Regulatory Hurdles:* The approval process for new treatments can be lengthy and complex, delaying the introduction of phenolic-based therapies (Contardi et al., 2021).

### 11.9.2 Future Directions

*Nanoformulations:* Advances in nanotechnology may enhance the delivery and stability of phenolic compounds, improving their therapeutic outcomes (Farhan, 2024).

*Clinical Trials:* More rigorous clinical studies are needed to validate the effectiveness of phenolic compounds in treating skin disorders (Gado et al., 2021).

*Combination Therapies:* Exploring the synergistic effects of phenolic compounds with other treatments could enhance their efficacy against skin diseases (Działo et al., 2016). While phenolic compounds show great promise, their limitations necessitate ongoing research and innovation to fully realize their potential in dermatological applications.

### 11.9.3 Delivery and Penetration of Phenolic Compounds into the Skin

It is a complicated issue because the distribution and penetration of phenolic compounds into the skin are greatly influenced by their physicochemical properties and the formulation methods employed. With a focus on the potential of phenolic compounds as medications and antioxidants, recent research identifies several innovative methods aimed at enhancing their skin absorption. It has been shown that a range of formulations, such as oleogels, emulsions, and double emulsions, greatly improve the delivery of these compounds; the type of formulation used is one of the crucial processes of penetration. For example, studies have demonstrated that oleogel with chlorogenic acid has potent antioxidant qualities and can penetrate the dermis and epidermis (Butkeviciute et al., 2022). Furthermore, it has become common practice to employ nanocarrier systems, such as liposomes and nanoemulsions; these lipid-based carriers enhance the stability and bioavailability of phenolic compounds, addressing issues with their high reactivity and poor solubility (Gugleva et al., 2021).

In terms of therapeutic applications, polyphenol mixtures delivered via double emulsions have shown promising results in treating conditions such as atopic dermatitis, with studies indicating significant reductions in inflammation markers (Marques et al., 2023). Additionally, the utilization of micro- and nanocarriers has proven effective in facilitating the transdermal delivery of resveratrol—a potent polyphenol—effectively bypassing barriers like the stratum corneum to deliver therapeutic benefits (Singh, 2020). Nevertheless, despite these advancements, challenges persist in optimizing formulations tailored to specific phenolic compounds, since their inherent physicochemical properties can limit effective skin penetration (Butkeviciute et al., 2022). Future research must continue to refine these delivery systems to maximize the therapeutic efficacy of phenolic compounds in dermatological applications.

### 11.9.4 Clinical Studies and Evidence-Based Applications

Because of their strong antioxidant, anti-inflammatory, and photoprotective qualities, phenolic compounds have become prospective therapeutic agents for a variety of skin conditions. These substances have attracted a lot of interest in clinical studies because of their potential to reduce symptoms and slow the development of skin conditions such as psoriasis, atopic dermatitis, and skin malignancies. Their potent antioxidant properties are essential for lowering inflammation and oxidative stress linked to skin conditions (Farhan, 2024; Gado et al., 2021). In addition to reducing cellular damage, phenolic chemicals also support general skin health by scavenging free radicals, which increases resilience and longevity (Nisa et al., 2024).

Some phenolic compounds have shown notable photoprotective effects against UV-induced skin damage in addition to their antioxidant qualities. In order to lower the incidence of melanoma and NMSCs, this defense mechanism is essential (Gado et al., 2021). These substances' relevance in oncological dermatology has been further cemented by clinical trials that have demonstrated their ability to prevent tumor growth and cause apoptosis in malignant cells.

Clinical applications of phenolic compound-rich sources, like effluent from olive oil mills, have been investigated for their potential to improve wound healing and treat inflammatory skin conditions (Carrara et al., 2021). The importance of dietary sources in maintaining skin health is further demonstrated by the association between fewer incidences of chronic skin problems and a Mediterranean diet strong in polyphenols (Di Salvo et al., 2023). Despite the potential of phenolic compounds in dermatological applications, further research is needed to elucidate their mechanisms of action and enhance their therapeutic applications for optimal efficacy. Several studies caution against relying exclusively on these drugs, suggesting that a more comprehensive approach that includes conventional medicines may provide additional benefits for the effective management of skin diseases.

## 11.10 ETHICAL CONSIDERATIONS AND SUSTAINABILITY

The ethical considerations and sustainability of phenolic compounds in the treatment of skin diseases are increasingly crucial topics within the field of dermatological research and therapy. As the demand for effective and safe skin treatments grows, clinicians and researchers must examine not only the efficacy of such compounds but also the broader implications of their sourcing and application. Phenolic compounds, which are often derived from natural sources like plants, fruits, and vegetables, present a unique opportunity for developing sustainable therapeutic options. Given their wide availability in nature, these compounds can be harvested in environmentally responsible ways that minimize ecological impact. However, the sustainable sourcing of such bioactive compounds necessitates adherence to ethical guidelines to prevent overharvesting and protect biodiversity.

The ethical implications extend beyond environmental sustainability to encompass issues related to sourcing practices, fair trade, and economic equity. For instance, many phenolic-rich plants are cultivated in regions where local communities depend on these natural resources for their livelihoods. It is paramount to ensure that the extraction and commercialization of phenolic compounds do not exploit these communities but rather provide them with fair compensation and support sustainable agricultural practices (Jimenez-Lopez et al., 2020). This approach can help foster a mutually beneficial relationship between industry and local populations by promoting economic sustainability alongside environmental stewardship.

Moreover, the application of phenolic compounds in dermatological treatments should consider patient safety and informed consent. The use of natural products can sometimes lead to variability in potency and efficacy, depending on the source and method of preparation. Therefore, stringent quality control measures and regulatory frameworks must be in place to ensure that dermatological products containing phenolic compounds are safe and efficacious, thus supporting ethical practice in healthcare (Fan et al., 2012).

In conclusion, while phenolic compounds offer promising therapeutic benefits for skin diseases, their ethical considerations and sustainability cannot be overlooked. A holistic approach that emphasizes responsible sourcing, community involvement, and rigorous safety standards is essential to maximize their potential while safeguarding both the environment and the well-being of local populations.

## 11.11 CONCLUSION AND FUTURE REMARKS

Polyphenols' anti-inflammatory and antioxidant properties make them useful in dermatology and medicine. Polyphenols are used in skincare products to protect the skin from the sun. Phenolic compounds offer a wide range of beneficial effects on the skin and can be used in a number of dermatological applications, including skincare products, treatments for skin conditions, and wound healing therapies. Because of their many benefits, phenolic compounds can be used to treat a wide range of skin-related issues and maintain the health of the skin. Healthy, glowing skin can be supported by incorporating foods and skincare products high in phenolics into everyday routines. Phenolic compounds' promise as a natural therapeutic agent for skin-related illnesses is becoming clearer as the scientific community works to understand the mechanisms underlying their impacts on skin health.

Due to their exceptional benefits for skin health, phenolic compounds—a broad class of bioactive molecules—have attracted a lot of interest in the dermatological community. They are useful in the prevention and treatment of a number of skin disorders, including photoaging, acne, dermatitis, and even skin malignancies, due to their antibacterial, anti-inflammatory, and antioxidant qualities.

These substances' potential for therapeutic uses is further highlighted by their capacity to alter cellular functions like proliferation and apoptosis.

It is increasingly evident that adding these natural ingredients to skincare products can improve their efficacy and safety as research continues to reveal the processes by which phenolic compounds affect skin health. In addition to improving skin health, using phenolics derived from plants satisfies customer demand for natural, eco-friendly, and clean cosmetics.

Future research on phenolic compounds presents promising opportunities for advances in dermatological science. Future studies ought to concentrate on: (1) targeted treatments for skin conditions will be developed with more focus on clarifying the precise biological mechanisms of action of particular phenolic compounds. (2) The creation of novel formulations containing phenolic compounds, like nanocarriers and microencapsulation, may enhance their efficacy, stability, and bioavailability in topical applications. (3) To verify the therapeutic claims linked to phenolic compounds and to guarantee their safety and effectiveness in various populations, thorough clinical evaluations are required. (4) As we get more knowledge about dermatogenetics, phenolic compounds may be used in customized skincare routines that are dependent on environmental and genetic variables. Lastly, a greater emphasis on the development of phenolic compounds from renewable resources and sustainable procurement might improve the ecological footprint of dermatological goods while fostering biodiversity.

## REFERENCES

Abeyrathne, E. D. N. S., Nam, K., Huang, X., & Ahn, D. U. (2022). Plant- and animal-based antioxidants' structure, efficacy, mechanisms, and applications: A review. *Antioxidants*, *11*(5), 1025. https://doi.org/10.3390/antiox11051025

Abozeid, D., Fawzy, G., Issa, M., Abdeltawab, N., & Soliman, F. (2022). Medicinal plants and their constituents in the treatment of acne vulgaris. *Biointerface Research in Applied Chemistry*, *13*(2), 189. https://doi.org/10.33263/BRIAC132.189

Abedi, A., Dianati, M., & Gilasi, H. (2021). Effect of Chamomile Extract on Prevention of Peristomal Skin Complications. *Medical - Surgical Nursing Journal, 9*(4). https://doi.org/10.5812/msnj.111795

Adekeye, T., Vandenakker, N., Tsakiroglou, P., & Klimis-Zacas, D. (2022). Phenolic extract(s) promote(s) wound remodeling. *Current Developments in Nutrition*, *6*, 260. https://doi.org/10.1093/cdn/nzac053.001

Ahmed, I. A., & Mikail, M. A. (2024). Diet and skin health: The good and the bad. *Nutrition*, *119*, 112350. https://doi.org/10.1016/j.nut.2023.112350

Aires, A. (2022). Polyphenols applications in pharmaceutic and cosmetic industries. In *Technologies to recover polyphenols from AgroFood by-products and wastes* (pp. 337–357). Elsevier. https://doi.org/10.1016/B978-0-323-85273-9.00009-0

Alemu, B. K., Misganaw, D., & Mengistu, G. (2020). Wound healing effect of acokanthera schimperi schweinf (Apocynaceae) methanol leaf extract ointment in mice and its *in-vitro* antioxidant activity. *Clinical Pharmacology: Advances and Applications*, *12*, 213–222. https://doi.org/10.2147/CPAA.S288394

Alkufeidy, R. M., Ameer Altuwijri, L., Aldosari, N. S., Alsakabi, N., & Dawoud, T. M. (2024). Antimicrobial and synergistic properties of green tea catechins against microbial pathogens. *Journal of King Saud University – Science*, *36*(8), 103277. https://doi.org/10.1016/j.jksus.2024.103277

Anantharaju, P. G., Gowda, P. C., Vimalambike, M. G., & Madhunapantula, S. V. (2016). An overview on the role of dietary phenolics for the treatment of cancers. *Nutrition Journal*, *15*(1), 1–16. https://doi.org/10.1186/s12937-016-0217-2

Aneja, R., Odoms, K., Denenberg, A. G., & Wong, H. R. (2004). Theaflavin, a black tea extract, is a novel anti-inflammatory compound. Critical Care Medicine, 32(10), 2097–2103. https://doi.org/10.1097/01.CCM.0000142661.73633.15

Anulika, P., Nwokeji, Ezenweani, R., Paul Anulika, N., Osamiabe Ignatius, E., Sunday Raymond, E., Osasere, O.-I., Hilda Abiola, A. (2016). The chemistry of natural product: plant secondary metabolites. *Int. J. Technol. Enhanc. Emerg. Eng. Res.* 4, 1. https://www.researchgate.net/publication/307858399

Arora, R., Aggarwal, G., Arora Dhingra, G., & Nagpal, M. (2019). Herbal active ingredients used in skin cosmetics. *Asian Journal of Pharmaceutical and Clinical Research*, *12*(9), 7–15. https://doi.org/10.22159/ajpcr.2019.v12i9.33620

Arts, I. C. W., & Hollman, P. C. H. (1998). Optimisation of a quantitative method for the determination of catechins in fruits and pulses. *Pathophysiology*, *5*, 268. https://doi.org/10.1016/s0928-4680(98)81342-4

Asfaw, T. B., Tadesse, M. G., Tessema, F. B., Woldemariam, H. W., Legesse, B. A., Esho, T. B., Bachheti, A., AL-Huqail, A. A., Taher, M. A., Zouidi, F., Saleh, K. A., Kumar, V., Širić, I., & Kumar, P. (2023). Optimization, Identification, and Quantification of Selected Phenolics in Three Underutilized Exotic Edible Fruits Using HPTLC. *Sustainability, 15*(12), 9568. https://doi.org/10.3390/su15129568

Asfaw, K. G., Liu, Q., Eghbalian, R., Purper, S., Akaberi, S., Dhakarey, R., Münch, S. W., Wehl, I., Bräse, S., Eiche, E., Hause, B., Bogeski, I., Schepers, U., Riemann, M., & Nick, P. (2022). The jasmonate biosynthesis Gene OsOPR7 can mitigate salinity induced mitochondrial oxidative stress. *Plant Science*, *316*, 111156. https://doi.org/10.1016/j.plantsci.2021.111156

Ayad, R., Lefahal, M., Makhloufi, E. H., & Akkal, S. (2024). Photoprotection strategies with antioxidant extracts: A new vision. *Physical Sciences Reviews*, *9*(6), 2273–2286. https://doi.org/10.1515/psr-2022-0313

Bae, J., Kim, N., Shin, Y., Kim, S.-Y., & Kim, Y.-J. (2020). Activity of catechins and their applications. *Biomedical Dermatology*, *4*(1), 8. https://doi.org/10.1186/s41702-020-0057-8

Bakrim, S., El Omari, N., El Hachlafi, N., Bakri, Y., Lee, L.-H., & Bouyahya, A. (2022). Dietary phenolic compounds as anticancer natural drugs: Recent update on molecular mechanisms and clinical trials. *Foods*, *11*(21), 3323. https://doi.org/10.3390/foods11213323

Baloghová, J., Michalková, R., Baranová, Z., Mojžišová, G., Fedáková, Z., & Mojžiš, J. (2023). Spice-derived phenolic compounds: Potential for skin cancer prevention and therapy. *Molecules*, *28*(17), 6251. https://doi.org/10.3390/molecules28176251

Barbosa, D. S. (2007). Green tea polyphenolic compounds and human health. *Journal Fur Verbraucherschutz Und Lebensmittelsicherheit*, *2*(4), 407–413. https://doi.org/10.1007/s00003-007-0246-z

Baron, G., Ferrario, G., Marinello, C., Carini, M., Morazzoni, P., & Aldini, G. (2021). Effect of extraction solvent and temperature on polyphenol profiles, antioxidant and anti-inflammatory effects of red grape skin by-product. *Molecules*, *26*(18), 5454. https://doi.org/10.3390/molecules26185454

Becker, M. M., Nunes, G. S., Ribeiro, D. B., Silva, F. E., Catanante, S., & Marty, G. (2019). Determination of the antioxidant capacity of red fruits by miniaturized spectrophotometry assays. *Journal of the Brazilian Chemical Society*, *30*(5), 1108–1114. https://doi.org/10.21577/0103-5053.20190003

Belayneh Asfaw, T., Getachew Tadesse, M., Beshah Tessema, F., Woldemichael Woldemariam, H., V. Chinchkar, A., Singh, A., Upadhyay, A., & Mehari, B. (2024). Ultrasonic-assisted extraction and UHPLC determination of ascorbic acid, polyphenols, and half-maximum effective concentration in *Citrus medica* and *Ziziphus spina-christi* fruits using multivariate experimental design. *Food Chemistry: X*, *22*, 101310. https://doi.org/10.1016/j.fochx.2024.101310

Berker, K. I., Güçlü, K., Tor, I., & Apak, R. (2007). Comparative evaluation of Fe(III) reducing power-based antioxidant capacity assays in the presence of phenanthroline, batho-phenanthroline, tripyridyltriazine (FRAP), and ferricyanide reagents. *Talanta*, *72*(3), 1157–1165. https://doi.org/10.1016/j.talanta.2007.01.019

Beya, M. M., Netzel, M. E., Sultanbawa, Y., Smyth, H., & Hoffman, L. C. (2021). Plant-based phenolic molecules as natural preservatives in comminuted meats: A review. *Antioxidants*, *10*(2), 263. https://doi.org/10.3390/antiox10020263

Bahramsoltani, R., Ahmadian, R., Daglia, M., & Rahimi, R. (2024). Petroselinum crispum (Mill.) Fuss (Parsley): An Updated Review of the Traditional Uses, Phytochemistry, and Pharmacology. *Journal of Agricultural and Food Chemistry, 72*(2), 956–972. https://doi.org/10.1021/acs.jafc.3c06429

Bhardawaj, H., Vasudeva, N., & Sharma, S. (2023). Phytochemistry and Pharmacological Aspects of Apigenin: A Review. *The Natural Products Journal, 13*(4). https://doi.org/10.2174/2210315512666220520150358

Bhupalam, V., Hetzel, J. D., Awad, N., & Nestor, M. S. (2024). Superficial radiation therapy in dermatology: A historical perspective. *Dermatological Reviews*, *5*(3). https://doi.org/10.1002/der2.229

Bickers, D. R., Lim, H. W., Margolis, D., Weinstock, M. A., Goodman, C., Faulkner, E., Gould, C., Gemmen, E., & Dall, T. (2006). The burden of skin diseases: 2004: A joint project of the American Academy of Dermatology Association and the Society for Investigative Dermatology. *Journal of the American Academy of Dermatology*, *55*(3), 490–500. https://doi.org/10.1016/J.JAAD.2006.05.048

Bondarenko, L., Kalachinskaya, M., Serhiichuk, N., Motronenko, V., & Biloshytska, O. (2023). Possibilities of plant preparations use for collagen structure and metabolism disturbances correction: Modern state of problem. *Innovative Biosystems and Bioengineering*, *7*(1), 3–13. https://doi.org/10.20535/ibb.2023.7.1.275743

Butkeviciute, A., Ramanauskiene, K., Kurapkiene, V., & Janulis, V. (2022). Dermal penetration studies of potential phenolic compounds ex vivo and their antioxidant activity in vitro. *Plants*, *11*(15), 1901. https://doi.org/10.3390/plants11151901

Cavanagh, H. M. A., & Wilkinson, J. M. (2005). Lavender essential oil: a review. *Australian Infection Control, 10*(1), 35–37. https://doi.org/10.1071/HI05035

Carlsen, M. H., Halvorsen, B. L., Holte, K., Bøhn, S. K., Dragland, S., Sampson, L., Willey, C., Senoo, H., Umezono, Y., Sanada, C., Barikmo, I., Berhe, N., Willett, W. C., Phillips, K. M., Jacobs, D. R., & Blomhoff, R. (2010). The total antioxidant content of more than 3100 foods, beverages, spices, herbs and supplements used worldwide. *Nutrition Journal, 9*(1). https://doi.org/10.1186/1475-2891-9-3

Carrara, M., Kelly, M. T., Roso, F., Larroque, M., & Margout, D. (2021). Potential of olive oil mill wastewater as a source of polyphenols for the treatment of skin disorders: A review. *Journal of Agricultural and Food Chemistry, 69*(26), 7268–7284. https://doi.org/10.1021/acs.jafc.1c00296

Chamcheu, J. C., Boateng, S. T., Roy, T., Agbo, M. E., Banang-Mbeumi, S., Chamcheu, R.-C. N., Bramwell, M., Pham, L. K., Jackson, K. E., Hill, R. A., Nagalo, B. M., Efimova, T., & Fotie, J. (2022). Identification of potential inhibitors of cutaneous melanoma and non-melanoma skin cancer cells through in-vitro and in-silico screening of a small library of phenolic compounds. *SSRN Electronic Journal*. https://doi.org/10.2139/ssrn.4102562

Chen, J., Liu, Y., Zhao, Z., & Qiu, J. (2021). Oxidative stress in the skin: Impact and related protection. *In International Journal of Cosmetic Science, 43*(5). https://doi.org/10.1111/ics.12728

Chi, C. (1994). Integrating traditional medicine into modern health care systems: Examining the role of Chinese medicine in Taiwan. *Social Science & Medicine, 39*(3), 307–321. https://doi.org/10.1016/0277-9536(94)90127-9

Chu, A. (2014). Antagonism by bioactive polyphenols against inflammation: A systematic view. *Inflammation & Allergy-Drug Targets, 13*(1), 34–64. https://doi.org/10.2174/1871528112666131119211002

Čižmárová, B., Hubková, B., Tomečková, V., & Birková, A. (2023). Flavonoids as promising natural compounds in the prevention and treatment of selected skin diseases. *International Journal of Molecular Sciences, 24*(7), 6324. https://doi.org/10.3390/ijms24076324

Contardi, M., Lenzuni, M., Fiorentini, F., Summa, M., Bertorelli, R., Suarato, G., & Athanassiou, A. (2021). Hydroxycinnamic acids and derivatives formulations for skin damages and disorders: A review. *Pharmaceutics, 13*(7). https://doi.org/10.3390/pharmaceutics13070999

Correia, P., Araújo, P., Ribeiro, C., Oliveira, H., Pereira, A. R., Mateus, N., de Freitas, V., Brás, N. F., Gameiro, P., Coelho, P., Bessa, L. J., Oliveira, J., & Fernandes, I. (2021). Anthocyanin-Related Pigments: Natural Allies for Skin Health Maintenance and Protection. *Antioxidants, 10*(7), 1038. https://doi.org/10.3390/antiox10071038

Damayanti, C. R. S. P., Purwanto, D. A., Endaryanto, A., Listiawan, M. Y., Wirohadidjoyo, Y. W., S., Soetjipto, Utomo, B., & Widjiati. (2023). The effect of topical epigallocatechin-3-gallate (EGCG) on collagen type-I, MMP-1 expression and dermal collagen count in photoaging prevention. *Natural and Life Sciences Communications, 22*(1). https://doi.org/10.12982/NLSC.2023.011

de Lima Cherubim, D. J., Buzanello Martins, C. V., Oliveira Fariña, L., & da Silva de Lucca, R. A. (2020). Polyphenols as natural antioxidants in cosmetics applications. *Journal of Cosmetic Dermatology, 19*(1), 33–37. https://doi.org/10.1111/jocd.13093

Deng, J., Li, J., Zhao, Y., & Wang, G. (2021). Effect and safety of cinnamaldehyde on immunosuppressed mice with invasive pulmonary candidiasis. *Chinese Journal of Integrative Medicine, 27*(4), 286–290. https://doi.org/10.1007/s11655-020-3075-x

Dhananjaya, B. L., Sudarshan, S., Dongol, Y., & More, S. S. (2016). The standard aqueous stem bark extract of *Mangifera indica* L. Inhibits toxic PLA2-NN-XIb-PLA2 of Indian cobra venom. *Saudi Pharmaceutical Journal, 24*(3), 371–378. https://doi.org/10.1016/j.jsps.2016.04.026

Di Salvo, E., Gangemi, S., Genovese, C., Cicero, N., & Casciaro, M. (2023). Polyphenols from Mediterranean plants: Biological activities for skin photoprotection in atopic dermatitis, psoriasis, and chronic urticaria. *Plants, 12*(20), 3579. https://doi.org/10.3390/plants12203579

Diaconeasa, Z., Ştirbu, I., Xiao, J., Leopold, N., Ayvaz, Z., Danciu, C., Ayvaz, H., Stănilă, A., Nistor, M., & Socaciu, C. (2020). Anthocyanins, Vibrant Color Pigments, and Their Role in Skin Cancer Prevention. *Biomedicines, 8*(9), 336. https://doi.org/10.3390/biomedicines8090336

dos Santos, Z. M. Q., dos Santos, M. Q., Zancanaro, V., Bellaver, E. H., Nardi, G. M., Gelinski, J. M. L., & Locatelli, C. (2019). Topical application of phenolic compounds suppresses *Propionibacterium acnes*-induced inflammatory responses in mice with ear edema. *Naunyn-Schmiedeberg's Archives of Pharmacology, 392*(5). https://doi.org/10.1007/s00210-018-01604-z

Dwevedi, D., & Srivastava, A. (2024). Molecular mechanisms of polyphenols in management of skin aging. *Current Aging Science, 17*(3), 180–188. https://doi.org/10.2174/0118746098287130240212085507

Działo, M., Mierziak, J., Korzun, U., Preisner, M., Szopa, J., & Kulma, A. (2016). The potential of plant phenolics in prevention and therapy of skin disorders. *International Journal of Molecular Sciences, 17*(2), 1–41. https://doi.org/10.3390/ijms17020160

Fan, T.-P., Deal, G., Koo, H.-L., Rees, D., Sun, H., Chen, S., Dou, J.-H., Makarov, V. G., Pozharitskaya, O. N., Shikov, A. N., Kim, Y. S., Huang, Y.-T., Chang, Y. S., Jia, W., Dias, A., Wong, V. C., & Chan, K. (2012). Future development of global regulations of Chinese herbal products. *Journal of Ethnopharmacology*, *140*(3), 568–586. https://doi.org/10.1016/j.jep.2012.02.029

Farhan, M. (2024). The promising role of polyphenols in skin disorders. *Molecules*, *29*(4), 865. https://doi.org/10.3390/molecules29040865

Fernandes, A., Rodrigues, P. M., Pintado, M., & Tavaria, F. K. (2023). A systematic review of natural products for skin applications: Targeting inflammation, wound healing, and photo-aging. *Phytomedicine*, *115*, 154824. https://doi.org/10.1016/j.phymed.2023.154824

Fernandes, I. de A. A., Maciel, G. M., Ribeiro, V. R., Rossetto, R., Pedro, A. C., & Haminiuk, C. W. I. (2021). The role of bacterial cellulose loaded with plant phenolics in prevention of UV-induced skin damage. *Carbohydrate Polymer Technologies and Applications*, *2*(June). https://doi.org/10.1016/j.carpta.2021.100122

Ferreira, I. G., Almeida, C. S., Bulcão, L. A., Ferreira, D. G., Weber, M. B., & Bonamigo, R. R. (2023). Hospital dermatology: Analysis of dermatological consultations in a tertiary teaching hospital. *Anais Brasileiros de Dermatologia*, *98*(5), 620–634. https://doi.org/10.1016/j.abd.2022.08.010

Foss, K., Przybyłowicz, K. E., & Sawicki, T. (2022). Antioxidant activity and profile of phenolic compounds in selected herbal plants. *Plant Foods for Human Nutrition*, *77*(3), 383–389. https://doi.org/10.1007/s11130-022-00989-w

Fukagawa, S., Haramizu, S., Sasaoka, S., Yasuda, Y., Tsujimura, H., & Murase, T. (2017). Coffee polyphenols extracted from green coffee beans improve skin properties and microcirculatory function. *Bioscience, Biotechnology, and Biochemistry, 81*(9), 1814–1822. https://doi.org/10.1080/09168451.2017.1345614

Gado, F., Digiacomo, M., Salsano, J. E., Macchia, M., & Manera, C. (2021). Phenolic compounds in prevention and treatment of skin cancers: A review. *Current Medicinal Chemistry*, *28*(33), 6730–6752. https://doi.org/10.2174/0929867328666210324160324

Galicka, A., & Nazaruk, J. (2007). Stimulation of collagen biosynthesis by flavonoid glycosides in skin fibroblasts of osteogenesis imperfecta type I and the potential mechanism of their action. *International Journal of Molecular Medicine*, *20*(6). https://doi.org/10.3892/ijmm.20.6.889

Gao, X., Xu, Z., Liu, G., & Wu, J. (2021). Polyphenols as a versatile component in tissue engineering. *Acta Biomaterialia*, *119*, 57–74. https://doi.org/10.1016/j.actbio.2020.11.004

Gendrisch, F., Esser, P. R., Schempp, C. M., & Wölfle, U. (2021). Luteolin as a modulator of skin aging and inflammation. *BioFactors*, *47*(2), 170–180. https://doi.org/10.1002/biof.1699

Ghafoor, K., Al-Juhaimi, F., & Choi, Y. H. (2011). Effects of grape (Vitis labrusca B.) peel and seed extracts on phenolics, antioxidants and anthocyanins in grape juice. *Pakistan Journal of Botany*, *43*(3), 1581–1586.

Ghazi, S. (2022). Do the polyphenolic compounds from natural products can protect the skin from ultraviolet rays? *Results in Chemistry*, *4*, 100428. https://doi.org/10.1016/j.rechem.2022.100428

Gomez-Molina, M., Albaladejo-Marico, L., Yepes-Molina, L., Nicolas-Espinosa, J., Navarro-León, E., Garcia-Ibañez, P., & Carvajal, M. (2024). Exploring phenolic compounds in crop by-products for cosmetic efficacy. *International Journal of Molecular Sciences*, *25*(11), 5884. https://doi.org/10.3390/ijms25115884

Gonzalez, T., & Sethi, A. (2012). Curcumin (turmeric) and its evolving role in skin health. In Handbook of diet, nutrition and the skin (pp. 332–348). Brill | Wageningen Academic. https://doi.org/10.3920/9789086867295_021

Gopinath, H., & Karthikeyan, K. (2018). Turmeric: A condiment, cosmetic and cure. *Indian Journal of Dermatology, Venereology and Leprology, 84*(1), 16. https://doi.org/10.4103/ijdvl.IJDVL_1143_16

Guan, C., Zhou, X., Li, H., Ma, X., & Zhuang, J. (2022). NF-κB inhibitors gifted by nature: The anticancer promise of polyphenol compounds. *Biomedicine & Pharmacotherapy*, *156*, 113951. https://doi.org/10.1016/j.biopha.2022.113951

Gugleva, V., Ivanova, N., Sotirova, Y., & Andonova, V. (2021). Dermal drug delivery of phytochemicals with phenolic structure via lipid-based nanotechnologies. *Pharmaceuticals*, *14*(9), 837. https://doi.org/10.3390/ph14090837

Guimarães, I., Baptista-Silva, S., Pintado, M., & L. Oliveira, A. L. (2021). Polyphenols: A promising avenue in therapeutic solutions for wound care. *Applied Sciences*, *11*(3), 1230. https://doi.org/10.3390/app11031230

Guo, J., Zhang, H., Lin, W., Lu, L., Su, J., & Chen, X. (2023). Signaling pathways and targeted therapies for psoriasis. *Signal Transduction and Targeted Therapy*, *8*(1), 437. https://doi.org/10.1038/s41392-023-01655-6

Hassan, F. I. (2024). Skin. In *Encyclopedia of toxicology, fourth edition: Volume 1-9, 8* (pp. 555–559). https://doi.org/10.1016/B978-0-12-824315-2.00135-4

Hasegawa, T., Shimada, S., Ishida, H., & Nakashima, M. (2013). Chafuroside B, an Oolong Tea Polyphenol, Ameliorates UVB-Induced DNA Damage and Generation of Photo-immunosuppression Related Mediators in Human Keratinocytes. *PLoS ONE, 8*(10), e77308. https://doi.org/10.1371/journal.pone.0077308

Hewlings, S., & Kalman, D. (2017). Curcumin: A Review of Its Effects on Human Health. *Foods, 6*(10), 92. https://doi.org/10.3390/foods6100092

Hopkins, R. (2019). Natural polyphenols for acne vulgaris: Clinical evidence and biological mechanisms. *Reactive Oxygen Species, 7*, 129–135. https://doi.org/10.20455/ros.2019.825

Hu, X., Zhang, H., & Chen, Z. (2024). Effects of four polyphenols on mouse wound healing and the gene expression profile of resveratrol action. *Histology and Histopathology, 39*(1), 105–116. https://doi.org/10.14670/HH-18-616

Husen, A. (2022). *Herbs, shrubs and trees of potential medicinal benefits*. CRC Press, Boca Raton, FL. https://doi.org/10.1201/9781003205067

Husen, A. (2023a). *Exploring poisonous plants: Medicinal values, toxicity response and other potential uses*. CRC Press, Boca Raton, FL. https://doi.org/10.1201/b23017

Husen, A. (2023b). *Antidiabetic medicinal plants and herbal treatments*. CRC Press, Boca Raton, FL. https://doi.org/10.1201/b23347

Husen, A. (2024a). *Antimalarial medicinal plants*. Taylor & Francis Group, LLC. ISBN: 9781032457154. https://doi.org/10.1201/9781003378396

Husen, A. (2024b). *Medicinal spices and condiment crops*. Taylor & Francis Group. ISBN: 9781032469478. https://doi.org/10.1201/9781003387046

Iacopetta, D., Ceramella, J., Scumaci, D., Catalano, A., Sinicropi, M. S., Tundis, R., Alcaro, S., & Borges, F. (2023). An Update on Recent Studies Focusing on the *Antioxidant Properties* of Salvia Species. Antioxidants, 12(12), 2106. https://doi.org/10.3390/antiox12122106

Ivarsson, J., Pecorelli, A., Lila, M. A., & Valacchi, G. (2023). Blueberry Supplementation and Skin Health. Antioxidants, 12(6), 1261. https://doi.org/10.3390/antiox12061261

Iwasaki, Y., Hirasawa, T., Maruyama, Y., Ishii, Y., Ito, R., Saito, K., Umemura, T., Nishikawa, A., & Nakazawa, H. (2011). Effect of interaction between phenolic compounds and copper ion on antioxidant and pro-oxidant activities. *Toxicology in Vitro, 25*(7), 1320–1327. https://doi.org/10.1016/j.tiv.2011.04.024

Jeong, S. H., Park, M. Y., Bhosale, P. B., Abusaliya, A., Won, C. K., Park, K., Il, Kim, E., Heo, J. D., Kim, H. W., Ahn, M., Seong, J. K., Kim, H. H., & Kim, G. S. (2023). Potential antioxidant and anti-inflammatory effects of *Lonicera japonica* and citri reticulatae pericarpium polyphenolic extract (LCPE). *Antioxidants, 12*(8), 1582. https://doi.org/10.3390/antiox12081582

Jimenez-Lopez, C., Fraga-Corral, M., Carpena, M., García-Oliveira, P., Echave, J., Pereira, A. G., Lourenço-Lopes, C., Prieto, M. A., & Simal-Gandara, J. (2020). Agriculture waste valorisation as a source of antioxidant phenolic compounds within a circular and sustainable bioeconomy. *Food & Function, 11*(6), 4853–4877. https://doi.org/10.1039/D0FO00937G

Kalasariya, H. S., & Pereira, L. (2022). Dermo-cosmetic benefits of marine macroalgae-derived phenolic compounds. *Applied Sciences, 12*(23), 11954. https://doi.org/10.3390/app122311954

Kaparekar, P. S., & Anandasadagopan, S. K. (2022). The potential role of bioactive plant-based polyphenolic compounds and their delivery systems—As a promising opportunity for a new therapeutic solution for acute and chronic wound healing. *Current Pharmacology Reports, 8*(5), 321–338. https://doi.org/10.1007/s40495-022-00296-7

Kapoor, M. P., Sugita, M., Fukuzawa, Y., Timm, D., Ozeki, M., & Okubo, T. (2021). Green Tea Catechin Association with Ultraviolet Radiation-Induced Erythema: A Systematic Review and Meta-Analysis. *Molecules, 26*(12), 3702. https://doi.org/10.3390/molecules26123702

Karağaç, M. S., & Ceylan, H. (2023). Effects of natural polyphenols against skin aging. *International Conference on Recent Academic Studies, 1*. https://doi.org/10.59287/icras.699

Kauffmann, A. C., & Castro, V. S. (2023). Phenolic compounds in bacterial inactivation: A perspective from Brazil. *Antibiotics, 12*(4), 645. https://doi.org/10.3390/antibiotics12040645

Ketnawa, S., Reginio, F. C., Jr., Thuengtung, S., & Ogawa, Y. (2022). Changes in bioactive compounds and antioxidant activity of plant-based foods by gastrointestinal digestion: A review. *Critical Reviews in Food Science and Nutrition, 62*(17), 4684–4705. https://doi.org/10.1080/10408398.2021.1878100

Khan, H. Y., Zubair, H., Faisal, M., Ullah, M. F., Farhan, M., Sarkar, F. H., Ahmad, A., & Hadi, S. M. (2014). Plant polyphenol induced cell death in human cancer cells involves mobilization of intracellular copper ions and reactive oxygen species generation: A mechanism for cancer chemopreventive action. *Molecular Nutrition & Food Research, 58*(3), 437–446. https://doi.org/10.1002/mnfr.201300417

Kim, E. B., Lee, E. K., Son, S. Y., & Lee, M. W. (2022). Antiacne and anti-inflammatory effects of phenolic compounds from *Quercus acutissima* Carruth. Leaves. *Genetics Research*, 2022. https://doi.org/10.1155/2022/9078475

Kim, H. H., Jeong, S. H., Park, M. Y., Bhosale, P. B., Abusaliya, A., Kim, H. W., Seong, J. K., Ahn, M., Park, K. Il, & Kim, G. S. (2023). Antioxidant effects of phenolic compounds in through the distillation of *Lonicera japonica* & Chenpi extract and anti-inflammation on skin keratinocyte. *Scientific Reports*, *13*(1), 20883. https://doi.org/10.1038/s41598-023-48170-w

Kim, Y. A., Gaidin, S. G., & Tarahovsky, Y. S. (2018). The influence of simple phenols on collagen type I fibrillogenesis in vitro. *Biophysics*, *63*(2), 162–168. https://doi.org/10.1134/S0006350918020148

Kumar, N., & Goel, N. (2019). Phenolic acids: Natural versatile molecules with promising therapeutic applications. *Biotechnology Reports*, *24*, e00370. https://doi.org/10.1016/j.btre.2019.e00370

Kunchana, K., Jarisarapurin, W., Chularojmontri, L., & Wattanapitayakul, S. K. (2021). Potential Use of Amla (Phyllanthus emblica L.) Fruit Extract to Protect Skin Keratinocytes from Inflammation and Apoptosis after UVB Irradiation. *Antioxidants, 10*(5), 703. https://doi.org/10.3390/antiox10050703

Lakey-Beitia, J., Burillo, A. M., La Penna, G., Hegde, M. L., & Rao, K. S. (2021). Polyphenols as potential metal chelation compounds against Alzheimer's disease. *Journal of Alzheimer's Disease*, *82*(s1). https://doi.org/10.3233/JAD-200185

Lee, E. H., Park, D. H., & Kwon, H. J. (2013). Nutritional and functional characterization of coffee cherry husk extract (CHE) for estimation of utility as raw material for cosmetics. *Korea Journal of Aesthetic Cosmetology*, *11*(3), 479–488.

Lee, J. H., Park, J., & Shin, D. W. (2022). The molecular mechanism of polyphenols with anti-aging activity in aged human dermal fibroblasts. *Molecules*, *27*(14), 4351. https://doi.org/10.3390/molecules27144351

Lephart, E. D. (2021). 4′,7-Isoflavandiol (equol) enhances human dermal fibroblast renewal and has effects similar to 17β-estradiol in stimulating collagen and elastin expression. Cell cycle and RT-PCR analysis without phenol red. *Cosmetics*, *8*(1), 5. https://doi.org/10.3390/cosmetics8010005

Liao, R., Parker, T., Bellerose, K., Vollmer, D., & Han, X. (2022). A green tea containing skincare system improves skin health and beauty in adults: An exploratory controlled clinical study. *Cosmetics*, *9*(5), 96. https://doi.org/10.3390/cosmetics9050096

Lima, M. C., Paiva de Sousa, C., Fernandez-Prada, C., Harel, J., Dubreuil, J. D., & de Souza, E. L. (2019). A review of the current evidence of fruit phenolic compounds as potential antimicrobials against pathogenic bacteria. *Microbial Pathogenesis, 130*(December 2018), 259–270. https://doi.org/10.1016/j.micpath.2019.03.025

Liu, H.-M., Cheng, M.-Y., Xun, M.-H., Zhao, Z.-W., Zhang, Y., Tang, W., Cheng, J., Ni, J., & Wang, W. (2023a). Possible mechanisms of oxidative stress-induced skin cellular senescence, inflammation, and cancer and the therapeutic potential of plant polyphenols. *International Journal of Molecular Sciences*, *24*(4), 3755. https://doi.org/10.3390/ijms24043755

Liu, R. H. (2003). Health benefits of fruit and vegetables are from additive and synergistic combinations of phytochemicals. *American Journal of Clinical Nutrition*, *78*(3 SUPPL), 3–6. https://doi.org/10.1093/ajcn/78.3.517s

Liu, S., Huang, X., Bin, Z., Yu, B., Lu, Z., Hu, R., & Long, C. (2023b). Wild edible plants and their cultural significance among the *Zhuang ethnic* group in Fangchenggang, Guangxi, China. *Journal of Ethnobiology and Ethnomedicine*, *19*(1), 52. https://doi.org/10.1186/s13002-023-00623-2

Liu, Y., Teng, J., Huang, R., Zhao, W., Yang, D., Ma, Y., Wei, H., Chen, H., Zhang, J., & Chen, J. (2024). Injectable plant-derived polysaccharide hydrogels with intrinsic antioxidant bioactivity accelerate wound healing by promoting epithelialization and angiogenesis. *International Journal of Biological Macromolecules*, *266*, 131170. https://doi.org/10.1016/j.ijbiomac.2024.131170

Luo, J., Si, H., Jia, Z., & Liu, D. (2021). Dietary anti-aging polyphenols and potential mechanisms. *Antioxidants*, *10*(2), 283. https://doi.org/10.3390/antiox10020283

Luong, H. X., Thanh, T. T., & Tran, T. H. (2020). Antimicrobial peptides – Advances in development of therapeutic applications. *Life Sciences*, *260*, 118407. https://doi.org/10.1016/j.lfs.2020.118407

Maan, A. A., Nazir, A., Khan, M. K. I., Ahmad, T., Zia, R., Murid, M., & Abrar, M. (2018). The therapeutic properties and applications of *Aloe vera*: A review. *Journal of Herbal Medicine*, *12*, 1–10. https://doi.org/10.1016/J.HERMED.2018.01.002

Mani, J. S., Johnson, J. B., Hosking, H., Ashwath, N., Walsh, K. B., Neilsen, P. M., Broszczak, D. A., & Naiker, M. (2021). Antioxidative and therapeutic potential of selected Australian plants: A review. *Journal of Ethnopharmacology*, *268*, 113580. https://doi.org/10.1016/j.jep.2020.113580

Majma Sanaye, P., Mojaveri, M. R., Ahmadian, R., Sabet Jahromi, M., & Bahramsoltani, R. (2022). Apigenin and its dermatological applications: A comprehensive review. *Phytochemistry, 203*, 113390. https://doi.org/10.1016/j.phytochem.2022.113390

Marko, M., & Pawliczak, R. (2023). Resveratrol and Its Derivatives in Inflammatory Skin Disorders—Atopic Dermatitis and Psoriasis: A Review. *Antioxidants, 12*(11), 19

Marchiosi, R., dos Santos, W. D., Constantin, R. P., de Lima, R. B., Soares, A. R., Finger-Teixeira, A., Mota, T. R., de Oliveira, D. M., Foletto-Felipe, M. de P., Abrahão, J., & Ferrarese-Filho, O. (2020). Biosynthesis and metabolic actions of simple phenolic acids in plants. *Phytochemistry Reviews, 19*(4). https://doi.org/10.1007/s11101-020-09689-2

Marques, M. P., Varela, C., Mendonça, L., & Cabral, C. (2023). Nanotechnology-based topical delivery of natural products for the management of atopic dermatitis. *Pharmaceutics, 15*(6), 1724. https://doi.org/10.3390/pharmaceutics15061724

Martic, I., Jansen-Dürr, P., & Cavinato, M. (2022). Effects of air pollution on cellular senescence and skin aging. *Cells, 11*(14), 2220. https://doi.org/10.3390/cells11142220

Martínez-Cuazitl, A., Gómez-García, M. del C., Hidalgo-Alegria, O., Flores, O. M., Núñez-Gastélum, J. A., Martínez, E. S. M., Ríos-Cortés, A. M., Garcia-Solis, M., & Pérez-Ishiwara, D. G. (2022). Characterization of polyphenolic compounds from *Bacopa procumbens* and their effects on wound-healing process. *Molecules, 27*(19), 6521. https://doi.org/10.3390/molecules27196521

Maver, T., Maver, U., Stana Kleinschek, K., Smrke, D. M., & Kreft, S. (2015). A review of herbal medicines in wound healing. *International Journal of Dermatology, 54*(7), 740–751. https://doi.org/10.1111/ijd.12766

Maya-Cano, D. A., Arango-Varela, S., & Santa-Gonzalez, G. A. (2021). Phenolic compounds of blueberries (*Vaccinium* spp) as a protective strategy against skin cell damage induced by ROS: A review of antioxidant potential and antiproliferative capacity. *Heliyon, 7*(2), e06297. https://doi.org/10.1016/j.heliyon.2021.e06297

Melguizo-rodríguez, L., de Luna-Bertos, E., Ramos-torrecillas, J., Illescas-montesa, R., Costela-Ruiz, V. J., & García-Martínez, O. (2021). Potential effects of phenolic compounds that can be found in olive oil on wound healing. *Foods, 10*(7), 1–18. https://doi.org/10.3390/foods10071642

Mendonça, B. S., Nascimento, L. M. M., Ferro, J. N. de S., & dos Santos Reis, M. D. (2024). The effect of plant-derived polyphenols on the immune system during aging: A systematic review. *Immunopharmacology and Immunotoxicology, 46*(5), 604–617. https://doi.org/10.1080/08923973.2024.2384911

Merecz-Sadowska, A., Sitarek, P., Kucharska, E., Kowalczyk, T., Zajdel, K., Cegliński, T., & Zajdel, R. (2021). Antioxidant properties of plant-derived phenolic compounds and their effect on skin fibroblast cells. *Antioxidants, 10*(5), 726. https://doi.org/10.3390/antiox10050726

Mir, R. H., & Masoodi, M. H. (2020). Anti-inflammatory plant polyphenolics and cellular action mechanisms. *Current Bioactive Compounds, 16*(6), 809–817. https://doi.org/10.2174/1573407215666190419205317

Mostafa, W. Z., & Hegazy, R. A. (2013). Vitamin D and the skin: Focus on a complex relationship: A review. *Journal of Advanced Research*, 6(6), 793–804). https://doi.org/10.1016/j.jare.2014.01.011

Nisa, R. U., Nisa, A. U., Tantray, A. Y., Shah, A. H., Jan, A. T., Shah, A. A., & Wani, I. A. (2024). Plant phenolics with promising therapeutic applications against skin disorders: A mechanistic review. *Journal of Agriculture and Food Research, 16*, 101090. https://doi.org/10.1016/j.jafr.2024.101090

Nobile, V., Schiano, I., Peral, A., Giardina, S., Spartà, E., & Caturla, N. (2021). Antioxidant and reduced skin-ageing effects of a polyphenol-enriched dietary supplement in response to air pollution: A randomized, double-blind, placebo-controlled study. *Food and Nutrition Research, 65*. https://doi.org/10.29219/fnr.v65.5619

Nowak, M., Tryniszewski, W., Sarniak, A., Wlodarczyk, A., Nowak, P. J., & Nowak, D. (2022). Concentration dependence of anti- and pro-oxidant activity of polyphenols as evaluated with a light-emitting $Fe^{2+}$-EGTA-$H_2O_2$ system. *Molecules, 27*(11), 3453. https://doi.org/10.3390/molecules27113453

Pebam, M., Sushma, M. V., Sankaranarayanan, S. A., Thanekar, A. M., Koyande, N., & Rengan, A. K. (2022). Antiviral perspectives of economically important Indian medicinal plants and spices. *Proceedings of the Indian National Science Academy, 88*(3), 392–416. https://doi.org/10.1007/s43538-022-00099-w

Pendota, S. C., Aremu, A. O., Slavětínská, L. P., Rárová, L., Grúz, J., Doležal, K., & Van Staden, J. (2018). Identification and characterization of potential bioactive compounds from the leaves of *Leucosidea sericea. Journal of Ethnopharmacology, 220*(January), 169–176. https://doi.org/10.1016/j.jep.2018.03.035

Pietta, P. G. (2000). Flavonoids as antioxidants. *Journal of Natural Products, 63*(7), 1035–1042. https://doi.org/10.1021/np9904509

Pintus, F., Floris, S., Fais, A., Era, B., Kumar, A., Gatto, G., Uriarte, E., & Matos, M. J. (2022). Hydroxy-3-phenylcoumarins as multitarget compounds for skin aging diseases: Synthesis, molecular docking and tyrosinase, elastase, collagenase and hyaluronidase inhibition, and sun protection factor. *Molecules, 27*(20), 6914. https://doi.org/10.3390/molecules27206914

Poomanee, W., Leelapornpisid, W., Trakoolpua, K., Salamon, I., & Leelapornpisid, P. (2022). Ameliorative effect of *Bouea macrophylla* Griffth seed extract against bacteria-induced acne inflammation: *In vitro* study. *Journal of Oleo Science, 71*(10), ess22168. https://doi.org/10.5650/jos.ess22168

Radić, B., Radić, S., Mašek, T., & Šuran, J. (2024). Anti-wrinkle efficacy of standardized phenolic acids polymer extract (PAPE) from propolis: Implications for antiaging and skin health. *Journal of Cosmetic Dermatology*, *23*(10), 3372–3381. https://doi.org/10.1111/jocd.16405

Rahman, M. M., Rahaman, M. S., Islam, M. R., Rahman, F., Mithi, F. M., Alqahtani, T., Almikhlafi, M. A., Alghamdi, S. Q., Alruwaili, A. S., Hossain, M. S., Ahmed, M., Das, R., Emran, T. B., Uddin, M. S. (2021). Role of phenolic compounds in human disease: Current knowledge and future prospects. *Molecules*, *27*(1), 233. https://doi.org/10.3390/molecules27010233

Rakotondrabe, T. F., Fan, M.-X., Muema, F. W., & Guo, M.-Q. (2023). Modulating inflammation-mediated diseases via natural phenolic compounds loaded in nanocarrier systems. *Pharmaceutics*, *15*(2), 699. https://doi.org/10.3390/pharmaceutics15020699

Rammohan, A., Zyryanov, G. V., Bhagath, Y. B., & Manjula, K. (2023). Antioxidants: Structure–activity of plant polyphenolics. *Vitamins and Hormones*, *121*, 395–411. https://doi.org/10.1016/bs.vh.2022.10.001

Ramsis, T., Refat, M., Selim, H. M., Elseedy, H., & Fayed, E. A. (2024). The role of current synthetic and possible plant and marine phytochemical compounds in the treatment of acne. *RSC Advances*, *14*(33), 24287–24321. https://doi.org/10.1039/D4RA03865G

Rinnerthaler, M., Bischof, J., Streubel, M. K., Trost, A., & Richter, K. (2015). Oxidative stress in aging human skin. *Biomolecules*, *5*(2), 545–589. https://doi.org/10.3390/biom5020545

Saad, N. M., Sekar, M., Gan, S. H., Lum, P. T., Vaijanathappa, J., & Ravi, S. (2020). Resveratrol: Latest scientific evidences of its chemical, biological activities and therapeutic potentials. *Pharmacognosy Journal*, *12*(6s), 1779–1791. https://doi.org/10.5530/pj.2020.12.240

Sajadimajd, S., Bahramsoltani, R., Iranpanah, A., Kumar Patra, J., Das, G., Gouda, S., Rahimi, R., Rezaeiamiri, E., Cao, H., Giampieri, F., Battino, M., Tundis, R., Campos, M. G., Farzaei, M. H., & Xiao, J. (2020). Advances on natural polyphenols as anticancer agents for skin cancer. *Pharmacological Research*, *151*, 104584. https://doi.org/10.1016/j.phrs.2019.104584

Saviuc, C., Ciubucă, B., Dincă, G., Bleotu, C., Drumea, V., Chifiriuc, M.-C., Popa, M., Pircalabioru, G., Marutescu, L., & Lazăr, V. (2017). Development and sequential analysis of a new multi-agent, anti-acne formulation based on plant-derived antimicrobial and anti-inflammatory compounds. *International Journal of Molecular Sciences*, *18*(1), 175. https://doi.org/10.3390/ijms18010175

Scarpa, E.-S., Antonelli, A., Balercia, G., Sabatelli, S., Maggi, F., Caprioli, G., Giacchetti, G., & Micucci, M. (2024). Antioxidant, anti-inflammatory, anti-diabetic, and pro-osteogenic activities of polyphenols for the treatment of two different chronic diseases: Type 2 diabetes mellitus and osteoporosis. *Biomolecules*, *14*(7), 836. https://doi.org/10.3390/biom14070836

Sharma, P., Dhiman, T., Negi, R. S., OC, A., Gupta, K., Bhatti, J. S., & Thareja, S. (2024). A comprehensive review of the molecular mechanisms driving skin photoaging and the recent advances in therapeutic interventions involving natural polyphenols. *South African Journal of Botany*, *166*, 466–482. https://doi.org/10.1016/j.sajb.2024.01.035

Silva, E. P. da, Herminio, V. L. de Q., Motta, D. N., Soares, M. B. P., Rodrigues, L. de A. P., Viana, J. D., Freitas, F. A. de, Silva, A. P. G. da, Souza, F. das C. do A., & Vilas Boas, E. V. de B. (2022). The role of phenolic compounds in metabolism and their antioxidant potential. *Research, Society and Development*, *11*(10). https://doi.org/10.33448/rsd-v11i10.31750

Singh, G. (2020). Resveratrol: Nanocarrier-based delivery systems to enhance its therapeutic potential. *Nanomedicine*, *15*(28), 2801–2817. https://doi.org/10.2217/nnm-2020-0289

Singh, N., & Yadav, S. S. (2022). A review on health benefits of phenolics derived from dietary spices. *Current Research in Food Science*, *5*(1), 1508–1523. https://doi.org/10.1016/j.crfs.2022.09.009

Soleymani, S., Farzaei, M. H., Zargaran, A., Niknam, S., & Rahimi, R. (2020). Promising plant-derived secondary metabolites for treatment of acne vulgaris: A mechanistic review. *Archives of Dermatological Research*, *312*(1), 5–23. https://doi.org/10.1007/s00403-019-01968-z

Stojiljković, D., Arsić, I., & Tadić, V. (2016). Extracts of wild apple fruit (*Malus sylvestris* (L.) Mill., Rosaceae) as a source of antioxidant substances for use in production of nutraceuticals and cosmeceuticals. *Industrial Crops and Products*, *80*, 165–176. https://doi.org/10.1016/j.indcrop.2015.11.023

Sun, M., Deng, Y., Cao, X., Xiao, L., Ding, Q., Luo, F., Huang, P., Gao, Y., Liu, M., & Zhao, H. (2022). Effects of natural polyphenols on skin and hair health: A review. *Molecules*, *27*(22). https://doi.org/10.3390/molecules27227832

Sun, W., & Shahrajabian, M. H. (2023). Therapeutic potential of phenolic compounds in medicinal plants—Natural health products for human health. *Molecules*, *28*(4), 1845. https://doi.org/10.3390/molecules28041845

Tariq, A., & Ahmed, A. (2024). Plant phenolics production: A strategy for biotic stress management. In *Plant phenolics in biotic stress management* (pp. 441–454). Springer Nature Singapore. https://doi.org/10.1007/978-981-99-3334-1_18

Tessema, F. B., Gonfa, Y. H., Asfaw, T. B., Tadesse, M. G., & Bachheti, R. K. (2023). Antioxidant activity of flavonoids and phenolic acids from *Dodonaea angustifolia* flower: HPLC profile and PASS prediction. *Journal of Chemistry, 2023*. https://doi.org/10.1155/2023/8315711

Truong, V.-L., & Jeong, W.-S. (2022). Antioxidant and anti-inflammatory roles of tea polyphenols in inflammatory bowel diseases. *Food Science and Human Wellness, 11*(3), 502–511. https://doi.org/10.1016/j.fshw.2021.12.008

Ulanowska, M., & Olas, B. (2021). Biological properties and prospects for the application of eugenol—A review. *International Journal of Molecular Sciences, 22*(7), 3671. https://doi.org/10.3390/ijms22073671

Ullah, S., Shoaib, R., Khan, S., & Masood, A. (2023). Phytochemicals; Targeted-based therapeutic approaches for pigmentation disorders. *Open Access Indonesian Journal of Medical Reviews, 3*(2), 368–381. https://doi.org/10.37275/oaijmr.v3i2.287

Vauzour, D., Rodriguez-Mateos, A., Corona, G., Oruna-Concha, M. J., & Spencer, J. P. E. (2010). Polyphenols and human health: Prevention of disease and mechanisms of action. *Nutrients, 2*(11), 1106–1131. https://doi.org/10.3390/nu2111106

Villegas-Aguilar, M. del C., Cádiz-Gurrea, M. de la L., Arráez-Román, D., & Segura-Carretero, A. (2022). Anti-aging effects of phenolic compounds. *Anti-Aging Pharmacology*. https://doi.org/10.1016/B978-0-12-823679-6.00017-5

Vivo, E., Antioxidant, T., In, A., de Lima Cherubim, D. J., Buzanello Martins, C. V., Oliveira Fariña, L., da Silva de Lucca, R. A., Lee, J. H., Park, J., Shin, D. W., Gugleva, V., Ivanova, N., Sotirova, Y., Andonova, V., Rispo, F., De Negri Atanasio, G., Demori, I., Costa, G., Marchese, E., … Suarez, M. (2021). Polyphenols as natural antioxidants in cosmetics applications. *Molecules, 10*(1), 1–16. https://doi.org/10.1111/jocd.13093

Vuolo, M. M., Lima, V. S., & Maróstica Junior, M. R. (2019). Phenolic compounds. In *Bioactive compounds* (pp. 33–50). Elsevier. https://doi.org/10.1016/B978-0-12-814774-0.00002-5

Wambe, H., Noubissi, P. A., Fokam Tagne, M. A., Foyet Fondjo, A., Fankem, G. O., Kamtchouing, I., Ngakou Mukam, J., Nguelefack, T. B., & Kamgang, R. (2019). Anti-shigellosis activity of *Cola anomala* water/ethanol pods extract on *Shigella flexneri*-induced diarrhea in rats. *BioMed Research International, 2019*. https://doi.org/10.1155/2019/6706230

Watzl, B. (2008). Anti-inflammatory effects of plant-based foods and of their constituents. *International Journal for Vitamin and Nutrition Research, 78*(6), 293–298. https://doi.org/10.1024/0300-9831.78.6.293

Woodby, B., Penta, K., Pecorelli, A., Lila, M. A., & Valacchi, G. (2020). Skin health from the inside out. *Annual Review of Food Science and Technology, 11*(1), 235–254. https://doi.org/10.1146/annurev-food-032519-051722

Xu, F.-W., Lv, Y.-L., Zhong, Y.-F., Xue, Y.-N., Wang, Y., Zhang, L.-Y., Hu, X., & Tan, W.-Q. (2021). Beneficial effects of green tea EGCG on skin wound healing: A comprehensive review. *Molecules, 26*(20), 6123. https://doi.org/10.3390/molecules26206123

Yar Khan, H., Zubair, H., Fahad Ullah, M., Ahmad, A., & Mumtaz Hadi, S. (2012). A prooxidant mechanism for the anticancer and chemopreventive properties of plant polyphenols. *Current Drug Targets, 13*(14), 1738–1749. https://doi.org/10.2174/138945012804545560

Yue, T., Bai, X., Zhang, H., & Yuan, Y. (2012). Fractionation and anti-inflammatory effects of polyphenol-enriched extracts from apple pomace. *Bangladesh Journal of Pharmacology, 7*(1). https://doi.org/10.3329/bjp.v7i1.10194

Zeb, A. (2020). Concept, mechanism, and applications of phenolic antioxidants in foods. *Journal of Food Biochemistry, 44*(9), 1–22. https://doi.org/10.1111/jfbc.13394

Zhang, X., Li, Z., Yang, P., Duan, G., Liu, X., Gu, Z., & Li, Y. (2021). Polyphenol scaffolds in tissue engineering. *Materials Horizons, 8*(1), 145–167. https://doi.org/10.1039/D0MH01317J

Zhao, H., Lou, Z., Chen, Y., Cheng, J., Wu, Y., Li, B., He, P., Tu, Y., & Liu, J. (2023). Tea polyphenols (TPP) as a promising wound healing agent: TPP exerts multiple and distinct mechanisms at different phases of wound healing in a mouse model. *Biomedicine & Pharmacotherapy, 166*, 115437. https://doi.org/10.1016/j.biopha.2023.115437

Zong, H., Wang, F., Fan, Q., & Wang, L. (2012). Curcumin inhibits metastatic progression of breast cancer cell through suppression of urokinase-type plasminogen activator by NF-kappa B signaling pathways. *Molecular Biology Reports, 39*(4), 4803–4808. https://doi.org/10.1007/s11033-011-1273-5

# 12 Gastrointestinal Health Benefits of Phenolic Compounds

*Karishma Singh and Pradeep Singh Negi**

## 12.1 INTRODUCTION

Phenols are a large group of chemicals featuring the phenolic structure of an aromatic hydrocarbon ring with a hydroxyl group (–OH) attached to it, widely known for its medicinal potential. Phenolics are classified as simple phenols (vanillin, gallic acid (GA), caffeic acid) or polyphenols (stilbenes, flavonoids, isoflavonoids, tannins) based on the number of phenol units in the molecular structure. Polyphenols are classified as plant secondary metabolites derived solely from the shikimate-derived phenylpropanoid and/or the polyketide pathway(s), comprising multiple phenolic rings and lacking nitrogen-based functional groups in their most fundamental structural form (Rashmi & Negi, 2020). They are the classes of phytochemicals widespread in plants and recognized for their bioactivities, such as antioxidant, anti-inflammatory, and antibacterial properties (Rana et al., 2022). Phenolic compounds may exert gastrointestinal benefits through their prebiotic potential, which can favourably alter the gut microbiota composition.

The impact of the gut microbiome on gastrointestinal and systemic infirmity is increasingly recognized. The human gastrointestinal tract is a highly active and dynamic environment that is exposed to ~100 trillion mutualistic microorganisms, collectively known as the microbiome. The nature of the various physiological and pathological processes that occur inside the human body is determined by the characteristics of their mutual coexistence (Hammerhøj, Gubatan, & Nielsen, 2024). Variations in host-inherited genetics and external interactions with the habituated environment (geographical region, diet, disease state, and lifestyle) are two significant elements that influence the gut microbiota at the genus and species level. Researchers have proposed a distinction between peripheral and core microbiomes. This concept suggests that the core microbiome encompasses gut bacterial genes present in most individuals. In contrast, the peripheral microbiome comprises functions observed in a subset of individuals but not all (Ogunrinola et al., 2020). The composition of an individual's gut evolves throughout life, with three primary scenarios: the gut microbial ecosystem first at birth and in infancy, then in adulthood, and finally in old age. As the host immune system develops soon after birth, bacteria colonise this environment from a very early stage in life. Studies have demonstrated that the microbiome profiles of newborns vary depending on their mode of birth and delivery method. Specifically, there is a lower presence of *Bifidobacterium* and *Bacteroides* genera and a notably high abundance of *Clostridium difficile* in children born via caesarean section (Nazzaro et al., 2020). A recently reported study suggests that the normalization of gut microbiota is possible in caesarean-delivered infants with continuous breastfeeding for six months as a microbial seeding source (Selma-Royo et al., 2024). Bacteria that constitute the majority of the adult gut microbiota include Firmicutes (79.4%) and Bacteroidetes (16.9%) being the two main phylotypes, along with Proteobacteria (1%), Verrucomicrobia (<1%), and Actinobacteria (2.5%). Additionally, adult microbiota is classified into three enterotypes based on the relative abundance

* Corresponding author: psnegi.cftri@csir.res.in

 DOI: 10.1201/9781003527671-12

of specific bacterial genera. The first enterotype is characterized by the presence of *Bacteroides*, the second by *Prevotella*, and the third by *Ruminococcus*, which is the most common. The enterotypes are responsible for the fermentation of fermentable substrates within the gut, facilitate food digestion and absorption, convert polyphenols and carbohydrates into biologically potent metabolites, including short-chain fatty acids (SCFAs), produce vitamins and minerals, preserve intestinal integrity, control host immunity, and both directly and indirectly guard against pathogens (Costea et al., 2018). As individuals progress through their lifespan, alterations in the composition of their gut microbiome are observed. Such alterations may be attributed to physiological factors peculiar to the elderly, such as chewing dysfunction, oedema, and gastrointestinal problems including decreased bowel motility and digestive difficulties. The changes in the microbial ecology observed in the elderly are linked to several factors, including immunosenescence, inflammatory ageing, and medical interventions (Nazzaro et al., 2020). However, variations in the microbiome's qualitative and quantitative makeup, despite age, can produce a range of diseases. These changes can also disrupt the synthesis of important bacterial metabolites, modify the activity of the intestinal mucosal barrier, and induce inflammation. GI tract microbial dysbiosis is frequently linked to autoimmune and intestinal diseases, including type 1 diabetes mellitus, inflammatory bowel disease (IBD), and other metabolic disorders.

## 12.2 PHENOLIC COMPOUNDS

The term "polyphenol" encompasses various distinct chemicals, classified under the number of connected phenol rings and the structural moieties that hold them together. Phenolics are secondary aromatic metabolites either produced by the polyketide acetate/malonate pathway or the shikimate/phenylpropanoid pathway, producing monomeric and polymeric phenolics. These metabolites are commonly found in plant-based sources, including fruits, vegetables, cereals, and herbs. They provide colour, flavour, and astringency and act in the organism's defence mechanism, protecting it against ultraviolet (UV) radiation, parasites, and harmful infections. The polyphenols are primarily classified into two major classes: flavonoids and non-flavonoids.

### 12.2.1 Flavonoids

The flavonoid backbone (C6–C3–C6) is composed of 15 carbon atoms and two phenolic subunits 1 and 2 and is further subdivided into seven subclasses based on subunit 3 (chromane or heterocyclic ring) that are the following: (1) Flavones (apigenin, luteolin, tangeritin, baicalein, diosmetin), (2) Flavanols/flavan-3-ols (catechin, epicatechin, epicatechin gallate, catechin gallate), (3) Flavonols (kaempferol, quercetin, myricetin, rutin, robinin), (4) Flavanones (pinocembrin, naringenin, hesperidin, eriodictyol, prunin), (5) isoflavones (daidzin, equol, genistein, glycitein), (6) anthocyanidins (delphinidin, malvidin, peonidin, cyanidin, petunidin, pelargonidin) and (7) chalcones (licochalcone A, licochalcone B, xanthohumol, pinostrobin). Flavonoids are abundantly found in dietary sources like citrus and non-citrus fruits, berries, green and black teas, and vegetable greens like celery, spinach, kale lettuce, and legumes (Pei, Liu, & Bolling, 2020).

### 12.2.2 Non-Flavonoids

The classification of non-flavonoid phenolic compounds includes phenolic acids, tannins, stilbenes, xanthones, and lignans. Among all non-flavonoid compounds, phenolic acids, which have a carboxyl group attached to a benzene ring, are the most abundant in fruits and vegetables. Fruits and vegetables have more free phenolic acids than cereals, where they are found in bound form as derivatives. They are derived from two primary phenolic compounds: benzoic acid and cinnamic acid. Hydroxybenzoic acid (C1–C6) derivatives include syringic acid, protocatechuic acid, p-hydroxybenzoic acid, ellagic acid, vanillic acid, GA, and salicylic acid. In contrast, hydroxycinnamic acid (C3–C6) derivatives include p-coumaric acid, caffeic acid, sinapic acid, and ferulic acid. Tannins are another non-flavonoid class with molecular weights ranging between 500 and 3000 Da. There are two types of tannins: hydrolyzable and non-hydrolyzable. As hydrolyzable tannins have hydroxyl or glucose groups, GA can esterify them to form gallotannins or hexahydroxydiphenic acid. Several plant parts in the families of Anacardiaceae, Leguminosae, Combretaceae, and Fabaceae contain hydrolyzable tannins. Composite polymers of flavan-3-ols, flavan-3,4-diols, or mixtures of the two are known as non-hydrolyzable tannins; examples include procyanidin. The most prevalent lignans are pinoresinol, medioresinol, enterodiol, enterolactone, syringaresinol, sinol, sesamol, lariciresinol, and secoisolariciresinol. Lignans are phenolic chemicals found in plants, including sesame and flaxseed, that are neither caloric nor nutritious. They are found in high concentrations in flaxseed (up to 8000 μg/g). Flax phenols offer protection against cancer and cardiovascular disease (CVD) and have antioxidant properties. Lignans are found to be reactive due to phenylpropanoid phenols that can easily pair with molecules, e.g., ether, carbon, or lactone (Oluwole et al., 2022).

## 12.3 INTERACTION OF PHENOLIC COMPOUNDS IN GASTROINTESTINAL TRACT

Growing research evidence highlights the prebiotic potential of polyphenols that allows them to alter the microbial composition of the gastrointestinal tract. Numerous studies on higher model organisms suggest that a multitude of polyphenols exert control over the diversity of the gut microbiota, increase the production of beneficial microbes of lactobacilli and *Bifidobacterium* species, and reduce the levels of pathogenic microorganisms, including *Clostridium histolyticum* and *Clostridium perfringens* (Rodríguez-Daza et al., 2021). Consequently, this process controls the synthesis of vitamins, branched-chain amino acids (BCAA), and SCFA. The microbial SCFAs butyrate, propionate, and acetate are typically produced in a molar ratio 60:20:20 (Whitman et al., 2024). Approximately 10% of the daily caloric intake is derived from SCFAs, which can regulate energy expenditure. Consequently, acetate and butyrate serve as substrates for hepatic and white adipose tissue (WAT) triglyceride (TG) synthesis, propionate is a precursor for hepatic gluconeogenesis, and propionate and acetate are precursors for cholesterol production. In addition, butyrate is the most preferred energy substrate for colonocyte growth and further differentiation. It also has a favourable impact on glucose and lipid metabolism, thereby improving the overall health of the host (Martinez et al., 2017). However, direct absorption is hampered by the majority of polyphenols being present as esters, glycosides, and polymers. As a result, these compounds undergo various degrees of metabolism prior to absorption. Dietary polyphenols combine with the oral microbiota and undergo initial modification in the mastication with the aid of different salivary enzymes. They then progressively pass through the oesophagus and reach the stomach (in an acidic environment), where they undergo mild enzymatic modification but no transformation. Approximately 5–10% polyphenols then travel through the small intestine, where they undergo further biotransformation processes and produce absorbable metabolites known as aglycones via sulfation, methylation, conjugation of glucuronidation, and enzymatic hydrolysis. Passive diffusion allows aglycones to infiltrate epithelial cells, and 90–95% of unabsorbed molecules travel to the colon (large intestine). Meanwhile, the assimilated molecules pass through the enterohepatic circulation to the liver, where they undergo stages I and II of biotransformation. Bile salts return previously biotransformed metabolites and are reabsorbed in the intestine. The metabolites and previously incompletely absorbed polyphenols are once more metabolized in the colon, but this time with the complete participation of gut bacteria. Low molecular weight polyphenol-derived metabolites were found to modulate the Firmicutes: *Bacteroides* ratio by increasing SCFA-producing bacteria, mainly *Akkermansia muciniphila*, in gut dysbiosis (Wang, Qi, & Zheng, 2022). Recently, Whitman et al. (2024) investigated the individual and combined impact of polyphenols (PP) and dietary fibre (FB) on gastrointestinal health. The results indicated that the incorporation of FB resulted in a significant increase in the levels of *Bifidobacterium* spp., *Lactobacillus* spp., and *Ruminococcus bromii*. Moreover, the production of SCFA was enhanced, while the concentrations of ammonia and indole were reduced. However, no impact was observed on the production of antioxidants. Furthermore, supplementation with PP resulted in increases in the concentrations of *Bifidobacterium*, Lactobacillus, and Dorea species, convoyed by elevations in free-radical scavenger levels and reductions in ammonia and indole production. However, the combined effect of both blends yielded positive results, indicating that PPFB has a beneficial impact on the gut microbiome, particularly when both PP and FB are included. Although there is still debate over what exactly defines a "healthy" gut microbiota, current research generally suggests that high microbial diversity is linked to a healthy phenotype, and the loss of diversity, along with an increase in pathogenic bacteria, corresponds with disease development. The presence of beneficial diversity in gut microbiota is responsible for the breakdown of polyphenols into useful metabolites. For example, flavonoid-converting bacteria are mainly found in the Actinobacteria, Bacteroidetes, and Firmicutes phyla, except for a few strains of Proteobacteria. *Lachnospiraceae* are known for their ability to deglycosylate and degrade various types of flavonoids. *Coriobacteriaceae* strains are responsible for the conversion of un-glycosylated isoflavones and flavan-3-ols. *Eubacterium ramulus* and *Flavonifractor plautii* have been extensively studied for their ability to convert flavonoids (Braune & Blaut, 2016). Recently, researchers have demonstrated the protective effect of GA on the pathogenic metabolite ammonia, which is known to be a primary cause of intestinal dysbiosis or IBD. The objective was to elucidate the anti-inflammatory mechanism in colitis induced by the known phenol GA by investigating its potential to scavenge the harmful byproduct ammonia and restore intestinal flora. Following oral administration of GA, an aminated complex was identified in the faeces of the model organism. This metabolite was subsequently characterized as 4-amino-substituted GA (4-$NH_2$-GA), which provides evidence for the ability of polyphenols to capture ammonia. The therapeutic effects of both GA and 4-$NH_2$-GA on dextran sodium sulphate (DSS)-induced IBD (ulcerative colitis) mice were then investigated, and it was reported that both active compounds succeeded in ameliorating the colitis state, showing that ammonia capturing had no detrimental effect on the originality of GA. Furthermore, both compounds significantly alleviated the gut dysbiosis triggered by DSS; subsequently, the excreta microbiota was transferred, demonstrating that the modified gut microbiota indirectly modulates the anti-colitis action of GA and aminated GA (Peng et al., 2024). Several studies are summarized in Table 12.1, showing the therapeutic effects of flavonoid and non-flavonoid polyphenols, changes in the microbiota, and possible metabolites produced in gastrointestinal disorders. Figure 12.1 shows the structure of selected polyphenolic compounds that act as a therapeutic agent in gastrointestinal disorders.

## 12.4 GASTROINTESTINAL HEALTH BENEFITS OF PHENOLIC COMPOUNDS

The therapeutic benefits of polyphenol-containing sources, along with their modulated individual factors and pathways, are presented in Tables 12.2 and 12.3, respectively, for alleviating diseased conditions in various disease models. Based on these studies, we have summarized the therapeutic effect of polyphenol intervention in gastrointestinal disorders in Figure 12.2.

### 12.4.1 Mechanism of Action

The gastrointestinal tract is frequently designated as the "second brain" of the complex human anatomy, as it performs a pivotal purpose in regulating a multitude of gut health axes, including the gut–immune, gut–brain, gut–muscle, gut–liver, gut–skin, and gut–hormone axis (Liang et al., 2024). The latest scientific literature has revealed that polyphenols from various dietary

**TABLE 12.1**
**Classification of Polyphenols, Sources, and Their Effects on Gut Microbiota in Dysbiosis**

| S. No. | Class | Subclass | Sources | Example with Effect on Microbial Population in Gut Dysbiosis and Their Resultant Metabolites | Reference |
|---|---|---|---|---|---|
| 1 | Phenolic acids | Hydroxybenzoic acids (ellagic acid, protocatechuic acid, gallic acid, vanillic acid, syringic Acid, salicylic acid) | Fruits (apple, grapes, and berries), vegetables (tomato, bell pepper, and spinach), and beverages (tea and wine), spices (nutmeg, cloves, cinnamon) | **Protocatechuic acid in colitis**<br>• ↑ *Bacteroidetes* and *Verrucomicrobiota*<br>**Ellagic acid in colitis**<br>• ↑*Bacteroidetes, Bacteroides*, and *Muribaculaceae;*<br>• ↓*Firmicutes*<br>**The gallic acid in colitis**<br>• ↑*Clostridiales, Enterobacterales,* and *Bacteroidales*<br>• ↑*Erysipelatoclostridium* and *Eubacterium*<br>• ↑*Akkermansia* and *Eubacterium coprostanoligenes*<br>**Vanillic ferulic acid in LPS inflammation**<br>• VA and FA-<br>↓*Prevotellaceae*; ↑*Lachnospira, Eubacterium eligens* and *Eubacterium xylanophilum;*<br>• VA-<br>↑*Firmicutes / Bacteroides ratio, Lachnospirceae*<br>**Metabolites**<br>Butyric acid, 4-$NH_2$-GA, catecholate, phenyl propionic acid | Yang et al. (2023), Li et al. (2023), Peng et al. (2024), Hu et al. (2022) |
| | | Hydroxycinnamic acids (cinnamic acid,chlorogenic acid, caffeic acid, ferulic acid) | Whole grains, fruits (apple, pears, blueberries, and plums), vegetables (artichokes, tomatoes, potatoes, and leafy greens), coffee, thyme, and oregano | **Chlorogenic acid in colitis**<br>• ↓*Enterobacteriaceae* and *Lachnospiraceae*<br>• ↑*Ligilactobacillus* and *Lachnospiraceae*<br>**Cinnamic acid in IBS-C**<br>• ↑*Akkermansia, Ruminococcus, Acinetobacter, Lachnoclostridium, Monoglobus* and *Verrucomicrobiota*<br>**Metabolites**<br>3-hydroxyphenylpropionic acid, phenylacetic acid, protocatechuic acid, hydroxytyrosol, vanillic acid, caffeoylquinic acids, valeric acid, butyric acid, and acetic acid | Xu et al. (2024), Jiang et al. (2023) |

| | | | | | |
|---|---|---|---|---|---|
| **2** | Flavonoids | Flavanones (naringenin, hesperidin and eriocitrin) | Citrus fruits, tomato and bell pepper, mint, thyme and green tea | **Hesperetin in colitis**<br>• ↓ *Proteobacteria* and *Gammaproteobacteri*↑*Rikenellaceae,* and *Rikenellaceae;*<br>• ↑ *Lachnospirales, Oscillospirales, Lachnospiraceae, Prevotellaceae*<br>• ↓ *Clostridia*<br>**Naringenin in colitis**<br>• ↑ *Firmicutes* and *Bacteroidetes;Lactobacillus* and *Bifidobacterium*<br>**Metabolites**<br>3-(4-hydroxyphenyl) propionic acid, eriodictyol, 3-(3-hydroxy-4-methoxyphenyl) propionic acid, hesperetin dihydrochalcone, butyrate | Wang et al. (2024a), Ma et al. (2024) |
| | | Flavanols or flavan-3-ols (catechins-catechin, epicatechin, epicatechin gallate, catechin gallate, theaflavins-**th**eaflavin and theaflavin-3-gallate, thearubigins) | Green tea, black tea, white tea, yellow tea, etc. | **Effect of flavanols on Colitis**<br>• ↓*Bacteroides, Escherichia Shigella, Parabacteroides,Turicibacter,* and *Alistipes*<br>• ↑*Saccharimonas*, *Akkermansia*, *Pseudomonas*, *Enterorhabdus*, *Dubosiella*, and *Lactobacillus,*Candidatus Saccharimonas<br>**Metabolites**<br>Butyric acid, threonine, isobutyric acid, and isovaleric acid | Lin et al. (2024) |
| | | flavonols (kaempferol, quercetin, myricetin, rutin, robinin) | | **Effect of flavonols on Colitis**<br>**Kaempferol**-<br>• ↓*Verrucomicrobiota, Campilobacterota,* and *Desulfobacterota;*<br>• ↑*Patescibacteria, Actinobacteria, Odoribacter,* and *Parasutterella*;<br>**Quercetin**-<br>• ↓*Bacteroidaceae* and *Bacteroides;*<br>• ↑*Lactobacillaceae, Lactobacillus,* and *Lactobacillales;*<br>• ↑*Akkermansia*<br>**Metabolites**<br>4-hydroxyphenylacetic acid, kaempferol, quercetin, quercetin-3-O-glucoside, 3,4-dihydroxyphenylacetic acid | Yu et al. (2023), Lv et al. (2024) |

*(Continued)*

**TABLE 12.1 *(Continued)***
**Classification of Polyphenols, Sources, and Their Effects on Gut Microbiota in Dysbiosis**

| S. No. | Class | Subclass | Sources | Example with Effect on Microbial Population in Gut Dysbiosis and Their Resultant Metabolites | Reference |
|---|---|---|---|---|---|
| | | Flavones (apigenin, luteolin) | Red pepper, mint, ginkgo biloba, chamomile, parsley, celery, oregano thyme | **Effect of Flavones on Colitis**<br>• ↑*Clostridia UCG-014, Enterorhabdus, Blautia;*<br>• ↓*Escherichia-Shigella, Streptococcus, Romboutsia, Clostridium sensu stricto 1, Turicibacter, Staphylococcus* and *Parasutterella;*<br>• ↓*Alistipes* and *Bilophila*<br>**Metabolites**<br>3-(4-hydroxyphenyl) propionic acid,3-(3,4-dihydroxyphenyl) propionic acid | Magadán-Corpas et al. (2024) |
| | | Isoflavones (genistein, daidzein, glycitein, formononetin, biochanin A, and irilone) | Soybeans, red clover, alfalfa | **Effect of isoflavones on IBD**<br>• ↑*Faecalibacterium prausnitzii* and *Lactobacillus rhamnosus;*<br>• ↓*Lactococcus lactis*, *Slackia equolifaciens*, and *Bacteroides fragilis*<br>**Metabolites**<br>6,7,4'-trihydroxyisoflavone, equol, dihydrodaidzein, O-demethylangolesin, daidzein, genistein, 2-(4-hydroxyphenyl) propionic acid | Wu, Sang, and Chang (2020) |
| | | Anthocyanins (delphinidin, maldivin, peonidin, cyanidin, petunidin, pelargonidin) | Bilberry, blackcurrant, blueberries, blackberry, blood orange, cranberry, concord grape, plum, raspberry, strawberries, pomegranate, sweet cherries | **Effect of anthocyanins on overall gut microbiota**<br>• ↑*Bifidobacterium* spp. and *Lactobacillus–Enterococcus* spp.<br>• ↓*Clostridium histolyticum* and *Bacteroides/Prevotella*<br>**Metabolites**<br>4-hydroxybenzoic acid, phloroglucinol aldehyde, 3-O-methyl gallic acid, protocatechuic acid, gallic acid, 3,4-dihydroxybenzaldehyde, syringic acid, 2,4,6-trihydroxybenzaldehyde, vanillic acid | Liang et al. (2024) |
| | | Chalcones (cardamonin, licochalcone A, xanthohumol, pinostrobin) | Fingerroot, apples, tomatoes, licorice root, basil, rhubarb, hops | **Effect of chalcones in IBS-D**<br>Licochalcone A-<br>• ↑*Allobaculum* and *Acinetobacter*<br>• ↑ Moraxellaceae, Bifidobacteriales, Akkermansiaceae<br>• ↓ Bacilliacea, Defluviitaleaceae, Streptococcaceae, Ruminococcaceae<br>• ↓*Enterococcus* and *prevotellaceae* | Zhang et al. (2021) |

| | | | | | |
|---|---|---|---|---|---|
| 3 | Stilbenes | Resveratrol | Grapes, peanuts, Redwine, berries, apples, soy, and Japanese knotweed | **Effect of Stilbenes in IBD**<br>• Resveratrol metabolites increase the population of *Lactobacillus* and *Bifidobacterium* and inhibit the adhesion of *Listeria monocytogenes, Salmonella typhimurium,* and *Escherichia coli* to colon cells in IBD<br>**Enzyme metabolites**<br>Resveratrol-4'-O-glucuronide, piceatannol, resveratrol-3-O-glucuronide, pterostilbene, trans-resveratrol-3-O-sulphate<br>**Gut microbiota metabolites**- lunularin, dihydroresveratril, piceid, resveratroloside, 3,4-dihydroxy-trans-stilbene<br>**RES intervention**<br>• ↑*Lactobacillus,Firmicutes* and *proteobacteria*<br>• ↓*Alistipes*<br>**Metabolites**<br>lysoPE (20:4(8Z,11Z,14Z,17Z)/0:0), 2-dodecylbenzene sulfonic acid, cholesterol sulphate, and 13-cis-retinoic acid increased in LPS-induced inflammation | M. Li et al. (2022, Ding et al. (2021) |
| | | Piceatannol (PIC)<br>Pterostilbene (PTS) | | • Piceatannol-<br>↑*Kineothrix*<br>• Pterostilbene-<br>↑*Lactobacillus intestinalis*<br>• PIC and PTS both-<br>↓*Spiroplasmataceae* and *Acholeplasmataceae;*↑*Akkermansiaceae* | Lin et al. (2023) |
| 4 | Lignans | Pinoresinol, lariciresinol, matairesinol, secoisolariciresinol, medioresinol, sesamol, and syringaresinol | Flax seeds, sesame seeds, cashews, bran of whole grains, and fruits vegetables, including citrus, grapes, and asparagus | • ↑*Bifidobacterium* and *Akkermansia*<br>• ↑ratio of *Firmicutes/Bacteroidetes*<br>• ↑*Lactobacillus;*↑abundance of *Coprococcuscus, Butyricicoccus,* and *Odoribacter*;<br>• ↓relative abundance of *Clostridium*<br>**Metabolite**<br>Enterodiol and enterolactone are two major metabolites. Lignan administration was found to increase SCFA (acids such as valeric, propionic, hexanoic, butyric, and acetic) production | Li et al. (2023a) |

*(Continued)*

**TABLE 12.1** *(Continued)*
**Classification of Polyphenols, Sources, and Their Effects on Gut Microbiota in Dysbiosis**

| S. No. | Class | Subclass | Sources | Example with Effect on Microbial Population in Gut Dysbiosis and Their Resultant Metabolites | Reference |
|---|---|---|---|---|---|
| 5 | Tannins | Hydrolysable tannins-Gallotannins (tannic acid, pentagalloyl glucose, ginnalin A, and gallic acid) and ellagitannins (punicalagin and ellagic acid) | Plants (oakwood, chestnut wood, gallnuts, tarapods, myrobalan, sumac), redwine, tea, and some fruits and nuts | • Tannic acid treatment reduced the population of *Ruminococcus & Bacteroides*, and enhanced the abundance of *Alloprevotella, Eubacterium siraeum, Enterorhabdus* and *Eubacterium xylanophilum* in UC model;<br>• Tannin enrichment in food-↓Bacteroides; ↑*Lachnospiraceae* and Ruminococcaceae families;↑*Akkermansia*, and *Faecalibacterium*<br>• Tannin supplementation increased the production level of essential metabolites such as butyric acid, acetic acid, and propionic acid compared to control subjects<br>**Metabolite**<br>Urolithins | Wang et al. (2024c), Molino et al. (2021) |
| | | Nonhydrolyzable/condensed tannins/proanthocyanidins (procyanidin C1) | Apple, grapes, berries (cranberries, blueberries, and strawberries), wine, cocoa, grapeseeds, cinnamon, mangosteen | **Effect of procyanidin C1 in IBD**<br>• ↑*Betaproteobacteria, Erysipelotrichia*, and *Verrucomicrobiae* at class level;<br>• ↑*Prevotellaceae, Sutterellaceae, Erysipelotrichaceae, Rikenellaceae, Erysipelotrichaceae* at family level;<br>• ↓*Helicobacter typhoons;*<br>• ↑*Akkermansia muciniphila*;<br>• Microbiota transplantation from the PC1 group treated model showed an increased abundance of species *Parabacteroides distasonis, Christensenella minuta, Enterorhabdus caecimuris, Faecalicoccus pleomorphus, Helicobacter hepaticus, Akkermansia muciniphila, Clostridium aldenense, Blautia producta Hydrogenoanaerobacterium*, and *Parabacteroides distasonis*<br>**Metabolite**<br>Valeric acid | Wang et al. (2024d) |

| | | | | | |
|---|---|---|---|---|---|
| **6** | Xanthones | Mangiferin, Mangostin, Garcinone, rutinosylxanthone | Mango<br>Mangosteen (*Garcinia mangostana*) | **Effect of xanthones on inflammation**<br>**Mangiferin**<br>• ↑ *Bacteroidota*<br>• ↓ *Firmicutes, Desulfobacterota,* Actinobacteriota, and Proteobacteria, Escherichia, Shigella, Pseudoalteromonas, Staphylococcus<br>**Mangostin and Garcinone**<br>• ↑ *Firmicutes: Bacteroidetes* ratio<br>• ↓ *Proteobacteria* colonization, *Rothia, Prevotella,* and *Aggregatibacter, Turicibacter*<br>**Metabolites**<br>Norathyriol, alpha-mangostin, Garcixanthone E | Chang et al. (2024), Nurrahma et al. (2022) |
| **7** | Anthraquinones | Emodin, aloe-emodin, chrysophanol, sennoside A, aloin A, and rhein | Plants- *Aloe barbadensis* (aloe vera), *Aloe ferox* (cape aloe), *Cassia obtusifolia* (sickle pod), *Cassia tora* (sickle senna), *Rhamnus purshiana* (cascara sagrada), *Rubia cordifolia* (Indian madder) | **Effect of anthraquinone in inflammation**<br>• ↑ Clostridiales and Lactobacillales; ↓Enterobacteriales; ↑Lactobacillus, Roseburia, and Akkermansia;<br>• ↓*Desulfovibrio* and *Lachnospiraceae*<br>**Metabolites**<br>ω-hydroxyemodin, 2-hydroxyemodin, aloe-emodin-9-O-glucuronide, aloe-emodin-8-O-glucuronide, aloe-emodin-3-O-sulphate, Aloe-emodin-8-O-sulphate, emodin, aloe-emodin (1,8-dihydroxy-3-hydroxymethyl anthraquinone), rheinanthrone, sennidin A and B, sennoside B, aloin B, aloe-emodin, emodin-8-O-β-D-glucoside, emodin, aloin A-glucuronide, rhein-9-O-glucuronide, rhein-2-O-sulphate, aloe-emodin-8-O-β-D-glucoside, rhein-9-O-glucuronide, rhein-3-O-sulphate, chrysophanol-8-O-β-D-glucoside | Wang et al. (2021), Cui et al. (2019) |

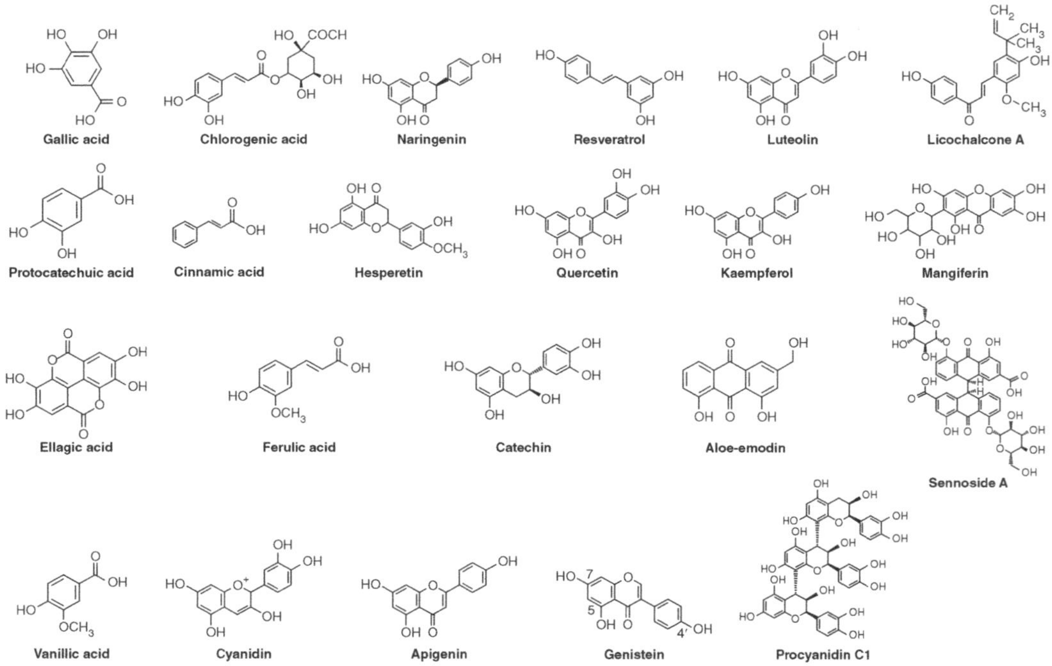


**FIGURE 12.1** Structures of major polyphenolic compounds that act as therapeutic agents in gastrointestinal disorders. (Compiled from Hu et al., 2022; Li et al., 2023b; Peng et al., 2024; Yang et al., 2023.)

and non-dietary sources can confer additional immunity by suppressing inflammatory and oxidative processes through multiple mechanisms. Oxidative stress and gut dysbiosis frequently arise as a consequence of external factors, including unhealthy lifestyles and the frequent use of certain medications. Furthermore, it is a substantial contributing factor to the development and progression of a range of pathological conditions, including cancer, neurological disorders, and infectious diseases. The endogenous production of reactive oxygen species (ROS) is attributable to various factors, including cytochrome P450-dependent metabolism, mitochondrial processes, cellular activity, and external exposure to UV, X-rays, and γ-rays. ROS can damage cellular macromolecules like proteins and lipids. Furthermore, examples of ROS include hypochlorous acid, lipid peroxides, superoxide anion radicals, per hydroxyl radicals, nitric oxide radicals, hydrogen peroxide, singlet oxygen, and hydroxyl radicals. Consequently, pro-inflammatory chemokines, cytokines, and enzymes, including tumour necrosis factor α (TNF-α), interleukin-1β (IL-1β), IL-6, and interferon-γ (IFN-γ), are released into the bloodstream over time, resulting in damage to the intestinal epithelium and an increase in intestinal overall permeability. The human body has evolved an intrinsic defence mechanism to safeguard cellular constituents from the detrimental effects of oxidative stress. A considerable number of hydroxyl groups within a polyphenol's structure endows it with the ability to scavenge radicals and mitigate oxidative stress effectively. Phenols are known to neutralize free radicals through the activation of antioxidant enzymes, including catalase (CAT), reduced glutathione (GSH), superoxide dismutase (SOD), and the inhibition of pro-inflammatory enzymes such as xanthine oxidase, cyclooxygenase-2 (COX-2), lipoxygenase (LOX), hyaluronidase, inducible nitric oxide synthase (iNOS), NF-κB, and activator protein (AP1) (Lee et al., 2021; Qin et al., 2022). Overall, the antioxidant properties of polyphenols provide a means to prevent free radical-induced oxidative damage, a recognized risk factor for chronic diseases.

**TABLE 12.2**
**The Therapeutic Effect of Polyphenol-Rich Sources on Gastrointestinal Diseases in In-Vitro Models**

| Therapeutic Sources | Active Components | Diseased Condition and Model of the Organism | Concentration | Activity in the Diseased Model after Treatment | Reference |
|---|---|---|---|---|---|
| **Grape seed phenolic extract** | Proanthocyanidin rich extract | Lipopolysaccharide (LPS)-stimulated macrophage (RAW264.7) -1*$10^5$ cells/mL | 10, 50, and 100 µg/mL | • ↓ levels of tumour necrosis factor-α, interleukin-6, interleukin-1β<br>• ↑ interleukin-10 levels | Chu et al. (2024) |
| **Vegetal extract (green artichoke-*Cynara scolymus*, fenugreek- *T. foenum-graecum*, and caigua-*Cyclanthera pedata*)** | Apigenin, luteolin, quinic acid, and derivatives of chrysin | pro-inflammatory cytokines induced IBD in caco-2 cell line | 50 and 100 µg/mL | • ↓ NFκB, P-Ikbα or Ikbα,IL-8 levels<br>• Extract intervention regulated cPLA2/AA/COX-2 pathway | Corbetta et al. (2024) |
| **Cinnamon (*Cinnamomum cassia*)** | Cinnamic acid | Caco-2 cell line | 100 µg/mL | • ↓ COX-2 (0.3-fold);<br>• ↓ mRNA levels of tumour necrosis factor-α, interleukin-6, interleukin-1β to 0.5,0.2 and 0.1-fold were observed<br>• ↑ ZO-1, claudin-1,3,4 | Park et al. (2023) |
| ***Camellia japonica*** | Traditional herbal medicine rich in various polyphenols | LPS-induced RAW264.7 cell | 100 and 200 µg/mL | • The treatment suppresses the production of ROS<br>• ↓ iNOS and TNF-α & JNK/p-38/MAPK expression | Nam, Nan, and Choo (2021) |
| **Herbal Mix (Shihosogan-tang, Yijung-tang, and Pyeongwi-san)** | Polyphenol-rich herbal extract | Pathogen *Helicobacter pylori-administered* dyspepsia in AGS (human gastric adenocarcinoma) cell line-1 × $10^5$ cells/well | 25, 50, and 100 µg/mL | • ↑ gene expression of occludin and claudin (tight junction protein) was observed<br>• ↓ cell vacuolation<br>• ↓ expression of TNF-α, COX-2, interleukin-6, interleukin-8, TLR-4, and interleukin-16; reduced ammonia production | Wang et al. (2024b) |
| **Flavonoid** | Quercetin | 125 µg/mL of p31-43 peptide was induced to mimic celiac disease in Caco-2 cell line | 50–100 µmol /L | • ↓Celiac toxicity<br>• ↑ GSH and SOD enzyme activity<br>• ↓ TG2 expression<br>• ↓ monolayer permeability<br>• ↓ Expression of MDA | Yu et al. (2024) |
| **Sour cherry** | anthocyanin-rich extract (cyanidin and rutinoside derivatives) | 1 mg/mL pepsin trypsin resistant gliadin (PT-gliadin)-induced celiac disease in caco-2 cells 2×$10^5$ cells/chamber | 100 µM AC | • ↓ ROS levels<br>• ↓ NF-κB activation pathway; ↓expression of TNF-α, interferon-γ, and interleukin-8<br>• ↑ TEER values even junctions were visible after treatment in the form of rearranged tightly packed occludin and ZO-1TJ proteins | Klusóczki et al. (2023) |

**TABLE 12.3**
**The Therapeutic Effects of Polyphenol-Rich Sources on Gastrointestinal Diseases in In-Vivo Models**

| Therapeutic Sources | Active Components | Diseased Condition and Model of the Organism | Concentration (mg/kg, Body Weight) | Activity in Diseased Model Organisms after Treatment | Reference |
|---|---|---|---|---|---|
| **Italian apple (*Limoncello Malus domestica*)** | Gallic acid (2.5%), procyanidin B1 (1.9%), catechin (16.5%), phlorizin (30.7%), chlorogenic acid (29.6%), procyanidin B2 (0.53%), epicatechin (9.6%), rutin (0.73%) | 150 (mg/kg) dinitrobenzenesulfonic acid (DNBS)-induced colitis in male CD1 mice (six to eight weeks old) | 3–300 (mg/kg) | • ↓IL-1β and IL-6↓FITC-dextran serum levels, ↑β-catenin<br>• ↓MPO-activity (intestinal permeability marker), ALT and AST<br>• Repression of nuclear factor-κB activation (Wnt/β-catenin signalling pathway) | Lama et al. (2024) |
| **Red fuji apple (peel and pomace extract)** | Epicatechin (3.13%), phlorizin- (6.55%), ellagic acid (0.05%), quercetin (0.29%), procyanidin B2 (4.05%), phloretin (1.07%), catechin (1.36%), caffeic acid (0.05%), chlorogenic acid (15.2%), rutin (0.04%), phloretin (1.07%), other (48.29%) | 3% dextran sodium sulphate (DSS) induced ulcerative colitis (UC) in C57BL/6 mice, six to eight weeks | 125 or 500 (mg/kg) | • ↓ CRP and IL-1β, ↑ colon length, and TUNEL-positive cells;↓spleen index<br>• ↑ zonula occludens-1(ZO-1), trefoil factor 3 (TFF3), mucin-2 (MUC-2), and several other occludin protein expressions<br>• ↓ pro-caspase-11, NLRP3, GSDMD, ASC, caspase-11, GSDMD- Ncaspase-1, caspase-11 protein expression | Liu et al. (2021) |
| **Grape seed phenolic extract** | Proanthocyanidin rich extract | 3% DSS induced colitis in male mice, six-week-old C57BL/6 | 50–100 (mg/kg) | • ↓TNF-α, IL-6, and IL-1β<br>• ↑IL-10 levels<br>↓ Expression of p-NF-κB and p-IκB in colon tissues; improved histological lesions | Chu et al. (2024) |
| **Wild banana (*Ensete superbum* (Roxb.) Cheesman)** | 3,5-dimethoxy-4-hydroxybenzoic acid, quercetin-3-O-rutinoside 4′,5,7-trihydroxy flavone | 1% 2,4,6- trinitrobenzene sulfonic acid ulcerative colitis (via rectal administration) in male Wistar model (n = 30, 200–250 g, rat, CFT strain) | 100, 250, and 500 mg/kg | • ↓Disease activity index (DAI) score, cellular infiltration, MDA levels, MPO activity<br>• ↑glutathione, glutathione peroxidase, and reductase enzyme in an increasing dose-dependent manner | Mathew et al. (2022) |

| | | | | | |
|---|---|---|---|---|---|
| ***Camellia Sinensis* at different fermenting stages (Green-GT, black-BT, yellow-YT, White-WT, oolong-OT, and dark-DT tea extract)** | Tea water extracts containing theaflavins, thearubigins, theabrownins, epicatechin (EC), epigallocatechin (EGC), epicatechin gallate (ECG), epigallocatechin gallate (EGCG), catechin (C) and gallocatechin (GC), catechin gallate (CG)and gallocatechin gallate (GCG) | Colitis and liver inflammation induced by 5% (w/v) DSS in C57BL/6 male mice | 600mg/kg, equivalent to 10–12g tea | • GT, WT and YT restored normal levels of goblet cells, villus crypt structure, and inflammatory infiltration while all six tea water extracts restored expression of KLF3 and Tff3 and reversed DSS-administered loss of intestinal proteins mucin and TJ, TWEs also reduced colonic inflammation by inhibiting NF-κB/NLRP3 pathway | Liu et al. (2023) |
| **Coffee Cherry husk phenols** | chlorogenic acid, ferulic acid, procyanidins, hydroxybenzoic protocatechuic acid, and epicatechin | 2.5% DSS (w/v)-induced colitis and neural inflammation in C57BL/6 J female mice | low conc.-10mg/kg and high conc.-30mg/kg | • Polyphenol treatment alleviates the upregulation of pNF-κB and pIκB-α, inhibiting TLR4/MyD88/ NF-signalling pathway.<br>• Downregulated harmful bacterial species (shigella, enterococcus pathogenic microbes) and upregulated required relative abundance of genus Bacteroides. | Li et al. (2022) |
| **Redwine** | monomeric procyanidins, oligomeric procyanidins, benzoic acids, hydroxycinnamic acids, catechin and monomeric anthocyanins | 2.5% TNBS induced colitis in CD-1 mice (male, ten-week old, 21–31 g body weight | 100 and 500 mg/kg | • ↓TNF-α, ALP and faecal haemoglobin in a dose-dependent manner<br>• PCE treatment reduced diarrhoea, moderate oedema, and retained colon length in the diseased group | Mateus et al. (2022) |
| **Turmeric (*Curcumae Longae Rhizoma*)** | An ethanolic extract containing curcumin | Acute reflux esophagitis condition in six-week-old male Sprague-Dawley rats (B.W. 180~200 g) | 50, 100, 200 mg/kg | • ↓ROS and ONOO in serum and oesophagus tissue;↓NADPH oxidase proteins (NOX4 and p22phox)<br>• ↑Nrf2 expression; ↑Antioxidant enzymes e.g., GPx-1/2, catalase, SOD, and HO-1<br>• ↓NF-κB (inducible nitric oxide synthase, cyclooxygenase-2, and cytokines- interleukin-1β, interleukin-6, and TNF-α) | Lee et al. (2021) |

*(Continued)*

**TABLE 12.3** *(Continued)*
**The Therapeutic Effects of Polyphenol-Rich Sources on Gastrointestinal Diseases in In-Vivo Models**

| Therapeutic Sources | Active Components | Diseased Condition and Model of the Organism | Concentration (mg/kg, Body Weight) | Activity in Diseased Model Organisms after Treatment | Reference |
|---|---|---|---|---|---|
| **Ginger** (***Zingiber officinale***) | Quinic acid, kaempferol caffeic acid, luteolin, rosmarinic acid, trans-ferulic acid, quercetin and p-coumaric acid, etc. | 3mg/kg loperamide (LP)-induced colonic dysmotility in seven-week-old male mice C57BL/6JRj strain weighing approximately 20–25 g | 75, 150, and 300 mg/kg | • ZOAE showed anti-constipation effect via gastric emptying in a dose-dependent manner with the highest no. of wet stool in 300 mg/kg dose. Thickness of mucosa and inflammatory infiltrate were maintained with lower oxidative markers | Abidi et al. (2022) |
| **Cinnamon (*Cinnamomum cassia*)** | Cinnamic acid | 5% DSS-induced colitis in five-week-old female BALB/c mice | 100, 500 mg/kg | • ↓TNF-α, IL-1β and IL-6<br>• ↓gut TJ barrier damage, crypt loss in the mucosa;<br>• ↑goblet cells | Park et al. (2023) |
| ***Camellia japonica*** | Traditional herbal medicine rich in various polyphenols | Ligation surgery administered reflux esophagitis, condition in seven-week-old model organism (male Sprague-Dawley rats, 180 ± 20 g) | 200 mg/kg | • ↓COX-2, TNF-α, and IL-1β expression in oesophagus tissues<br>• ↓NF-κB and IκBα<br>• Improvement in oesophageal-damaged tissues | Nam and Choo (2021) |
| **Tongxieyaofang Decoction (*Saposhnikoviae Radix*-siler, *Citri Reticulatae Pericarpium*-tangerine peel, *Atractylodis macrocephala*-Bai Zhu, Rhizoma, and *Paeoniae Radix Alba* – white peony root)** | Nobiletin, tangeritin, hesperidin, naringenin, quinic acid | wrap restraint stress (WRS) induced IBS in seven-week-old male C57BL/6 laboratory mice | 11.2 g/Kg | • ↓ZO-1 and occludin expression<br>• ↓TNF-α, AKT, MMP9, and PTGS2<br>• ↓CHRM3 and acetylcholine<br>• ↓PLC and MLCK<br>• ↓diarrhoea and loss in body weight<br>• Overall suppression of NF-κB/MLCK pathway | Feng et al. (2024) |

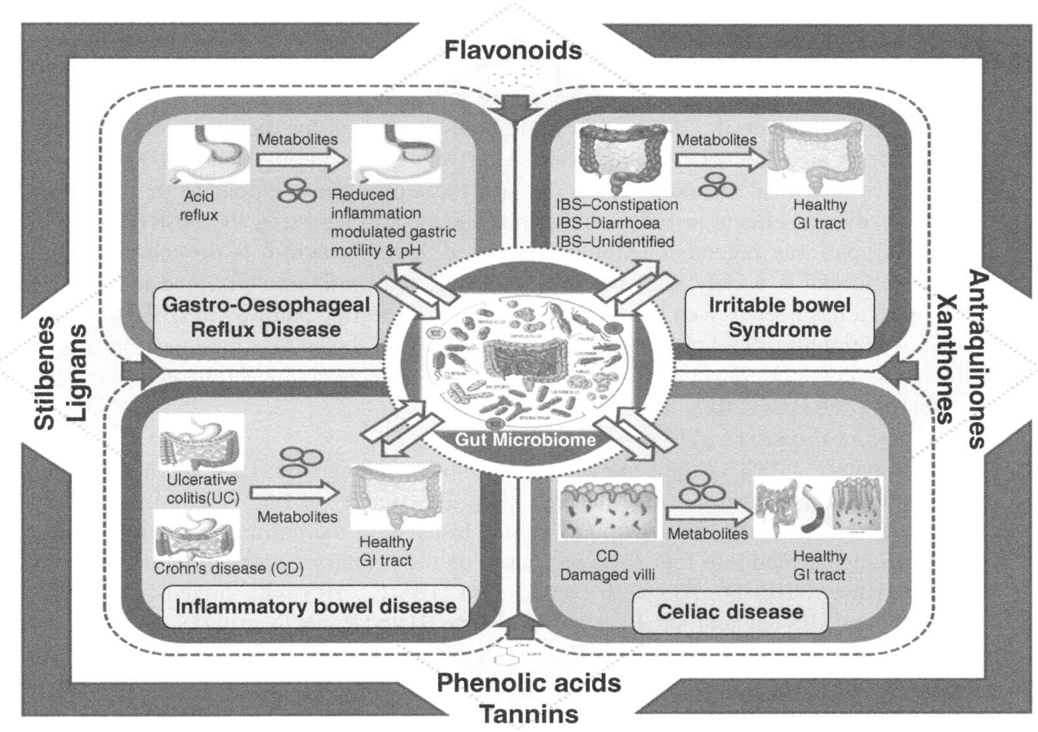


**FIGURE 12.2** Therapeutic effect of polyphenol intervention in gastrointestinal disorders. (Authors original work.)

### 12.4.2 Protective Effect of Phenols in GERD, IBS, IBD, and Celiac Disease

The term "gastrointestinal disorders" refers to a broad range of conditions that impact the stomach, intestines, and liver, among other organs. Although these conditions are crucial to human health, they also provide treatment challenges. These illnesses are becoming a significant concern for people's health and a major strain on healthcare systems, especially after the COVID-19 pandemic. Herbal medicine, or phytotherapy, is regarded as a significant alternative to conventional medicine in treating various conditions.

a. *Gastro-oesophageal reflux disease (GERD) and Ulcers*
GERD is one of the persistent diseases that is increasing in prevalence worldwide. It is associated with mucosal damage and epithelial metaplasia. It is caused by acid reflux from the stomach and duodenum and appears with or without regurgitation. This condition typically includes symptoms like dysphagia, heartburn, regurgitation, nausea, and chest pain. GERD often precedes Barrett's oesophagus and oesophageal cancer and is also associated with hypotensive lower oesophageal sphincter, frequent transient relaxation of the lower oesophageal sphincter, and oesophageal hypersensitivity. Peptic ulcer disease is characterized by lesions in the internal wall lining of the stomach and duodenum, causing irritation and inflammation of the oesophagus. Polyphenolic compounds are found to be highly protective in reflux and peptic ulcer disease. Geraniin, an ellagitannin primarily found in sources such as geranium, rhubarb, cinnamon, green tea, pomegranate, and grapes, has shown excellent anti-inflammatory and gastroprotective activity in several tested models by regulating various pathways (Nam & Choo, 2021). Flavonoids from the herb *Euphorbia*

*hirta* have shown antioxidant and radical-scavenging properties, as well as they have ability to stimulate the secretion of gastric mucus, block H+ and K+ ATPase, and inhibit the effects of histamine. The standardized *E. hirta* plant extract, including quercetin, rutin, and kaempferol, was found to be highly active in suppressing the release of stomach acid and mucus (Gupta et al., 2017). Two traditional treatments for digestive problems used for centuries are aloe vera and turmeric. Aloe vera gel is mainly rich in anthraquinone derivatives such as aloin and aloe-emodin, and these phenolic compounds are responsible for its medicinal effects in reflux and peptic ulcer via readjusting the NLRP3/GSDMD signalling pathway outcomes (Mohamed et al., 2024). Curcumin is the main flavonoid of turmeric and is widely known for its antiseptic, anti-inflammatory, and antibacterial activities. It has been proven to regulate the NF-κB signalling pathway in GERD management. Additionally, these phenolic compounds have been shown to exhibit gastroprotective effects in alleviating the symptoms of GERD and peptic ulcers by acting as proton pump inhibitors (Lee et al., 2021).

b. *Irritable bowel syndrome (IBS)*

Approximately 7–18% of the general population is suffering from IBS, and it is the most common digestive disorder affecting the day-to-day quality of life. Abdominal pain and associated bowel dysfunction without organic lesions are the hallmarks of IBS. Patients with IBS are divided into four subtypes based on the primary pattern of their stools: IBS with diarrhoea (IBS-D), IBS with constipation (IBS-C), IBS with mixed bowel habits (IBS-M), and IBS unclassified. Based on the Rome III and Rome IV criteria, the prevalence rates of IBS worldwide are 9.2 and 3.8%, respectively. According to a global meta-analysis, IBS subtypes D, C, M, and U accounted for 23.4, 22.0, 24.0, and 22.2% of individuals with IBS, respectively (Oka et al., 2020; Zeeshan et al., 2022). Polyphenol-rich decoctions may help relieve IBS symptoms and modulate various pathways such as the NF-κB/MLCK signalling mechanism.

c. *Inflammatory bowel disease (IBD)*

IBD is a disorder that encompasses chronic IBD, including two significant conditions: 1. ulcerative colitis and 2. Crohn's disease. These conditions are characterized by intestinal injury and persistent inflammation. Abdominal pain, diarrhoea, bloody stools, exhaustion, and excessive weight loss are possible symptoms. Natural as well as herbal remedies are also being extensively researched for the treatment of IBD patients, in addition to pharmaceutical medications.

d. *Celiac Disease (CD)*

One in 100–300 people worldwide are affected by CD, a known chronic, T-cell-mediated enteropathy with triggered autoimmune characteristics caused by the ingestion of gluten-containing grains (wheat, rye, barley) in genetically vulnerable individuals. One of the most significant constituents of gluten is gliadin, a protein that also contains high proportions of amino acids proline and glutamine. These amino acids undergo only partial degradation by enzymes like human proteases, forming substantial immunogenic peptides. In the small intestine lumen, the digestion of gliadin is incomplete, leading to the production of various peptide fragments. Among these is the pepsin-trypsin-resistant gliadin peptides (PT-gliadin), which retain their structure even when subjected to the digestive enzymes of both pepsin and trypsin. Immunogenic gluten peptides produced during gastrointestinal digestion cause this autoimmune disease, and researchers have found that stimulating covalent protein–polyphenol interactions may be a promising strategy to stop the synthesis of immunogenic gluten peptides. The amount of immunogenic protein peptides was observed following a laboratory digestion setup. Gluten was claimed to interact with polyphenol-rich green tea extract (GTE) at alkaline pH to increase covalent gluten–polyphenol interactions. Increase in thermal properties, a decrease in the number of amino groups, and an increase in antioxidant capacity were all obtained by treating gluten with

GTE. However, after treating gluten along with 2% extract for 2 hours at a moderate temperature (50°C) and alkaline (pH 9) conditions, immunogenic gluten peptide release was remarkably eliminated (up to 73%). All this demonstrated that GTE polyphenols effectively modified gluten under the given conditions. These results are crucial for the development of novel approaches to produce low- or no-gluten foods with reduced immunogenicity (Aksoy, Hamzalıoğlu, & Gökmen, 2024).

## 12.5 CONCLUSION

This chapter summarizes the prebiotic medicinal potential of various polyphenolic compounds in gastrointestinal diseases or disorders. A substantial growing body of research evidence supports the idea that dietary and non-dietary polyphenols can improve host health by interacting with the gastrointestinal microbial community. Numerous in vivo and in vitro studies have demonstrated how polyphenols regulate various disease-related factors and pathways, thereby improving overall host health. The mechanisms underlying these interactions may include changes in the gut environment's composition, the synthesis of potent therapeutic molecules by the microbiota, modulation of gut barrier function, downregulation of inflammatory pathways, and the biotransformation of immunogenic peptides into interacting compounds. However, more insights are necessary to fully understand these complex mechanisms, particularly those involving metabolic pathways, as they may provide new therapeutic targets in the future. Considering the multitude of elements that affect the absorption and bioavailability of polyphenols, there is considerable scope for further research in this aspect. The bioavailability of polyphenols is known to be affected by chemical interactions at the molecular level with the plant matrix, the chemical and physical structural properties of the polyphenols, gastrointestinal pH, anabolism and catabolism reactions involving hepatic and intestinal enzymes, and the gut microbiota. There is immense potential for research to further improve the bioavailability of polyphenolic compounds, thereby resulting in higher GI health benefits.

## COMPETING INTERESTS

The authors declare that they have no competing interest.

## DECLARATION OF GENERATIVE AI AND AI-ASSISTED TECHNOLOGIES IN THE WRITING PROCESS

During the preparation of this work the book chapter author(s) have not taken any assistance from AI, and the book chapter author(s) reviewed and edited the content as needed and take(s) full responsibility for the content of the publication.

## REFERENCES

Abidi, C., Rtibi, K., Boutahiri, S., Tounsi, H., Abdellaoui, A., Wahabi, S., Gressier, B., Eto, B., & Sebai, H. (2022). Dose-dependent action of *Zingiber officinale* on colonic dysmotility and ex vivo spontaneous intestinal contraction modulation. *Dose-Response*, *20*(3), 15593258221127556. https://doi.org/10.1177/15593258221127556

Aksoy, M., Hamzalıoğlu, A., & Gökmen, V. (2024). Investigating the formation of in vitro immunogenic gluten peptides after covalent modification of their structure with green tea phenolic compounds under alkaline conditions. *Journal of Agricultural and Food Chemistry*, *72*(24), 13898–13905. https://doi.org/10.1021/acs.jafc.4c00334

Braune, A., & Blaut, M. (2016). Bacterial species involved in the conversion of dietary flavonoids in the human gut. *Gut Microbes*, *7*(3), 216–234. https://doi.org/10.1080/19490976.2016.1158395

Chang, B., Wang, Y., Tu, W., Zhang, Z., Pu, Y., Xie, L., Yuan, F., Gao, Y., Xu, N., & Yao, Q. (2024). Regulatory effects of mangiferin on LPS-induced inflammatory responses and intestinal flora imbalance during sepsis. *Food Science & Nutrition*, *12*(3), 2068–2080. https://doi.org/10.1002/fsn3.3907

Chu, L., Zhang, S., Wu, W., Gong, Y., Chen, Z., Wen, Y., Wang, Y., & Wang, L. (2024). Grape seed proanthocyanidin extract alleviates inflammation in experimental colitis mice by inhibiting NF-κB signaling pathway. *Environmental Toxicology*, *39*(5), 2572–2582. https://doi.org/10.1002/tox.24129

Corbetta, P., Lonati, E., Pagliari, S., Mauri, M., Cazzaniga, E., Botto, L., Campone, L., Palestini, P., & Bulbarelli, A. (2024). Flavonoids-enriched vegetal extract prevents the activation of NFκB downstream mechanisms in a bowel disease in vitro model. *International Journal of Molecular Sciences*, *25*(14), Article 14. https://doi.org/10.3390/ijms25147869

Costea, P. I., Hildebrand, F., Arumugam, M., Bäckhed, F., Blaser, M. J., Bushman, F. D., de Vos, W. M., Ehrlich, S. D., Fraser, C. M., Hattori, M., Huttenhower, C., Jeffery, I. B., Knights, D., Lewis, J. D., Ley, R. E., Ochman, H., O'Toole, P. W., Quince, C., Relman, D. A., … Bork, P. (2018). Enterotypes in the landscape of gut microbial community composition. *Nature Microbiology*, *3*(1), 8–16. https://doi.org/10.1038/s41564-017-0072-8

Cui, H.-X., Zhang, L.-S., Luo, Y., Yuan, K., Huang, Z.-Y., & Guo, Y. (2019). A purified anthraquinone-glycoside preparation from rhubarb ameliorates type 2 diabetes mellitus by modulating the gut microbiota and reducing inflammation. *Frontiers in Microbiology*, *10*, 1423. https://doi.org/10.3389/fmicb.2019.01423

Ding, S., Jiang, H., Fang, J., & Liu, G. (2021). Regulatory effect of resveratrol on inflammation induced by lipopolysaccharides via reprograming intestinal microbes and ameliorating serum metabolism profiles. *Frontiers in Immunology*, *12*. https://doi.org/10.3389/fimmu.2021.777159

Feng, T., Zhou, Y., Lv, B., & Cai, L. (2024). Tongxieyaofang decotion alleviates IBS by modulating CHRM3 and gut barrier. *Drug Design, Development and Therapy*, *18*, 3191–3208. https://doi.org/10.2147/DDDT.S455497

Gupta, S. S., Azmi, L., Mohapatra, P. K., & Rao, C. V. (2017). Flavonoids from whole plant of *Euphorbia hirta* and their evaluation against experimentally induced gastroesophageal reflux disease in rats. *Pharmacognosy Magazine*, *13*(1), S127–S134. https://doi.org/10.4103/0973-1296.203987

Hammerhøj, A., Gubatan, J. M., & Nielsen, O. H. (2024). Gut microbiome and disorders of the gastrointestinal tract. *Microorganisms*, *12*(3), 576. https://doi.org/10.3390/microorganisms12030576

Hu, R., Wu, S., Li, B., Tan, J., Yan, J., Wang, Y., Tang, Z., Liu, M., Fu, C., Zhang, H., & He, J. (2022). Dietary ferulic acid and vanillic acid on inflammation, gut barrier function and growth performance in lipopolysaccharide-challenged piglets. *Animal Nutrition*, *8*, 144–152. https://doi.org/10.1016/j.aninu.2021.06.009

Jiang, J.-G., Luo, Q., Li, S.-S., Tan, T.-Y., Xiong, K., Yang, T., & Xiao, T.-B. (2023). Cinnamic acid regulates the intestinal microbiome and short-chain fatty acids to treat slow transit constipation. *World Journal of Gastrointestinal Pharmacology and Therapeutics*, *14*(2), 4–21. https://doi.org/10.4292/wjgpt.v14.i2.4

Klusóczki, Á., Oláh, B., Hosszú, D., Fenyvesi, F., Remenyik, J., Homoki, J., Gyöngyösi, A., Bácskay, I., & Váradi, J. (2023). Effectiveness of anthocyanin-rich sour cherry extract on gliadin-induced caco-2 barrier damage. *Nutrients*, *15*(18), Article 18. https://doi.org/10.3390/nu15184022

Lama, S., Pagano, E., Borrelli, F., Maisto, M., Tenore, G. C., Nanì, M. F., Chacon-Millan, P., Novellino, E., & Stiuso, P. (2024). Polyphenol-rich extract from 'Limoncella' apple variety ameliorates dinitrobenzene sulfonic acid-induced colitis and linked liver damage. *International Journal of Molecular Sciences*, *25*(6), Article 6. https://doi.org/10.3390/ijms25063210

Lee, J. A., Shin, M.-R., Kim, M. J., Lee, J. H., Park, H.-J., & Roh, S.-S. (2021). Protective effects of inflammation of curcumae longae rhizoma 30% EtOH extract on acute reflux esophagitis rats. *BioMed Research International*, *2021*, 8854945. https://doi.org/10.1155/2021/8854945

Li, D., Luo, F., Guo, T., Han, S., Wang, H., & Lin, Q. (2023a). Targeting NF-κB pathway by dietary lignans in inflammation: Expanding roles of gut microbiota and metabolites. *Critical Reviews in Food Science and Nutrition*, *63*(22), 5967–5983. https://doi.org/10.1080/10408398.2022.2026871

Li, M., Li, P., Tang, R., & Lu, H. (2022). Resveratrol and its derivates improve inflammatory bowel disease by targeting gut microbiota and inflammatory signaling pathways. *Food Science and Human Wellness*, *11*, 22–31. https://doi.org/10.1016/j.fshw.2021.07.003

Li, S., Chen, Y., Zhang, Y., Lv, H., Luo, L., Wang, S., & Guan, X. (2022). Polyphenolic extracts of coffee cherry husks alleviated colitis-induced neural inflammation via NF-κB signaling regulation and gut microbiota modification. *Journal of Agricultural and Food Chemistry*, *70*(21), 6467–6477. https://doi.org/10.1021/acs.jafc.2c02079

Li, X., Xu, L., Peng, X., Zhang, H., Kang, M., Jiang, Y., Shi, H., Chen, H., Zhao, C., Yu, Y., Ma, R., Li, X., & Cao, Y. (2023b). The alleviating effect of ellagic acid on DSS-induced colitis via regulating gut microbiomes and gene expression of colonic epithelial cells. *Food & Function*, *14*(16), 7550–7561. https://doi.org/10.1039/D3FO01226C

Liang, A., Leonard, W., Beasley, J. T., Fang, Z., Zhang, P., & Ranadheera, C. S. (2024). Anthocyanins-gut microbiota-health axis: A review. *Critical Reviews in Food Science and Nutrition*, *64*(21), 7563–7588. https://doi.org/10.1080/10408398.2023.2187212

Lin, W.-S., Chueh, T.-L., Nagabhushanam, K., Ho, C.-T., & Pan, M.-H. (2023). Piceatannol and 3′-hydroxypterostilbene alleviate inflammatory bowel disease by maintaining intestinal epithelial integrity and regulating gut microbiota in mice. *Journal of Agricultural and Food Chemistry*, *71*(4), 1994–2005. https://doi.org/10.1021/acs.jafc.2c08170

Lin, Z., Dai, W., Hu, S., Chen, D., Yan, H., Zeng, L., & Lin, Z. (2024). Stored white tea ameliorates DSS-induced ulcerative colitis in mice by modulating the composition of the gut microbiota and intestinal metabolites. *Food & Function*, *15*(8), 4262–4275. https://doi.org/10.1039/D3FO05176E

Liu, F., Wang, X., Cui, Y., Yin, Y., Qiu, D., Li, S., & Li, X. (2021). Apple polyphenols extract (APE) alleviated dextran sulfate sodium induced acute ulcerative colitis and accompanying neuroinflammation via inhibition of apoptosis and pyroptosis. *Foods (Basel, Switzerland)*, *10*(11), 2711. https://doi.org/10.3390/foods10112711

Liu, H., Chen, R., Wen, S., Li, Q., Lai, X., Zhang, Z., Sun, L., Sun, S., & Cao, F. (2023). Tea (*Camellia sinensis*) ameliorates DSS-induced colitis and liver injury by inhibiting TLR4/NF-κB/NLRP3 inflammasome in mice. *Biomedicine & Pharmacotherapy*, *158*, 114136. https://doi.org/10.1016/j.biopha.2022.114136

Lv, Y., Peng, J., Ma, X., Liang, Z., Salekdeh, G. H., Ke, Q., Shen, W., Yan, Z., Li, H., Wang, S., & Ding, X. (2024). Dietary quercetin protects against dextran sodium sulfate (DSS)-induced colitis mice by restoring intestinal barrier, reducting oxidative stress and inflammation via modulating gut microbiota. https://doi.org/10.21203/rs.3.rs-4536789/v1

Ma, W., Zhao, Y., Wang, Q., Gao, L., Huang, Z., Wang, J., Ma, W., Li, P., Liang, Z., & Duan, H. (2024). IDDF2024-ABS-0317 Mechanistic study of naringenin to alleviate DSS-induced IBD as revealed by 16s rDNA-based and metabolomics analysis. *Gut*, *73*(Suppl 2), A201–A204. https://doi.org/10.1136/gutjnl-2024-IDDF.126

Magadán-Corpas, P., Pérez-Valero, Á., Ye, S., Sordon, S., Huszcza, E., Popłoński, J., Villar, C. J., & Lombó, F. (2024). Gut microbiota and inflammation modulation in a rat model for ulcerative colitis after the intraperitoneal administration of apigenin, luteolin, and xanthohumol. *International Journal of Molecular Sciences*, *25*(6), Article 6. https://doi.org/10.3390/ijms25063236

Martinez, K. B., Mackert, J. D., & McIntosh, M. K. (2017). Polyphenols and intestinal health. In R. R. Watson (Ed.), *Nutrition and Functional Foods for Healthy Aging* (pp. 191–210). Academic Press. https://doi.org/10.1016/B978-0-12-805376-8.00018-6

Mateus, V., Estarreja, J., Silva, I., Gonçalves, F., Teixeira-Lemos, E., & Pinto, R. (2022). Effect of aqueous extract of phenolic compounds obtained from red wine in experimental model of colitis in mice. *Current Issues in Molecular Biology*, *44*(6), Article 6. https://doi.org/10.3390/cimb44060188

Mathew, N. S., Muthukumar, S. P., Bettadaiah, B. K., & Negi, P. S. (2022). *Ensete superbum* Roxb Cheesman (wild banana) peel extract ameliorates 2, 4, 6-trinitrobenzenesulfonic acid-induced colitis in rats. https://doi.org/10.21203/rs.3.rs-1505416/v1

Mohamed, A. O., Abd-Elghaffar, S., Kh., Mousa, R. A., & Kamel, A. A. (2024). Aloe vera gel confers therapeutic effect by reducing pyroptosis in ethanol-induced gastric ulcer rat model: Role of NLRP3/GSDMD signaling pathway. *Molecular Biology Reports*, *51*(1), 401. https://doi.org/10.1007/s11033-024-09329-4

Molino, S., Lerma-Aguilera, A., Jiménez-Hernández, N., Gosalbes, M. J., Rufián-Henares, J., & Francino, M. P. (2021). Enrichment of food with tannin extracts promotes healthy changes in the human gut microbiota. *Frontiers in Microbiology*, *12*. https://doi.org/10.3389/fmicb.2021.625782

Nam, H., Nan, L., & Choo, B. (2021). Inhibitory effects of *Camellia japonica* on cell inflammation and acute rat reflux esophagitis. *Chinese Medicine*, *16*. https://doi.org/10.1186/s13020-020-00411-0

Nam, H. H., & Choo, B. K. (2021). *Geranium koreanum*, a medicinal plant Geranii Herba, ameliorate the gastric mucosal injury in gastritis-induced mice. *Journal of Ethnopharmacology*, *265*, 113041. https://doi.org/10.1016/j.jep.2020.113041

Nazzaro, F., Fratianni, F., De Feo, V., Battistelli, A., Da Cruz, A. G., & Coppola, R. (2020). Polyphenols, the new frontiers of prebiotics. *Advances in Food and Nutrition Research*, *94*, 35–89. https://doi.org/10.1016/bs.afnr.2020.06.002

Nurrahma, B. A., Yeh, T.-H., Hsieh, R.-H., Tsao, S.-P., Chen, C.-W., Lee, Y.-P., Pan, C.-H., & Huang, H.-Y. (2022). Mangosteen pericarp extract supplementation boosts antioxidant status via rebuilding gut microbiota to attenuate motor deficit in 6-OHDA-induced Parkinson's disease. *Antioxidants*, *11*(12), Article 12. https://doi.org/10.3390/antiox11122396

Ogunrinola, G. A., Oyewale, J. O., Oshamika, O. O., & Olasehinde, G. I. (2020). The human microbiome and its impacts on health. *International Journal of Microbiology*, *2020*, 8045646. https://doi.org/10.1155/2020/8045646

Oka, P., Parr, H., Barberio, B., Black, C. J., Savarino, E. V., & Ford, A. C. (2020). Global prevalence of irritable bowel syndrome according to Rome III or IV criteria: A systematic review and meta-analysis. *The Lancet Gastroenterology & Hepatology*, *5*(10), 908–917. https://doi.org/10.1016/S2468-1253(20)30217-X

Oluwole, O., Fernando, W. B., Lumanlan, J., Ademuyiwa, O., & Jayasena, V. (2022). Role of phenolic acid, tannins, stilbenes, lignans and flavonoids in human health – A review. *International Journal of Food Science & Technology*, *57*(10), 6326–6335. https://doi.org/10.1111/ijfs.15936

Park, T. G., Kim, Y. R., Park, S., Choi, K., Kim, K. J., & Kim, J. Y. (2023). Cinnamon (*Cinnamomum cassia*) hot water extract improves inflammation and tight junctions in the intestine in vitro and in vivo. *Food Science and Biotechnology*, *32*(13), 1925–1933. https://doi.org/10.1007/s10068-023-01292-3

Pei, R., Liu, X., & Bolling, B. (2020). Flavonoids and gut health. *Current Opinion in Biotechnology*, *61*, 153–159. https://doi.org/10.1016/j.copbio.2019.12.018

Peng, J., Liu, T., Meng, P., Luo, Y., Zhu, S., Wang, Y., Ma, M., Han, J., Zhou, J., Su, X., Li, S., Ho, C.-T., & Lu, C. (2024). Gallic acid ameliorates colitis by trapping deleterious metabolite ammonia and improving gut microbiota dysbiosis. *mBio*, *15*(2), e0275223. https://doi.org/10.1128/mbio.02752-23

Qin, C., Wang, J., Zhao, W., Pi, D., Yan, X., & Nie, G. (2022). Effects of dietary bitter melon extract on growth performance, antioxidant capacity, inflammatory cytokines expression, and intestinal microbiota in common carp (*Cyprinus carpio* L.). *Aquaculture Nutrition*, *2022*(1). https://doi.org/10.1155/2022/3705943

Rana, A., Samtiya, M., Dhewa, T., Mishra, V., & Aluko, R. E. (2022). Health benefits of polyphenols: A concise review. *Journal of Food Biochemistry*, *46*(10), e14264. https://doi.org/10.1111/jfbc.14264

Rashmi, H. B., & Negi, P. S. (2020). Phenolic acids from vegetables: A review on processing stability and health benefits. *Food Research International (Ottawa, Ont*, *136*, 109298. https://doi.org/10.1016/j.foodres.2020.109298

Rodríguez-Daza, M. C., Pulido-Mateos, E. C., Lupien-Meilleur, J., Guyonnet, D., Desjardins, Y., & Roy, D. (2021). Polyphenol-mediated gut microbiota modulation: Toward prebiotics and further. *Frontiers in Nutrition*, *8*, 689456. https://doi.org/10.3389/fnut.2021.689456

Selma-Royo, M., Dubois, L., Manara, S., Armanini, F., Cabrera-Rubio, R., Valles-Colomer, M., González, S., Parra-Llorca, A., Escuriet, R., Bode, L., Martínez-Costa, C., Segata, N., & Collado, M. C. (2024). Birthmode and environment-dependent microbiota transmission dynamics are complemented by breastfeeding during the first year. *Cell Host & Microbe*, *32*(6), 996–1010.e4. https://doi.org/10.1016/j.chom.2024.05.005

Wang, D., Wang, X.-H., Yu, X., Cao, F., Cai, X., Chen, P., Li, M., Feng, Y., Li, H., & Wang, X. (2021). Pharmacokinetics of anthraquinones from medicinal plants. *Frontiers in Pharmacology*, *12*, 638993. https://doi.org/10.3389/fphar.2021.638993

Wang, J., Yao, Y., Yao, T., Shi, Q., Zeng, Y., & Li, L. (2024a). Hesperetin alleviated experimental colitis via regulating ferroptosis and gut microbiota. *Nutrients*, *16*(14), 2343. https://doi.org/10.3390/nu16142343

Wang, J.-H., Han, S.-Y., Kim, J., Lim, S., Jeong, C., Wu, L., & Kim, H. (2024b). Comparative assessment of the anti-helicobacter pylori activity and gastroprotective effects of three herbal formulas for functional dyspepsia in vitro. *Cells*, *13*(11), Article 11. https://doi.org/10.3390/cells13110901

Wang, M., Xu, X., Sheng, M., Zhang, M., Wu, F., Zhao, Z., Guo, M., Fang, B., & Wu, J. (2024c). Tannic acid protects against colitis by regulating the IL17—NFκB and microbiota—Methylation pathways. *International Journal of Biological Macromolecules*, *274*, 133334. https://doi.org/10.1016/j.ijbiomac.2024.133334

Wang, X., Qi, Y., & Zheng, H. (2022). Dietary polyphenol, gut microbiota, and health benefits. *Antioxidants*, *11*(6), Article 6. https://doi.org/10.3390/antiox11061212

Wang, X., Wang, P., Li, Y., Guo, H., Wang, R., Liu, S., Qiu, J., Wang, X., Hao, Y., Zhao, Y., Liao, H., Zou, Z., Thinwa, J., & Liu, R. (2024d). Procyanidin C1 modulates the microbiome to increase FOXO1 signaling and valeric acid levels to protect the mucosal barrier in inflammatory bowel disease. *Engineering*. https://doi.org/10.1016/j.eng.2023.10.016

Whitman, J. A., Doherty, L. A., Pantoja-Feliciano de Goodfellow, I. G., Racicot, K., Anderson, D. J., Kensil, K., Karl, J. P., Gibson, G. R., & Soares, J. W. (2024). In vitro fermentation shows polyphenol and fiber blends have an additive beneficial effect on gut microbiota states. *Nutrients*, *16*(8), 1159. https://doi.org/10.3390/nu16081159

Wu, Z.-Y., Sang, L.-X., & Chang, B. (2020). Isoflavones and inflammatory bowel disease. *World Journal of Clinical Cases*, *8*(11), 2081–2091. https://doi.org/10.12998/wjcc.v8.i11.2081

Xu, P., Chen, S., Fu, Q., Zhu, S., Wang, Z., & Li, J. (2024). Amelioration effects of chlorogenic acid on mice colitis: Anti-inflammatory and regulation of gut flora. *Food Bioscience*, *61*, 104942. https://doi.org/10.1016/j.fbio.2024.104942

Yang, X., Sun, X., Zhou, F., Xiao, S., Zhong, L., Hu, S., Zhou, Z., Li, L., & Tan, Y. (2023). Protocatechuic acid alleviates dextran-sulfate-sodium-induced ulcerative colitis in mice via the regulation of intestinal flora and ferroptosis. *Molecules*, *28*(9), Article 9. https://doi.org/10.3390/molecules28093775

Yu, R., Zhou, Q., Liu, T., Liu, P., Li, H., Bian, Y., & Liu, Z. (2023). Kaempferol relieves the DSS-induced chronic colitis in C57BL/6J mice, alleviates intestinal angiogenesis, and regulates colonic microflora structure. *Journal of Functional Foods*, *107*, 105646. https://doi.org/10.1016/j.jff.2023.105646

Yu, T., Xie, Y., Wang, Z., Li, J., Shen, Y., Yuan, J., Gao, J., Fakruddin, M., Wu, Y., & Chen, H. (2024). Quercetin ameliorates celiac-related intestinal inflammation caused by wheat gluten through modulating oxidative stress, Th1/Th2/Treg balance, and intestinal microflora structure. *Food & Function*. https://doi.org/10.1039/D4FO03025G

Zeeshan, M. H., Vakkalagadda, N. P., Sree, G. S., Anne, K., kishore, devi, S., Parkash, O., Fawwad, S. B. U., Haider, S. M. W., Mumtaz, H., & Hasan, M. (2022). Irritable bowel syndrome in adults: Prevalence and risk factors. *Annals of Medicine and Surgery*, *81*, 104408. https://doi.org/10.1016/j.amsu.2022.104408

Zhang, J., Cao, L., Sun, Y., Qing, D.-G., Xu, X.-Q., Wang, J.-C., Si, J.-Y., & Li, N. (2021). The regulatory effects of licochalcone a on the intestinal epithelium and gut microbiota in murine colitis. *Molecules*, *26*(14), Article 14. https://doi.org/10.3390/molecules26144149

# 13 Phenolic Compounds in Respiratory Health and Asthma Management

*Manish Dev Sharma, Ruchi, Sonika Kalia, Indra Rautela, Vaibhavi Thapliyal, and Sheetal Tyagi**

## 13.1 INTRODUCTION

The term "chronic respiratory disease" (CRD) encompasses asthma, pneumoconiosis, interstitial lung disease (ILD), chronic obstructive pulmonary disease (COPD), pulmonary sarcoidosis, and other conditions that impact the lungs and airways. As the world's third-leading cause of death in 2019, CRD carries a significant financial burden (Syamlal, 2020; Vos et al., 2020).

A third fewer premature deaths from non-communicable diseases (NCDs), such as CRDs, are to occur by 2030, according to Sustainable Development Goal (SDG) 3.4. To achieve this aim, a strong, evidence-based research approach is required that focuses on identifying, preventing, and managing modifiable risk factors, including tobacco use, indoor and outdoor air pollution, occupational exposures, and lifestyle-related variables. Personalized treatment plans, scalable public health initiatives, and the creation of early diagnostic biomarkers are among the top research goals. Implementation research and longitudinal epidemiological studies are also essential for addressing health disparities among diverse communities, monitoring progress, and understanding disease burden (Assembly, 2015). Achieving universal health coverage (UHC), strengthening global defenses against health emergencies, and empowering an extra billion people to achieve improved health and well-being are the three main goals of the World Health Organization's (WHO's) 2019–2023 Strategic Plan. These priorities are supported by evidence-based strategies that aim to improve access to essential services, strengthen health systems, and promote resilience to public health threats. The WHO is a key player in directing global health policy, promoting international cooperation, and tracking advancements in sustainable development and health equity as the primary coordinating body for implementing the SDGs related to health. Apart from its responsibility to track health-related metrics, the WHO has developed the SDG3 GAP (Global Action Plan for Healthy Lives and Well-being for All) to enhance cooperation among key stakeholders in the multilateral system. Through the promotion of coordinated action, resource alignment, and the integration of evidence-based methods across global health efforts, this initiative aims to accelerate progress toward achieving healthy SDG targets (World Health Organization, 2018, 2025).

Although the overall goal of these programs is to enhance health, CRD-specific mitigating measures have also been implemented. The Global Alliance against Chronic Respiratory Diseases (GARD) of the World Health Organization (World Health Organization, 2007) has targeted the Global Initiatives for COPD (GOLD) (Pauwels, 2001) and Asthma (GINA) to lessen the burden of CRDs (Reddel et al., 2019). According to Soriano et al. (2020), the Global Burden of Disorders (GBD), Injuries, and Risk Factors Study 2017 provided the most recent thorough evaluation of the prevalence and related health burden of chronic respiratory disorders (CRDs) worldwide. By quantifying morbidity, mortality, and disability-adjusted life years (DALYs) using standardized

* Corresponding author: sheetalty@gmail.com

 DOI: 10.1201/9781003527671-13

procedures, this study provides a robust epidemiological framework for understanding the global impact of CRDs. The improved GBD 2019 provides more accurate estimates due to new data sources, locations, and several risk factors, as well as specific analytical adjustments.

The main risk factors for CRDs are smoking, occupational and environmental hazards, geographic location, culture, age, and sex. Policymakers can create and effectively target risk modification strategies that can reduce disability and premature mortality by having a better understanding of the trends in these risk factors and identifying the populations that are at risk. We characterized the burden of CRDs and related risk factors by age, sex, and socio-demographic index (SDI) at the national, regional, and worldwide levels, as well as their changes from 1990 to 2019, using the GBD 2019 survey. The goal of this report is to provide a broad assessment of the existing CRD burden. Under the direction of the GBD protocol, we wrote this manuscript as members of the GBD Collaborator Network. The ultimate goal is to draw attention to the most significant risk factors and vulnerable groups, enabling policymakers and caregivers to develop targeted risk reduction strategies effectively. All previous GBD estimations of CRDs are updated in this paper (GBD 2015 Chronic Respiratory Disease Collaborators, 2017; Soriano et al., 2020).

In the US, between 4 and 8% of adults suffer from asthma (Dodge & Burrows, 1980). Between 14 and 15 million individuals in the US suffer from asthma, and the condition is thought to be the cause of over 5,000 fatalities each year (Bascom, 1996). Approximately 20% of asthma cases occur in individuals over 45, despite the majority typically developing in childhood or early adulthood. About 30% of asthmatics experience moderate or severe symptoms, whereas 70% have mild symptoms (Martinez, 1995). Many studies conducted over the past few decades have significantly advanced our understanding of asthma as a complex and diverse chronic inflammatory airway disease. The primary pathogenic characteristics include recurring episodes of bronchial inflammation, mucus hypersecretion, airway hyper-responsivity, and reversible airflow obstruction. These symptoms are caused by a complex interplay between environmental exposures, genetic predisposition, and immune system dysregulation, particularly type 2 (T2) inflammation, which is mediated by cytokines such as interleukin-4 (IL-4), IL-5, and IL-13. Despite advancements in characterizing these systems, research is ongoing to determine the exact relationships between underlying molecular pathways, observable physiological changes, and clinical symptoms.

Phenotyping and endotyping asthma has been the focus of recent research in an effort to increase diagnostic precision and direct individualized treatment plans. Examples of these endotypes include T2-high and T2-low asthma, which react differentially to corticosteroids and biologic treatments, according to research employing transcriptome and proteomic analyses. Furthermore, to clarify the development of illness, treatment response, and potential biomarkers for early diagnosis and therapeutic targeting, systems biology techniques and longitudinal cohort studies are being employed (Yeh & Schwartzstein, 2009).

The pathophysiology of asthma is complicated, and its symptoms can be perplexing. It can resemble other diseases and coexist with them. Because asthma is episodic, patients may have routine testing and examinations in between episodes. When someone has both reversible and episodic airway obstruction, the National Heart, Lung, and Blood Institute advises diagnosing asthma (Teichtahl, Buckmaster, & Pertnikovs, 1997). Pulmonary function testing and other diagnostic investigations are essential. Patients with asthma frequently exhibit wheezing, a high-pitched sound made during breathing, as their earliest symptom. However, a diagnosis of asthma is not always ruled out by the absence of wheezing. Because asthma is an episodic condition, wheezing is not always a symptom. Furthermore, a severe asthma attack may be indicated by a decrease in wheezing due to severe airway resistance. When interpreting a lack of wheezing as an improvement in asthma, care should be taken, as it could be a sign of deteriorating respiratory failure and airway obstruction.

GERD, postnasal drip, and asthma are the leading causes of chronic cough, which is frequently accompanied by cough-variant asthma (Irwin & Curley, 1990). Asthma is often diagnosed and treated with bronchodilator medication. GERD and asthma have a complicated relationship, though, with postnasal drip exhibiting the allergic phenotype and acid reflux increasing airway resistance. Individuals with a persistent cough should be evaluated for asthma. Asthma frequently manifests as dyspnea and shortness of breath, but it can be challenging to differentiate it from other respiratory conditions, such as COPD. Asthma patients' subjective perceptions of dyspnea vary and are not always consistent with objective assessments of lung function (Teeter & Bleecker, 1998).

Despite having higher airway resistance, patients with significant airway obstruction may experience a "blunted" feeling of dyspnea. There is no correlation between symptoms and objective measurements of lung function. Since shortness of breath in various respiratory disorders is characterized by unique features, such as chest pain and cardiovascular disease, it is more beneficial to understand the language of dyspnea. Despite having higher airway resistance, patients with significant airway obstruction may experience a "blunted" feeling of dyspnea. There is no correlation between symptoms and objective measurements of lung function. Since shortness of breath in various respiratory disorders is characterized by unique features, such as chest pain and cardiovascular disease, it is more beneficial to understand the language of dyspnea (Jeffery, 2000). Whereas COPD is linked to bronchiolitis or inflammation in the small airways, asthma primarily affects the large airways (Bousquet et al., 2000). The central airways, peripheral airways, and lung parenchyma are among the several lung compartments affected by COPD, and each has unique pathological characteristics that overlap (Saetta et al., 2001). Studies have demonstrated that the accumulation of activated macrophages and $CD8^+$ cytotoxic T-lymphocytes, which lead to epithelial damage and mucus hypersecretion, is indicative of chronic inflammation in the central airways. In addition to comparable inflammatory mechanisms, peripheral airway disease also involves substantial airway remodeling, characterized by myofibroblast differentiation and fibroblast proliferation. The molecular mechanisms via which profibrotic mediators, including fibronectin, insulin-like growth factors (IGF-1/2), endothelin-1, transforming growth factor-beta (TGF-β), and platelet-derived growth factor (PDGF), control the deposition of extracellular matrix and fibrosis are currently being studied. These signaling cascades are being characterized using experimental models and translational research initiatives to identify potential therapeutic targets that can prevent or reverse airway remodeling. To further guide individualized therapies in the management of COPD, biomarker identification and sophisticated imaging techniques are being used to measure the amount of remodeling and forecast disease progression (Rennard, 1999). Patients with COPD experience varied degrees of pathological alterations in their lung parenchyma and airways.

Proteins, DNA, and lipids can undergo oxidation as a result of oxidative stress, a condition characterized by an imbalance between oxidants and antioxidants that can lead to lung damage or cellular dysfunction (Gutteridge & Halliwell, 2000).

Cell division, mitochondrial respiration, extracellular matrix remodeling, and lung defense mechanisms can all be impacted by this stress. Additionally, it plays a key role in inflammatory reactions by triggering the transcription of transcription factors and the expression of pro-inflammatory mediators (Rahman & MacNee, 1998). Natural substances are essential for maintaining respiratory health, as they provide multifaceted protection against a range of lung conditions. As antioxidants, anti-inflammatory agents, bronchodilators, mucolytics, and antimicrobials, a variety of plant-derived compounds, including flavonoids, alkaloids, terpenoids, and polyphenols, reduce airway inflammation, clear mucus, neutralize free radicals (FRs), and prevent respiratory infections. Herbs such as vasaka (*Adhatoda vasica*), licorice, ginger, turmeric, and garlic have been used for a long time to treat ailments like bronchitis, asthma, and COPD. These natural substances are crucial for managing and preventing respiratory diseases, as they often have fewer adverse effects than manufactured medications and boost immunity. The goal of this book chapter is to understand the function of phenolic compounds in respiratory health and asthma therapy.

## 13.2 PHENOLIC COMPOUNDS: STRUCTURE AND CLASSIFICATION

The large class of secondary metabolites known as plant-derived phenolic compounds is mainly composed of different phenolic acids and flavonoids, which include flavonols, flavones, flavanols, flavanones, anthocyanins, chalcones, and isoflavones. According to Wen et al. (2021), more than 5,000 structurally unique flavonoids have been identified and extracted from a variety of plant species. The phenylalanine ammonia-lyase (PAL) enzyme initiates the biosynthesis of these hydroxylated polyphenolic compounds by converting phenylalanine to cinnamic acid, which triggers further enzymatic changes. By preventing the growth of pathogens, scavenging reactive oxygen species (ROS) produced under ultraviolet (UV) light exposure, and modifying signaling pathways associated with plant stress responses, flavonoids perform vital biological functions, including antimicrobial defense. Additionally, flavonoids play a significant role in flower colors by affecting pollinator attractiveness through the accumulation of anthocyanins, which contribute to pigmentation. Their ecological and evolutionary significance is further highlighted by the fact that they interact with other secondary metabolites to influence the sensory attributes of plants, including flavor and scent.

Flavonoids, a broad class of plant polyphenolic compounds, are consumed in huge amounts by humans. Their intricate molecular makeup enables them to perform a variety of biological tasks within the human body, including providing UV protection, anti-aging properties, antibacterial effects, and antiradical capabilities. Recently, microalgae have garnered interest as a promising yet understudied source of flavonoids, which are well-known secondary metabolites of polyphenols primarily examined in terrestrial plants (Ferdous & Yusof, 2021). Flavonoids are found in plants and serve a variety of physiological roles, including defense against pathogens, regulation of growth and development, and protection against UV radiation. Crucially, these substances possess strong antibacterial, anti-inflammatory, and antioxidant properties that have been linked to several beneficial effects on human health, including a lower incidence of chronic illnesses. With their rapid growth rates and ability to adapt to various conditions, microalgae offer a scalable and sustainable platform for flavonoid production. Nevertheless, in contrast to higher plants, the biosynthetic pathways, diversity, and bioactivity of flavonoids in microalgae are still poorly understood. This suggests that further biochemical and pharmacological research is necessary to investigate their potential applications in pharmaceuticals, functional foods, and nutraceuticals.

One of the most structurally varied and physiologically important families of secondary metabolites present in higher plants is flavonoids. Flavonoids are widely distributed across almost all plant tissues, including leaves, stems, roots, fruits, and seeds, although they are most typically identified as floral colors in many angiosperm species (Dewick, 2002). These substances affect not only the color of flowers but also the taste of foods and drinks made from plants (Harahap et al., 2024).

Flavonoids are characterized chemically by a 15-carbon skeleton (C6–C3–C6) that is made up of two aromatic rings (A and B) connected by a three-carbon bridge to form a core heterocyclic ring (C-ring). A prominent structural characteristic that adds to the stereochemical diversity of flavonoid compounds is the existence of a chiral center at the heterocyclic ring's C2 position (Oliveira et al., 2022). The six main subclasses of flavonoids are (1) flavanones, (2) flavonols, (3) flavones, (4) isoflavones, (5) flavan-3-ols, and (6) anthocyanidins. These subclasses are distinguished by differences in the degree of hydroxylation, methoxylation, glycosylation, and the location of double bonds and oxygen-containing functional groups (Santhakumar, Battino, & Alvarez-Suarez, 2018). These structural variations are directly linked to their unique biological functions and activities in plant physiology, including signaling, defense responses, and UV protection.

Aloe species are particularly rich in flavonoids, known for their therapeutic potential in managing microbial infections, cancers, diabetes, and age-related conditions. Flavonoid biosynthesis involves the shikimate pathway, where coumaroyl-CoA forms rings B and C, while malonyl-CoA forms ring A. The process begins with phosphoenolpyruvic acid and D-erythrose 4-phosphate, ultimately producing chalcone, which is then converted into flavanone and subsequently diversified into other flavonoid classes. Flavonoids combat the FR damage through direct scavenging, stabilizing ROS by neutralizing radicals via their hydroxyl groups. Hanasaki et al. highlighted that flavonoids, such as epicatechin and rutin, effectively scavenge superoxides and peroxynitrite, with the potency of rutin linked to its inhibition of the enzyme XO. Kerry and Abbey noted that this radical-scavenging ability can inhibit low-density lipoprotein (LDL) oxidation, potentially protecting against atherosclerosis (Hanasaki, Ogawa, & Fukui, 1994; Kerry & Abbey, 1997; Korkina & Afanasev, 1997).

The pharmacological potential of quercetin, a well-known member of the flavonol subclass of flavonoids, has garnered much interest, especially in relation to the prevention and treatment of chronic diseases. Numerous investigations conducted in vitro, in vivo, and in clinical settings have demonstrated that quercetin exhibits a range of beneficial biological effects, primarily by modulating intracellular signaling pathways and gene expression associated with inflammation, oxidative stress, and apoptosis (Wang et al., 2016). According to Lee, Bode, and Dong (2011) and Kim et al. (2013), quercetin has demonstrated the capacity to prevent the growth of different types of cancer cells, such as colorectal, prostate, liver, pancreatic, and lung cancer cells, by disrupting important cellular functions like cell cycle progression, apoptosis regulation, and signaling pathways linked to metastasis. The compound's ability to activate caspase and modify pathways such as PI3K/Akt, mitogen-activated protein kinase (MAPK), and nuclear factor kappa beta (NF-κB) is frequently associated with these anticancer actions (Table 13.1). There are many dietary sources of quercetin, including leafy green vegetables (such as spinach and kale), apples (*Malus domestica*), berries (like cranberries and blueberries), onions (*Allium cepa*), and medicinal plants like *Hypericum perforatum* and *Ginkgo biloba*. Quercetin is currently being investigated as a lead compound for the development of innovative medicinal agents and functional foods due to its abundance in everyday foods and its encouraging bioactivity.

## 13.3 MECHANISM OF ACTION OF PHENOLIC COMPOUNDS

### 13.3.1 Antioxidant Activity

The pathophysiology of aging and numerous human diseases, including cancer, cardiovascular disease, rheumatoid arthritis, and Alzheimer's disease, is caused by oxidative damage to biomolecules, which is primarily caused by FR species like hydroxyl radical, hydrogen peroxide, superoxide, nitric oxide (NO), peroxynitrite, and others (De Groot, 1994; Devasagayam et al., 2004). Therefore, FR-induced oxidative stress avoidance is important for both disease prevention and treatment. Low concentrations of phenolic compounds can function as antioxidants, shielding biomolecules (proteins, nucleic acids, lipids, and carbohydrates) against oxidative damage caused by FR-mediated processes (Leopoldini, Russo, & Toscano, 2011). Through metal-ion chelation, enzyme inhibition, and direct quenching of ROS, these substances can prevent oxidizing chain reactions (Vaya & Aviram, 2001). The molecular structure of phenolic compounds, specifically the quantity and presence of hydroxyl groups, as well as conjugation and resonance effects, is linked to their antioxidant qualities. H-atom transfer and one-electron transfer are the two suggested ways by which phenolic chemicals, as major antioxidants, might carry out their protective function. In the former, the antioxidant loses a hydrogen atom to resonance stabilization, which makes the antioxidant a less reactive radical. The antioxidant can transfer an electron to the FR in the second pathway, turning it into a stable radical cation (Leopoldini et al., 2004). The potential for phenolic compounds to chelate transition metal ions ($Fe^{3+}$, $Cu^{+}$, and $Mn^{+}$) and produce stable, complexed molecules gives rise to another antioxidant mechanism. The metals' entrapment stops them from participating in the processes that produce FR (Perron & Brumaghim, 2009).

Several studies have shown that several phenolic compounds have positive effects on the cardiovascular system, mainly because of their strong antioxidant and FR scavenging properties. It has been demonstrated that these substances lower blood pressure, enhance endothelial function, inhibit platelet aggregation, and prevent LDL cholesterol from undergoing oxidative modification (Vauzour et al., 2010). According to Zou et al. (2003) and Quiñones, Miguel, and Aleixandre (2013), these effects are significant in the prevention and treatment of several cardiovascular diseases, including atherosclerosis, hypertension, congestive heart failure, and hypercholesterolemia, which are frequently associated with oxidative stress and vascular dysfunction. Some phenolic compounds show their sources and biological activity in Table 13.2. Strong inhibitors of LDL peroxidation, a crucial

**TABLE 13.1**
**Major Class, Sources, Pharmacological Activity, and Mechanism of Action of Phenolic Compounds**

| Major Class | Natural Sources | Pharmacological Activity | Mechanism of Action | Reference |
|---|---|---|---|---|
| **Flavonols** (e.g., kaempferol, myricetin, quercetin) | Onions, kale, spinach, broccoli, apples, grapes, tea leaves, capers, berries, cilantro, dill, *Ginkgo biloba* | Anti-inflammatory, antioxidant, cardiovascular health | modulating inflammatory pathways like NF-κB, ROS | Mahmud et al. (2023), Rajendran et al. (2014) |
| **Flavones** (e.g., apigenin, luteolin) | Parsley, celery, chamomile, basil, thyme, oregano, bell peppers, mint, citrus peels | Neuroprotection, anticancer, anti-inflammatory | MAPK, NF-κB, and STAT3 | Che et al. (2020), Kashyap et al. (2022) |
| **Flavanones** (e.g., hesperidin, naringenin, eriodictyol) | Citrus fruits (oranges, lemons, grapefruit, tangerines), bergamot, tomatoes | Skin protection, heart health, metabolic regulation | activating the Nrf2 pathway, modulating inflammatory cytokines like IL-6 and TNF-α, and reducing oxidative stress | Mahmoud et al. (2019), Madureira et al. (2023), Jain et al. (2021) |
| **Isoflavones** (e.g., genistein, daidzein, glycitein) | Soybeans, chickpeas, lentils, peanuts, red clover, alfalfa | Estrogenic effects, bone health, menopausal relief | bind to estrogen receptors (ERs), inhibit tyrosine kinase activity, mitogen-activated protein kinase (MAPK) pathway, phosphoinositide 3-kinase/Akt (PI3K/Akt) pathway | Wu et al. (2025), Pan et al. (2001), Xiang et al. (2025) |
| **Anthocyanins** (e.g., cyanidin, delphinidin, malvidin) | Blueberries, blackberries, cherries, red grapes, cranberries, pomegranates, plums, eggplants, purple sweet potatoes, red cabbage, hibiscus | Anti-diabetic, Anticancer, angiogenesis, cardiovascular health, eye health | by blocking MAPK pathways and the epidermal growth factor receptor (EGFR) | Saulite et al. (2019), Lin et al. (2017), Meiers et al. (2001) |
| **Chalcones** (e.g., phloretin, isoliquiritigenin) | Apples, licorice root, hops, tomatoes, tea, ginger, turmeric | Anti-inflammatory, antimicrobial, anticancer | Enhancing Nrf2 activity and suppressing STAT3 signaling, apoptosis in cancer cells, activating the AMPK/mTOR pathway, inhibition of the p38/mTOR/STAT3 | Kwon et al. (2025), Qiu et al. (2024), Ruan et al. (2023), Yang et al. (2022), Tuli et al. (2022) |
| **Flavan-3-ols** (catechins, proanthocyanidins) | Green and black tea, cocoa fruit, dark chocolate, berries, pears, red wine, apples cinnamon | Antioxidant, brain health, anti-aging | Free radical scavenging, metal-ion chelation, GLUT4, and PI3/AKT pathway activation | Bernatoniene and Kopustinskiene (2018), Özduran, Becer, and Vatansever (2023), Patanè et al. (2023) |

**TABLE 13.2**
**Natural Sources and Biological Activity of Some Phenolic Compounds**

| Phenolic Compounds | Natural Sources | Biological Activity | References |
|---|---|---|---|
| | | **Flavonoids** | |
| Quercetin | Apples, berries, onions, spinach | Antioxidant, anti-inflammatory | Kelly (2011) |
| Apigenin | Celery, basil, parsley, oranges, and lemons | Anti-inflammatory, neuroprotective, skin protective, anticancer | Salehi et al. (2019 |
| Catechin | Fava beans, green tea, cocoa beans, grapes | Antioxidant potential, antimicrobial, and antidiabetic | Isemura (2019) |
| Cyanidin | Berries, red cabbage, cherries | Antidiabetic and antiobesity, neuroprotective, and vision health | Stintzing et al. (2002) |
| Genistein | Legumes, red clover, soybeans | Phytoestrogenic, osteoprotective, epigenetic modulation | Dixon and Ferreira (2002) |
| Eriodictyol | Citrus fruits | Antiallergic, hepatoprotective, anticancer potential | El-Taher et al. (2020) |
| | | **Stilbenes** | |
| Astringin | Bark of source and rhubarb | Potent antioxidant, cardioprotective, antimicrobial | Mérillon et al. (1997) |
| Piceatannol | Blueberries, white tea, passion fruit seeds, rhubarb | Specific kinase inhibition, anticancer, induces apoptosis, anti-adipogenic | Piotrowska, Kucinska, and Murias (2012) |
| Resveratrol | Grapes, blueberries, mulberries, red wine, peanuts | Anticancer, antioxidant, antidiabetic, and metabolic | Walle (2011) |
| Pterostilbene | Grapes, blueberries, and the heartwood of sandalwood trees | Potent antioxidant, cardioprotective, enhanced bioavailability | Estrela et al. (2013) |
| Gnetol | Gnetum species | Potent antioxidant, antimicrobial, anti-inflammatory | Remsberg et al. (2015) |
| | | **Phenolic acids** | |
| Caffeic acid | Apples, pears, thyme, coffee | Anticancer potential, antidiabetic, and immunomodulatory | Murtaza et al. (2014) |
| Gallic acid | Green tea, gallnuts, berries | Anticancer, anti-inflammatory, antimicrobial | Choubey et al. (2015) |
| Vanillic acid | Vanilla beans, olives | Antimicrobial, antidiabetic, neuroprotective | Pang et al. (2015) |
| Ellagic acid | Raspberries, strawberries, pomegranates | Potent antioxidant, chemopreventive | Vattem and Shetty (2005) |
| | | **Lignans** | |
| Secoisolariciresinol | Flaxseed, sesame seed, lentils, apricots | Antioxidant, anticancerous, cardioprotective, anti-inflammatory | Adolphe et al. (2010) |
| Pinoresinol | Broccoli, kale, rye, sesame seeds, forsythia fruit | Anti-inflammatory, anticancer potent | Markulin et al. (2019) |
| Matairesinol | Rye, barley, pumpkin seeds, cranberries, strawberries | Antioxidant properties, anti-inflammatory | Broomhead et al. (1991) |
| | | **Tannins** | |
| Gallotanins | Berries, oak galls, berries, walnuts | Antioxidant, antimicrobial, astringent | Kiss and Piwowarski (2018) |
| Eckol | Brown algae, *Ascophyllum nodosum* | Antidiabetic, photoprotective, antiallergic | Manandhar et al. (2019) |

step in the pathophysiology of atherosclerotic plaque development and progression, are phenolic substances. According to Fuhrman and Aviram (2001), these bioactive molecules help maintain vascular homeostasis and may lower the risk of cardiovascular illnesses by reducing oxidative damage to LDL particles. Notably, both experimental and clinical research have shown that phenolic components are present in extra-virgin olive oil.

Olive oil, including hydroxytyrosol, tyrosol, and oleuropein, has cardioprotective benefits. These benefits support their involvement in lowering cardiovascular risk, as evidenced by improvements in lipid profiles, a drop in oxidative stress indicators, and a decrease in LDL atherogenicity (Covas, De La Torre, & Farré-Albaladejo, 2006; Hernáez, Remaley, & Farràs, 2015).

### 13.3.2 Anti-Inflammatory Activity

The body uses inflammation as an immunological defense mechanism to combat infection or damage caused by bacteria, viruses, and other pathogens (Larsen & Henson, 1983). Furthermore, it is becoming more widely recognized that inflammation plays a role in the emergence of several chronic illnesses, including cancer, obesity, diabetes, neurodegenerative diseases, and arteriosclerosis. Acute inflammation has been treated with aspirin and other non-steroidal anti-inflammatory Drugs (NSAIDs) since the 1970s (Santangelo et al., 2007). Nevertheless, these medications have not proven totally effective in treating long-term inflammatory conditions and often come with unanticipated adverse effects.

Several studies support the notion that various phenolic compounds, including flavonoids and phenolic acids, inhibit the inflammatory process (Dos Santos et al., 2006; Sur et al., 2008; Tipoe et al., 2007). Through their modulation of several molecular targets and signaling pathways associated with inflammation, oxidative stress, and cell survival, phenolic compounds have a substantial biological impact. One of their primary methods of action is by activating the peroxisome proliferator-activated receptor gamma (PPAR-γ), which leads to the downregulation of important pro-inflammatory mediators, including inducible NO synthase (iNOS), lipoxygenase (LOX), and cyclooxygenase-2 (COX-2). Furthermore, phenolic substances block important pro-inflammatory and proliferative signaling molecules such as c-Jun, phosphoinositide 3-kinase (PI3K), tyrosine kinases, and NF-κB. Together with vital kinases engaged in cellular defense and survival mechanisms, they also promote the activation of phase II antioxidant and detoxification enzymes, including serine/threonine protein kinase B (PKB/Akt), MAPKs, and protein kinase C (PKC). Furthermore, these substances alter the expression of genes that control the cell cycle, proliferation, and survival (Yoon & Baek, 2005). Phenolic chemicals regulate many physiological and pathological processes through these complex molecular interactions, underscoring their promise as therapeutic agents for the treatment of disorders linked to oxidative stress and chronic inflammation.

### 13.3.3 Immunomodulatory Actions

Increased leukocyte infiltration into the air passages, particularly granulocytes, is a hallmark of asthma. Airway hyperresponsiveness (AHR), bronchoconstriction, and increased mucus production are frequently brought on by inflammation (Djukanovic et al., 1990). An allergen, such as pollen, causes a Th2-dominated reaction in the airways by attracting and activating inflammatory cells, thereby increasing IL-4, IL-13, and IL-5 (Bisset & Schmid-Grendelmeier, 2005). Resveratrol was found to suppress asthmatic symptoms in an animal model of the disease by inhibiting the production of Th2 cytokines, such as IL-4 and IL-5, in the plasma and bronchoalveolar lavage fluid. It has also suppressed eosinophilia, mucus hypersecretion, and AHR (Lee, Bode, & Dong, 2011). By altering the Th1/Th2 immunological balance and inhibiting the synthesis of key cytokines, including IL-4 and interferon-gamma (IFN-γ), quercetin has been shown to improve the pathophysiology of asthma (Park et al., 2009). Additionally, by suppressing mast cell-mediated inflammation and downregulating the expression of pro-inflammatory mediators such as tumor necrosis factor-alpha

(TNF-α), IL-1β, IL-6, Th2 cytokines, extracellular signal-regulated kinase (ERK), and MAPK, a polyphenol-rich ethanolic extract of *Boehmeria nivea* has shown antiallergic effects in murine models. Numerous bioactive polyphenols, including caffeic acid, epicatechin, catechin, β-sitosterol, rutin, luteolin-7-glucoside, naringin, hesperidin, chlorogenic acid (CGA), and tangeretin, were discovered to be present in the extract. Sesamin, a lignan molecule rich in flavonoids, has also been shown to reduce allergic inflammation and AHR in asthmatic mice, suggesting that it may be used therapeutically to treat asthma (Lin et al., 2014).

### 13.3.4 Bronchodilatory and Smooth Muscle Relaxation Effects

To induce bronchodilatory and smooth muscle-relaxing effects, various methods can be achieved using phenolic chemicals. One significant method is the regulation of calcium channels. These substances prevent calcium ions from entering smooth muscle cells, which is necessary for muscular contraction (Khan et al., 2016). The cyclic AMP (cAMP) signaling cascade is activated through an alternative route. For example, a study has demonstrated that quercetin increases the synthesis of cAMP, which subsequently triggers the activation of cAMP-dependent protein kinase (PKA). This kinase reduces smooth muscle contraction by inhibiting myosin light chain kinase (MLCK) activity (Wani, Iqbal, & Basir, 2017). Furthermore, several polyphenols regulate or modulate the production of NO. NO is a well-known mediator of smooth muscle relaxation and vasodilation (Mohammed & Al-Habib, 2021). Additionally, by blocking contraction mediators such as histamine and acetylcholine that cause airway smooth muscle constriction, phenolic chemicals facilitate bronchodilation (Franova et al., 2011). By interfering with cellular processes that regulate muscle contraction, such as blocking calcium channels, triggering cAMP-dependent pathways, releasing NO, or lessening acetylcholine and histamine-induced contractions, phenolic compounds can produce bronchodilatory and smooth muscle relaxant effects. By various phenolic compounds, such as quercetin, caffeic acid, and gingerols, these effects have been demonstrated in a variety of tissues, including tracheal and bronchial smooth muscle. These attributes may be used to treat respiratory and other disorders involving smooth muscles (de Alencar Silva et al., 2020). Phenolic chemicals use a variety of molecular mechanisms to produce their smooth muscle relaxant actions. When considered collectively, these processes demonstrate the therapeutic potential of phenolic compounds in the treatment of bronchoconstriction-related respiratory diseases.

## 13.4 PROMINENT PHENOLIC COMPOUNDS IN ASTHMA AND LUNG DISORDERS

The immunomodulatory, antioxidant, and anti-inflammatory properties of some phenolic compounds, as well as their structure, as shown in Figure 13.1, have made them promising medicines for the treatment of respiratory disorders, including asthma.

### 13.4.1 Quercetin

As a potent natural substance that reduces inflammatory cytokines and stabilizes mast cells, quercetin plays a crucial role in the treatment of asthma and other pulmonary conditions. Key signaling pathways, including PI3K/AKT/mTOR, NF-κB, and MAPK, are modulated by quercetin (Zhang et al., 2025) to produce its anti-tumor effects, which repress the growth and spread of cancer. Its effectiveness in inhibiting tumor growth in a variety of cancer types, such as breast, colon, liver, and stomach cancer, has been shown in several in vitro and in vivo investigations. It shows utility in controlling inflammation, angiogenesis, and the tumor microenvironment. In comprehensive cancer therapy regimens, quercetin frequently offers promising benefits as an adjuvant due to its multitarget effects and mild side-effect profile (Thangasamy, Sittadjody, & Burd, 2009). Quercetin decreases the production of histamine and tryptase, two important mediators of allergic and inflammatory

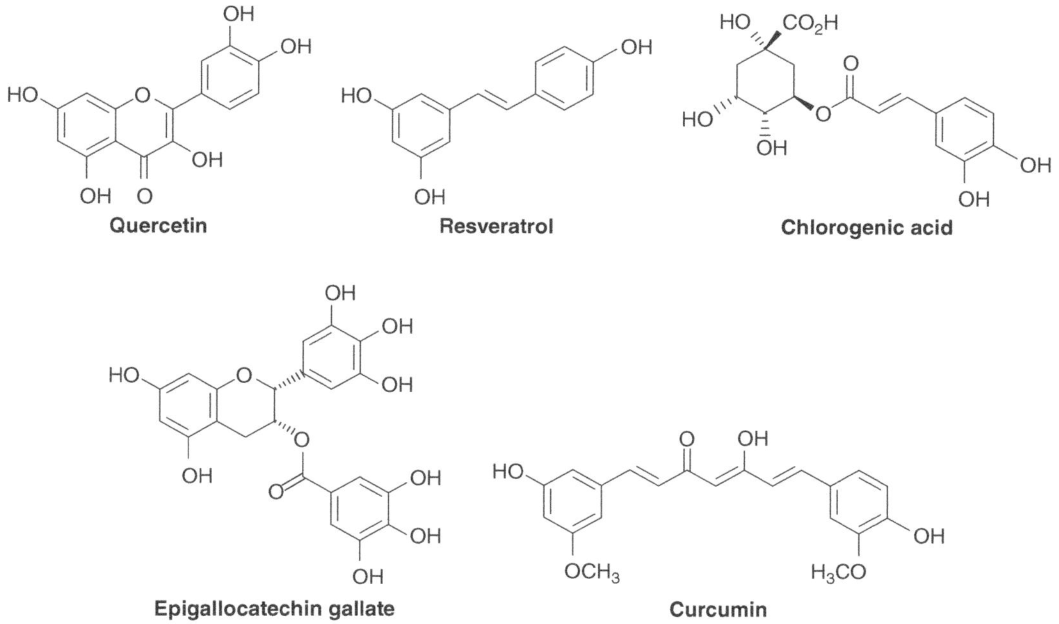


**FIGURE 13.1** Structure of few Phenolic compounds.

reactions in asthma and other lung disorders, and inhibits mast cell activation. It stops excessive inflammation and tissue damage in the lungs and other connective tissues by stabilizing mast cells. It can sometimes prevent the release of cytokines more effectively than conventional mast cell stabilizers, such as cromolyn. It has been demonstrated that quercetin is more effective than cromolyn in preventing human mast cells from releasing pro-inflammatory cytokines, such as TNF and IL-8. Both substances prevent the release of leukotrienes, histamine, and prostaglandin D2, although quercetin suppresses mast cell cytokines more effectively. Quercetin works as a preventative measure, but cromolyn quickly loses its effectiveness and only works when combined with a mast cell trigger. Quercetin significantly reduced the severity of contact dermatitis and increased resistance to skin photosensitivity in human trials, two conditions that are typically unresponsive to traditional treatments. With benefits for upcoming treatment formulations, quercetin shows promise as a mast cell inhibitor for inflammatory and allergic skin conditions (Weng et al., 2012).

Quercetin prevents lung tissue from producing inflammatory cytokines, including TNF-$\alpha$, IL-1$\beta$, and IL-6 (Aghababaei & Hadidi, 2023). Asthma and lung injury pathogenesis depend on these cytokines. In conditions such as pulmonary fibrosis, paraquat-induced lung injury, and hyperoxic/hypoxic lung injury, quercetin reduces lung tissue damage by reducing oxidative stress and inflammatory cell infiltration (Huang et al., 2024). It prevents emphysema and enhances lung function in experimental rats exposed to dangerous chemicals like cigarette smoke (Araújo et al., 2022). Due to its ability to stabilize mast cells and decrease cytokines, quercetin is a promising adjuvant for treating inflammatory lung diseases, including asthma.

### 13.4.2 Resveratrol

Resveratrol is said to activate SIRT1 and prevent airway remodeling and has a substantial positive impact on lung conditions, including asthma. The pathophysiology of asthma is closely linked to sirtuins (SIRT1–SIRT7), which are NAD$^+$-dependent enzymes with various functions in inflammation, energy metabolism, and cellular control. While SIRT2 and SIRT7 may encourage pro-inflammatory and proliferative processes, SIRT3 and SIRT6 also show protective benefits against airway inflammation and remodeling. By stimulating SIRT1 and associated pathways, sirtuin-targeting modulators, such as resveratrol, can reduce asthma symptoms. Although sirtuin modulators show promise as therapeutic targets, further investigation is needed to elucidate their precise functions, mechanisms, and clinical value in the treatment of asthma (Liu & Shi, 2022).

The study demonstrated that resveratrol significantly suppresses airway inflammation and remodeling by upregulating PTEN expression in a mouse model of asthma induced by ovalbumin. Following resveratrol treatment, PTEN levels, which had been lowered in asthmatic mice, were raised, which was consistent with a decrease in airway remodeling indicators. Since SIRT1 inhibition reduced PTEN expression, resveratrol's elevation of PTEN was mediated by SIRT1 activation. The results suggest that the SIRT1/PTEN signaling pathway is linked to resveratrol's anti-asthmatic properties. Overall, by addressing inflammation and structural alterations in the airways through PTEN regulation, resveratrol exhibits promise as a treatment for asthma (Chen et al., 2015).

The study demonstrated that in a chronic asthma mouse model induced by ovalbumin (OVA) exposure, resveratrol significantly reduces airway inflammation and remodeling. Resveratrol decreased collagen deposition in lung tissues, goblet cell hyperplasia, inflammatory cell infiltration, smooth muscle thickening, and airway hyper-responsivity. The reduction of the TGF-β1/Smad signaling pathway, which is essential for airway remodeling and the epithelial–mesenchymal transition (EMT), mediates the inhibitory effects. In bronchial epithelial cells, resveratrol administration reduced the expression of TGF-β1, phosphorylated Smad2/3, and mesenchymal markers, including Snail and Slug. According to these results, resveratrol may be able to reverse structural alterations in the airways that are usually unresponsive to traditional asthma treatments. All things considered, resveratrol shows promise as a therapeutic agent that targets airway remodeling in chronic asthma by modulating key molecular pathways (Lee et al., 2017).

### 13.4.3 Curcumin

The primary mechanism by which curcumin influences immunological responses in lung diseases, including asthma, is by modulating Th2 and Th17 responses. It inhibits Th2 cytokines, including IL-4, IL-5, and IL-13, and it lowers mucus formation and eosinophilic inflammation in the airways. Curcumin also restores the equilibrium between Th17 and Treg cells by promoting anti-inflammatory Treg cells and inhibiting the development and the activity of pro-inflammatory Th17 cells. These effects are mediated by pathways involving NF-κB, mTOR, and STAT3. Ultimately, it results in less airway remodeling and inflammation. All things considered, curcumin's ability to modulate Th2/Th17 responses enhances its promise as an anti-inflammatory and anti-asthmatic agent in lung diseases (Chen et al., 2018).

### 13.4.4 Epigallocatechin Gallate (EGCG)

Epigallocatechin gallate (EGCG) is the main phenolic ingredient of green tea. It demonstrates an antioxidant effect by reducing inflammation and oxidative stress in conditions such as asthma. These effects are demonstrated by controlling the TGF-β1 signaling pathway and reducing pro-inflammatory cytokines (TNF-α and IL-6), while also increasing anti-inflammatory cytokines (such as IL-10). EGCG reduces airway inflammation. Furthermore, it reduces eosinophil and neutrophil infiltration in lung tissues by restoring the balance of Th17/Treg cells. It also prevents excessive mucus production and lowers airway remodeling, including collagen deposition. All of these properties work together to preserve mucous membranes and reduce inflammation in respiratory conditions, making EGCG a potentially effective treatment for lung conditions, including asthma (Shan et al., 2018).

### 13.4.5 Chlorogenic Acid

It is a common phenolic molecule that shows promise in the treatment of lung diseases and asthma. By preventing the production and effects of ROS, CGA demonstrates potent antioxidant qualities and reduces oxidative stress in respiratory tissues. Through the suppression of pro-inflammatory cytokines and the regulation of signaling pathways like NF-κB, MAPK, and TLR4, CGA modifies immunological responses and reduces airway inflammation. To reduce allergy and inflammatory symptoms in the respiratory system, it also balances T cell responses and controls Th17 cell activity, which lowers IL-17 levels. The immunomodulatory and antioxidant properties of CGA reduce immune cell infiltration, prevent airway remodeling, and shield lung tissue from harm. Chologenic acid is considered a viable natural medicinal agent for the treatment of asthma and chronic lung problems due to its versatile and combined activities (Bagdas et al., 2020). In this context, one study shows that by controlling Th17 immune responses, CGA dramatically reduces the symptoms of allergic rhinitis in mice. It showed regulation of Th17 differentiation by boosting IFN-γ and decreasing serum levels of OVA-IgE, IL-17A, and the transcription factor ROR-gamma-t. CGA reduced the percentage of $CD4^{+}IL\text{-}17^{+}Th17$ cells in peripheral blood, as shown by flow cytometry. These results suggest the therapeutic potential of CGA in treating AR by inhibiting Th17-mediated immune regulation and inflammation (Shi et al., 2020).

### 13.4.6 Dietary Sources and Nutraceutical Formulations

Phosphorus-rich foods are crucial for respiratory health due to their antibacterial, bronchodilatory, anti-inflammatory, and antioxidant properties. They reduce inflammation, relax bronchial muscles, neutralize FRs, and prevent infections. Additionally, flavonoids and phenolic acids are found in a variety of foods, including vegetables, fruits, coffee, tea, herbs, spices, red wine, olives, and dark chocolate. By improving lung function and reducing the risk of respiratory diseases, including infections, asthma, and COPD, these nutrients work synergistically to increase overall respiratory resilience.

Dietary fibers, on the other hand, are polymers of carbohydrates primarily present in a variety of foods, such as cereals, fruits, vegetables, and nuts. They are complex, heterogeneous polysaccharides that are resistant to the gut's digestive enzymes and exhibit a variety of physicochemical and physiological characteristics (Cui et al., 2019). The gut microbiome uses different types of dietary fibers, including starch, pectin, cellulose, hemicellulose, dextrin, and lignin, as substrates. The microbiota may then transform the fiber into metabolites such as short-chain fatty acids (SCFAs), which display a range of immunomodulatory properties (Ratajczak et al., 2019).

However, the link between dietary fiber consumption and improved lung health is not yet well supported by the available data. Lung cancer incidence and dietary fiber intake were found to be inversely correlated in a pooled analytic research involving 1,445,850 people from the US, Europe, and China. Notably, among nonsmokers who eat yogurt and high-fiber foods, the risk of lung cancer is significantly lower (Yang et al., 2020). Furthermore, a high-fiber diet has been shown to considerably enhance lung function, whereas other research has linked a low-fiber diet to decreased lung function (Hanson et al., 2016).

High dietary fiber consumption has also been linked to a lower incidence of asthmatic symptoms, including inflammation, airway hyper-responsivity, airway contraction, coughing, and wheezing, according to observational studies (Dreher, 2018; Williams, Scott, & Wood, 2019). Taken together, these findings highlight the importance of dietary fiber and phosphorus-rich foods in supporting respiratory health.

Numerous preclinical investigations have also demonstrated a link between dietary fiber consumption and the improvement of respiratory conditions. A low-fiber diet was linked to a significant increase in inflammatory parameters (eosinophil and lymphocyte infiltration, levels of IL-4, IL-5, IL-13, and Th17 (IL-17A) cytokines) in the lungs of mice sensitized to house dust mites, according to a study by Trompette et al. (2014). When mice on a high-fiber diet were compared to those fed a regular diet, the same study also showed that the symptoms of asthma caused by house dust mites were diminished (Trompette et al., 2014). According to a different study, mice given a high-fiber diet showed less emphysema caused by cigarette smoke by having lower levels of IL-6 and IFN-γ in their blood and BALF, as well as a lower mean linear intercept (MLI) of their alveolar wall (Jang et al., 2020). Mechanistically, it has been shown that dietary fiber reduces inflammation by producing SCFAs, which are byproducts of the gut microbiota that interact with the G-protein-coupled free-fatty acid receptors GPR43, GPR41, and GPR109a (Brown et al., 2003). When SCFAs bind to these free fatty acid receptors, they restrict adenyl cyclase activity and further diminish a number of inflammatory processes, including reduced arachidonic acid synthesis, decreased calcium influx, and activation of NF-κB, which helps to improve the condition (Williams, Scott, & Wood, 2019).

Although some chemical medications are used to treat respiratory conditions, they are not always successful and can have adverse side effects. Herbal therapy has demonstrated promise in treating respiratory conditions and provides an alternative treatment option with fewer side effects. Lung disorders may be treated with the use of some medicinal plants, including ashwagandha (*Withania somnifera*), moringa (*Moringa oleifera*), hawthorn (*Crataegus rhipidophylla*), and garlic (*Allium sativum*). Plant-based natural substances such as quercetin, ligustrazine, salidroside, resveratrol, and apple polyphenol can also aid in symptom relief. By decreasing cell proliferation, combating oxidative stress, reducing inflammation, halting tumor growth, enhancing blood flow, and calming the airways, these herbs and chemicals act (Campbell, 2025). The majority of herbalists and ancient people believed that adding crushed peppercorns to green tea helped alleviate asthma (Abbasi et al., 2010). Kim and Lee (2009) found that administering piperine to mice orally in varying amounts reduced their production of histamine, immunoglobulin E, IL-4, and IL-5. This, in turn, reduced airway inflammation, hyperresponsiveness, and eosinophil infiltration (Kim et al., 2017).

Furthermore, TGF-β products were better in the piperine-treated group than in the control group. As a result, piperine can be used to treat asthma. Piperine may have the ability to suppress immunomodulatory Th2 cytokines (Kim et al., 2017). Phenolic compounds play a vital role in exerting antioxidant and anti-inflammatory effects, contributing significantly to respiratory health by modulating oxidative stress, inflammatory responses, and immune regulation (Figure 13.2).

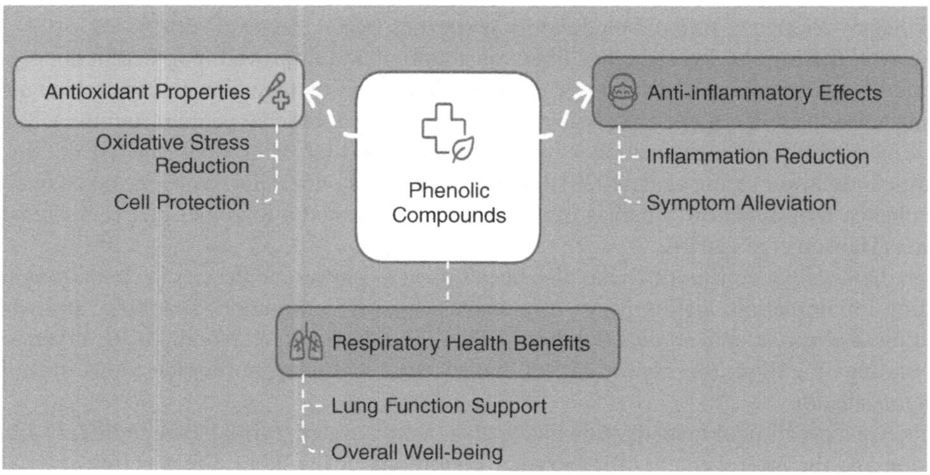


**FIGURE 13.2** Impact of Phenolic compounds on health.

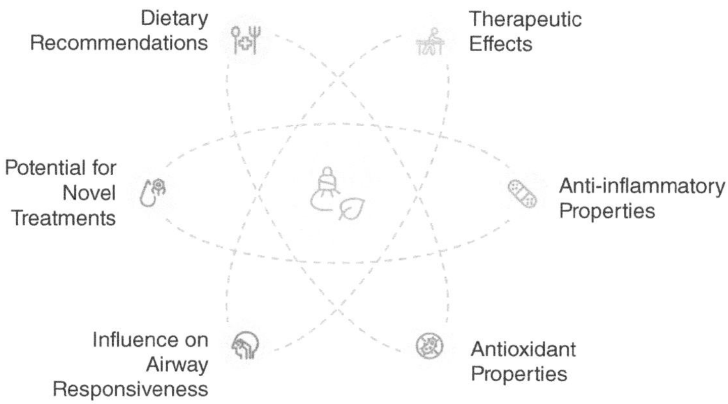


**FIGURE 13.3** Importance of Phenolic compounds in asthma management.

## 13.5 SAFETY AND DOSAGE

Depending on their type, quantity, and duration of exposure, phenolic chemicals, including flavonoids, phthalates, bisphenols, and their derivatives, can have both beneficial and harmful effects on lung conditions like asthma (Emran et al., 2024). Flavonoids and other phenolic compounds, such as apigenin and fisetin, have been studied in animal models at doses ranging from 2 to 20 mg/kg and have been shown to possess anti-inflammatory and anti-asthmatic properties. They also display low toxicity. In one such study, phenolic-rich plant extracts, such as *Mentha longifolia*, were evaluated in acute and subacute oral toxicity trials. In these trials, they exhibited comparatively low toxicity in animal models.

On the other hand, environmental phenols such as parabens, phthalates, and bisphenol A (BPA) are associated with an increased risk of respiratory issues and asthma. Their toxicity is mainly caused by inflammation and endocrine disruption (Tourabi et al., 2024). Further studies on human asthma have revealed that pycnogenol, at doses of 100–200 mg daily, is safe, with only mild adverse effects such as occasional gastrointestinal distress (Tanaka & Takahashi, 2013). There are two environmental phenolics—BPA and phthalates—that may worsen asthma. Because of the potential for endocrine disruption, the EPA has established a reference dose of 0.05 mg/kg/day for BPA and phthalates (Li et al., 2025). There are no confirmed safe exposure limits for phthalates; even minimal amounts might cause respiratory problems, particularly in those with asthma (Ramos et al., 2021). When consumed at normal dietary levels, dietary plant polyphenols are generally well tolerated and may protect against airway inflammation (Franova et al., 2011). Therefore, by modifying inflammatory pathways, reducing oxidative stress, and controlling immunological responses involved in the pathogenesis of the disease, as illustrated in Figure 13.3, phenolic compounds have significant therapeutic potential in regulating and managing asthma.

## 13.6 CONCLUSION

With their potent anti-inflammatory, immunomodulatory, and antioxidant properties, phenolic compounds have considerable potential for promoting lung health and treating diseases such as asthma. Phenolic-rich foods and plant-based therapies may give supplementary benefits that enhance overall quality of life, even as modern medicine continues to provide patients with vital respite. However, the majority of the available data comes from experimental research, and well-planned human trials are needed to translate these results into clinical practice. Phenolic chemicals have the potential to become a crucial component of holistic approaches to respiratory treatment by bridging the gap between traditional knowledge and scientific validation, offering safer, more sustainable, and natural alternatives for the future.

## COMPETING INTERESTS

The authors declare that they have no competing interest.

## DECLARATION OF GENERATIVE AI AND AI-ASSISTED TECHNOLOGIES IN THE WRITING PROCESS

During the preparation of this work, the book chapter author(s) have not taken any assistance from AI, and the book chapter author(s) reviewed and edited the content as needed and take(s) full responsibility for the content of the publication.

## REFERENCES

Abbasi, B. H., Ahmad, N., Fazal, H., & Mahmood, T. (2010). Conventional and modern propagation techniques in *Piper nigrum*. *Journal of Medicinal Plants Research*, *4*(1), 7–12.

Adolphe, J. L., Whiting, S. J., Juurlink, B. H., Thorpe, L. U., & Alcorn, J. (2010). Health effects with consumption of the flax lignan secoisolariciresinol diglucoside. *British Journal of Nutrition*, *103*(7), 929–938.

Aghababaei, F., & Hadidi, M. (2023). Recent advances in potential health benefits of quercetin. *Pharmaceuticals*, *16*(7), 1020.

Araújo, N. P. D. S., de Matos, N. A., Oliveira, M., de Souza, A. B. F., Castro, T. D. F., Machado-Júnior, P. A., … & Bezerra, F. S. (2022). Quercetin improves pulmonary function and prevents emphysema caused by exposure to cigarette smoke in male mice. *Antioxidants*, *11*(2), 181.

Assembly, U. G. (2015). Transforming our world: The 2030 Agenda for Sustainable Development.

Bagdas, D., Gul, Z., Meade, J. A., Cam, B., Cinkilic, N., & Gurun, M. S. (2020). Pharmacologic overview of chlorogenic acid and its metabolites in chronic pain and inflammation. *Current Neuropharmacology*, *18*(3), 216–228.

Bascom, R. (1996). Committee of the Environmental and Occupational Health Assembly of the American Thoracic Society Health effects of outdoor air pollution (part 1 of 2 parts). *American Journal of Respiratory and Critical Care Medicine*, *153*, 3–50.

Bernatoniene, J., & Kopustinskiene, D. M. (2018). The role of catechins in cellular responses to oxidative stress. *Molecules*, *23*(4), 965.

Bisset, L. R., & Schmid-Grendelmeier, P. (2005). Chemokines and their receptors in the pathogenesis of allergic asthma: Progress and perspective. *Current Opinion in Pulmonary Medicine*, *11*(1), 35–42.

Bousquet, J., Jeffery, P. K., Busse, W. W., Johnson, M., & Vignola, A. M. (2000). Asthma: From bronchoconstriction to airways inflammation and remodeling. *American Journal of Respiratory and Critical Care Medicine*, *161*, 1720–1745.

Broomhead, A. J., Rahman, M. M., Dewick, M., Jackson, P., & Lucas, D. E. (1991). Matairesinol as precursor of *Podophyllum lignans*. *Phytochemistry*, *30*(5), 1489–1492.

Brown, A. J., Goldsworthy, S. M., Barnes, A. A., Eilert, M. M., Tcheang, L., Daniels, D., Muir, A. I., Wigglesworth, M. J., Kinghorn, I., Fraser, N. J., et al. (2003). The Orphan G protein-coupled receptors GPR41 and GPR43 are activated by propionate and other short chain carboxylic acids. *Journal of Biological Chemistry*, *278*(13), 11312–11319. doi: 10.1074/jbc.M211609200.

Che, D. N., Cho, B. O., Kim, J. S., Shin, J. Y., Kang, H. J., & Jang, S. I. (2020). Effect of luteolin and apigenin on the production of IL-31 and IL-33 in lipopolysaccharides-activated microglia cells and their mechanism of action. *Nutrients*, *12*(3), 811.

Chen, B. L., Chen, Y. Q., Ma, B. H., Yu, S. F., Li, L. Y., Zeng, Q. X., … & Li, C. W. (2018). Tetrahydrocurcumin, a major metabolite of curcumin, ameliorates allergic airway inflammation by attenuating Th2 response and suppressing the IL-4Rα-Jak1-STAT6 and Jagged1/Jagged2-Notch1/Notch2 pathways in asthmatic mice. *Clinical & Experimental Allergy*, *48*(11), 1494–1508.

Chen, G., Tang, J., Ni, Z., Chen, Q., Li, Z., Yang, W., … & Wang, X. (2015). Antiasthmatic effects of resveratrol in ovalbumin-induced asthma model mice involved in the upregulation of PTEN. *Biological and Pharmaceutical Bulletin*, *38*(4), 507–513.

Choubey, S., Varughese, L. R., Kumar, V., & Beniwal, V. (2015). Medicinal importance of gallic acid and its ester derivatives: A patent review. *Pharmaceutical Patent Analyst*, *4*(4), 305–315.

Covas, M. I., De La Torre, K., Farré-Albaladejo, M. et al. (2006). Postprandial LDL phenolic content and LDL oxidation are modulated by olive oil phenolic compounds in humans. *Free Radical Biology and Medicine*, *40*(4), 608–616.

Cui, J., Lian, Y., Zhao, C., Du, H., Han, Y., Gao, W., Xiao, H., & Zheng, J. 2019. Dietary fibers from fruits and vegetables and their health benefits via modulation of gut microbiota. *Comprehensive Reviews in Food Science and Food Safety*, *18*(5), 1514–1532. doi: 10.1111/1541-4337.12489.

de Alencar Silva, A., Pereira-de-Morais, L., da Silva, R. E. R., de Menezes Dantas, D., Milfont, C. G. B., Gomes, M. F., … & Barbosa, R. (2020). Pharmacological screening of the phenolic compound caffeic acid using rat aorta, uterus and ileum smooth muscle. *Chemico-Biological Interactions*, *332*, 109269.

De Groot, H. (1994). Reactive oxygen species in tissue injury. *Hepato-Gastroenterology*, *41*(4), 328–332.

Devasagayam, T., Tilak, J., Boloor, K., Sane, K. S., Ghaskadbi, S. S., & Lele, R. (2004). Free radicals and antioxidants in human health: Current status and future prospects. *Journal of the Association of Physicians of India*, *52*, 794–804.

Dewick, P. M. (2002). *Medicinal natural products: A biosynthetic approach*. John Wiley & Sons.

Dixon, R. A., & Ferreira, D. (2002). Genistein. *Phytochemistry*, *60*(3), 205–211.

Djukanovic, R., Roche, W., Wilson, J., Beasley, C., Twentyman, O., Howarth, P., & Holgate, S. (1990). Mucosal inflammation in asthma. *American Review of Respiratory Disease*, *142*, 434–457.

Dodge, R. R., & Burrows, B. (1980). The prevalence and incidence of asthma and asthma-like symptoms in a general population sample. *American Review of Respiratory Disease*, *122*(4), 567–575.

Dos Santos, M. D., Almeida, M. C., Lopes, N. P., & De Souza, G. E. P. (2006). Evaluation of the anti-inflammatory, analgesic and antipyretic activities of the natural polyphenol chlorogenic acid. *Biological and Pharmaceutical Bulletin*, *29*(11), 2236–2240.

Dreher, M. L. 2018. Whole fruits and fruit fiber emerging health effects. *Nutrients*, *10*(12), 1833. doi: 10.3390/nu10121833.

El-Taher, E. M., El-Sherei, M. M., El Dine, R. S., El Naggar, D. M., Khalil, W. K., Eshak, M. G., & Kassem, M. E. (2020). Chemical profiling of *Acacia auriculiformis* (*A. Cunn.*) ex Benth. Leaves extract and its impact on gene expression and apoptotic DNA damage in CCl4-induced hepatotoxicity in male rats. *Tropical Journal of Natural Product Research*, *4*(11), 936–944.

Emran, T. B., Eva, T. A., Zehravi, M., Islam, F., Khan, J., Kareemulla, S., … & Kim, B. (2024). Polyphenols as therapeutics in respiratory diseases: Moving from preclinical evidence to potential clinical applications. *International Journal of Biological Sciences*, *20*(8), 3236.

Estrela, J. M., Ortega, A., Mena, S., Rodriguez, M. L., & Asensi, M. (2013). Pterostilbene: Biomedical applications. *Critical Reviews in Clinical Laboratory Sciences*, *50*(3), 65–78.

Ferdous, U. T., & Yusof, Z. N. B. (2021). Insight into potential anticancer activity of algal flavonoids: Current status and challenges. *Molecules*, *26*, 6844. https://doi.org/10.3390/molecules26226844.

Franova, S., Joskova, M., Sutovska, M., Novakova, E., Adamicova, K., Pechanova, O., & Nosalova, G. (2011). The efficiency of polyphenolic compounds on allergen induced hyperreactivity of the airways. *Biomedicine & Preventive Nutrition*, *1*(4), 232–235.

Fuhrman, B., & Aviram, M. (2001). Flavonoids protect LDL from oxidation and attenuate atherosclerosis. *Current Opinion in Lipidology*, *12*(1), 41–48.

GBD 2015 Chronic Respiratory Disease Collaborators. (2017). Global, regional, and national deaths, prevalence, disability-adjusted life years, and years lived with disability for chronic obstructive pulmonary disease and asthma, 1990–2015: A systematic analysis for the Global Burden of Disease Study 2015. *The Lancet. Respiratory Medicine*, *5*(9), 691.

Gutteridge, J. M., & Halliwell, B. (2000). Free radicals and antioxidants in the year 2000. A historical look to the future. *Annals of the New York Academy of Sciences*, *899*, 136–147.

Hanasaki, Y., Ogawa, S., & Fukui, S. (1994). The correlation between active oxygens scavenging and antioxidative effects of flavonoids. *Free Radical Biology and Medicine*, *16*, 845–850.

Hanson, C., Lyden, E., Rennard, S., Mannino, D. M., Rutten, E. P., Hopkins, R., & Young, R. (2016). The relationship between dietary fiber intake and lung function in the national health and nutrition examination surveys. *Annals of the American Thoracic Society*, *13*(5), 643–650. doi: 10.1513/AnnalsATS.201509-609OC.

Harahap, U., Syahputra, R. A., Ahmed, A., Nasution, A., Wisely, W., Sirait, M. L., Dalimunthe, A., Zainalabidin, S., Taslim, N. A., & Nurkolis, F. et al. (2024). Current insights and future perspectives of flavonoids: A promising antihypertensive approach. *Phytotherapy Research*, *38*, 3146–3168.

Hernáez, Á., Remaley, A. T., & Farràs, M. et al. (2015). Olive oil polyphenols decrease LDL concentrations and LDL atherogenicity in men in a randomized controlled trial. *The Journal of Nutrition*, *145*(8), 1692–1697.

Huang, M., Liu, X., Ren, Y., Huang, Q., Shi, Y., Yuan, P., & Chen, M. (2024). Quercetin: A flavonoid with potential for treating acute lung injury. *Drug Design, Development and Therapy*, 5709–5728.

Irwin, R. S., & Curley, F. J. (1990). FrenchCL. Chronic cough. The spectrum and frequency of causes, key components of the diagnostic evaluation and outcome of specific therapy. *American Review of Respiratory Disease*, *141*(3), 640–647.

Isemura, M. (2019). Catechin in human health and disease. *Molecules*, *24*(3), 528.

Jain, N., Mohan, S. C., & Sumathi, S. (2021). Mechanisms of action of flavonoids in the management of diabetes mellitus. *Journal of Drug Delivery & Therapeutics*, *11*.

Jang, Y. O., Lee, S. H., Choi, J. J., Kim, D.-H., Choi, J.-M., Kang, M.-J., Oh, Y.-M., Park, Y.-J., Shin, Y., & Won Lee, S. (2020). Fecal microbial transplantation and a high fiber diet attenuates emphysema development by suppressing inflammation and apoptosis. *Experimental & Molecular Medicine*, *52*(7), 1128–1139. doi: 10.1038/s12276-020-0469-y.

Jeffery, P. K. (2000). Comparison of the structural and inflammatory features of chronic obstructive pulmonary disease and asthma. Giles F. Filley lecture. *Chest*, *117*, 251S–260S.

Kashyap, P., Shikha, D., Thakur, M., & Aneja, A. (2022). Functionality of apigenin as a potent antioxidant with emphasis on bioavailability, metabolism, action mechanism and in vitro and in vivo studies: A review. *Journal of Food Biochemistry*, *46*(4), e13950.

Kelly, G. S. (2011). Quercetin. *Alternative Medicine Review*, *16*(2), 172–194.

Kerry, N., & Abbey, M. (1997). Red wine and fractionated phenolic compounds prepared from red wine inhibit low density lipoprotein oxidation. Atherosclerosis,135(1),93–102 after oxidation

Khan, T., Ali, S., Qayyum, R., Hussain, I., Wahid, F., & Shah, A. J. (2016). Intestinal and vascular smooth muscle relaxant effect of *Viscum album* explains its medicinal use in hyperactive gut disorders and hypertension. *BMC Complementary and Alternative Medicine*, *16*(1), 251.

Kim, B. H., Choi, J. S., Yi, E. H., Lee, J. K., Won, C., Ye, S. K. et al. (2013). Relative antioxidant activities of quercetin and its structurally related substances and their effects on NF-κB/CRE/AP-1 signaling in murine macrophages. *Molecules and Cells*, *35*, 410–420.

Kim, J. E., Park, K. M., Lee, S. Y., Seo, J. H., Yoon, I. S., Bae, C. S., & Park, D. H. (2017). Anti-inflammatory effect of *Allium hookeri* on carrageenan-induced air pouch mouse model. *PLoS One*, *12*(12), e0190305.

Kiss, A. K., & Piwowarski, J. P. (2018). Ellagitannins, gallotannins and their metabolites-the contribution to the anti-inflammatory effect of food products and medicinal plants. *Current Medicinal Chemistry*, *25*(37), 4946–4967.

Korkina, L., & Afanasev, I. (1997). Antioxidant and chelating properties of flavonoids. *Advances in Pharmacology*, *38*, 151–163.

Kwon, M. J., Raut, P. K., Jang, J. H., & Chun, K. S. (2025). Isoliquiritigenin induces apoptosis via ROS-mediated inhibition of p38/mTOR/STAT3 pathway in human melanoma cells. *Biomolecules & Therapeutics*, *33*(2), 378.

Larsen, G. L., & Henson, P. M. (1983). Mediators of inflammation. *Annual Review of Immunology*, *1*(1), 335–359.

Lee, H. Y., Kim, I. K., Yoon, H. K., Kwon, S. S., Rhee, C. K., & Lee, S. Y. (2017). Inhibitory effects of resveratrol on airway remodeling by transforming growth factor-β/Smad signaling pathway in chronic asthma model. *Allergy, Asthma & Immunology Research*, *9*(1), 25–34.

Lee, K. W., Bode, A. M., & Dong, Z. (2011). Molecular targets of phytochemicals for cancer prevention. *Nature Reviews Cancer*, *11*(3), 211–218.

Lee, M., Kim, S., Kwon, O. K., Oh, S. R., Lee, H. K., & Ahn, K. (2009). Anti-inflammatory and anti-asthmatic effects of resveratrol, a polyphenolic stilbene, in a mouse model of allergic asthma. *International Immunopharmacology*, *9*(4), 418–424.

Leopoldini, M., Marino, T., Russo, N., & Toscano, M. (2004). Antioxidant properties of phenolic compounds: H-atom versus electron transfer mechanism. *The Journal of Physical Chemistry A*, *108*(22), 4916–4922.

Leopoldini, M., Russo, N., & Toscano, M. (2011). The molecular basis of working mechanism of natural polyphenolic antioxidants. *Food Chemistry*, *125*(2), 288–306.

Li, R., Zheng, B., Zhang, Y., He, L., Ren, C., Guan, L., … & Zhang, Z. (2025). The impact of phthalates on asthma and chronic obstructive pulmonary disease: A comprehensive analysis based on network toxicology and molecular docking. *Frontiers in Pharmacology*, *16*, 1566965.

Lin, B. W., Gong, C. C., Song, H. F., & Cui, Y. Y. (2017). Effects of anthocyanins on the prevention and treatment of cancer. *British Journal of Pharmacology*, *174*(11), 1226–1243.

Liu, Y., & Shi, G. (2022). Roles of sirtuins in asthma. *Respiratory Research*, *23*(1), 251.

Madureira, M. B., Concato, V. M., Cruz, E. M. S., Bitencourt de Morais, J. M., Inoue, F. S. R., Concimo Santos, N., … & Pavanelli, W. R. (2023). Naringenin and hesperidin as promising alternatives for prevention and co-adjuvant therapy for breast cancer. *Antioxidants*, *12*(3), 586.

Mahmoud, A. M., Hernandez Bautista, R. J., Sandhu, M. A., & Hussein, O. E. (2019). Beneficial effects of citrus flavonoids on cardiovascular and metabolic health. *Oxidative Medicine and Cellular Longevity*, *2019*(1), 5484138.

Mahmud, A. R., Ema, T. I., Siddiquee, M. F. R., Shahriar, A., Ahmed, H., Mosfeq-Ul-Hasan, M., … & Mizan, M. F. R. (2023). Natural flavonols: Actions, mechanisms, and potential therapeutic utility for various diseases. *Beni-Suef University Journal of Basic and Applied Sciences*, *12*(1), 47.

Manandhar, B., Paudel, P., Seong, S. H., Jung, H. A., & Choi, J. S. (2019). Characterizing eckol as a therapeutic aid: A systematic review. *Marine Drugs*, *17*(6), 361.

Markulin, L., Corbin, C., Renouard, S., Drouet, S., Gutierrez, L., Mateljak, I., … & Lainé, E. (2019). Pinoresinol–lariciresinol reductases, key to the lignan synthesis in plants. *Planta*, *249*(6), 1695–1714.

Martinez, F. (1995). Definitions, risk factors, and early natural history. *American Journal of Respiratory and Critical Care Medicine*, *151*, S2–S2.

Meiers, S., Keméņy, M., Weyand, U., Gastpar, R., von Angerer, E., & Marko, D. (2001). The anthocyanidins cyanidin and delphinidin are potent inhibitors of the epidermal growth-factor receptor. *Journal of Agricultural and Food Chemistry*, *49*(2), 958–962.

Mérillon, J. M., Fauconneau, B., Teguo, P. W., Barrier, L., Vercauteren, J., & Huguet, F. (1997). Antioxidant activity of the stilbene astringin, newly extracted from *Vitis vinifera* cell cultures. *Clinical Chemistry*, *43*(6), 1092–1093.

Mohammed, A. A., & Al-Habib, O. A. (2021). Bronchodilator effect of *Crataegus azarolus var aronia* unripped fruit extracts on Rat's tracheal smooth muscle. *Indian Journal of Forensic Medicine & Toxicology*, *15*(4), 827.

Murtaza, G., Karim, S., Akram, M. R., Khan, S. A., Azhar, S., Mumtaz, A., & Bin Asad, M. H. H. (2014). Caffeic acid phenethyl ester and therapeutic potentials. *BioMed Research International*, *2014*(1), 145342.

Oliveira, A. K. D. S., de Oliveira e Silva, A. M., Pereira, R. O., Santos, A. S., Barbosa Junior, E. V., Bezerra, M. T., Barreto, R. S., Quintans-Junior, L. J., & Quintans, J. S. (2022). Antiobesity properties and mechanism of action of flavonoids: A review. *Critical Reviews in Food Science and Nutrition*, *62*(28), 7827–7848.

Özduran, G., Becer, E., & Vatansever, H. S. (2023). The role and mechanisms of action of catechins in neurodegenerative diseases. *Journal of the American Nutrition Association*, *42*(1), 67–74.

Pan, W., Ikeda, K., Takebe, M., & Yamori, Y. (2001). Genistein, daidzein and glycitein inhibit growth and DNA synthesis of aortic smooth muscle cells from stroke-prone spontaneously hypertensive rats. *The Journal of Nutrition*, *131*(4), 1154–1158.

Pang, C., Zhang, J., Zhang, Q., Wu, G., Wang, Y., & Ma, J. (2015). Novel vanillic acid-based poly (ether–ester) s: From synthesis to properties. *Polymer Chemistry*, *6*(5), 797–804.

Park, H. J., Lee, C. M., Jung, I. D., Lee, J. S., Jeong, Y. I., Chang, J. H., … & Park, Y. M. (2009). Quercetin regulates Th1/Th2 balance in a murine model of asthma. *International Immunopharmacology*, *9*(3), 261–267.

Patanè, G. T., Putaggio, S., Tellone, E., Barreca, D., Ficarra, S., Maffei, C., … & Laganà, G. (2023). Catechins and proanthocyanidins involvement in metabolic syndrome. *International Journal of Molecular Sciences*, *24*(11), 9228.

Pauwels, R. (2001). Global initiative for chronic obstructive lung diseases (GOLD): Time to act. *European Respiratory Journal*, *18*(6), 901–902.

Perron, N. R., & Brumaghim, J. L. (2009). A review of the antioxidant mechanisms of polyphenol compounds related to iron binding. *Cell Biochemistry and Biophysics*, *53*(2), 75–100.

Piotrowska, H., Kucinska, M., & Murias, M. (2012). Biological activity of piceatannol: Leaving the shadow of resveratrol. *Mutation Research/Reviews in Mutation Research*, *750*(1), 60–82.

Qiu, M., Ma, K., Zhang, J., Zhao, Z., Wang, S., Wang, Q., & Xu, H. (2024). Isoliquiritigenin as a modulator of the Nrf2 signaling pathway: Potential therapeutic implications. *Frontiers in Pharmacology*, *15*, 1395735.

Quiñones, M., Miguel, M., & Aleixandre, A. (2013). Beneficial effects of polyphenols on cardiovascular disease. *Pharmacological Research*, *68*(1), 125–131.

Rahman, I., & MacNee, W. (1998). Role of transcription factors in inflammatory lung diseases. *Thorax*, *53*, 601–612.

Rajendran, P., Rengarajan, T., Nandakumar, N., Palaniswami, R., Nishigaki, Y., & Nishigaki, I. (2014). Kaempferol, a potential cytostatic and cure for inflammatory disorders. *European Journal of Medicinal Chemistry*, *86*, 103–112.

Ramos, R. L., Moreira, V. R., Lebron, Y. A., Santos, A. V., Santos, L. V., & Amaral, M. C. (2021). Phenolic compounds seasonal occurrence and risk assessment in surface and treated waters in Minas Gerais—Brazil. *Environmental Pollution*, *268*, 115782.

Ratajczak, W., Rył, A., Mizerski, A., Walczakiewicz, K., Sipak, O., & Laszczynska, M. (2019). Immunomodulatory potential of gut micro- biome-derived short-chain fatty acids (SCFAs). *Acta Biochimica Polonica*, *66*, 1–12.

Reddel, H. K., FitzGerald, J. M., Bateman, E. D., Bacharier, L. B., Becker, A., Brusselle, G., … & Boulet, L. P. (2019). GINA 2019: A fundamental change in asthma management: Treatment of asthma with short-acting bronchodilators alone is no longer recommended for adults and adolescents. *European Respiratory Journal*, *53*(6).

Remsberg, C. M., Martinez, S. E., Akinwumi, B. C., Anderson, H. D., Takemoto, J. K., Sayre, C. L., & Davies, N. M. (2015). Preclinical pharmacokinetics and pharmacodynamics and content analysis of gnetol in foodstuffs. *Phytotherapy Research*, *29*(8), 1168–1179.

Rennard, S. I. (1999). Inflammation and repair processes in chronic obstructive pulmonary disease. *American Journal of Respiratory and Critical Care Medicine*, *160*, S12–S16.

Ruan, Q., Wen, C., Jin, G., Yuan, Z., Yang, X., Wen, Z., … & Bai, Y. (2023). Phloretin-induced STAT3 inhibition suppresses pancreatic cancer growth and progression via enhancing Nrf2 activity. *Phytomedicine*, *118*, 154990.

Saetta, M., Turato, G., Maestrelli, P., Mapp, C. E., & Fabbri, L. M. (2001). Cellular and structural bases of chronic obstructive pulmonary disease. *American Journal of Respiratory and Critical Care Medicine*, *163*, 1304–1309.

Salehi, B., Venditti, A., Sharifi-Rad, M., Kręgiel, D., Sharifi-Rad, J., Durazzo, A., & Martins, N. (2019). The therapeutic potential of apigenin. *International Journal of Molecular Sciences*, *20*(6), 1305.

Santangelo, C., Varì, R., Scazzocchio, B., Di Benedetto, R., Filesi, C., & Masella, R. (2007). Polyphenols, intracellular signalling and inflammation. *Annali-istituto Superiore di Sanita 43*(4):394.

Santhakumar, A. B., Battino, M., & Alvarez-Suarez, J. M. (2018). Dietary polyphenols: Structures, bioavailability and protective effects against atherosclerosis. *Food and Chemical Toxicology*, *113*, 49–65.

Saulite, L., Jekabsons, K., Klavins, M., Muceniece, R., & Riekstina, U. (2019). Effects of malvidin, cyanidin and delphinidin on human adipose mesenchymal stem cell differentiation into adipocytes, chondrocytes and osteocytes. *Phytomedicine*, *53*, 86–95.

Shan, L., Kang, X., Liu, F., Cai, X., Han, X., & Shang, Y. (2018). Epigallocatechin gallate improves airway inflammation through TGF-β1 signaling pathway in asthmatic mice. *Molecular Medicine Reports*, *18*(2), 2088–2096.

Shi, Z., Jiang, W., Chen, X., Xu, M., Wang, J., Lai, Y., & Zha, D. (2020). Chlorogenic acid ameliorated allergic rhinitis-related symptoms in mice by regulating Th17 cells. *Bioscience Reports*, *40*(11), BSR20201643.

Soriano, J. B., Kendrick, P. J., Paulson, K. R., Gupta, V., Abrams, E. M., Adedoyin, R. A., … & Moradi, M. (2020). Prevalence and attributable health burden of chronic respiratory diseases, 1990–2017: A systematic analysis for the Global Burden of Disease Study 2017. *The Lancet Respiratory Medicine*, *8*(6), 585–596.

Stintzing, F. C., Stintzing, A. S., Carle, R., Frei, B., & Wrolstad, R. E. (2002). Color and antioxidant properties of cyanidin-based anthocyanin pigments. *Journal of Agricultural and Food Chemistry*, *50*(21), 6172–6181.

Sur, R., Nigam, A., Grote, D., Liebel, F., & Southall, M. D. (2008). Avenanthramides, polyphenols from oats, exhibit anti-inflammatory and anti-itch activity. *Archives of Dermatological Research*, *300*(10), 569–574.

Syamlal, G. (2020). Medical expenditures attributed to asthma and chronic obstructive pulmonary disease among workers—United States, 2011–2015. *MMWR. Morbidity and Mortality Weekly Report*, *69*.

Tanaka, T., & Takahashi, R. (2013). Flavonoids and asthma. *Nutrients*, *5*(6), 2128–2143.

Teeter, J. G., & Bleecker, E. R. (1998). Relationship between airway obstruction and respiratory symptoms in adult asthmatics. *Chest*, *113*(2), 272–277.

Teichtahl, H., Buckmaster, N., & Pertnikovs, E. (1997). The incidence of respiratory tract infection in adults requiring hospitalization for asthma. *Chest*, *112*(3), 591–596.

Thangasamy, T., Sittadjody, S., & Burd, R. (2009). Quercetin: A potential complementary and alternative cancer therapy. *Complementary and Alternative Therapies and the Aging Population*, 563–584.

Tipoe, G. L., Leung, T.-M., Hung, M.-W., & Fung, M.-L. (2007). Green tea polyphenols as an antioxidant and anti-inflammatory agent for cardiovascular protection. *Cardiovascular and Haematological Disorders – Drug Targets*, *7*(2), 135–144.

Tourabi, M., Ghouizi, A. E., Nouioura, G., Faiz, K., Elfatemi, H., El-Yagoubi, K., … & Derwich, E. (2024). Phenolic profile, acute and subacute oral toxicity of the aqueous extract from Moroccan *Mentha longifolia* L. aerial part in Swiss Albino mice model. *Journal of Ethnopharmacology*, *319*, 117293.

Trompette, A., Gollwitzer, E. S., Yadava, K., Sichelstiel, A. K., Sprenger, N., Ngom-Bru, C., Blanchard, C., Junt, T., Nicod, L. P., Harris, N. L., et al. 2014. Gut microbiota metabolism of dietary fiber influences allergic airway disease and hematopoiesis. *Nature Medicine*, *20*(2), 159–166. doi: 10.1038/nm.3444.

Tuli, H. S., Rath, P., Chauhan, A., Ramniwas, S., Vashishth, K., Varol, M., … & Sak, K. (2022). Phloretin, as a potent anticancer compound: From chemistry to cellular interactions. *Molecules*, *27*(24), 8819.

Vattem, D. A., & Shetty, K. (2005). Biological functionality of ellagic acid: A review. *Journal of Food Biochemistry*, *29*(3), 234–266.

Vauzour, D., Rodriguez-Mateos, A., Corona, G., OrunaConcha, M. J., & Spencer, J. P. (2010). Polyphenols and human health: Prevention of disease and mechanisms of action. *Nutrients*, *2*(11), 1106–1131.

Vaya, J., & Aviram, M. (2001). Nutritional antioxidants mechanisms of action, analyses of activities and medical applications. *Current Medicinal Chemistry Immunology, Endocrine and Metabolic Agents*, *1*(1), 99–117.

Vos, T., Lim, S. S., Abbafati, C., Abbas, K. M., Abbasi, M., Abbasifard, M., … & Bhutta, Z. A. (2020). Global burden of 369 diseases and injuries in 204 countries and territories, 1990–2019: A systematic analysis for the Global Burden of Disease Study 2019. *The Lancet*, *396*(10258), 1204–1222.

Walle, T. (2011). Bioavailability of resveratrol. *Annals of the New York Academy of Sciences*, *1215*(1), 9–15.

Wang, W., Sun, C., Mao, L., Ma, P., Liu, F., Yang, J., & Gao, Y. (2016). The biological activities, chemical stability, metabolism and delivery systems of quercetin: A review. *Trends in Food Science & Technology*, *56*, 21–38.

Wani, S. A., Iqbal, H., & Basir, S. F. (2017). Mechanism of flavonoids action in smooth muscle relaxation. *WJPPS*, *6*, 514–550.

Wen, K., Fang, X., Yang, J., Yao, Y., Nandakumar, K. S., Salem, M. L., & Cheng, K. (2021). Recent research on flavonoids and their biomedical applications. *Current Medicinal Chemistry*, *28*, 1042–1066.

Weng, Z., Zhang, B., Asadi, S., Sismanopoulos, N., Butcher, A., Fu, X., … & Theoharides, T. C. (2012). Quercetin is more effective than cromolyn in blocking human mast cell cytokine release and inhibits contact dermatitis and photosensitivity in humans. *PLoS One*, *7*(3), e33805.

Williams, L. M., Scott, H. A., & Wood, L. G. (2019). Soluble fibre as a treatment for inflammation in asthma. *Journal of Nutrition & Intermediary Metabolism*, *18*(100108), 100108. doi: 10.1016/j.jnim.2019.100108.

World Health Organization. (2007). Global surveillance, prevention and control of chronic respiratory diseases: a comprehensive approach.

World Health Organization. (2018). Towards a global action plan for healthy lives and well-being for all: Uniting to accelerate progress towards the health-related SDGs. In *Towards a global action plan for healthy lives and well-being for all: Uniting to accelerate progress towards the health-related SDGs*.

World Health Organization. (2025). *World health statistics 2025: Monitoring health for the SDGs, sustainable development goals*. World Health Organization.

Wu, H., Feng, Y., Xie, X., & Zhang, Y. (2025). Glycitin in soy beans: A mine of their structures, functions, and potential applications. *Foods*. *14*(17), 2940

Xiang, Z., Ma, B., Pei, X., Wang, W., & Gong, W. (2025). Mechanism of action of genistein on breast cancer and differential effects of different age stages. *Pharmaceutical Biology*, *63*(1), 141–155.

Yang, J. J., Yu, D., Xiang, Y. B., Blot, W., White, E., Robien, K., Sinha, R., Park, Y., Takata, Y., Lazovich, D., et al. 2020. Association of dietary fiber and yogurt consumption with lung cancer risk: A pooled analysis. *JAMA Oncology*, *6*, e194107. doi: 10.1001/jamaoncol.2019.4107

Yang, L., Wang, D., Zhang, Z., Jiang, Y., & Liu, Y. (2022). Isoliquiritigenin alleviates diabetic symptoms via activating AMPK and inhibiting mTORC1 signaling in diet-induced diabetic mice. *Phytomedicine*, *98*, 153950.

Yeh, S. Y., & Schwartzstein, R. (2009). Asthma: Pathophysiology and diagnosis. In *Asthma, health and society: A public health perspective* (pp. 19–42).

Yoon, J.-H., & Baek, S. J. (2005). Molecular targets of dietary polyphenols with anti-inflammatory properties. *Yonsei Medical Journal*, *46*(5), 585–596.

Zhang, J., Guo, J., Qian, Y., Yu, L., Ma, J., Gu, B., … & Wu, W. (2025). Quercetin induces apoptosis through downregulating P4HA2 and inhibiting the PI3K/Akt/mTOR axis in hepatocellular carcinoma cells: An in vitro study. *Cancer Reports*, *8*(5), e70220.

Zou, J.-G., Wang, Z.-R., Huang, Y.-Z., Cao, K.-J., & Wu, J. M. (2003). Effect of red wine and wine polyphenol resveratrol on endothelial function in hypercholesterolemic rabbits. *International Journal of Molecular Medicine*, *11*(3), 317–320.

# 14 Pharmacokinetics and Bioavailability of Phenolic Compounds

*Monica Butnariu**

## 14.1 INTRODUCTION

Phenolic compounds (vegetable phenolic substances) are chemical compounds essential for maintaining health. They are particularly known for their antioxidant characteristics, but their benefits are much more numerous. Structurally, vegetable phenolic substances are segregated into different classes, the most significant of which are flavonoids, phenolic acids, stilbenes, and tannins. Phenolic acids are a cluster of phytochemicals from plant-based foods that exhibit antioxidant, anti-inflammatory, prebiotic, and anticancer characteristics. They can help reduce the risk of chronic diseases, which is of interest to both the food industry and the pharmaceutical industry [Moazzen et al., 2022]. Investigations on the vegetable phenolic substance outline of cereal bran have revealed that ferulic acid is one of the most significant substances, being closely accompanied by lignin and polysaccharides in cell walls. Other significant substances involve caffeic acid, p-coumaric acid, p-hydroxybenzoic acid, syringic acid, and vanillic acid.

The reduced bioavailability of vegetable phenolic substances such as curcumin obstructs its use as a therapeutic agent. However, literature data have reported the existence of some chemical agents that can positively influence (piperine) its absorption and metabolism [Durazzo et al., 2019].

Like curcumin, (–)-epigallocatechin-3-gallate (EGCG) is a polyphenol that is well known for its beneficial effects on the body, being the main active ingredient in green tea.

Vegetable phenolic substances are a category of organic substances very widespread in nature, differentiated predominantly by the existence of several phenolic-category –OH (hydroxyl) substitutes. Vegetable phenolic substances are very abundant in vegetable organisms and comprise diverse structural categories. The most significant categories of vegetable phenolic substances are phenolic acids, flavonoids, and lignans. Vegetable phenolic substances are segregated into three subclasses for organic substances: flavonoids, phenolic acids, and vegetable phenolic substances. Vegetable phenolic substances, also known as bioflavonoids, construct a very substantial cluster of natural substances. Like most vegetable phenolic substances, flavonoids are found predominantly in fruits and vegetable organisms. They have several functions: antioxidant effect, defending against harmful insects and fungi, and are responsible for plant colors [Singla et al., 2019]. The most significant vegetable phenolic substances are the categories of substances from Table 14.1.

Vegetable phenolic substances can be supplied to the body with the help of food supplements.

Vegetable phenolic substances emblematize the substantial cluster of natural vegetable phenolic substances.

They are predominantly found in bark, flowers, fruits, leaves, seeds, stems, and other parts of vegetable organisms. Professionals speak about their anticancer potentialities by interfering with the work capacity of enzymes involved in the neovascularization mechanisms. Vegetable phenolic substances are accompanied by estrogen biosynthesis in women, obstruct infections, and have antiallergic

* Corresponding author: monicabutnariu@yahoo.com

 DOI: 10.1201/9781003527671-14

**TABLE 14.1**
**Categories of Vegetable Phenolic Substances**

| | | |
|---|---|---|
| Anthocyanidins | They are found in vegetable organisms with a pink – dark purple or even marine blue color. They play a protective role in the body. | Günther et al. (2020) |
| Isoflavones | They are found, among others, in legumes and soybeans. They regulate the hormonal background and obstruct sebum secretion. | Nakai, Fujita, and Kamei, (2020) |
| Flavonols | Have antioxidant and anti-inflammatory characteristics, found in black grapes, hawthorn, onions, walnuts, or red wine. | Wang et al. (2021) |
| Catechins | Found in tea, cocoa, fruits, and vegetables (especially fruit skins). They have antibacterial, antioxidant, antiviral, cancer obstruction, anti-inflammatory, antifungal, anticoagulant characteristics. | Katz, Doughty, and Ali, (2011) |
| Flavanones | They lower cholesterol, have anti-atherosclerotic characteristics, protect the skin, and obstruct ulcers. The highest concentration is found in citrus fruits. | Barreca et al. (2017) |
| Bioflavonoids | They have many health benefits. They support the functions of the immune system, have antioxidant characteristics, and regulate cholesterol levels. In addition, flavonoids have a diuretic effect and seal blood vessels, obstructing the formation of varicose veins and atherosclerosis. | Al-Aubaidy et al. (2021) |

characteristics. Choleretic, hepatoprotective, and antiulcer characteristics differentiate vegetable phenolic substances. Phenolic acids are also part of the polyphenol cluster. They are also known as carboxylic acids. This cluster of vegetable phenolic substances is primarily composed of tannins (chemical substances used in medicine). Phenolic acids are found in many plant species. Significant quantities can be found in dried fruits. Tannins are known for their anti-inflammatory, astringent, and hemostatic characteristics. Like all vegetable phenolic substances, they also have a strong antioxidant effect [Pais et al., 2024]. A substantial cluster of biologically active substances is phenolics. They constitute secondary/specialized plant metabolites containing benzene rings with one or more –OH (hydroxyl) groups, ranging from simple phenolic compounds to very complex arrangements.

Vegetable phenolic substances have the strongest antioxidant and antiradical characteristics of all secondary metabolites, as they react with a wide range of free radicals, including –OH (hydroxyl) anions, superoxide, and other organic and non-organic radicals.

Vegetable phenolic substances may also enhance the action of other antioxidants, including fat-soluble vitamins and low molecular weight water-soluble substances. The nutritional value of fruits depends on the quality, chemical composition, and quantity of constituent substances, which determine their therapeutic and prophylactic effect on the human body. Biologically active substances from seasonal fruits are well known and widely applied in the cosmetic, food, nutraceutical, and medicine industries.

The main appreciated substances are phenolics (anthocyanins, leucoanthocyanins, catechins, flavonols, etc.), vitamins, and minerals [Hasnat et al., 2024].

These seasonally grown vegetable raw materials are carried in various forms, including fresh, preserved (such as jams), dried, frozen, as infusions, soft drinks, and concentrated extracts. According to some research, different techniques can be employed to analyze the content of polyphenolic substances in experimental specimens. These techniques are summarized in the following: spectroscopic procedures (NIR and UV absorbance by Folin-Ciocalteu index, polyphenolic index, etc.); chromatographic procedures (gas chromatography, high-performance liquid chromatography, ultra-high-performance liquid chromatography (UHPLC)); and other analytical possibilities (network-attached storage (NAS) and electronic language) [Gamella et al., 2006].

The most accurate procedure is UHPLC, while the spectroscopic procedure is uncomplicated but less selective [Ajila et al., 2011]. Bioavailability emblematizes the ratio between the quantity of a vegetable phenolic substance and the speed with which it is released and absorbed in the body, which exceeds the site of action and demonstrates its biological consequence. Bioavailability is stipulated as the part (percentage) of a prescribed quantity of vegetable phenolic substances that exceeds the circulatory network unchanged (systemic circulation). In all situations of using a vegetable phenolic substance, it is compulsory that active compounds like polyphenolic compounds, known as the "active pharmaceutical compound" (APC), can reach the body. To possess a therapeutic consequence, nevertheless, it is not sufficient for the effective compound to enter the body [Karakaya, 2004]. The effective compound must be accessible in the correct quantity in the body where it can have an effect. This is what the field of the body is designated as the "*target area*". Also, the effective compound must reach the target field inside an unquestionable time frame and continue to exist there for an unquestionable duration. In the appearance of intravenous (IV) infusion inoculation, i.e., when the inoculation is prescribed directly into the circulatory system, the bioavailability is 100%. After inoculation, the effective compound exceeds the target field after traveling a complex route through the circulatory network [Muller et al., 2019]. To assess bioavailability, blood specimens are aggregated, and the agglomeration of the effective compound in the blood (systemic circulation) is established.

Consequently, the bioavailability will be 100% immediately after the inoculation, because the vegetable phenolic compounds are absorbed directly into the blood.

Consequently, if 75 mg of the vegetable phenolic compounds are injected into the circulatory network, 100% bioavailability corresponds to the quantity of 75 mg of polyphenolic compound. When the effective compound/vegetable phenolic substances circulate in the blood, some of them will be metabolized or eliminated by excretion; consequently, the agglomeration of polyphenolic substances in the body will decrease over time [Zeb, 2021].

The bioavailability outline is assessed and differentiated from that of other vegetable phenolic substances by observing the area under the curve, which emblematizes the total body absorption of the effective compound. The moment when the culminating agglomeration of the effective compound is found in the blood is designated $T_{max}$, and the maximum agglomeration of the effective compound found in the blood is designated $C_{max}$.

If the identical effective compound indicated is prescribed by another route, for example, as an oral lozenge (swallowed), the bioavailability will be less than 100%. Oral bioavailability represents the percentage of the effective compound that is absorbed after swallowing a lozenge, measured over 15 hours. ASC is hatched [Liu et al., 2016].

$T_{max}$ emblematizes the moment at which the culminating agglomeration of the phenolic compound is found in the blood, and $C_{max}$ emblematizes the maximum agglomeration of the vegetable phenolic substances found in the blood [de Beer et al., 2023]. Lower bioavailability of the oral route of administration differentiated from IV infusion inoculation.

After swallowing a vegetable phenolic substance or a lozenge, it reaches the stomach in 1–2 minutes. In the stomach, the vegetable phenolic substances or lozenge dissolve, and part of the effective compound is absorbed into the circulatory network.

The constituents are passed on to the small intestine once assimilation is complete. Assimilation from the gastrointestinal network can vary significantly. Low bioavailability may result from reduced or no assimilation in the stomach and intestines; therefore, this stage is a significant one that can significantly influence availability [Dantas et al., 2023].

Once the effective compound is absorbed, it first exceeds the hepatic portal vein and is transmitted to the liver. This is the first metabolism of the effective compound in the liver, known as "first-pass metabolism". In this first pass, some effective compounds are metabolized to a greater extent than others. The non-metabolized part of the effective compound, which will ordinarily be less than 100%, will reach the systemic circulation through the medium of the hepatic vein.

The quantity of effective compound that exceeds the systemic circulation is designated as "absolute bioavailability".

Absolute bioavailability differentiates the bioavailability of APC in the systemic circulation following non-IV infusion administration from the bioavailability of the identical vegetable phenolic substances prescribed IV infusion [Catteuw et al., 2019].

This emblematizes the quantity of APC absorbed by administration other than IV infusion, expressed as a percentage of the quantity absorbed from the identical vegetable phenolic substances prescribed IV infusion.

In other words, the standard transferred for absolute bioavailability is IV infusion administration.

Relative bioavailability estimates the bioavailability of one assemblage (A) of a given vegetable phenolic substance, differentiated from that of another assemblage (B) of the identical vegetable phenolic substances – usually a standard set different from IV administration, i.e., prescribed by a route other than IV infusion [Arce-Reynoso et al., 2023].

Bioavailability is also affected by other individual ingredients. See the attached data sheet for some examples of bioavailability. Bioequivalence is the interconnection between two assemblages of identical vegetable phenolic substances, with identical construction of presentation in identical dosing, which have indistinguishable bioavailability.

Relative bioavailability is not only used for comparing different formulas but also when a comparison is required among two polyphenolic substances (or any other polyphenolic compounds with the same formula) having the same effective compound manufactured by different pharmaceutical organizations. Vegetable phenolic compound A is a generic vegetable phenolic substance, unlike the reference lozenge from compound B (vegetable phenolic substances marketed under a brand name). To determine whether there is bioequivalence between vegetable phenolic substance A and vegetable phenolic substance B, the bioavailability rates of the two are differentiated [Manach et al., 2005].

The –OH (hydroxyl) substances are structurally related to water, given the fact that in their molecular arrangement, an organic radical (alkyl or aryl) appears instead of a hydrogen atom in the water molecule. As in water, the oxygen atom is $sp^3$ hybridized, resulting in a tetrahedral geometry.

The roles of the four hybrid orbitals of oxygen are as follows:

- One orbital intertwines with the s orbital of the hydrogen atom, forming the O–H bond;
- the second $sp^3$ orbital interpenetrates with an $sp^3$ or $sp^2$ orbital of the saturated or aromatic carbon, to which the –OH cluster is bound;
- The other two $sp^3$ hybrid orbitals of oxygen are occupied by a pair of non-participating electrons [Zhang et al., 2016].

The characteristic characteristics of –OH (hydroxyl) substances are due to the existence in their molecules of the heteroatom (in this case, oxygen) with high electronegativity.

Consequently, in alcohols, the C–O and O–H bonds are polar, the $\sigma$ electrons being moved toward the oxygen atom, which makes the electron density for both the hydrogen atom and the alkyl residue to be low, while the oxygen atom is negatively polarized.

The existence of a high electron density at the oxygen atom, respectively, a low one at the hydrogen atom has two significant consequences: the appearance of electrostatic attractions among the positive hydrogen of an –OH cluster and the electronegative oxygen atom of the –OH (hydroxyl) cluster from another molecule of alcohol; consequently, in liquid and solid state, alcohols are accompanying by hydrogen bonds, existing in the construction of molecular associations $(ROH)_n$ formed by several n molecules that vary permanently [Biela et al., 2023].

They have the character of weak acids, being able to give up the proton of the –OH (hydroxyl) cluster in the existence of alkali metals and, in the appearance of phenols, of alkali hydroxides.

Alkyl substitutes in alcohols, through the +I electron donor consequence, increase the electron density at the oxygen atom (at the identical moment giving alcohols a weak basicity – they can fix a proton – see, for example, the reaction of removing water in an acidic environment), strengthening the O–H bond and consequently reducing the acidity of alcohols compared to water [Breniaux et al., 2018].

Therefore, alcohols are weaker acids than water, having a pKa between 16 and 19 ($Ka = 10^{-16}$–$10^{-19}$), differentiated from water, whose pKa is 15.6.

In the appearance of vegetable phenolic substances, the situation is different due to the existence of π electrons of the aromatic nucleus. Consequently, in the phenol molecule, conjugation takes place among the p electrons (non-participating) of oxygen and the π electrons of the aromatic nucleus, because of which the electron density at the nucleus increases while the electron density at the oxygen atom decreases. As a result, on the one hand, the attraction of the oxygen atom for the σ electrons in the O–H bond increases, its stability decreases, which favors the easier release of hydrogen in the construction of $H^+$; on the other hand, the aromatic nucleus of phenols, enriched in electrons, reacts more trippingly with electrophilic reactants (E+) than an unsubstituted benzene [Godos et al., 2024].

In conclusion, phenols are stronger acids than water and alcohols (having the pKa of $8^{-10}$) and are reactive substrates in SEAr (aromatic electrophilic substitution mechanism) reactions. The existence of attractive substituents or electron donors in the aromatic nucleus will bring about changes in the acidity of the respective phenol.

Consequently, an electro-attracting cluster, by decreasing the electron density at the oxygen atom, will increase the acidity. In contrast, an electron-donating cluster will have the opposite consequence, decreasing the ability to donate the proton.

The existence of the OH functional cluster and the hydrocarbon residue establishes the water solubility of alcohols. If the –OH (hydroxyl) cluster confers solubility in water and in polar solvents, the remaining hydrocarbon, due to its hydrophobic character, increases the insolubility in the identical solvents (water and polar solvents), but allows the solubilization of alcohols in non-polar solvents (hydrocarbons, halogenated derivatives, ethers) [Alinaghi et al., 2023].

The existence of the benzene nucleus, non-polar and strongly hydrophobic, makes the water solubility of these –OH (hydroxyl) derivatives low. Solubility can be improved by increasing the temperature (at approximately 65°C, phenol is miscible with water in any proportion – because at this temperature, phenol is liquid). The existence of a substantial number of –OH (hydroxyl) substitutes also increases the solubility of phenols – vegetable phenolic substances are more soluble in water than monohydroxy phenols [Furmanek et al., 2024].

## 14.2 MICROENCAPSULATION OF BIOLOGICALLY ACTIVE VEGETABLE PHENOLIC SUBSTANCES

With the advent of suitable biodegradable polymers, microencapsulation allowed the development of directed launching arrangements. These thorough arrangements allow for monitoring the speed, course, and administration of vegetable phenolic substances.

With these arrangements, microparticles sensitive to the biological environment are designed to yield an effective phenolic compound in an unquestionable field of the body.

One of the main advantages of these arrangements is to protect vegetable phenolic substances from sudden environmental changes (pH) and to enhance compliance by reducing the need for medical attention [Charoenprasert & Mitchell, 2012].

Microencapsulation can be thought of as a technological process in which vegetable phenolic substances in solid, liquid, or gaseous states of small sizes are surrounded by individual polymer shells.

In a broader sense, microencapsulation can be stipulated as a process by which individual entities of solid, liquid, or gas are embedded in a polymeric material so that the incorporated compounds are released at an unquestionable rate and/or under unquestionable conditions.

Microencapsulated vegetable phenolic substances are helpful in various fields in this way, achieving a series of objectives such as slow, prolonged launching of the effective compound; easier handling of some products; simplifying the problems related to the assemblage by separating the constituents that could interact; protecting trippingly degradable materials from the action of environmental ingredients; the transition of a liquid into a solid state; camouflage the disagreeable taste of some compounds; the possibility of creating multi-particulate arrangements by encapsulating the constituents [Bergesse et al., 2023].

Incorporating bioactive compounds, such as antioxidants, bioactive peptides, probiotics, and vitamins, into food products is a straightforward way to develop new products that may offer physiological benefits or reduce disease risk. Developing foods by adding bioactive compounds, however, induces many technological challenges. Sometimes, nevertheless, the compounds or ingredients incorporated in a food can negatively influence its sensory characteristics or change because of processing ingredients.

As a result, to successfully incorporate various constituents into a food network, encapsulation matrices with a complex arrangement must be designed, which ensure physical stability, defend the ingredients against chemical degradation, and allow control over the launching of the encapsulated compounds during chewing and digestion to maximize assimilation.

The encapsulation is a complex process through which the inclusion and isolation of solid, liquid, or gaseous compounds in closed, well-stipulated spaces, covered with various protective materials, and their launch under controlled conditions are sought [Paulo & Santos, 2021].

Microencapsulation offers food semi-cooked (processors), a method to defend susceptible constituents, obstruct nutritional losses, efficiently use unstable ingredients that are launching at the right moment, preserve flavor and aroma, or transform liquids into solid ingredients that are trippingly handled.

The results from the process can be categorized into categories conformable to their morphology: mononuclear, polynuclear, or matrix type.

Microencapsulation ensures:

- protection of ingredients and bioactive constituents against the aggression of physical–chemical and biological ingredients (oxygen, light, humidity, temperature, pH, enzymatic degradation);
- avoiding the mixing of constituents that can react and change the sensory characteristics of food;
- solubilization of compounds hardly soluble in polar or non-polar solvents;
- the slow and directed launching of the ingredients, either by destroying the protective wall (mechanical action, dissolution, melting, biodegradation) or by the diffusion of molecules through the protective wall;
- camouflage the disagreeable taste and smell of some constituents;
- the transition of some constituents from a liquid state to a solid state in powder construction;
- Extending the life of bioactive constituents and maintaining their nutritional capacity [Ždero Pavlović et al., 2023].

Just as many alcohols are found in nature, there are also many natural phenols (especially vegetable phenolic compounds and their byproducts), widespread in the vegetable kingdom, often responsible for the specific smell or color of the respective vegetable organisms and flowers. Some of them are transferred as food flavors.

## 14.3 CALCULATION MODELS FOR THE PHENOLIC COMPOUND'S DISTRIBUTION

The FDA's Critical Path Initiative has emphasized that efforts are greatly needed to enhance the speed of bringing new medications to patients with serious illnesses. For example, since 1995, when the BCS (*Biopharmaceutics Classification System*) was inaugurated, it has a significant consequence on the regulatory process and application of vegetable phenolic compounds [Truzzi et al., 2021].

For an oral construction with immediate launching in the effective quantity, the speed and degree of assimilation are established by its solubility in water and permeability in the free gastrointestinal tract. That is why the BCS emblematizes a new and powerful representation for bioequivalence investigations based on physiological parameters and the physicochemical characteristics of the drug molecule.

BCS, in the construction appropriated by the WHO (*World Health Organization*), categorizes the medicinal compounds recorded in the *essential medicines list* (EML) according to their solubilization and permeability characteristics in four different categories [Davit et al., 2016].

Therefore, unquestionable categories of drugs can be considered for biowaiver, i.e., approval of products based on *in vitro* dissolution tests of vegetable phenolic substances instead of bioequivalence data on humans, a costly task for medication manufacturers.

Such representations have significantly enhanced the speed and decreased the cost of bringing vegetable phenolic substances to market through oral administration.

The BCS network enables *in vivo* bioequivalence testing of immediate-release solid pharmaceutical formulations for category 1 medications, while waivers for category three medications are only registered based on knowledge from research.

The categorization of vegetable phenolic substances conformable to BCS needs acquaintance from solubility and acquaintance from permeability [Papich & Martinez, 2015].

The estimation of the permeability of vegetable phenolic substances is ordinarily established on well-stipulated experimental data of permeability. This acquaintance is only accessible for a small fraction of the medications recorded in the EML [Scott, Cox, & Njardarson, 2022].

A biowaiver representation can fluently be required for oral effective constructions with immediate launching (with a launching greater than or equal to 85% in 30 minutes), which contain medication compounds with an enhanced solubility over a pH range between 1 and 7.5 (quantity/solubility ratio smaller than 250 mL) and an enhanced permeability (absorbed fraction greater than or equal to 90%) if the excipients transferred in the assemblage of the product do not intervene in the assimilation strategy of the medicine.

The compound with a narrow therapeutic index and the medications created for oral administration cannot be examined for a biowaiver.

Nevertheless, the basic idea of the BCS categorization network is to predict the *in vivo* pharmacokinetic work capacity of the pharmaceutical product depending on the solubilization and permeability characteristics *in vitro* (of the medication) [Hofsäss & Dressman, 2019].

Recently, 167 medicinal compounds/vegetable phenolic substances categorized with BCS were extensively examined, and it was observed that the existence of food, assimilation, transporters, efflux transporters, and elimination routes (renal/biliary) were significant factors in assimilation and general bioavailability for pharmaceutical formulations administered orally with immediate release.

Nevertheless, they suggested that the categorization of molecules conformable to metabolism is less ambiguous than that conformable to permeability and assimilation [Fagerholm, 2007].

This categorization may also increase the number of Category 1 medications that become viable for biowaivering.

The *Biopharmaceutics Drug Disposition Classification System* (BDDCS), like the BCS, proposes categorizing medication molecules into four categories, defining the broad metabolic criterion as greater than or equal to 50% (approximately 10%) of the metabolism of an oral quantity *in vivo* (in humans).

Established on this criterion, some medications that were initially in BCS category 1 were reclassified as BDDCS category 3 and, therefore, no longer viable for biowaivering [Charalabidis et al., 2019].

One study looked at least eight BDDCS category 1 medications that were viable for bioremoval.

Considering the overall significance, the BDDCS approach can be helpful in the successful categorization of vegetable phenolic substances in category 1, consequently increasing the viability for biowaivering.

A challenge for both BCS and BDDCS is the categorization of vegetable phenolic substances in a manner conforming to the *in vitro* data required for metabolism, solubilization, or permeability. Nevertheless, considerable research has been conducted in the last decade on in silico or computational procedures for estimating assimilation, distribution, metabolism, and excretion (ADME) [Darwich et al., 2010].

The study aimed to facilitate an uncomplicated and rapid categorization of BDDCS by utilizing enhanced computational representations to predict the BDDCS category based on the characteristics of the molecules.

The calculation representations were programmed around a dataset related to 165 medications, which served as a reference point and, in turn, was established on BDDCS data. The calculation representations were programmed around a dataset related to 165 medications, which served as a reference point and, in turn, was established on BDDCS data. To test and verify these calculation representations, another 56 medications recorded in WHO/ELM were added that were not categorized in the BDDCS lists but for which the specialized literature provides sufficient data to allow their categorization [Heimbach et al., 2013].

*Definition of solubility*: Data related to the solubility of vegetable phenolic substances compulsory for categorization were obtained from standard references and are expressed in mg/mL. If solubility data were not accessible or unclear, benchmarks were ensured by data on maximum quantities from the WHO Essential Medicines core list, expressed in mg [Bustos et al., 2018].

*Calculation of the amount of quantities*: The number of quantity ($D_0$) was deliberated to utilize the formula $D0=(M_0/V_0)/C_s$, where $M_0$ is the maximum quantity (mg), Cs is the solubility (mg/mL), and $V_0$=250 mL [Khan et al., 2018].

## 14.4 VALIDATION CRITERIA

The three criteria, producer accuracy, consumer accuracy, and absolute accuracy, indicate the efficiency of the estimation of a calculation representation. Percent accuracy is the percentage of all vegetable phenolic substances correctly recorded. Absolute accuracy is the percentage of all vegetable phenolic substances in a category that are rightly estimated to be in that category. The occurrence of category I (false positive) and category II (false negative) errors is indicated, depending on the correct belonging category [Huang et al., 2002].

Consumer accuracy is formed and established on absolute accuracy, continues to "penalize" category I error but overlooks category II errors and mitigates the consequence of such errors to defend the consumer's interest. Analogous to the consumer's accuracy, the producer's accuracy tries to consider its interest in evaluating an estimative representation.

Consumer accuracy insists on avoiding "false positives", while producer accuracy emphasizes avoiding "false negatives". Each validation criterion partially diminishes the importance of the error category on which it is not focused. For any set of estimated data, it is expected that the accuracy of the consumer differs from that of the producer, as the rates of category I and category II errors differ. The accuracy of the consumer and the accuracy of the producer exceed the absolute accuracy, except for the appearance, where the category I errors are equal to the category II ones when the three criteria are identical. In general, a medication manufacturer and a consumer will have differing needs regarding the accuracy of estimates for BDDCS categorizations [DiNicolantonio, McCarty, & O'Keefe, 2022].

The best representation of support vector machines (SVM) still presents a lower accuracy for the consumer on the test set than that of the producer, consequently increasing the risk of incorrect categorizations from the consumer's point of view [Patil et al., 2023].

## 14.5 THE CONCEPT OF BIOAVAILABILITY

The postulation of bioavailability is more precisely stipulated as the speed and extent of assimilation of a phenolic compound from the pharmaceutical formulation into the systemic circulation. It is known that the therapeutic effect of a medication is determined by its distribution in plasma or blood. The importance of bioavailability in medication therapy lies in the fact that the speed and extent of absorption of this medication can affect the pharmacological response of the patient.

It was examined that the therapeutic performance is ensured only by the existence of the declared quantity of vegetable phenolic substances in the medicinal product. Nevertheless, it was found that therapeutic failure is due to variations in the bioavailability of the medication, being rather harmonized with some sick person variability [Anand et al., 2007].

Several ingredients can cause variations in the bioavailability of a medication, which can be harmonized with the individual or the pharmaceutical formulation.

The ingredients related to the pharmaceutical construction and which can produce big differences in the bioavailability of solid vegetable phenolic substances for oral use can be ingredients related to the vegetable phenolic substances (physical, physicochemical, or chemical characteristics), ingredients related to the nature, quantity, and reactivity of the excipients transferred, or technological ingredients.

Obviously, the administration of a vegetable phenolic substance in various pharmaceutical constructions (granules, capsules, lozenges, coated) will lead to variations in terms of the size and speed of assimilation. The realization of vegetable phenolic substances with a reproducible bioavailability is related to the estimation of the quantity and speed of assimilation of vegetable phenolic substances from the pharmaceutical construction [Stockley et al., 2012].

The equal bioavailability of two pharmaceutical products, assessed under standardized conditions in an identical cluster of healthy volunteers, defines bioequivalence.

The estimation of bioequivalence is made for generic products in comparison with the inventor's products to ensure the identical biological quality of both varieties.

The existence of bioequivalence in some assemblages, unlike manufacturers, which contain identical chemical substances that induce pharmacological work capacity, allows for methods to be examined interchangeably because it is expected to produce identical therapeutic consequences [Tatipamula & Kukavica, 2021].

The kinetic (speed) strategies that take place when the vegetable phenolic substances are unquestionably absorbed in the body are stipulated by unquestionable specific parameters established because of the pharmacokinetic analysis.

The analysis is based on determining the plasma agglomerations of the vegetable phenolic substances or their effective metabolites, depending on the time elapsed since administration.

The parameters obtained from the analysis can be transferred later for the pharmacokinetic description of the compound, for comparing the quality of certain pharmaceutical products, or for establishing a treatment scheme or individualizing it in accordance with the physiological peculiarities of the body [An et al., 2020].

In 1992, the FDA established clear guidelines for discovering new chiral vegetable phenolic substances, which were later applied in other countries, including European ones, Canada, Australia, and Japan. These new regulations may lead, in the moment, to the disappearance of racemic vegetable phenolic substances. In this context, the development of stereoselective procedures for synthesizing chiral molecules with sometimes particularly complex arrangements and efficient procedures for separating new or old chiral vegetable phenolic substances is of great significance. The second objective of the present work was to investigate and develop biocatalytic procedures for the resolution of enantiomers of bioactive substances. Bioavailability emblematizes the extent to which an ingested nutrient is absorbed, assimilated by body tissues, and transferred for its specific functions [Matlin et al., 1988].

## 14.6 BIOSYNTHESIS AND CLASSIFICATION

The PAL (phenylalanine ammonia-lyase) catalyzes the release of ammonia from phenylalanine and leads to the formation of a carbon–carbon (C=C) double bond, with the formation of cinnamic acid.

In some vegetable organisms and herbs, tyrosine is transformed into 4-hydroxycinnamic acid by the action of TAL (tyrosine ammonia-lyase) [Shiragaki et al., 2020].

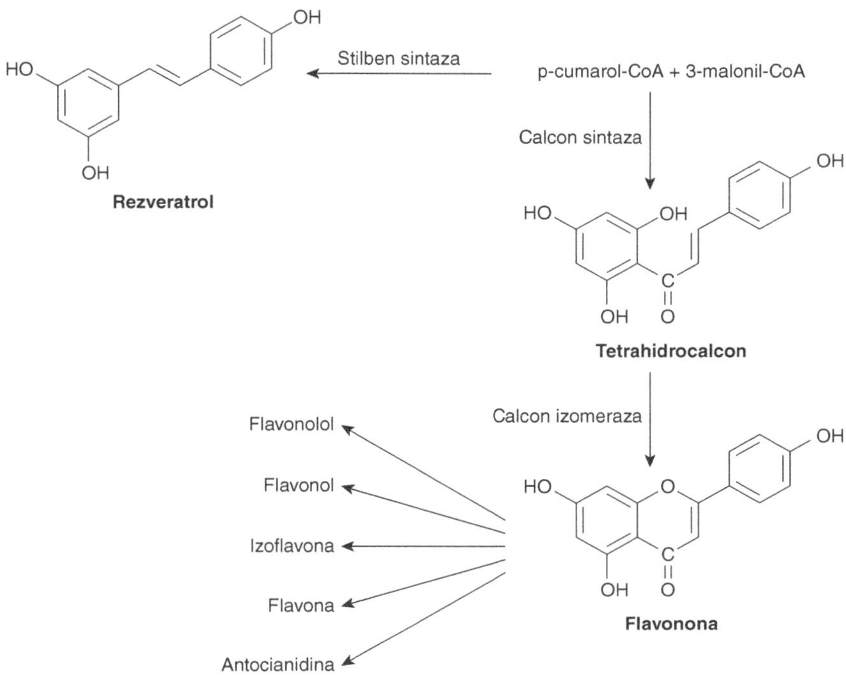


**FIGURE 14.1** Conversion to Flavanones (Cyclization).

The introduction of a –OH (hydroxyl) cluster in the para location in the phenolic ring of cinnamic acid is catalyzed by cytochrome P 450 monooxygenase as an oxygen binding site.

The p-coumaric acid that is consequently formed can be further hydroxylated at positions 3 and 5 by hydroxylases. It can also be feasibly methylated through the o-methyl transferase pathway, using S-adenosylmethionine as a methyl donor, which leads to the formation of caffeic, ferulic, and sinapic acids (Figure 14.1).

All these molecules contain a phenyl ring in the C3 (C6) position and a side chain; consequently, they are designated phenylpropanes, serving as precursors for the synthesis of lignins and many other molecules.

The benzoic acid byproducts are manufactured by the loss of two carbon fragments from phenylpropane. Salicylic acid is a byproduct of benzoic acid, which acts as a signal compound. After infection or UV radiation, the content of salicylic acid increases in many plant species, which can induce the biosynthesis of defense molecules [Zhang & Liu, 2015].

Fragments like the phenylpropane series, the hydroxylation, and possible methylation of hydroxybenzoic acid lead to the constitution of dihydroxybenzoic, vanillic, syringic, and gallic acids. Hydroxybenzoic acids are frequently present in food products; they are often constituents of complex arrangements, such as hydrolyzable tannins and lignins.

They are also present in the construction of organic acids and carbohydrate byproducts. Conventionally, propylpropanes (e.g., the cinnamic acid class) and benzoic acid byproducts are collectively referred to as phenolic acids in the references [Mouafon et al., 2023].

The decarboxylation of benzoic acid and phenylpropane byproducts leads to the constitution of uncomplicated phenols. Thermal degradation of lignin or microbial metamorphosis can also produce uncomplicated phenols in food products. Consequently, substituted vinyl-phenols can be manufactured by decarboxylation of hydroxycinnamic acids.

Nevertheless, the reaction of 4-vinylguaiacol with oxygen leads to the formation of vanillin. Several uncomplicated phenols, such as phenols, o-cresol, 4-ethylphenol, guaiacol, 4-vinylguaiacol, and eugenol, are found in food products of plant origin. Coumarins are lactones of cis-o-hydroxycinnamic acid byproducts, which are also present in unquestionable plant foods and exist free or as glycosides [Tao et al., 2023].

The most significant of these are uncomplicated coumarins, furanocoumarins (also designated psoralens), and pyranocoumarins, which are present both in free construction and in glycosylated construction in food products. This cluster of molecules acts as phytoalexins in response to infections. It is worth noting that most investigations on coumarins have been conducted on *Umbelliferae* and the *Rataceae* family. The existence of psoralens in the various parts of these vegetable organisms was confirmed.

Unlike flavonoids, flavones, isoflavones, and anthocyanidins are formed by the condensation of a phenylpropane (C6–C3) with the participation of three molecules of malonyl coenzyme A, which leads to the constitution of chalcones that are subsequently cyclized [Mlakić et al., 2023], in an acidic environment (Figure 14.1). Consequently, flavonoids have the basic arrangement of diphenylpropane (C6–C3–C6), with unlike oxidation states at the central pyran nucleus. This is also true for stilbenes, but in this case, after the introduction of the second phenyl moiety, one carbon atom of the phenylpropane is removed. Stilbenes are potential fungicides in vegetable organisms (e.g., tocopherol in grape wine). In the appearance of flavonoids and isoflavonoids, depending on the substitution configuration and the unsaturation of the molecules, flavones, flavonones, flavonols, and flavononols as well as flavan-3-ols and related molecules can be formed.

Surrounded by flavonoids, anthocyanins, and catechins known as flavans due to the lack of the carbonyl cluster in location 3, the following are significant: flavan-3-ols, 3-flavans, and 4-diols belonging to this category [Teng et al., 2007].

Anthocyanins are on the border between glycosides and anthocyanidins (Figure 14.2) and are present in many flowers and fruits.

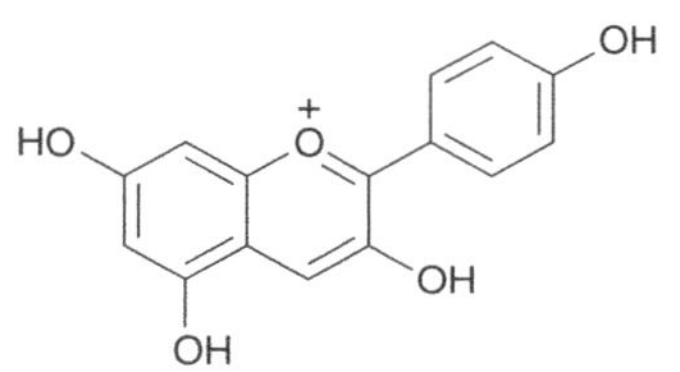

**Pelargonidin**

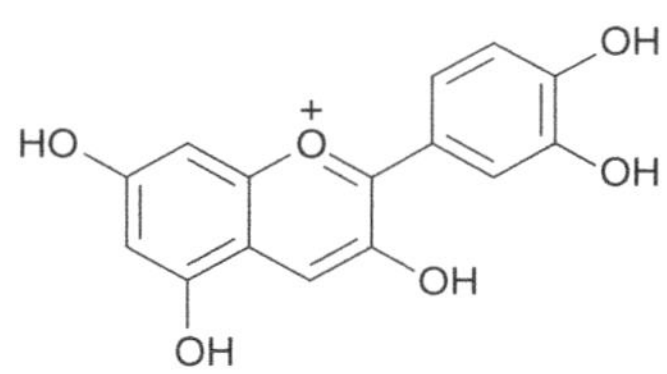

**Cianidin**

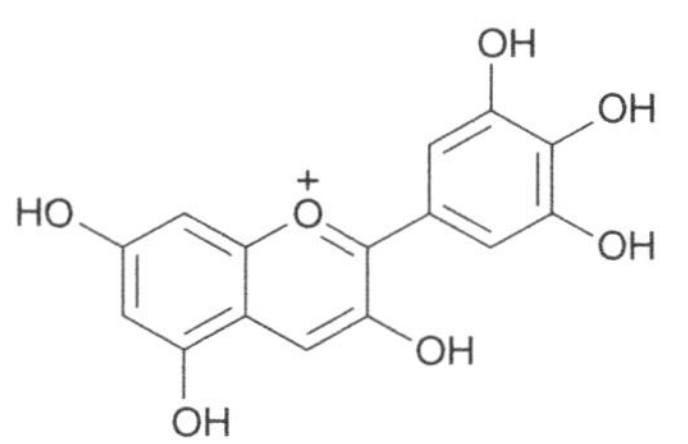

**Delfinidin**

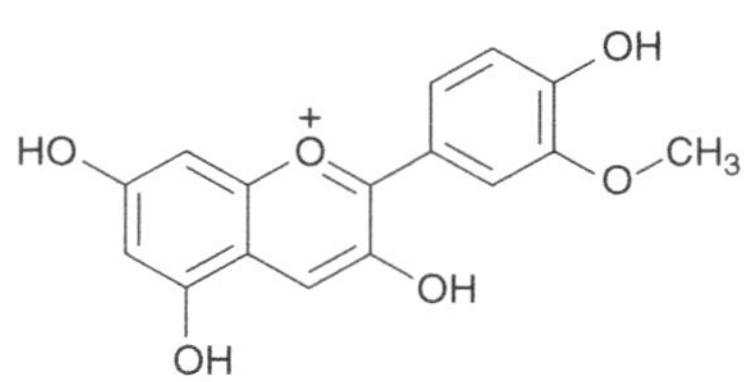

**Peonidin**

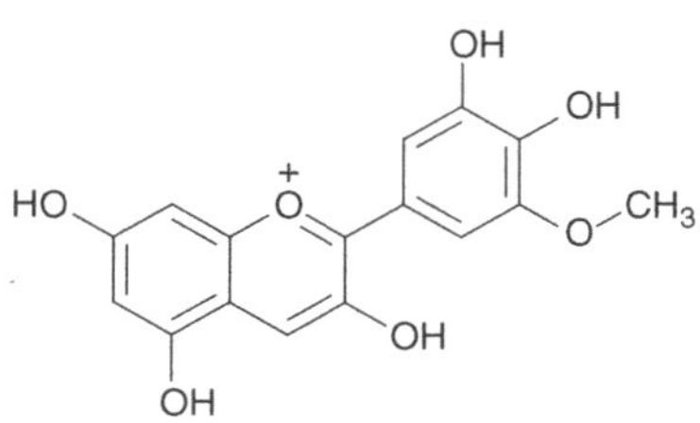

**Petunidin**

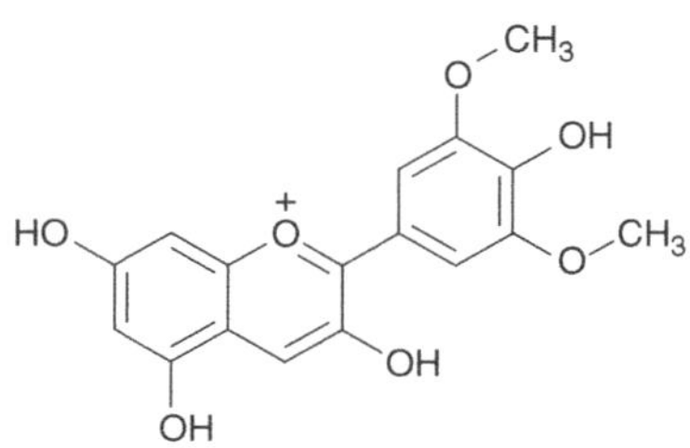

**Malvidin**

**FIGURE 14.2** The anthocyanins are on the border between glycosides and anthocyanidins, mentioned are as follows: Pelargonidin, Cianidin, Delfinidin, Peonidin, Petunidin and Malvidin.

Chalcones and flavones are yellow; anthocyanins are water-soluble pigments responsible for the red, blue, and purple colors of flowers, fruits, and other foods.

Consequently, the red color of radishes, the red color of potato skins, and the dark color of egg-plants are due to the existence of anthocyanins. Gooseberries, raspberries, blueberries, cherries, raisins, red grapes, and plums contain anthocyanins. Almost 200 kinds of anthocyanins have been recognized in vegetable organisms. The anthocyanins extracted from geranium contain pelargonin as a chromophore (color compound) [Asad et al., 2023].

The introduction of one or more –OH (hydroxyl) substitutes in the 3′ and 5′ location in the phenyl residue and their successive methylation leads to the constitution of five anthocyanins, unlike in color. Most anthocyanins are present as monoglycosides and digicosides of pelargonidin, cyanidin, peonidin, delphinidin, petunidin, and malvidin. At the same time, anthocyanins have different colors when subjected to pH variations in solutions.

Furthermore, catechins and epicatechins, found in various vegetable organisms, in substantial quantities in green tea leaves, are indistinguishable in arrangement to anthocyanidins; nevertheless, they are colorless. Tannin is a collective term used to describe a variety of vegetable organisms containing vegetable phenolic substances transferred in the tanning of raw hides. This cluster of molecules involves oligomers and polymers as constituents. Tannins are vegetable phenolic substances capable of precipitating proteins from aqueous solutions and are widely distributed in vegetable organisms. They are found in substantial quantities, especially in the bark of some trees, such as oak wood, and in the gall bladder [Bravo, 1998].

Tannins also construct complex structures with an unquestionable category of polysaccharides, nucleic acids, and alkaloids.

Depending on their arrangements, tannins are classified as hydrolyzed (Figure 14.3) or condensed (Figure 14.4), also known as proanthocyanidins.

Condensed tannins are oligomers and polymers of flavonoids, especially 3 flavan-oils, while hydrolyzed tannins are glycosylated with gallic acids. Many of these gallic acids are bound to sugar molecules. The molecular weight of hydrolyzable tannins ranges from 500 to 2800 Da or more. Based on their hydrolysis products, hydrolyzable tannins involve gallotannins and ellagitannins.

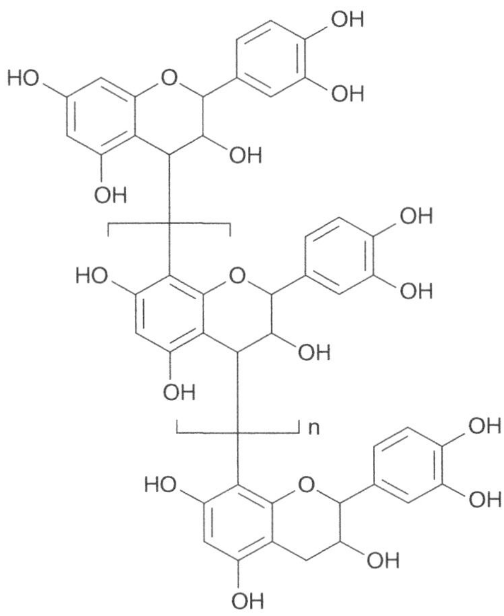


**FIGURE 14.3** The hydrolyzed tannins are glycosylated with gallic acids.

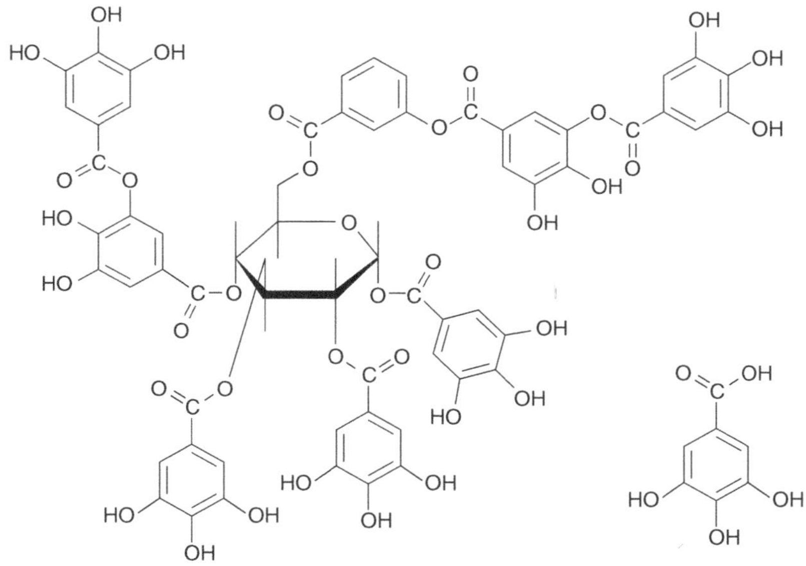

**FIGURE 14.4** Condensed tannins are oligomers and polymers of flavonoids, especially 3 flavanols.

Condensed tannins involve dimers, oligomers, and polymers of 3-flavan-ols. About 50 procyanidins wandering from dimers to hexamers have been recognized. Established on the hydroxylation pattern at the A and B rings, proanthocyanidins can be segregated into procyanidins, propelargonidins, and prodelphinidins. Successive components of condensed tannins are linked by interflavonoid obligations between C4 and C8 or C6. The molecular weight of proanthocyanidins isolated from fruits varies between 2000 and 4000 Da. The phenolic substitutes in tannins are very closely linked with NH– substitutes of peptides and proteins and consequently obstruct their hydrolysis and digestion in the stomach, making tannins anti-nutritional in nature [Dixon, Xie, & Sharma, 2005].

In the tanning strategy, the tannins bind the collagen in the animal fur, so the leather manufactured can resist the degrading attack of microorganisms. Because they aggressively inactivate enzymes, tannins also defend vegetable organisms against attack by microorganisms.

### 14.6.1 Enzymatic Tests

Enzymatic procedures are established on the potentiality of tannins to form intricate complexes with enzymes, consequently interfering with their work capacity. Vegetable phenolic substances can affect enzymes either by forming insoluble protein–phenolic complexes and decreasing the enzyme agglomeration in the mixture reaction or by forming inactive protein–phenolic complexes. Consequently, the competitive and noncompetitive kinetic reaction, as well as the mixed kinetics for enzyme inhibition with vegetable phenolic substances, must be examined.

It has been suggested that the results of the enzyme assay are difficult to interpret because the interconnection between enzyme work capacity and the constitution of insoluble complexes is not fully understood [Zhou et al., 2022].

Later, the inhibition of β-galactosidase work capacity was stilted with the help of polyphenolic substances. They illustrated that the trans 3-(4-hydroxyphenyl) cluster, common to 4′-hydroxyflavone acids and p-hydroxyphenols, is significant for the inhibition of enzyme work capacity. They found that the addition of tannic acid to a standard trypsin solution leads to the formation of an insoluble complex. Nevertheless, trypsin remaining in solution has an enhanced rate of autolysis. The conformational changes of the enzyme brought about by the constitution of insoluble complexes with tannic acid may be responsible. Nevertheless, the phenomenon was not perceived when excess vegetable phenolic substances were added to the network containing trypsin and tannic acid. An enzymatic test was developed for the quantification of tannins.

The test utilizes the potential of tannins to form complex structures with more than one protein and is established to determine the work capacity of alkaline phosphatase activities.

The tannins in this test are sandwiched between two phenolic molecules and two alkaline phosphatases. A linear interconnection between the agglomeration of tannins and the work capacity of the bound alkaline phosphatase was discovered. The procedure involves immobilizing vegetable phenolic substances on polystyrene microplates.

The quantity of immobilized vegetable phenolic substances is small, differentiated from the quantity of added tannin [Velmurugan et al., 2014].

Unbound tannins are removed, and the microplates are incubated with an alkaline phosphatase solution. As a result, unbound enzymes are removed, and the bound alkaline phosphatase work capacity is established utilizing p-nitrophenyl phosphate as substrate.

Taking into account the noteworthiness of flavonoids both for the organoleptic characteristics and for human health, as well as the reactivity of vegetable phenolic substances and the role played by these metamorphosis about the organoleptic characteristics of food products, during processing, storage, and aging, the specialized literature offers a glimpse overview of the findings regarding the semisynthesis, reactivity, and enzymatic oxidation of some flavonoids and the mechanisms complicated in some processes as well as the chemical arrangement of the resulting products [Lu et al., 2021].

The free radical elimination work capacity of some synthesized molecules is also presented in detail in the specialized literature, as well as discussions regarding the chemical arrangement–work capacity interconnection. In the semisynthesis of changed flavonols, the starting material was (+)-catechin, while two molecules without a hydroxy cluster at the C3 atom, namely, chrysin and luteolin, as well as three flavonols with a hydroxy cluster at C3, presenting various hydroxylation patterns on the ring lor B, namely, kaempferol, quercetin, and myricetin, were transferred for enzymatic oxidation.

The first research, regarding the *synthesis of modified byproducts of (+)-catechin*, aimed at the synthesis of new monomeric units derived from catechin with the objective of exploring the consequence of the substitution product of the A ring vis-à-vis the biological characteristics.

For this study, six 8-substituted byproducts of flavan-3-ols (3–8) besides taxifolin (9) were synthesized, and their antioxidant work capacity was investigated [Sarmiento et al., 2018].

Figure 14.5 shows the byproduct monomers changed by flavan-3-ol, and the synthesis routes are revealed in Figure 14.6.

It can be perceived that the stilted molecules were synthesized in their benzylated constructions (11,12,14–16,23). Their free constructions were assured by deprotection in MeOH/THF medium in a hydrogen atmosphere and in the presence of the palladium catalyst.

Molecule 11 was prepared by the action of trifluoroacetic anhydride on tetrabenzylated catechin 10, following a Friedel-Crafts reaction on the C8 nucleophilic location.

Further reduction with $NaBH_4$ led to the constitution of molecule 12. Molecule 14 was assured starting from pentabenzylated catechin 13. After the formylation reaction conformable to the classic Vilsmeier reaction, the assured molecule 14 was decreased with $LiAlH_4$, giving rise to the hydroxymethylated byproduct 15. Further reduction of the latter generated the target molecule 16 in good yield [Hayes et al., 2006]. The brominated adduct 17 was obtained from 13 by the action of N-bromosuccinimide. Quantities of taxifolin (9) of the order of grams were prepared from (+)-catechin through reactions involving oxidation processes conformable to the procedure described in detail in the references [Jiang et al., 2021].

The arrangements of these changed catechin byproducts were established by assimilation spectroscopy in the UV (ultraviolet) spectral range, mass spectroscopy ("Mass Spectrometry" – MS), and nuclear magnetic resonance spectroscopy ("Nuclear Magnetic Resonance" – NMR).

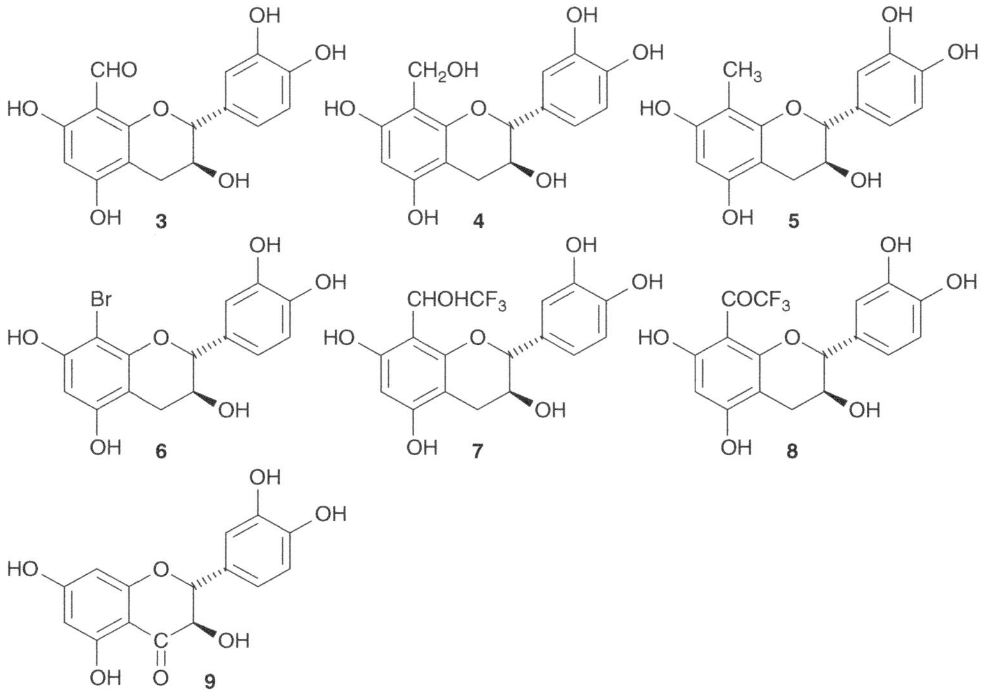

**FIGURE 14.5** The byproduct monomers changed by flavan-3-ol.

Arrangement clarification of molecule 16 is revealed as an example. The mass spectrum with "electro-spray" ionization ("Electro-Spray Ionization Mass Spectroscopy" – ESI–MS) recorded in positive-ion mode shows signals located at m/z 754, 771, and 776 corresponding, respectively, to the ions $[M+H]^+$, $[M+NH4]^+$, and $[M+Na]^+$ and indicating a molecular mass of 754 atomic mass units (aum), in accordance with the arrangement molecule 16 [Elshorbagy et al., 2024].

It is also perceived that the characteristic ADR fragmentation of flavan-3-ols was at m/z 423 $[M+H-332]^+$, which corresponds to the protonated cluster A (Figure 14.7).

The remaining interrogation that needed to be solved was related to the location of the methyl cluster on the A ring of the flavonol. This is the most frequently encountered difficulty in the structural description of flavonols. Since locations 6 and 8 are almost magnetically equivalent, they cannot be distinguished based on NMR chemical shift.

In the present appearance, if the substitution takes place in the more nucleophilic locations of the flavonol skeleton, i.e., 6 or 8, as confirmed by ESI–MS spectrometry, the estimation of the residual proton ($H_6$ or $H_8$) cannot be made establish only on the displacement its chemistry, instead, the use of two-dimensional NMR analysis (2D-NMR) is required. The location of the $CH_3$ cluster on the aromatic ring A was annotated by the long-range carbon–proton correlation established by 2D NMR HMBC (heteronuclear multiple-bond correlation spectroscopy) experiments by the following reasoning. The usual $H_4$ protons on the pyran ring [(2.1 ppm, dd, J= 16.7 and 6.0 Hz; 4.0 ppm) (dd, J= 17.1 and 8.9 Hz)], $H_3$ (6.7 ppm, m), and $H_2$ (8.0 ppm, d, J= 9.0 Hz) were trippingly allocated by 1H-NMR analysis. The three protons on ring B were perceived to be between 7.0 and 7.1 ppm [García Marrero et al., 2014].

For the aromatic ring A, only the signal of a proton, appearing as a singlet at 6.3 ppm, was present, indicating a monosubstitution. The existence of the methyl cluster was confirmed by 1H NMR analysis, indicating a singlet at 2.2 ppm. The chemical shifts of the protonated carbons were allocated by HSQC NMR analysis. The decisive clarification of the arrangement of molecule

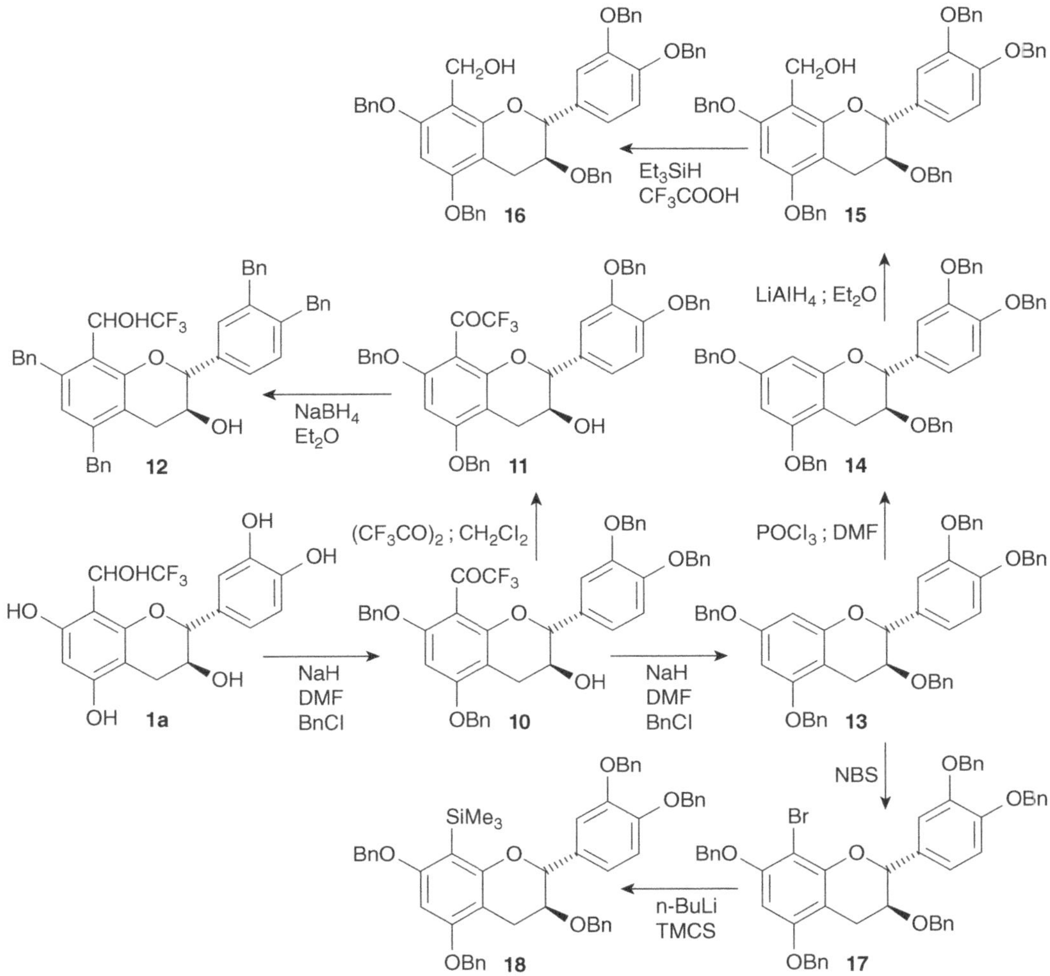


**FIGURE 14.6** The synthesis routes of byproduct monomers changed by flavan-3-ol.

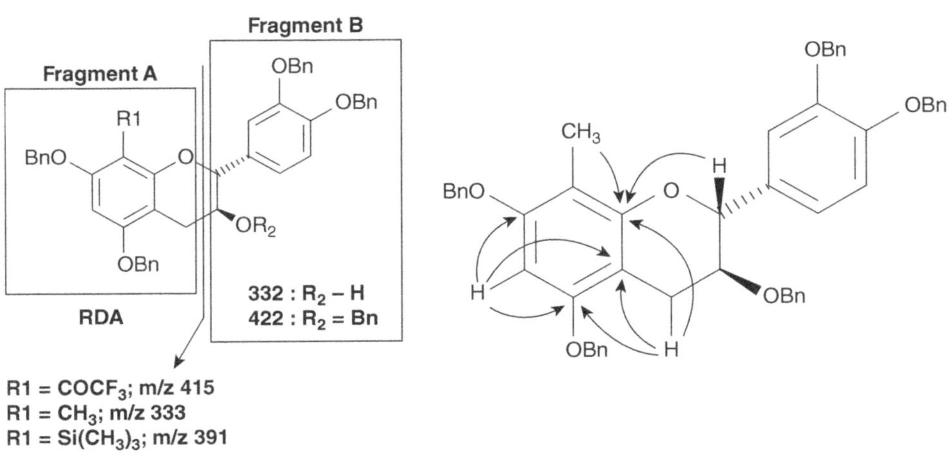


**FIGURE 14.7** The characteristic ADR fragmentation of flavan-3-ols at m/z 423 [M+H-332]+, corresponds to the protonated cluster A.

16 was achieved experimentally by HMBC, establishing the exact location of all carbon and hydrogen atoms. Furthermore, to the correlations with $C_2$ (79.8 ppm) and with $C_3$ (74.0 ppm), the $H_4$ protons (2.9 and 2.9 ppm) correlate with three carbon atoms at 101.8, 152.9, and 155.0 ppm. Carbon atoms $C_{4a}$, $C_8$, and $C_5$ are in appreciative locations to generate such correlations. The signal perceived at 103.0 ppm was attributed to carbon $C_{4a}$ due to its chemical location differentiated from $C_{8a}$ and $C_5$, which are linked by an oxygen atom.

The carbon signal located at 152.8 ppm was also harmonized with $H_2$ bound to $C_{8a}$, and the residual signal, observed at 155.1 ppm, was assigned to the $C_5$ atom. The signal attributed to the carbon $C_{8a}$ did not show any correlation with the residual protons of the aromatic nucleus A, which is, therefore, $H_6$ [Pérez-Navarro et al., 2019].

This was also confirmed by the existence of the correlation between the $C_5$ carbon and the residual proton of the aromatic nucleus and between the protons of the methyl cluster and the $C_8$ carbon. Consequently, location 8 of the methyl cluster on ring A was illustrated.

### 14.6.2 The Epigallocatechin Gallate

The epigallocatechin gallate (EGCG) is a polyphenol contained in unquestionable vegetable organisms and foods (especially in *Camellia sinensis* green tea) with numerous pharmacological characteristics, among which we note the antioxidant, antibacterial, antimutagenic, antiresorptive, and finally, antitumor ones. Starting from the observation that tumor cells are more susceptible to the action of EGCG compared to intact ones, these vegetable phenolic substances have been successfully transferred in the obstruction as well as in the treatment of many categories of cancer [Singh, Shankar, & Srivastava, 2011].

Moreover, *in vitro* investigations have illustrated additive or synergistic interactions between EGCG and other chemotherapeutics such as erlotinib, capecitabine, docetaxel, and gemcitabine. Interference with cellular mechanisms is predominantly through the pro- or antioxidant character of EGCG, depending on the quantity.

The annihilation of reactive oxygen or nitrogen species and interaction with bivalent metals such as Fe or Cu validate its antioxidant behavior, but in the presence of Fe or Cu, EGCG can undergo oxidation processes of the phenolic substitutes in its location, generating reactive oxygen species [Wei et al., 2024].

### 14.6.3 The Curcumin

The curcumin is the effective molecule in turmeric, extracted from the root of the *Curcuma longa* plant, and is a safe, affordable, and effective agent. Curcumin is a lipophilic polyphenol, insoluble in water, but stable in gastric acid. Investigations have illustrated curcumin's rapid metabolism but also limited the systemic potentiality of bioavailability due to rapid elimination from the body. Curcumin is the only antioxidant that contains a variety of functional substitutes, including β-diketone substitutes, C=C double bonds, and phenyl substitutes containing varying quantities of –OH (hydroxyl) and methoxy substituents [Kocaadam & Şanlier, 2017].

Curcumin exhibits anticancer characteristics due to distinct chemical features that allow its interaction with numerous extracellular and intracellular molecules, consequently intervening with cancer progression. The administration of curcumin illustrated the normalization of lipid peroxidation but also of the changes made in the tissues of Wistar guinea pigs. An adjuvant consequence of curcumin with piperine and quercetin has been perceived in the treatment of hyperlipidemia. The antioxidant property of curcumin was illustrated by oral administration of an agglomeration of 90 mg of curcumin or in the construction of a placebo, inside a 2-hour interval, before and after physical exertion. After curcumin administration, an attenuation of exercise-induced oxidative stress was perceived by increasing the antioxidant capacity of the blood [Pourhabibi-Zarandi et al., 2024].

One of the most significant characteristics of curcumin is its antitumor work capacity, illustrated over the years in numerous investigations. Due to the low solubility of curcumin in water, various assemblages of encapsulated curcumin nanoparticles have been developed in the construction of soluble copolymers for their use in antitumor therapy. Curcumin has been revealed to selectively induce apoptosis in six tumor cells in the G2 phase by regulating TP53 gene expression and the initiation of the mitochondrial apoptotic pathway by increasing BAX gene expression and cytochrome C release. Despite the positive consequences of curcumin, its use in clinical investigations *in vivo* has been related due to low bioavailability, which is attributed to low assimilation, accelerated metabolism, and rapid elimination from the body. The poor bioavailability of curcumin was established by the low levels of glucuronide and sulfate [El-Hack et al., 2021]. The result assured following the administration of 8 g of curcumin daily indicated a clinical response in the appearance of two sick persons, and in one sick person, a slight tumor regression was recorded simultaneously with the enhancement in the level of cytokines IL-6, IL-8, IL-10, and IL-1, in serum. On the other hand, curcumin regulated the expression of nuclear ingredient NF-κB, COX-2, and STAT3 (signal transducer and activator of transcription 3) protein in peripheral blood mononuclear cells obtained from sick persons [Kunnumakkara et al., 2017]. Curcumin was prescribed with a rest every two months.

### 14.6.4 The 3-O-Epigallocatechin Gallate

The 3-O-EGCG emblematizes 38–40% of the total catechins present in green tea, being the most abundant. The catechins in green tea are part of the flavonoid family that have strong antioxidant capacity on the human body. From a chemical point of view, 3-O-EGCG contains three heterocyclic rings, A, B, and C, and two aromatic arrangements. The two aromatic arrangements are joined to the B ring by three bridge carbon arrangements (C6–C3–C9) together with the –OH (hydroxyl) cluster. The low bioavailability of EGCG must be examined when extrapolating results obtained *in vitro* to *in vivo* experiments. To administer correct and effective quantities of EGCG, it is compulsory to consider pharmacokinetic parameters [Ramesh & Mandal, 2019]. The effective ingredient in black pepper, piperine, is a potential modulator of EGCG bioavailability, by intervening with its glucuronide conjugation in the small intestine, as well as intervening with gastrointestinal transit, which can lead to enhanced assimilation. Conformable to investigations conducted over time, EGCG possesses numerous anticancer characteristics due to its potential to destroy free radicals, avoiding the destruction of the cellular arrangement caused by them.

The beneficial consequences that EGCG demonstrates are not achieved through a single mechanism; it acts through several signaling pathways. Green tea consumption has been reported to have beneficial consequences on the gastrointestinal network by improving blood levels of EGCG, leading to inhibition of cellular glucose uptake. In conclusion, following the experiments performed, the cytotoxic role of curcumin was perceived as more pronounced than that of EGCG, both in the appearance of divulgence of A-431 malignant cells to curcumin individually and in the presence of EGCG [Tao, Forester, & Lambert, 2014].

The objectives of the scientific research were aimed at highlighting the cytotoxic consequences of the two vegetable phenolic substances, evaluating the synergistic consequences of curcumin and EGCG, as well as highlighting the absence of cytotoxicity of curcumin and EGCG on a normal cell line. To assure additional results regarding the evaluation of the synergistic consequence between curcumin and EGCG, experiments can be performed to utilize several agglomerations of the two vegetable phenolic substances. Also, although the experiments transferred out in this chapter contribute to an improvement in the knowledge of the mode of action of the two vegetable phenolic substances *in vitro*, it is essential to carry out *in vivo* investigations, followed by clinical investigations, to discover an anticancer treatment establish on phytochemical molecules, it is known that they do not present harmful side effects on the human body [Hu et al., 2023].

### 14.6.5 The Vegetable Phenolic Substances as Molecular Systems with Characteristics that Can Be Thermally Modulated

Thermosensitive arrangements can be complicated in the category of "smart" materials that have applicability in the biomedical field through their use as medication carriers and the directed launching of chemical compound which induces pharmacological work capacity's/vegetable phenolic substances from a pharmaceutical assemblage, ensuring the bioavailability of medications by increasing the solubility of medications that have a pronounced lipophilic character [Wang et al., 2023].

Thermosensitive molecular arrangements could also help decrease non-specific bioadministration, i.e., the medication exceeding the desired tissue where it is administered, to avoid other adverse consequences. The response of smart materials to physical stimuli, such as temperature, and chemical stimuli, such as pH and ionic strength, involves changes in their initial characteristics, serving as an environmental response to external stimuli. The answers lie in the possibility of changing the molecular conformation and, implicitly, the shape, the water solubility, and the sol-gel transition.

In particular, the characteristics of thermosensitive arrangements will be changed by temperature variation, but this is not the only ingredient. The phase transition of the network from the tense state to the initial state involves a relaxation by removing the stimulus that generated the change at the molecular level [Van de Velde et al., 2022]. The reversible consequence is also known as the memory consequence.

Between thermosensitive arrangements, we can list micelles, supramolecular gels, or biphasic materials that generate hydrogels. The generated arrangements contain amphiphilic, pluronic polymeric surfactants in which sol-micelle or micelle-gel transitions can be perceived.

Encapsulation of molecules with the formation of "host-guest" arrangements in supramolecular chemistry exemplifies another category of thermosensitive networks. It is established on the interaction of the two molecules through the constitution of non-covalent interactions between them in the absence or presence of external stimuli. The constitution of these complexes is the result of several weak interactions of a physical nature, among which we can mention those of the ion–ion, ion–dipole, dipole–dipole, van der Waals, hydrogen obligations, etc., also involving steric consequences in many appearances [Li et al., 2023].

## 14.7 CONCLUSION AND FUTURE REMARKS

The bioavailability of vegetable phenolic substances differs, but there is no interconnection between the quantity of vegetable phenolic substances in food and their bioavailability in the human body. Following the works mentioned above and following a thorough analysis of the specialized literature, especially of the most recent investigations, we came to discover the great potentialities that vegetable phenolic substances can offer in the treatment or improvement of chronic diseases with a high incidence in the human population, with the added benefit of milder side effects compared to fluently accessible treatments. Nevertheless, we also discovered gaps and limitations in current knowledge about anthocyanins. A significant aspect, which is still under debate, is the role of the 3-OH cluster. *In vitro* investigations have revealed that this hydroxy cluster contributes to the antioxidant potentialities. Indeed, blocking the 3-OH cluster, as in rutin, or removing it, as in luteolin, significantly decreases the work capacity. The influence of the 3-OH cluster was also proven during the metabolism of quercetin, as it is assumed that quercetin coordinatively binds to the copper-containing consequence site of 2,3-dioxygenase. A significant aspect of the investigations transferred out is the use of laccase as an oxidizing agent. Indeed, numerous investigations frequently report the oxidation of flavonoids either with chemical radicals, initiators such as DPPH, or with enzymatic arrangements such as PPO polyphenol oxidase. The possibility of a di-electronic transfer can favor the oxidation of diphenol by forming semiquinone and quinone, and the thermodynamics of the reaction would rather be under the control of the stability of the possible final products than under the control of the stability of the radical species resulting from the splitting of a single hydrogen atom. In the appearance of laccase, even if the complete mechanisms of the reaction leading to the reduction of oxygen are not clearly established, the monoelectronic transfer is the more privileged mechanism. Most significant of all is the effort to develop economically sustainable treatments with the potential to maintain the highly volatile chemical stability of anthocyanins and, consequently, their beneficial function. For this reason, we would like to continue investigating anthocyanins and incorporating these biomolecules into nanoscale polymeric delivery arrangements as a targeted delivery therapy for melanoma, aiming for more precise and faster action, while preserving the arrangement and its chemical functions.

## COMPETING INTERESTS

The authors declare that they have no competing interests.

## DECLARATION OF GENERATIVE AI AND AI-ASSISTED TECHNOLOGIES IN THE WRITING PROCESS

During the preparation of this work, the book chapter author(s) have not taken any assistance from AI, and the book chapter author(s) reviewed and edited the content as needed and take(s) full responsibility for the content of the publication.

## REFERENCES

Ajila, C. M., Brar, S. K., Verma, M., Tyagi, R. D., Godbout, S., & Valéro, J. R. (2011). Extraction and analysis of polyphenols: Recent trends. *Critical Reviews in Biotechnology*, *31*(3), 227–249. https://doi.org/10.3109/07388551.2010.513677.

Al-Aubaidy, H. A., Dayan, A., Deseo, M. A., Itsiopoulos, C., Jamil, D., Hadi, N. R., & Thomas, C. J. (2021). Twelve-week mediterranean diet intervention increases citrus bioflavonoid levels and reduces inflammation in people with type 2 diabetes mellitus. *Nutrients*, *13*(4), 1133. https://doi.org/10.3390/nu13041133.

Alinaghi, A., Zhang, Q., Williams, D. B., Mackenzie, L., & Roberts, M. S. (2023). Thermodynamic and kinetic analysis of human epidermal penetration of phenolic compounds: I. Stratum corneum solubility and partitioning. *International Journal of Pharmaceutics*, *630*, 122424. https://doi.org/10.1016/j.ijpharm.2022.122424.

An, J., Liu, J., Liang, Y., Ma, Y., Chen, C., Cheng, Y., Peng, P., Zhou, N., Zhang, R., Addy, M., Chen, P., Liu, Y., Huang, G., Ren, D., & Ruan, R. (2020). Characterization, bioavailability and protective effects of phenolic-rich extracts from almond hulls against pro-oxidant induced toxicity in Caco-2 cells. *Food Chemistry*, *322*, 126742. https://doi.org/10.1016/j.foodchem.2020.126742.

Anand, P., Kunnumakkara, A. B., Newman, R. A., & Aggarwal, B. B. (2007). Bioavailability of curcumin: Problems and promises. *Molecular Pharmaceutics*, *4*(6), 807–818. https://doi.org/10.1021/mp700113r.

Arce-Reynoso, A., Mateos, R., Mendivil, E. J., Zamora-Gasga, V. M., & Sáyago-Ayerdi, S. G. (2023). Bioavailability of bioactive compounds in *Hibiscus sabdariffa* beverage as a potential anti-inflammatory. *Food Research International (Ottawa, Ont.)*, *174*(Pt 1), 113581. https://doi.org/10.1016/j.foodres.2023.113581.

Asad, N., Deodato, D., Asad, N., Gore, S., & Dore, T. M. (2023). Multi-photon-sensitive chromophore for the photorelease of biologically active phenols. *ACS Chemical Neuroscience*, *14*(23), 4163–4175. https://doi.org/10.1021/acschemneuro.3c00552.

Barreca, D., Gattuso, G., Bellocco, E., Calderaro, A., Trombetta, D., Smeriglio, A., Laganà, G., Daglia, M., Meneghini, S., & Nabavi, S. M. (2017). Flavanones: Citrus phytochemical with health-promoting properties. *BioFactors (Oxford, England)*, *43*(4), 495–506. https://doi.org/10.1002/biof.1363.

Bergesse, A. E., Asensio, C. M., Quiroga, P. R., Ryan, L. C., Grosso, N. R., & Nepote, V. (2023). Microencapsulation of phenolic compounds extracted from soybean seed coats by spray-drying. *Journal of Food Science*, *88*(11), 4457–4471. https://doi.org/10.1111/1750-3841.16775.

Biela, M., Kleinová, A., Uhliar, M., & Klein, E. (2023). Investigation of substituent effect on O-C bond dissociation enthalpy of methoxy group in meta- and para-substituted anisoles. *Journal of Molecular Graphics & Modelling*, *122*, 108465. https://doi.org/10.1016/j.jmgm.2023.108465.

Bravo, L. (1998). Polyphenols: Chemistry, dietary sources, metabolism, and nutritional significance. *Nutrition Reviews*, *56*(11), 317–333. https://doi.org/10.1111/j.1753-4887.1998.tb01670.x.

Breniaux, M., Dutilh, L., Petrel, M., Gontier, E., Campbell-Sills, H., Deleris-Bou, M., Krieger, S., Teissedre, P. L., Jourdes, M., Reguant, C., & Lucas, P. (2018). Adaptation of two groups of *Oenococcus oeni* strains to red and white wines: The role of acidity and phenolic compounds. *Journal of Applied Microbiology*, *125*(4), 1117–1127. https://doi.org/10.1111/jam.13946.

Bustos, A. S., Håkansson, A., Linares-Pastén, J. A., Penarrieta, J. M., & Nilsson, L. (2018). Interaction between phenolic compounds and lipase: The influence of solubility and presence of particles in the $IC_{50}$ value. *Journal of Food Science*, *83*(8), 2071–2076. https://doi.org/10.1111/1750-3841.14217.

Catteuw, A., Broekaert, N., De Baere, S., Lauwers, M., Gasthuys, E., Huybrechts, B., Callebaut, A., Ivanova, L., Uhlig, S., De Boevre, M., De Saeger, S., Gehring, R., Devreese, M., & Croubels, S. (2019). Insights into *in vivo* absolute oral bioavailability, biotransformation, and toxicokinetics of zearalenone, α-zearalenol, β-zearalenol, zearalenone-14-glucoside, and zearalenone-14-sulfate in pigs. *Journal of Agricultural and Food Chemistry*, *67*(12), 3448–3458. https://doi.org/10.1021/acs.jafc.8b05838.

Charalabidis, A., Sfouni, M., Bergström, C., & Macheras, P. (2019). The biopharmaceutics classification system (BCS) and the biopharmaceutics drug disposition classification system (BDDCS): Beyond guidelines. *International Journal of Pharmaceutics*, *566*, 264–281. https://doi.org/10.1016/j.ijpharm.2019.05.041.

Charoenprasert, S., & Mitchell, A. (2012). Factors influencing phenolic compounds in table olives (*Olea europaea*). *Journal of Agricultural and Food Chemistry*, *60*(29), 7081–7095. https://doi.org/10.1021/jf3017699.

Dantas, A. M., Fernandes, F. G., Magnani, M., & da Silva Campelo Borges, G. (2023). Gastrointestinal digestion assays for evaluating the bioaccessibility of phenolic compounds in fruits and their derivates: An overview. *Food Research International (Ottawa, Ont.)*, *170*, 112920. https://doi.org/10.1016/j.foodres.2023.112920.

Darwich, A. S., Neuhoff, S., Jamei, M., & Rostami-Hodjegan, A. (2010). Interplay of metabolism and transport in determining oral drug absorption and gut wall metabolism: A simulation assessment using the "Advanced Dissolution, Absorption, Metabolism (ADAM) model". *Current Drug Metabolism*, *11*(9), 716–729.. https://doi.org/10.2174/138920010794328913.

Davit, B. M., Kanfer, I., Tsang, Y. C., & Cardot, J. M. (2016). BCS biowaivers: Similarities and differences among EMA, FDA, and WHO requirements. *The AAPS Journal*, *18*(3), 612–618. https://doi.org/10.1208/s12248-016-9877-2.

de Beer, D., Beelders, T., Human, C., & Joubert, E. (2023). Assessment of the stability of compounds belonging to neglected phenolic classes and flavonoid sub-classes using reaction kinetic modeling. *Critical Reviews in Food Science and Nutrition*, *63*(33), 11802–11829. https://doi.org/10.1080/10408398.2022.2096561.

DiNicolantonio, J. J., McCarty, M. F., & O'Keefe, J. H. (2022). Nutraceutical activation of Sirt1: A review. *Open Heart*, *9*(2), e002171. https://doi.org/10.1136/openhrt-2022-002171.

Dixon, R. A., Xie, D. Y., & Sharma, S. B. (2005). Proanthocyanidins – A final frontier in flavonoid research?. *The New Phytologist*, *165*(1), 9–28. https://doi.org/10.1111/j.1469-8137.2004.01217.x.

Durazzo, A., Lucarini, M., Souto, E. B., Cicala, C., Caiazzo, E., Izzo, A. A., Novellino, E., & Santini, A. (2019). Polyphenols: A concise overview on the chemistry, occurrence, and human health. *Phytotherapy Research: PTR*, *33*(9), 2221–2243. https://doi. org/10.1002/ptr.6419.

El-Hack, Abd., El-Saadony, M. E., Swelum, M. T., Arif, A. A., Abo Ghanima, M., Shukry, M. M., Noreldin, M., Taha, A., & El-Tarabily, A. E. (2021). Curcumin, the active substance of turmeric: Its effects on health and ways to improve its bioavailability. *Journal of the Science of Food and Agriculture*, *101*(14), 5747–5762. https://doi.org/10.1002/jsfa.11372.

Elshorbagy, A. M., Fayed, M. A. A., Sallam, A., & Badria, F. A. (2024). Metabolic profiling, GC-MS, LC-ESI-MS/MS analysis, phenolics isolation and biological evaluation of the aerial parts extracts of *Felicia abyssinica* L. *Chemistry & Biodiversity*, *21*(3), e202301347. https://doi.org/10.1002/cbdv.202301347.

Fagerholm, U. (2007). Evaluation and suggested improvements of the biopharmaceutics classification system (BCS). *The Journal of Pharmacy and Pharmacology*, *59*(6), 751–757. https://doi.org/10.1211/jpp.59.6.0001.

Furmanek, L., Czarnota, P., Tekiela, A., Kapusta, I., & Seaward, M. R. D. (2024). A spectrophotometric analysis of extracted water-soluble phenolic metabolites of lichens. *Planta*, *260*(2), 40. https://doi.org/10.1007/s00425-024-04474-3.

Gamella, M., Campuzano, S., Reviejo, A. J., & Pingarrón, J. M. (2006). Electrochemical estimation of the polyphenol index in wines using a laccase biosensor. *Journal of Agricultural and Food Chemistry*, *54*(21), 7960–7967. https://doi.org/10.1021/jf061451r.

García Marrero, D. E., Glasser, W. G., Pizzi, A., Paczkowski, S., & Laborie, M. P. (2014). Substitution pattern elucidation of hydroxypropyl *Pinus pinaster* (Ait.) bark polyflavonoid derivatives by ESI(–)-MS/MS. *Journal of Mass Spectrometry: JMS*, *49*(10), 1050–1058. https://doi.org/10.1002/jms.3431.

Godos, J., Micek, A., Mena, P., Del Rio, D., Galvano, F., Castellano, S., & Grosso, G. (2024). Dietary (poly)phenols and cognitive decline: A systematic review and meta-analysis of observational studies. *Molecular Nutrition & Food Research*, *68*(1), e2300472. https://doi.org/10.1002/mnfr.202300472.

Günther, C. S., Dare, A. P., McGhie, T. K., Deng, C., Lafferty, D. J., Plunkett, B. J., Grierson, E. R. P., Turner, J. L., Jaakola, L., Albert, N. W., & Espley, R. V. (2020). Spatiotemporal modulation of flavonoid metabolism in blueberries. *Frontiers in Plant Science*, *11*, 545. https://doi.org/10.3389/fpls.2020.00545.

Hasnat, H., Shompa, S. A., Islam, M. M., Alam, S., Richi, F. T., Emon, N. U., Ashrafi, S., Ahmed, N. U., Chowdhury, M. N. R., Fatema, N., Hossain, M. S., Ghosh, A., & Ahmed, F. (2024). Flavonoids: A treasure house of prospective pharmacological potentials. *Heliyon*, *10*(6), e27533. https://doi.org/10.1016/j.heliyon.2024.e27533.

Hayes, C. J., Whittaker, B. P., Watson, S. A., & Grabowska, A. M. (2006). Synthesis and preliminary anticancer activity studies of C4 and C8-modified derivatives of catechin gallate (CG) and epicatechin gallate (ECG). *The Journal of Organic Chemistry*, *71*(26), 9701–9712. https://doi.org/10.1021/jo061740e.

Heimbach, T., Xia, B., Lin, T. H., & He, H. (2013). Case studies for practical food effect assessments across BCS/BDDCS class compounds using in silico, *in vitro*, and preclinical *in vivo* data. *The AAPS Journal*, *15*(1), 143–158. https://doi.org/10.1208/s12248-012-9419-5.

Hofsäss, M. A., & Dressman, J. B. (2019). The discriminatory power of the BCS-based biowaiver: A retrospective with focus on essential medicines. *Journal of Pharmaceutical Sciences*, *108*(9), 2824–2837. https://doi.org/10.1016/j.xphs.2019.04.030.

Hu, M., Huang, Y., Du, X., Liu, G., Qi, B., & Li, Y. (2023). The synergistic effect of epigallocatechin-3-gallate and quercetin co-loaded hydrogel beads on inflammatory bowel disease. *Food & Function*, *14*(10), 4539–4551. https://doi.org/10.1039/d2fo04029h.

Huang, D., Ou, B., Hampsch-Woodill, M., Flanagan, J. A., & Deemer, E. K. (2002). Development and validation of oxygen radical absorbance capacity assay for lipophilic antioxidants using randomly methylated beta-cyclodextrin as the solubility enhancer. *Journal of Agricultural and Food Chemistry*, *50*(7), 1815–1821. https://doi.org/10.1021/jf0113732.

Jiang, C., Wang, L., Huang, X., Zhu, S., Ma, C., & Wang, H. (2021). Structural characterization and antioxidant property of enzymatic-transesterification derivatives of (–)-epigallocatechin-3-O-gallate and vinyl laurate. *Journal of Food Science*, *86*(10), 4717–4729. https://doi.org/10.1111/1750-3841.15894.

Karakaya, S. (2004). Bioavailability of phenolic compounds. *Critical Reviews in Food Science and Nutrition*, *44*(6), 453–464. https://doi.org/10.1080/10408690490886683.

Katz, D. L., Doughty, K., & Ali, A. (2011). Cocoa and chocolate in human health and disease. *Antioxidants & Redox Signaling*, *15*(10), 2779–2811. https://doi.org/10.1089/ars.2010.3697.

Khan, B. A., Mahmood, T., Menaa, F., Shahzad, Y., Yousaf, A. M., Hussain, T., & Ray, S. D. (2018). New perspectives on the efficacy of gallic acid in cosmetics & nanocosmeceuticals. *Current Pharmaceutical Design*, *24*(43), 5181–5187. https://doi.org/10.2174/1381612825666190118150614.

Kocaadam, B., & Şanlier, N. (2017). Curcumin, an active component of turmeric (*Curcuma longa*), and its effects on health. *Critical Reviews in Food Science and Nutrition*, *57*(13), 2889–2895. https://doi.org/10.1080/10408398.2015.1077195.

Kunnumakkara, A. B., Bordoloi, D., Padmavathi, G., Monisha, J., Roy, N. K., Prasad, S., & Aggarwal, B. B. (2017). Curcumin, the golden nutraceutical: multitargeting for multiple chronic diseases. *British Journal of Pharmacology*, *174*(11), 1325–1348. https://doi.org/10.1111/bph.13621.

Li, L., Wang, Q., Liu, C., Hong, J., & Zheng, X. (2023). Effect of oven roasting on major chemical components in cereals and its modulation on flour-based products quality. *Journal of Food Science*, *88*(7), 2740–2757. https://doi.org/10.1111/1750-3841.16625.

Liu, W., Zhai, Y., Heng, X., Che, F. Y., Chen, W., Sun, D., & Zhai, G. (2016). Oral Bioavailability of curcumin: Problems and advancements. *Journal of Drug Targeting*, *24*(8), 694–702. https://doi.org/10.3109/1061186X.2016.1157883.

Lu, Y., Xu, Y., Song, M. T., Qian, L. L., Liu, X. L., Gao, R. Y., Han, R. M., Skibsted, L. H., & Zhang, J. P. (2021). Promotion effects of flavonoids on browning induced by enzymatic oxidation of tyrosinase: Structure-activity relationship. *RSC Advances*, *11*(23), 13769–13779. https://doi.org/10.1039/d1ra01369f.

Manach, C., Williamson, G., Morand, C., Scalbert, A., & Rémésy, C. (2005). Bioavailability and bioefficacy of polyphenols in humans. I. Review of 97 bioavailability studies. *The American Journal of Clinical Nutrition*, *81*(1 Suppl), 230S–242S. https://doi. org/10.1093/ajcn/81.1.230S.

Matlin, S. A., Zhou, R. H., Belenguer, A., Tyson, R. G., & Brookes, A. N. (1988). Large-scale resolution of gossypol enantiomers for biological evaluation. *Contraception*, *37*(3), 229–237. https://doi.org/10.1016/0010-7824(88)90025-x.

Mlakić, M., Perinić, H., Vušak, V., Horváth, O., Sampedro, D., Losantos, R., Odak, I., & Škorić, I. (2023). Photochemical transformations of diverse biologically active resveratrol analogs in batch and flow reactors. *Molecules (Basel, Switzerland)*, *29*(1), 201. https://doi.org/10.3390/molecules29010201.

Moazzen, A., Öztinen, N., Ak-Sakalli, E., & Koşar, M. (2022). Structure-antiradical activity relationships of 25 natural antioxidant phenolic compounds from different classes. *Heliyon*, *8*(9), e10467. https://doi.org/10.1016/j.heliyon.2022.e10467.

Mouafon, I. L., Mountessou, B. Y. G., Lateef, M., Tchamgoue, J., Shaiq Ali, M., Tchouankeu, J. C., Green, I. R., Ngadjui, B. T., & Kouam, S. F. (2023). Atricephenols A and B, two phenolic compounds from *Indigofera atriceps* Hook.f. (Fabaceae). *Natural Product Research*, *37*(14), 2319–2326. https://doi.org/10.1080/14786419.2022.2041007.

Muller, A. G., Sarker, S. D., Saleem, I. Y., & Hutcheon, G. A. (2019). Delivery of natural phenolic compounds for the potential treatment of lung cancer. *Daru: Journal of Faculty of Pharmacy, Tehran University of Medical Sciences*, *27*(1), 433–449. https:/ /doi.org/10.1007/s40199-019-00267-2.

Nakai, S., Fujita, M., & Kamei, Y. (2020). Health promotion effects of soy isoflavones. *Journal of Nutritional Science and Vitaminology*, *66*(6), 502–507. https://doi.org/10.3177/jnsv.66.502.

Pais, A. C. S., Coscueta, E. R., Pintado, M. M., Silvestre, A. J. D., & Santos, S. A. O. (2024). Exploring the bioaccessibility and intestinal absorption of major classes of pure phenolic compounds using *in vitro* simulated gastrointestinal digestion. *Heliyon*, *10*(7), e28894. https://doi.org/10.1016/j.heliyon.2024.e28894.

Papich, M. G., & Martinez, M. N. (2015). Applying biopharmaceutical classification system (BCS) criteria to predict oral absorption of drugs in dogs: Challenges and pitfalls. *The AAPS Journal*, *17*(4), 948–964. https://doi.org/10.1208/s12248-015-9743-7.

Patil, S. S., Deshannavar, U. B., Gadekar-Shinde, S. N., Gadagi, A. H., & Kadapure, S. A. (2023). Optimization studies on batch extraction of phenolic compounds from *Azadirachta indica* using genetic algorithm and machine learning techniques. *Heliyon*, *9*(11), e21991. https://doi.org/10.1016/j.heliyon.2023.e21991.

Paulo, F., & Santos, L. (2021). Deriving valorization of phenolic compounds from olive oil by-products for food applications through microencapsulation approaches: A comprehensive review. *Critical Reviews in Food Science and Nutrition*, *61*(6), 920–945. https://doi.org/10.1080/10408398.2020.1748563.

Pérez-Navarro, J., Izquierdo-Cañas, P. M., Mena-Morales, A., Martínez-Gascueña, J., Chacón-Vozmediano, J. L., García-Romero, E., Hermosín-Gutiérrez, I., & Gómez-Alonso, S. (2019). Phenolic compounds profile of different berry parts from novel *Vitis vinifera* L. red grape genotypes and Tempranillo using HPLC-DAD-ESI-MS/MS: A varietal differentiation tool. *Food Chemistry*, *295*, 350–360. https://doi.org/10.1016/j.foodchem.2019.05.137.

Pourhabibi-Zarandi, F., Rafraf, M., Zayeni, H., Asghari-Jafarabadi, M., & Ebrahimi, A. A. (2024). The efficacy of curcumin supplementation on serum total antioxidant capacity, malondialdehyde, and disease activity in women with rheumatoid arthritis: A randomized, double-blind, placebo-controlled clinical trial. *Phytotherapy Research: PTR*, *38*(7), 3552–3563. https://doi.org/10.1002/ptr.8225.

Ramesh, N., & Mandal, A. K. A. (2019). Pharmacokinetic, toxicokinetic, and bioavailability studies of epigallocatechin-3-gallate loaded solid lipid nanoparticle in rat model. *Drug Development and Industrial Pharmacy*, *45*(9), 1506–1514. https://doi.org/10.1080/03639045.2019.1634091.

Sarmiento, V., Ramirez-Sanchez, I., Moreno-Ulloa, A., Romero-Perez, D., Chávez, D., Ortiz, M., Najera, N., Correa-Basurto, J., Villarreal, F., & Ceballos, G. (2018). Synthesis of novel (–)-epicatechin derivatives as potential endothelial GPER agonists: Evaluation of biological effects. *Bioorganic & Medicinal Chemistry Letters*, *28*(4), 658–663. https://doi.org/10.1016/j.bmcl.2018.01.025.

Scott, K. A., Cox, P. B., & Njardarson, J. T. (2022). Phenols in pharmaceuticals: Analysis of a recurring motif. *Journal of Medicinal Chemistry*, *65*(10), 7044–7072. https://doi.org/10.1021/acs.jmedchem.2c00223.

Shiragaki, K., Nakamura, R., Nomura, S., He, H., Yamada, T., Marubashi, W., Oda, M., & Tezuka, T. (2020). Phenylalanine ammonia-lyase and phenolic compounds are related to hybrid lethality in the cross *Nicotiana suaveolens×N. tabacum*. *Plant Biotechnology (Tokyo, Japan)*, *37*(3), 327–333. https://doi.org/10.5511/plantbiotechnology.20.0606a.

Singh, B. N., Shankar, S., & Srivastava, R. K. (2011). Green tea catechin, epigallocatechin-3-gallate (EGCG): mechanisms, perspectives and clinical applications. *Biochemical Pharmacology*, *82*(12), 1807–1821. https://doi.org/10.1016/j.bcp.2011.07.093.

Singla, R. K., Dubey, A. K., Garg, A., Sharma, R. K., Fiorino, M., Ameen, S. M., Haddad, M. A., & Al-Hiary, M. (2019). Natural polyphenols: Chemical classification, definition of classes, subcategories, and structures. *Journal of AOAC International*, *102*(5), 1397–1400. https://doi.org/10.5740/jaoacint.19-0133.

Stockley, C., Teissedre, P. L., Boban, M., Di Lorenzo, C., & Restani, P. (2012). Bioavailability of wine-derived phenolic compounds in humans: A review. *Food & Function*, *3*(10), 995–1007. https://doi.org/10.1039/c2fo10208k.

Tao, L., Forester, S. C., & Lambert, J. D. (2014). The role of the mitochondrial oxidative stress in the cytotoxic effects of the green tea catechin, (–)-epigallocatechin-3-gallate, in oral cells. *Molecular Nutrition & Food Research*, *58*(4), 665–676. https://doi.org/10.1002/mnfr.201300427.

Tao, M., Liu, S., Li, Y., Liu, A., Tian, J., Liu, Y., Fu, H., & Zhu, W. (2023). Molecular characterization of a feruloyl-CoA 6′-hydroxylase involved in coumarin biosynthesis in Clematis terniflora DC. *Plant Physiology and Biochemistry: PPB*, *196*, 162–170. https://doi.org/10.1016/j.plaphy.2023.01.046.

Tatipamula, V. B., & Kukavica, B. (2021). Phenolic compounds as antidiabetic, antiinflammatory, and anticancer agents and improvement of their bioavailability by liposomes. *Cell Biochemistry and Function*, *39*(8), 926–944. https://doi.org/10.1002/cbf.3667.

Teng, M., Zhu, J., Johnson, M. D., Chen, P., Kornmann, J., Chen, E., Blasina, A., Register, J., Anderes, K., Rogers, C., Deng, Y., Ninkovic, S., Grant, S., Hu, Q., Lundgren, K., Peng, Z., & Kania, R. S. (2007). Structure-based design and synthesis of (5-arylamino-2H-pyrazol-3-yl)-biphenyl-2',4'-diols as novel and potent human CHK1 inhibitors. *Journal of Medicinal Chemistry*, *50*(22), 5253–5256. https://doi.org/10.1021/jm0704604.

Truzzi, F., Tibaldi, C., Zhang, Y., Dinelli, G., & Amen, D. (2021). An overview on dietary polyphenols and their biopharmaceutical classification system (BCS). *International Journal of Molecular Sciences*, *22*(11), 5514. https://doi.org/10.3390/ijms22115514.

Van de Velde, F., Vignatti, C., Paula Méndez-Galarraga, M., Gomila, M., Fenoglio, C., Donda Zbinden, M., & Élida Pirovani, M. (2022). Intestinal and colonic bioaccessibility of phenolic compounds from fruit smoothies as affected by the thermal processing and the storage conditions. *Food Research International (Ottawa, Ont.)*, *155*, 111086. https://doi.org/10.1016/j.foodres.2022.111086.

Velmurugan, P., Singam, E. R., Jonnalagadda, R. R., & Subramanian, V. (2014). Investigation on interaction of tannic acid with type I collagen and its effect on thermal, enzymatic, and conformational stability for tissue engineering applications. *Biopolymers*, *101*(5), 471–483. https://doi.org/10.1002/bip.22405.

Wang, R., Li, M., Brennan, M. A., Dhital, S., Kulasiri, D., Brennan, C. S., & Guo, B. (2023). Complexation of starch and phenolic compounds during food processing and impacts on the release of phenolic compounds. *Comprehensive Reviews in Food Science and Food Safety*, *22*(4), 3185–3211. https://doi.org/10.1111/1541-4337.13180.

Wang, Y., Gao, X. T., Li, H. Q., Lu, H. C., He, L., Peng, W. T., Chen, W., Li, S. D., Li, S. P., Duan, C. Q., & Wang, J. (2021). Microcliamte changes caused by black inter-row mulch decrease flavonoids concentrations in grapes and wines under semi-arid climate. *Food Chemistry*, *361*, 130064. https://doi.org/10.1016/j.foodchem.2021.130064.

Wei, F., Li, D., Chen, X., Li, Y., Zeng, Y., Cai, Y., Zeng, Y., Chen, Y., Ma, X., & Zeng, J. (2024). Therapeutic effects of epigallocatechin-3-gallate for inflammatory bowel disease: A preclinical meta-analysis. *Phytomedicine: International Journal of Phytotherapy and Phytopharmacology*, *128*, 155408. https://doi.org/10.1016/j.phymed.2024.155408.

Ždero Pavlović, R., Blagojević, B., Kukrić, T., Mocko Blažek, K., Lazzara, G., Četojević-Simin, D., & Popović, B. M. (2023). Phenolic compounds from sour cherry pomace: Microencapsulation, *in vitro* digestion, and cell growth activities. *Plant Foods for Human Nutrition (Dordrecht, Netherlands)*, *78*(1), 221–227. https://doi.org/10.1007/s11130-023-01046-w.

Zeb, A. (2021). A comprehensive review on different classes of polyphenolic compounds present in edible oils. *Food research international (Ottawa, Ont.)*, *143*, 110312. https://doi.org/10.1016/j.foodres.2021.110312.

Zhang, N., Geronimo, I., Paneth, P., Schindelka, J., Schaefer, T., Herrmann, H., Vogt, C., & Richnow, H. H. (2016). Analyzing sites of OH radical attack (ring vs. side chain) in oxidation of substituted benzenes via dual stable isotope analysis (δ(13)C and δ(2)H). *The Science of the Total Environment*, *542*(Pt A), 484–494. https://doi.org/10.1016/j.scitotenv.2015.10.075.

Zhang, X., & Liu, C. J. (2015). Multifaceted regulations of gateway enzyme phenylalanine ammonia-lyase in the biosynthesis of phenylpropanoids. *Molecular Plant*, *8*(1), 17–27. https://doi.org/10.1016/j.molp.2014.11.001.

Zhou, M., Chen, J., Bi, J., Li, X., & Xin, G. (2022). The roles of soluble poly and insoluble tannin in the enzymatic browning during storage of dried persimmon. *Food Chemistry*, *366*, 130632. https://doi.org/10.1016/j.foodchem.2021.130632.

# 15 Synergistic Interactions of Phenolic Compounds with Other Natural Compounds

*Sudeshna Shyam Choudhury*[*], *Arkajyoti Mahanta, and Aranya Hazari*

## 15.1 INTRODUCTION

Phenolic compounds, a diverse group of secondary metabolites found extensively in fruits, vegetables, cereals, and medicinal plants, have significant implications for human health due to their antioxidant, anti-inflammatory, and antimicrobial properties (Vauzour et al., 2010). These compounds are characterized by one or more aromatic rings that contain hydroxyl groups, which allow them to effectively scavenge reactive oxygen species (ROS) and reactive nitrogen species (RNS), chelate metal ions, and inhibit oxidative enzymes (Rice-Evans, Miller, & Paganga, 1997). Historically, the study of phenolic compounds has concentrated on their individual biological impacts; however, in natural environments and dietary contexts, phenolics coexist with various other bioactive substances, such as vitamins (for instance, vitamins C and E), carotenoids, terpenoids, alkaloids, and additional polyphenols, resulting in potential synergistic interactions that may enhance their bioefficacy beyond mere additive effects (Lotito & Frei, 2006; Tagliazucchi et al., 2010). Synergistic interactions occur when the collective activity of compounds exceeds the sum of their individual effects, often attributed to mechanisms such as the regeneration of active radical forms, modulation of signal transduction pathways, and enhanced cellular uptake and bioavailability (Hayek et al., 2013; Wang et al., 2011). For instance, the combination of flavonoids with vitamin C promotes redox cycling, thereby sustaining antioxidant capacity within biological systems. Their co-presence with carotenoids bolsters protection against lipid peroxidation, which is essential for alleviating oxidative stress-related cellular damage. In a similar vein, the amalgamation of phenolic compounds with terpenoids and essential oils found in plant extracts has been shown to enhance antimicrobial and anti-inflammatory properties, indicating synergistic or potentiating effects against pathogens and inflammatory pathways.

Furthermore, synergistic interactions can influence the bioavailability of phenolic compounds, a vital consideration given their low absorption rates and rapid metabolism within the human body (Manach et al., 2004). The co-ingestion of phenolics alongside other dietary components can affect gastrointestinal stability, enzymatic hydrolysis, and transport across enterocytes, leading to enhanced systemic availability and bioactivity (Del Rio et al., 2013). For example, specific phenolic acids and flavonoids may inhibit glucuronidation and sulfation, thereby extending the presence of their bioactive forms and those of coexisting compounds in circulation (Wang et al., 2011). Such interactions are particularly significant when formulating functional foods, nutraceuticals, and phytopharmaceuticals aimed at addressing oxidative stress, inflammation, cardiovascular health,

[*] Corresponding author: sudeshna.s.choudhury@sxccal.edu


DOI: 10.1201/9781003527671-15

neurodegeneration, and microbial infections (López-Alarcón & Denicola, 2013). Understanding the synergistic relationships between phenolic compounds and other natural substances is crucial for maximizing their therapeutic efficacy, particularly in the context of multifactorial chronic diseases where oxidative stress and inflammation play a pivotal role. This review aims to examine the origins and chemical characteristics of phenolic compounds and other bioactive substances, emphasize their potential synergistic counterparts, clarify the mechanisms that drive synergy, and investigate the consequences of these interactions in food systems, nutraceutical innovation, and therapeutic applications.

## 15.2 SOURCES AND NATURE OF PHENOLIC COMPOUND

Natural sources of phenolic compounds include both natural and artificial sources. Natural sources include plants and microbes.

### 15.2.1 Natural Sources of Phenolic Compounds from Plants

Fruits are rich in phenolic acids, flavonoids, and tannins, such as blueberries (high in anthocyanin), apples (rich in quercetin and chlorogenic acid), grapes (high in resveratrol), citrus fruits (which provide Flavonoids), and cherries and plums (Dai & Mumper, 2010). Onions (rich in quercetin), broccoli and spinach (contain hydroxycinnamic acid), whole grains, such as wheat and barley, contain high levels of ferulic acid and lignans. Legumes, such as soybeans, lentils, and chickpeas, are rich in isoflavones and phenolic acids. Beverage such as tea conatins contain high catechins (epigallocatechin gallate), theaflavins, etc., coffee (rich in chlorogenic acid and caffeic acid), and red wine (contains flavanols and anthocyanins). Plants such as ginseng, ginger, turmeric, neem, and tulsi are rich in several phenolic compounds and possess antioxidant properties (Balasundram, Sundram, & Samman, 2006).

### 15.2.2 Natural Sources of Phenolic Compounds from Microbes

Filamentous fungi such as *Aspergillus* and *Penicillium* produce phenolic acid (gallic acid). Yeast, such as *Saccharomyces cerevisiae*, produces phenolic derivatives during the fermentation process (Huang, Cai, & Zhang, 2010). Bacteria like *Streptomyces* produce phenolic antibiotics, such as chloramphenicol, and phenolic pigments. *Lactobacillus* releases free phenolic acids during fermentation. Algae such as chlorella produce phenolic compounds with antioxidant activity (Goiris et al., 2014).

### 15.2.3 Artificial Sources of Phenolic Compounds

Chemical synthesis, through targeted production and nano-technology integration, enables the production of artificial polyphenols.

## 15.3 NATURE OF NATURAL PHENOLIC COMPOUNDS

Phenolic compounds are a diverse group of organic compounds characterized by the presence of one or more hydroxyl groups (–OH) bonded to an aromatic (often benzene ring) (Tsao, 2010).

### 15.3.1 Physicochemical Nature

According to solubility, low molecular weight phenolic compounds exhibit solubility in polar solvents such as water, methanol, and ethanol; in contrast, high molecular weight polyphenols, specifically tannins, show limited solubility in water but can dissolve in organic polar solvents

(Dai & Mumper, 2010). For ultraviolet (UV) absorption, owing to their conjugated π-electron systems, these compounds are capable of absorbing UV light, which facilitates detection through UV-spectrophotometry.

### 15.3.2 Biological Nature

Polyphenols play protective roles in plants against UV radiation, pathogens, herbivores, and oxidative stress (Crozier, Jaganath, & Clifford, 2009). Polyphenols play a role in the flavor, color, and astringency of foods (for instance, tannins). They possess bioactive effects in humans, encompassing antioxidant, anti-inflammatory, antimicrobial, and anticancer characteristics. They exhibit analgesic, stimulant, anti-malarial, and other antiparasitic activities as well.

### 15.3.3 Chemical Nature

Polyphenolic groups of compounds have unique redox properties. They possess the ability to donate hydrogen atoms or electrons, thereby stabilizing free radicals and chelating metal ions, which enhances their antioxidant properties. They have unique reactivity also. These compounds interact with proteins and carbohydrates, forming complexes and engaging in reactions such as esterification, glycosylation, and methylation.

Diversified chemical natures are there for different polyphenolic groups of compounds.

- *Phenolic Acids:* These compounds are characterized by a phenolic ring with one or more carboxylic acid groups. Examples include gallic acid, caffeic acid, and ferulic acid (Cardona et al., 2013) (Figure 15.1).
- *Flavonoids:* A large family of phenolic compounds with a characteristic three-ring structure. They include flavonols (e.g., quercetin), flavanones, and anthocyanins (Figure 15.1).
- *Tannins:* Complex phenolic compounds known for their astringent taste. They can be further divided into hydrolyzable and condensed tannins (Balasundram, Sundram, & Samman, 2006). Theaflavin and thearubigins are tannins found in tea (Figure 15.1).
- *Lignans:* Polymers formed from phenylpropane units, found in wood and other plant materials. They are often present in flaxseed and buckwheat. In sesame seeds, sesamin and episesamin are found, which are lignans (Figure 15.1).
- *Stilbenes:* Phenolic compounds with a stilbene structure, like resveratrol, which is found in grapes and berries (Figure 15.1).
- *Alkaloids:* Nitrogen-containing compounds, often with medicinal properties, like morphine (Figure 15.1) and quinine. Morphine is obtained from *Papaver somniferum*, caffeine is obtained from *Camellia sinensis*, quinine is obtained from *Cinchona spp.*, and nicotine is obtained from *Nicotiana tabacum.*
- *Terpenoids:* A diverse group derived from isoprene units, including essential oils, carotenoids, and sterols. Limonene (Figure 15.1) is obtained from citrus peels, menthol is obtained from mint, carotenoids (β-carotene is found in carrots), and Taxol is obtained from the Pacific yew tree (Wink, 2010).
- *Saponins:* Ginsenosides are obtained from *Panax ginseng*, soyasaponins (Figure 15.1) are obtained from soybeans, and *Quillaja saponins* are obtained from *Quillaja saponaria* bark.
- *Glycosides:* Cardiac glycosides, such as digoxin, are obtained from *Digitalis* spp. Anthraquinone glycosides, like aloe-emodin, are found in *Aloe* spp.
- *Essential Oils*: Eugenol, which is obtained from clove oil, thymol, which is obtained from thyme, and linalool (Figure 15.1), which is obtained from lavender.
- *Sulfur-Containing Compounds:* Allicin (Figure 15.1) is obtained from garlic, while glucosinolates are obtained from broccoli and cabbage.

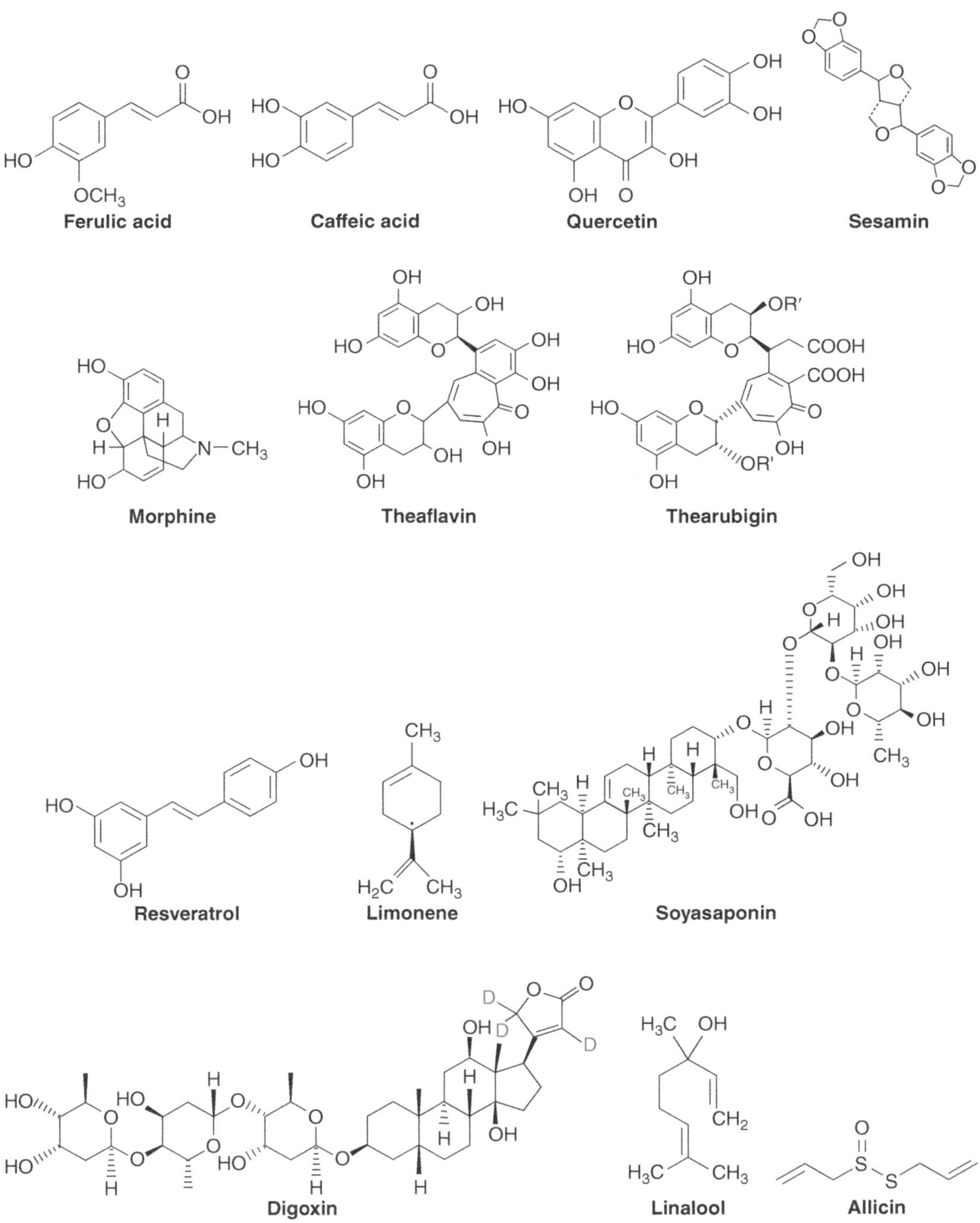


**FIGURE 15.1** Structures of different Phenolic compounds from natural sources.

## 15.4 BIOLOGICAL ACTIVITIES OF SYNERGISM OF PHENOLICS AND OTHER NATURAL COMPOUNDS

Numerous biological activities are associated with various polyphenolic compounds. Through synergistic actions, such activities are increasing.

### 15.4.1 Antimicrobial Activity

The synergistic antimicrobial effect is observed when phenolic compounds interact with other natural substances (such as essential oils, alkaloids, and terpenoids) to amplify their bactericidal and fungicidal properties (Bachheti et al. 2023; Gutiérrez et al., 2008). The mechanism involves enhanced disruption of cell membranes, inhibition of enzymes, and interference with microbial metabolism, resulting in a higher rate of microbial death than that achieved by the individual compounds (Burt, 2004). Phenolic acids, such as gallic acid, combined with essential oils like thymol and carvacrol, exhibit improved antimicrobial properties against *Escherichia coli*, *Staphylococcus aureus*, and *Salmonella typhimurium* (Bajpai, Baek, & Kang, 2013). The combination of flavonoids and terpenoids enhances the permeability of bacterial cell walls, thereby facilitating the entry of phenolic compounds and amplifying their antimicrobial effects. When phenolics are used in conjunction with alkaloids, such as caffeine, they can reduce the required dosage of antibiotics and delay the development of antibiotic resistance when targeting pathogens (Hemaiswarya, Kruthiventi, & Doble, 2008).

### 15.4.2 Antioxidant Activity

The synergistic antioxidant activity occurs when phenolic compounds interact with other natural substances, such as ascorbic acid, flavonoids, carotenoids, and terpenoids, thereby enhancing the scavenging of free radicals, metal chelation, and the inhibition of lipid peroxidation (Tapiero et al., 2004). This synergism decreases the concentration needed for antioxidant effectiveness, which in turn reduces potential toxicity while preserving efficacy. The combined action provides enhanced protection for biomolecules, including lipids, proteins, and DNA, against oxidative damage within biological systems. Phenolics combined with ascorbic acid (vitamin C) can regenerate oxidized phenolic radicals, thereby extending antioxidant activity and enhancing cellular protection (Rice-Evans, Miller, & Paganga, 1997). Flavonoids, particularly combinations such as quercetin and catechin, exhibit a greater capacity for DPPH and ABTS radical scavenging compared to individual flavonoids, suggesting a genuine synergistic effect. Phenolics, in conjunction with carotenoids such as β-carotene, exhibit synergistic properties in lipid-rich environments, thereby enhancing the inhibition of lipid peroxidation (Tapiero et al., 2004). The combination of phenolics and terpenoids, including essential oils, enhances antioxidant defense by merging the radical scavenging capabilities of phenolics with the singlet oxygen quenching properties of terpenoids.

### 15.4.3 Anticancer Activity

The synergistic anticancer effects occur when phenolic compounds interact with other natural substances, such as curcumin, flavonoids, alkaloids, and terpenoids, resulting in increased apoptosis, decreased proliferation, and inhibited angiogenesis in cancer cells. These synergistic combinations can improve the effectiveness of chemotherapeutic agents and mitigate drug resistance in cancer therapies. Practical applications encompass their incorporation in the development of nutraceuticals and supplementary cancer treatments aimed at enhancing therapeutic results while minimizing adverse effects. Phenolics combined with curcumin, derived from turmeric, demonstrate a greater ability to inhibit cancer cell proliferation and more effectively induce apoptosis compared to either compound used independently (Khan & Mukhtar, 2007). Green tea catechins, specifically EGCG, in conjunction with quercetin, display synergistic effects against prostate and breast cancer cells by influencing cell signaling pathways and reducing the levels of anti-apoptotic proteins (Shankar, Ganapathy, & Srivastava, 2007). The combination of phenolic acids and alkaloids, such as caffeine, promotes cell cycle arrest and enhances the susceptibility of cancer cells to apoptosis triggered by oxidative stress.

### 15.4.4 Antidiabetic Activity

Synergistic antidiabetic effects arise when phenolic compounds interact with various natural substances (such as flavonoids, alkaloids, terpenoids, and vitamins) to enhance glucose metabolism, increase insulin sensitivity, and bolster antioxidant defense in the management of diabetes. Phenolics and flavonoids (such as quercetin and kaempferol) work together to inhibit carbohydrate-digesting enzymes ($\alpha$-amylase and $\alpha$-glucosidase), thereby decreasing postprandial hyperglycemia (Tundis, Loizzo, & Menichini, 2010). Phenolic acids and alkaloids (for instance, berberine) promote insulin secretion, enhance the sensitivity of insulin receptors, and regulate glucose uptake pathways, which supports glycemic control. Green tea catechins (EGCG) and vitamin C bolster antioxidant defenses, mitigating oxidative stress linked to diabetes and helping to prevent complications (Bahadoran, Mirmiran, & Azizi, 2013).

### 15.4.5 Anti-Inflammatory Activity

The synergistic anti-inflammatory effect is observed when phenolic compounds interact with various natural substances, such as flavonoids, terpenoids, vitamins, and alkaloids, thereby reducing inflammation markers. This synergistic action is more effective in diminishing leukocyte infiltration and edema in inflammation models compared to the use of individual compounds. The combination of these compounds aids in inhibiting lipid peroxidation in inflamed tissues, thereby mitigating damage associated with chronic inflammatory conditions (Vauzour et al., 2008). Phenolics and flavonoids (such as quercetin and kaempferol) work together to inhibit pro-inflammatory cytokines (TNF-$\alpha$, IL-1$\beta$, IL-6) and decrease the levels of inflammatory mediators like COX-2 and iNOS (Tapas et al., 2008). Phenolics and terpenoids (for example, curcumin and thymol) enhance the suppression of NF-$\kappa$B and MAPK signaling pathways, thereby reducing inflammatory responses in cells. Phenolic acids, when combined with vitamin E, enhance antioxidant defenses, resulting in a reduction of oxidative stress associated with chronic inflammation.

Flavonoids and alkaloids (such as berberine) exhibit a synergistic effect in modulating the activity of inflammatory enzymes while also safeguarding tissues from oxidative damage.

### 15.4.6 Anti-Obesity Activity

The synergistic anti-obesity effect is observed when phenolic compounds interact with various natural substances such as flavonoids, alkaloids, terpenoids, and vitamins, leading to enhanced fat metabolism, decreased lipid accumulation, and increased energy expenditure.

This synergistic interaction facilitates lipolysis, which is the breakdown of fats, while simultaneously inhibiting lipogenesis, thereby aiding in the regulation of body weight and fat mass. The combination of these compounds influences the AMPK and PI3K/Akt signaling pathways, promoting glucose uptake and fatty acid oxidation, which ultimately mitigates metabolic disorders associated with obesity. Phenolics and flavonoids (such as quercetin and catechins) work together to inhibit adipogenesis by downregulating PPAR$\gamma$ and C/EBP$\alpha$, which leads to a decrease in fat cell formation (Hsu & Yen, 2007). Green tea catechins (EGCG) combined with caffeine (an alkaloid) exhibit a synergistic effect in boosting thermogenesis and fat oxidation, thereby facilitating weight loss and increasing energy expenditure (Dulloo et al., 2009). Phenolics and terpenoids (for example, essential oils like limonene) contribute to the reduction of lipid accumulation in adipocytes and influence the activity of lipid metabolism enzymes. Phenolic acids, along with vitamin C, improve the antioxidant defense mechanisms in adipose tissues, which helps to mitigate inflammation and oxidative stress associated with obesity.

### 15.4.7 Sun Protection Activity

The synergistic activity of sun protection occurs when phenolic compounds interact with various natural substances, such as flavonoids, carotenoids, vitamins, and terpenoids, thereby enhancing

UV absorption, bolstering antioxidant defenses, and protecting the skin from photoaging (Lin & Lin, 2005). This synergistic effect diminishes UV-induced DNA damage and inhibits the activity of MMPs, which break down skin collagen, thereby preventing the development of wrinkles. The combination of these compounds also enhances the modulation of melanin production, offering natural photoprotection (Nichols & Katiyar, 2010). Phenolics combined with flavonoids (such as quercetin and kaempferol) work together to absorb UVA and UVB radiation, thereby minimizing skin damage and erythema (Sanchez, Llorens, & Luque-Garcia, 2011). Phenolics in conjunction with carotenoids (for instance, β-carotene and lycopene) bolster protection against oxidative stress induced by UV rays by scavenging free radicals and decreasing lipid peroxidation in skin cells. Phenolics, along with vitamin E and vitamin C, collaborate to neutralize ROS produced by UV exposure, thereby safeguarding skin collagen and elastin (Lin & Lin, 2005). Phenolics, paired with terpenoids (such as essential oils like limonene), enhance the antioxidant defense mechanisms in the skin, leading to a reduction in inflammation and photodamage.

## 15.5 MODE OF ACTION OF POLYPHENOLIC GROUP OF COMPOUNDS

Various modes of action are observed for the biological functioning of the polyphenolic groups of compounds.

### 15.5.1 Structure–Activity-Based Relationship of Polyphenolic Group of Compounds

The quantity and arrangement of hydroxyl groups on the aromatic ring (Rice-Evans, Miller, & Paganga, 1996), the extent of polymerization and conjugation, the presence of glycosides, methoxy, or other substituent groups, and all these structural features collectively enable them to exhibit biological functioning or bioactivities.

### 15.5.2 Synergistic Action of the Polyphenolic Groups of Compounds

Synergism denotes the occurrence in which the collective action of two or more substances results in a biological effect that exceeds the total of their separate effects. This interaction may occur between similar and dissimilar classes of substances, resulting in enhanced antioxidant, antimicrobial, anti-inflammatory, or therapeutic properties. The synergistic effect is especially noteworthy in relation to natural substances, where phenolic compounds, flavonoids, terpenoids, vitamins, and various other phytochemicals collaborate to offer increased health benefits (Lotito & Frei, 2006).

The benefit of synergism is its ability to increase bioefficacy while reducing the necessary dosage of each compound, thereby lowering the likelihood of side effects and toxicity (Scalbert, Johnson, & Saltmarsh, 2005). Synergistic interactions can enhance the stability and bioavailability of phenolic compounds, facilitating better absorption and utilization within biological systems. Synergism can aid in overcoming microbial resistance when phenolic compounds are combined with essential oils or alkaloids, thereby enhancing their antimicrobial efficacy against pathogenic bacteria (Burt, 2004). Vitamin C can regenerate oxidized phenolic radicals, extending antioxidant activity and cellular protection. The synergistic effect of natural compounds presents a promising approach for the development of effective preventive and therapeutic agents in human health management.

### 15.5.3 Major Chemical Interactions and Mechanisms of Action of Polyphenolic Compounds to Show Biological Activities

Polyphenols, a category of natural substances found in plants, can modulate the redox potential by functioning as both antioxidants and pro-oxidants, thereby impacting cellular redox homeostasis. This modulation may prove advantageous in the prevention and treatment of numerous diseases by influencing redox-dependent signaling pathways and cellular activities such as proliferation and apoptosis (Ozdal, Capanoglu, & Altay, 2013).

#### 15.5.3.1 Redox Potential Modulation

Phenolic compounds demonstrate biological activity by modulating redox potential, an effect that is amplified when they are combined with other natural substances like flavonoids, carotenoids, terpenoids, vitamins, and alkaloids. Redox potential refers to a compound's ability to either donate or accept electrons in oxidation–reduction reactions, which is crucial for neutralizing ROS and maintaining cellular redox homeostasis (Pisoschi & Pop, 2015).

#### 15.5.3.2 Electron and Hydrogen Atom Transfer

Phenolic compounds can donate electrons or hydrogen atoms, effectively neutralizing free radicals. Synergistically, co-compounds such as vitamin C can regenerate the oxidized forms of phenolics (for instance, flavonoid radicals) back to their active antioxidant state, thereby extending their antioxidant activity (Rice-Evans, Miller, & Paganga, 1997).

#### 15.5.3.3 Regeneration Cycles

When combined with ascorbic acid and tocopherol, phenolic compounds engage in antioxidant recycling. In this process, ascorbic acid reduces oxidized flavonoids, while tocopherol radicals are also reduced by ascorbic acid, thus sustaining a continuous antioxidant defense system (Pisoschi & Pop, 2015).

#### 15.5.3.4 Modulation of Cellular Antioxidant Enzymes

The combination of phenolics and other compounds can enhance the expression of endogenous antioxidant enzymes such as superoxide dismutase (SOD), catalase (CAT), and glutathione (GSH) peroxidase, thereby improving the cellular ability to cope with redox stress.

#### 15.5.3.5 Metal Chelation Synergism

Phenolic compounds and flavonoids possess the capability to chelate pro-oxidant metal ions ($Fe^{2+}$, $Cu^{2+}$), which helps to prevent Fenton-type reactions. The presence of co-compounds further enhances this chelating ability, thereby reducing the generation of ROS at their source.

#### 15.5.3.6 Reduction of Lipid Peroxidation

The synergistic actions of these compounds inhibit the propagation of lipid radical chains within cellular membranes by breaking these radical chains and stabilizing the integrity of the membranes.

#### 15.5.3.7 Modulation of Redox-Sensitive Signaling Pathways

The synergistic effects of phenolics modulate redox-sensitive transcription factors (such as Nrf2 and NF-κB), which enhance cellular defense mechanisms and diminish oxidative stress-related inflammatory pathways.

#### 15.5.3.8 Lowering Redox Potential

Certain synergistic combinations can lower the redox potential necessary for phenolic compounds to function as antioxidants, thereby increasing their effectiveness in scavenging highly reactive radicals, including hydroxyl radicals.

#### 15.5.3.9 Complex Formation

The formation of complexes serves as a fundamental mechanism through which phenolic compounds synergistically engage with various natural substances (such as flavonoids, alkaloids, metal ions, proteins, and terpenoids) to improve bioactivity, stability, and bioavailability.

##### *15.5.3.9.1 Hydrogen Bonding and π–π Stacking*

Phenolic compounds contain hydroxyl groups and aromatic rings that establish hydrogen bonds and π–π interactions with other polyphenols, flavonoids, and alkaloids, leading to the formation of stable complexes that improve solubility and offer protection against degradation (Bourassa et al., 2016).

##### *15.5.3.9.2 Metal Ion Chelation Complexes*

Phenolics (such as gallic acid and catechins) are capable of chelating metal ions ($Fe^{2+}$, $Cu^{2+}$, $Zn^{2+}$) through their hydroxyl and carbonyl groups, resulting in stable complexes that diminish the pro-oxidant activity of metal ions and inhibit Fenton reactions, thereby reducing the generation of ROS (Santos et al., 2017).

##### *15.5.3.9.3 Phenolic–Protein Complexes*

Phenolics interact with proteins through hydrogen bonding, hydrophobic interactions, and covalent bonding, resulting in the formation of either soluble or insoluble complexes. This interaction can influence enzyme activity (for instance, by inhibiting digestive enzymes), stabilize proteins, and safeguard phenolics from degradation during the digestive process (Ozdal, Capanoglu, & Altay, 2013).

##### *15.5.3.9.4 Complexes with Vitamins*

Phenolics create non-covalent complexes with vitamins (such as vitamins C and E), which enhance the recycling, stability, and activity of antioxidants within biological systems.

#### 15.5.3.10 Synergistic Phytochemical Complexes

Phenolics and flavonoids can form complexes with other secondary metabolites from plants (such as terpenoids and carotenoids), which stabilize each other against oxidation and enhance their bioefficacy in antioxidant and antimicrobial applications.

#### 15.5.3.11 Increased Solubility and Bioavailability

The formation of complexes enhances the aqueous solubility of hydrophobic phenolics (for example, quercetin), thereby improving their absorption and bioavailability in the gastrointestinal tract.

### 15.5.4 Modulation of Biological Activity by Synergistic Action of Polyphenols and Their Mechanism of Action

Complexes can modify the redox potential of phenolic compounds, thereby enhancing their antioxidant, antimicrobial, anti-inflammatory, and anticancer properties by increasing their stability and reactivity through *in vitro* and *in vivo* biochemical activities.

#### 15.5.4.1 Biochemical Interaction *(In Vitro)*

##### *15.5.4.1.1 Enzyme Inhibition*

Phenolic compounds, when combined with other natural compounds (such as flavonoids, alkaloids, terpenoids, and vitamins), exhibit synergistic enzyme inhibition, thereby enhancing biological activities, including antioxidant, antidiabetic, anti-inflammatory, and anticancer effects.

*Direct Binding to Enzyme Active Sites:* Phenolic compounds interact with amino acid residues in enzyme active sites through hydrogen bonding, hydrophobic interactions, and π–π stacking, thereby obstructing substrate access and impairing enzyme functionality.

*Allosteric Modulation:* Phenolics can bind to allosteric sites, inducing conformational changes in the enzyme's structure, which reduce catalytic efficiency.

*Metal Ion Chelation:* Numerous enzymes necessitate metal cofactors ($Fe^{2+}$, $Cu^{2+}$, $Zn^{2+}$) for their activity. Phenolics chelate these metal ions, depriving enzymes of crucial cofactors, resulting in enzyme inactivation (Santos et al., 2017).

*Synergistic Inhibition with Other Compounds:* Phenolics and flavonoids demonstrate additive or synergistic inhibition of carbohydrate-digesting enzymes (α-amylase, α-glucosidase), contributing to antidiabetic effects by mitigating postprandial glucose spikes (Tundis, Loizzo, & Menichini, 2010). Phenolics and alkaloids, when combined, can inhibit acetylcholinesterase, supporting neuroprotection and enhancing memory (Orhan et al., 2007). Phenolics and terpenoids exhibit increased inhibition of lipase enzymes, thereby decreasing fat digestion and absorption in anti-obesity approaches (Hsu & Yen, 2007).

*Oxidative Modulation of Enzyme Activity:* Phenolics and their synergistic partners mitigate oxidative stress, protecting cellular enzymes from oxidative inactivation while selectively inactivating pro-oxidant enzymes, such as xanthine oxidase.

*Protein Precipitation:* Phenolic compounds can induce enzyme precipitation by forming insoluble complexes, which diminishes their catalytic activity, a phenomenon frequently observed in microbial enzyme inhibition during antimicrobial actions (Daglia, 2012).

##### *15.5.4.1.2 Membrane Interaction*

Phenolic compounds, when combined with other natural substances such as flavonoids, terpenoids, alkaloids, and essential oils, exhibit synergistic interactions with membranes, thereby enhancing their antimicrobial, antioxidant, and therapeutic effects.

*Disruption of Membrane Integrity:* Phenolic compounds penetrate microbial or cellular membranes owing to their amphipathic characteristics, which enhance membrane fluidity and permeability, resulting in the leakage of cellular contents (Daglia, 2012). The co-administration of terpenoids (such as thymol and carvacrol) amplifies this disruption, demonstrating synergistic antimicrobial properties (Burt, 2004).

*Interaction with Membrane Lipids:* Phenolics engage with phospholipid bilayers, disrupting lipid organization and inducing structural changes that modify membrane potential and ion gradients.

*Modification of Membrane Proteins:* Phenolics and their synergistic counterparts can interact with membrane-bound enzymes and transport proteins, influencing nutrient absorption and cellular signaling.

*Induction of Membrane Depolarization:* The combination of phenolics and essential oils can lead to the depolarization of microbial membranes, hindering ATP production and energy metabolism, ultimately resulting in cell death.

*Modulation of Lipid Rafts:* Phenolic compounds can influence lipid raft domains within membranes, affecting receptor-mediated signaling and enzymatic activity associated with the membrane.

*Protection of Membranes from Oxidative Damage:* Phenolics, when combined with vitamins (such as vitamins E and C), safeguard cell membranes against lipid peroxidation caused by ROS, thereby preserving membrane integrity and functionality (Rice-Evans, Miller, & Paganga, 1997).

*Synergistic Permeabilization:* Phenolics and essential oils demonstrate enhanced permeability effects on microbial membranes, resulting in increased antimicrobial efficacy at reduced concentrations. Phenolics and alkaloids can synergistically augment membrane disruption, thereby enhancing both antimicrobial and anticancer effects.

*Enhanced Transport Across Membranes:* Interactions with membranes can facilitate the absorption of phenolics, thereby increasing their bioavailability and systemic impact.

#### 15.5.4.2 Biochemical Interaction *(In Vivo)*

Phenolic compounds, when integrated with various natural substances (such as flavonoids, alkaloids, terpenoids, vitamins, and minerals), demonstrate synergistic biological interactions in vivo, thereby amplifying their antioxidant, anti-inflammatory, antidiabetic, anticancer, cardioprotective, and neuroprotective properties.

###### 15.5.4.2.1 *Improved Bioavailability*

Numerous phenolic compounds exhibit low bioavailability due to inadequate absorption and swift metabolism. The co-administration of other substances (such as piperine and vitamin C) enhances absorption, minimizes degradation, and boosts systemic availability (Manach et al., 2004).

###### 15.5.4.2.2 *Combined Antioxidant Defense*

*In vivo*, phenolics and vitamins (C, E) work together to strengthen the antioxidant defense system by eliminating ROS in blood and tissues, restoring oxidized antioxidants (for instance, ascorbate restoring flavonoids), increasing the activity of antioxidant enzymes (SOD, CAT, GPx), thereby diminishing oxidative stress in organs (Pisoschi & Pop, 2015).

###### 15.5.4.2.3 *Modulation of Anti-Inflammatory Pathways*

Phenolics and their synergistic compounds inhibit NF-κB, COX-2, and pro-inflammatory cytokines (TNF-α, IL-6) in vivo, resulting in a reduction of systemic and tissue-specific inflammation in models of arthritis, colitis, and cardiovascular inflammation.

###### 15.5.4.2.4 *Enzyme Modulation*

*In vivo*, phenolics work synergistically with other substances to inhibit: Carbohydrate-hydrolyzing enzymes (which contribute to antidiabetic effects), lipases (which help combat obesity), and acetylcholinesterase (providing neuroprotective benefits), thereby enhancing therapeutic outcomes.

###### 15.5.4.2.5 *Modulation of Gut Microbiota*

Phenolics engage with gut microbiota, and when paired with prebiotics, they boost the production of advantageous metabolites (such as SCFAs), thereby promoting gut health and overall immune function (Cardona et al., 2013).

###### 15.5.4.2.6 *Hormonal and Signal Modulation*

Compounds that work in synergy with phenolics influence insulin signaling pathways (including AMPK and PI3K/Akt), lipid metabolism pathways (notably PPARγ), and apoptosis pathways relevant to cancer (such as caspases and the Bcl-2 family), thereby facilitating disease management.

###### 15.5.4.2.7 *Tissue-Specific Accumulation and Protection*

Phenolics are stored in specific tissues (such as the liver, brain, and heart) where they, along with co-compounds, offer protection against oxidative and inflammatory damage while influencing cellular pathways.

###### 15.5.4.2.8 *Reduction of Drug Resistance*

The combination of phenolics and alkaloids (for instance, curcumin and piperine) in vivo can suppress drug efflux transporters, thereby enhancing the effectiveness of chemotherapeutic and antimicrobial treatments (Anand et al., 2007).

## 15.6 TECHNIQUES TO IDENTIFY THE SYNERGISM OF NATURAL PHENOLICS WITH OTHER NATURAL COMPOUNDS

### 15.6.1 *In Vivo* Techniques

*In vivo* studies play a crucial role in confirming the synergistic effects of phenolic compounds in conjunction with other natural bioactives. These investigations offer valuable insights into the biological significance of these interactions within physiological contexts. Various in vivo methodologies can be utilized to evaluate synergism, typically concentrating on antioxidant activity, anti-inflammatory properties, metabolic modulation, and disease prevention. The following are important in vivo techniques and models that are commonly employed.

#### 15.6.1.1 Animal Models for Oxidative Stress and Antioxidant Evaluation

The synergistic antioxidant properties of phenolic compounds in conjunction with other natural substances can be assessed in animal models subjected to chemically induced oxidative stress (for instance, $CCl_4$, $H_2O_2$, or doxorubicin). To evaluate the antioxidant response, markers such as malondialdehyde (MDA), SOD, CAT, and GSH are quantified (El-Shitany, El-Haggar, & El-Desoky, 2014).

For instance, the combined administration of quercetin and resveratrol in rats resulted in a significant decrease in MDA levels and an enhancement in SOD and CAT activity compared to the administration of either compound alone, thereby demonstrating synergistic effects.

#### 15.6.1.2 *In Vivo* Anti-Inflammatory Models

Rodent models, including those induced by carrageenan and formalin, are extensively utilized. The co-administration of phenolic-rich extracts alongside other natural anti-inflammatory substances (such as curcumin and gingerols) can demonstrate a synergistic effect in decreasing edema, cytokines (TNF-$\alpha$, IL-6), and prostaglandin levels (Banerjee et al., 2012). For instance, a synergistic anti-inflammatory effect of curcumin and epigallocatechin gallate (EGCG) has been documented in a mouse model of arthritis.

#### 15.6.1.3 Disease-Specific Animal Models

*In vivo disease-specific animal models* (e.g., diabetes, cancer, and neurodegeneration) can be observed. *In vivo* synergism can also be examined in disease models by observing disease progression, biomarkers, and histopathological alterations (Liu et al., 2017).

Diabetes model rats with streptozotocin-induced diabetes, treated with a combination of ferulic acid and berberine, exhibited improved hypoglycemic and insulin-sensitizing effects compared to the effects of individual treatments (Zhou et al., 2019). Cancer Models as Xenograft models utilizing mice are employed to investigate the reduction of tumor volume through phenolic-based combinations (such as resveratrol and genistein), demonstrating enhanced apoptosis and diminished tumor growth (Liu et al., 2017).

##### *15.6.1.3.1 Pharmacokinetic and Bioavailability Studies*

Phenolic compounds often exhibit enhanced absorption or prolonged circulation when combined with other natural substances, thereby indirectly promoting synergy (Shoba et al., 1998). For instance, piperine, derived from black pepper, boosts the bioavailability of curcumin, thereby augmenting its therapeutic efficacy in vivo.

##### *15.6.1.3.2 Biomarker and Gene Expression Analysis*

*In vivo* synergy is further corroborated by examining the expression of genes and proteins associated with antioxidant defense, inflammation, or apoptosis (e.g., Nrf2, NF-$\kappa$B, caspases) using techniques such as RT-PCR, ELISA, or Western blotting. For example, the combination of polyphenol treatments resulted in a more significant upregulation of Nrf2 and a more pronounced downregulation of NF-$\kappa$B compared to monotherapies in rat brain tissue (Sgarbossa, Giacomazza, & Di Carlo, 2015).

### 15.6.2 *In Vitro* Techniques

*In vitro* techniques provide a regulated and economical platform for the preliminary evaluation of the synergistic effects between natural phenolic compounds and other natural bioactive substances. These methods assist in identifying combinations that exhibit improved biological activity (such as antioxidant, antimicrobial, anti-inflammatory, and anticancer properties) and are crucial prior to advancing to in vivo validation.

Below is a comprehensive overview of frequently utilized *in vitro* techniques for investigating such synergism.

#### 15.6.2.1 Antioxidant Assays

The synergistic effect of antioxidant activity is often evaluated by merging phenolic compounds with other substances and analyzing their impact in relation to the cumulative effects of each component.

##### *15.6.2.1.1 DPPH (2,2-Diphenyl-1-Picrylhydrazyl) Assay*

It assesses the scavenging of free radicals. A combination that demonstrates a significantly higher scavenging effect than the anticipated additive effect suggests synergism. The synergistic antioxidant activity of rosmarinic acid and ascorbic acid is determined by the DPPH assay (Akinmoladun et al., 2010).

##### *15.6.2.1.2 ABTS, FRAP, ORAC Assays*

There are numerous other chemical reactions through which antioxidant activities can be assessed. All these assess the antioxidant capacity through various mechanisms. A range of assays helps confirm synergism across diverse radical systems: ABTS (2,2′-azino-bis(3-ethylbenzothiazoline-6-sulfonic acid)), FRAP (ferric reducing antioxidant power), ORAC (oxygen radical absorbance capacity). The synergistic improvement in ABTS scavenging achieved by the combination of curcumin and catechin (Kumar et al., 2013).

#### 15.6.2.2 Antimicrobial Assays

The synergistic effects of antimicrobial activity can be evaluated through standardized microbiological techniques.

##### *15.6.2.2.1 Checkerboard Assay*

It employs a two-dimensional array of concentrations for two substances. The fractional inhibitory concentration index (FICI) is determined: FICI $< 0.5$ designates synergism, FICI $= 0.5$–1 indicates additivity, and FICI $> 1$ indicates antagonism. The example includes the synergistic antimicrobial effect of eugenol and gallic acid on *S. aureus* (Bouarab-Chibane et al., 2019).

##### *15.6.2.2.2 Time-Kill Assay*

It assesses the viability of bacteria over time across various treatments. Synergistic effects are indicated by a more significant logarithmic reduction in colony count.

#### 15.6.2.3 Cytotoxicity and Anticancer Activity

There are so many ways in which anticancer assays could be done.

##### *15.6.2.3.1 MTT or XTT Assay*

MTT (3-(4,5-dimethylthiazol-2-yl)-2,5-diphenyl-2H-tetrazolium bromide) and XTT (2,3-bis (2-methoxy-4-nitro-5-sulfophenyl)-2H-tetrazolium-5-carboxanilide) are used as substrates for anticancer assays. The cell viability assay and cell proliferation assay are employed in both cancerous and normal cell lines. By measuring the reduction of a yellow tetrazolium salt (MTT) into a purple formazan product by metabolically active cells, the concept of cellular viability is established. The combination index (CI) is determined using the Chou–Talalay Method. A CI value of less than 1 signifies synergism. Tools such as CompuSyn may be utilized for accurate analysis. The combined cytotoxic impacts of resveratrol and curcumin on human colon cancer cells had been observed (Park, Kim, & Lee, 2014).

##### *15.6.2.3.2 Fluorimetric and Colorimetric Assay*

Spectrofluorimetrically and colourimetrically measuring the lactate dehydrogenase (LDH) activity, the cell proliferation and viability could be assessed with and without the polyphenolic extracts.

#### 15.6.2.4 Enzyme Inhibition Assays

Natural phenolic compounds frequently interact with enzymes, including the following:

- α-glucosidase and α-amylase (which are antidiabetic),
- Acetylcholinesterase (which provides neuroprotection),
- Lipoxygenase and COX-2 (which have anti-inflammatory properties).

Dual or Triple Enzyme Inhibition Models involve assessing whether a combination demonstrates greater enzyme inhibition compared to individual compounds.

For instance, the combined phenolics from tea and cinnamon exhibited synergistic α-glucosidase inhibition (Adisakwattana et al., 2012).

#### 15.6.2.5 Gene Expression and Molecular Mechanism (Optional *In Vitro*)

qPCR, Western blot, and ELISA are capable of detecting alterations in gene or protein expression (for instance, NF-κB, COX-2, Nrf2, Bcl-2).

Molecular synergy is established when synergistic compounds collectively influence pathways to a greater extent than they do separately.

Example: The synergistic enhancement of Nrf2 and HO-1 through the combination of quercetin and sulforaphane (Tanigawa, Fujii, & Hou, 2007).

#### 15.6.2.6 Lipid Peroxidation Assay (TBARS/MDA)

This assay is utilized for assessing oxidative damage in cellular or tissue homogenates.

The combinations of phenolic compounds that exhibit enhanced inhibition of MDA production indicate a synergistic effect.

### 15.6.3 *In Silico* Technique

*In silico* methods have become essential in contemporary pharmacognosy and drug discovery, particularly in exploring the synergistic interactions among natural phenolic compounds and other natural substances. These computational techniques facilitate cost-efficient, high-throughput screening, prediction of mechanisms of action, target binding, and interaction potential, thereby assisting in the guidance and validation of in vitro and in vivo studies.

#### 15.6.3.1 Molecular Docking Investigations

To forecast the binding affinity and orientation of compounds toward biological targets (such as enzymes, receptors, and DNA), phenolic compounds are docked individually and in conjunction with other natural molecules onto pertinent protein targets. The synergism indicator is designated as when the combined binding energies are lower than the total of individual energies, or when co-binding occurs at different active/allosteric sites, resulting in enhanced activity. The docking of resveratrol and berberine demonstrated co-binding to DNA topoisomerase, exhibiting a superior binding affinity compared to each compound separately (Saad et al., 2014).

#### 15.6.3.2 Molecular Dynamics (MD) Simulations

To assess the stability and behavior of compound-target complexes over time at the atomic scale, it aids in validating docking outcomes. Capable of identifying synergistic stabilization of protein–ligand complexes in co-binding situations, the combination of quercetin and curcumin with the COX-2 enzyme demonstrated enhanced stability and decreased RMSD fluctuations when compared to the individual compounds (Khan et al., 2019).

#### 15.6.3.3 Network Pharmacology and Systems Biology

This technique illustrates multi-target interactions and forecasts synergistic pathways by utilizing protein–protein interaction (PPI) networks and gene-pathway mapping. It evaluates the impact of phenolic compounds and others on various disease-related targets simultaneously. It emphasizes overlapping or complementary pathways to uncover synergistic potential. An analysis of network pharmacology concerning EGCG and sulforaphane revealed shared targets within the Nrf2-mediated antioxidant pathway (Li et al., 2014).

#### 15.6.3.4 Quantitative Structure–Activity Relationship (QSAR)

It employs statistical models to establish a correlation between chemical structure and biological activity. It can forecast the cumulative activity of two or more phenolics based on their descriptors (e.g., logP, H-bonding, molecular weight). The potential for synergy is deduced from the comparison of predicted and actual activities. Quantitative structure–activity relationship (QSAR) modeling predicted an increased α-glucosidase inhibitory activity of a flavonoid-alkaloid combination (Roy, Kar, & Das, 2015).

#### 15.6.3.5 Artificial Intelligence (AI) and Machine Learning (ML) in Synergy Prediction

By utilizing big data of machine learning (ML) for data-driven predictions of compound combinations that may exhibit synergy. It was trained using datasets related to bioactivity, molecular fingerprints, and docking studies. It enables the prediction of synergistic combinations of drugs (drug–drug or compound–compound). ML models developed on databases of phenolic compounds successfully predicted pairs that demonstrate known antioxidant and antimicrobial properties (Sun et al., 2020).

#### 15.6.3.6 ADMET and Toxicity Profiling

ADMET is to forecast absorption, distribution, excretion and toxicity to confirm that synergistic combinations are pharmacologically feasible. It ensures that the synergy does not elevate toxicity or diminish bioavailability. The combination of curcumin and EGCG demonstrated synergistic anti-inflammatory effects without heightened hepatotoxicity, as indicated by in silico ADMET evaluation (Daina, Michielin, & Zoete, 2017).

## 15.7 IMPLICATIONS OF SYNERGISTIC ACTIONS OF PHENOLICS WITH OTHER COMPOUNDS

There is a wide range of implications of synergistic actions of phenolics with other compounds.

### 15.7.1 Drug Designing (Pharmaceutical Aspect)

The concept of synergy between phenolic compounds and various natural or synthetic bioactives holds significant implications for the design and development of pharmaceutical drugs. These synergistic interactions enhance therapeutic effectiveness, minimize toxicity, and counteract drug resistance, thereby establishing natural phenolic-based combinations as valuable assets in contemporary pharmacotherapy (Wang et al., 2017).

### 15.7.2 Enhanced Therapeutic Efficacy

Synergism enables the use of lower doses of each compound while still achieving greater or equivalent biological effects compared to higher doses of individual agents. This is especially significant in cancer therapy: Phenolics such as resveratrol, curcumin, and quercetin, when used in conjunction with chemotherapeutic agents (e.g., paclitaxel, cisplatin), promote apoptosis and decrease tumor growth. In metabolic disorders, the synergistic inhibition of enzymes such as α-amylase and α-glucosidase, achieved through the combination of phenolic and alkaloid compounds, contributes to improved glycemic control. Curcumin works synergistically with doxorubicin, enhancing its effectiveness in breast cancer treatment while minimizing cardiotoxicity (Wang et al., 2017).

### 15.7.3 Minimizing Side Effects and Toxicity

By reducing the necessary dosage of traditional medications through synergy with phenolics, side effects can be lessened. Anti-inflammatory medications: The combined use of NSAIDs with flavonoids such as quercetin or EGCG can decrease gastric toxicity while preserving anti-inflammatory effects. Flavonoids amplify the anti-inflammatory efficacy of aspirin without exacerbating gastric mucosal injury (Rauf et al., 2018).

### 15.7.4 Overcoming Drug Resistance

A significant challenge in the pharmaceutical field is the issue of multidrug resistance (MDR), which is found in both pathogens and cancer cells. Phenolic compounds play a role in modulating proteins related to resistance and efflux pumps, thereby improving the effectiveness of standard medications. To combat cancer, phenolics reduce the expression of MDR genes (such as P-glycoprotein), which leads to increased accumulation of drugs within cells. The antibiotic resistance is observed when the combination of phenolics and antibiotics (for instance, gallic acid paired with tetracycline) reinstates effectiveness against bacteria that have developed resistance (Cushine et al., 2018).

### 15.7.5 Multi-Target Drug Design (Polypharmacology)

Phenolic compounds naturally interact with various molecular targets, which corresponds with the developing polypharmacological strategy in drug design. Combination therapy utilizing phenolics provides multi-target modulation, proving particularly effective in complex diseases such as neurodegenerative disorders (e.g., Alzheimer's, Parkinson's), inflammatory diseases, and metabolic syndromes. A blend of curcumin and EGCG addresses oxidative stress, neuroinflammation, and amyloid aggregation in Alzheimer's disease (Singh et al., 2020).

#### 15.7.5.1 Pharmacokinetic Enhancement and Bioavailability

Some phenolic compounds enhance the bioavailability of concurrently administered medications by inhibiting metabolic enzymes (such as CYP450), inhibiting efflux transporters (like P-gp), and improving intestinal permeability. Piperine, a naturally occurring alkaloid, markedly boosts the bioavailability of curcumin (by approximately 2000%) (Shoba et al., 1998).

#### 15.7.5.2 Development of Innovative Pharmaceutical Formulations

Synergistic combinations of phenolic compounds motivate the creation of fixed-dose combinations (FDCs), nanoformulations (including liposomes and solid lipid nanoparticles), co-crystals, or prodrugs that integrate phenolics with synthetic medications. Nanoparticles loaded with curcumin and resveratrol demonstrate improved anti-inflammatory effects and prolonged release (Yallapu et al., 2014).

#### 15.7.5.3 Challenges in Pharmaceutical Application

Despite the significant potential, the conversion of synergistic phenolics into pharmaceuticals presents several challenges, such as standardization and optimization of dosage, issues related to stability (sensitive to light, pH, and temperature), complicated pharmacokinetics, and regulatory constraints concerning botanical drug combinations.

#### 15.7.5.4 Health Promotions

The synergistic interactions among phenolic compounds and various natural bioactives hold significant implications for health promotion, particularly through preventive measures, functional foods, and nutraceuticals. Such interactions frequently enhance the bioactivity of individual compounds, leading to improved physiological outcomes, more effective disease prevention, and a general enhancement of overall well-being (Singh et al., 2020).

#### *15.7.5.4.1 Improved Antioxidant Protection*

One of the key health advantages of phenolic compounds lies in their antioxidant properties. When they are synergistically combined with other natural substances (such as vitamins, alkaloids, and terpenes), they offer exceptional defense against oxidative stress. Phenolics neutralize free radicals, restore other antioxidants (such as vitamin C or E), and stimulate the activity of endogenous antioxidant enzymes (SOD, CAT). The simultaneous intake of quercetin and vitamin C results in a greater reduction in lipid peroxidation and oxidative DNA damage (Lotito & Frei, 2006).

#### *15.7.5.4.2 Cardiovascular Health*

Synergistic combinations of phenolic compounds with various nutraceuticals (such as omega-3 fatty acids, garlic compounds, and coenzyme Q10) assist in regulating blood pressure, cholesterol levels, and endothelial function. Grape seed proanthocyanidins combined with resveratrol enhance vascular relaxation and lower blood pressure more effectively than either component alone (Pérez-Jiménez et al., 2010).

#### *15.7.5.4.3 Cognitive and Neuroprotective Benefits*

Synergistic interactions among phenolic compounds contribute to brain health by promoting neurogenesis, reducing neuroinflammation, and protecting neurons from oxidative damage. The combination of curcumin and EGCG offers synergistic defense against β-amyloid toxicity in models of Alzheimer's disease (Ahmed, Gilani, & Abdollahi, 2017).

#### *15.7.5.4.4 Metabolic Health and Obesity Management*

Synergistic combinations of phenolic compounds enhance glucose metabolism, regulate lipid levels, and improve energy expenditure. The combination of green tea catechins and chlorogenic acid has been shown to significantly reduce body fat, improve insulin sensitivity, and impact gut microbiota (Hsu et al., 2008).

#### *15.7.5.4.5 Immunomodulation and Anti-Inflammatory Action*

Synergistic interactions between phenolics and terpenoids, flavonoids, or sulfur-containing compounds can influence immune responses and mitigate chronic inflammation. The combined anti-inflammatory effect of quercetin and sulforaphane in decreasing cytokine release (IL-6, TNF-α) in immune cells (Dinkova-Kostova & Kostov, 2012).

#### *15.7.5.4.6 Gut Microbiome and Digestive Health*

Phenolic compounds possess prebiotic properties, and when paired with dietary fibers or probiotics, they enhance the diversity of gut microbiota, reduce inflammation, and strengthen gut barrier function. Berries rich in polyphenols combined with inulin work together to boost beneficial bacteria such as *Bifidobacterium* and *Lactobacillus* (González-Sarrías et al., 2017).

#### *15.7.5.4.7 Skin Health and Anti-Aging*

Topical or dietary combinations of phenolic antioxidants with vitamins (such as E and C), coenzyme Q10, or carotenoids can work together to enhance their effects (Grether-Beck et al., 2012). They are effective in minimizing UV-induced damage, enhancing collagen production, and postponing the visible signs of aging.

#### *15.7.5.4.8 Cancer Prevention*

Phenolic compounds, when combined with other phytochemicals, have the potential to (Lambert et al., 2007). They can trigger apoptosis, prevent angiogenesis, and diminish proliferation.

### 15.7.6 Food Preservation

Natural phenolic compounds are gaining recognition as effective, safe, and sustainable substitutes for synthetic food preservatives. When combined with other natural substances (such as essential oils, organic acids, flavonoids, and terpenes), phenolics exhibit improved antimicrobial, antioxidant, and shelf-life-extending characteristics, providing a multifunctional approach to food preservation.

#### 15.7.6.1 Improved Antimicrobial Efficacy

Phenolic compounds compromise microbial membranes, hinder enzyme activity, and bind to vital metal ions. Their cooperative effect with other natural substances often results in a broader antimicrobial range, lower minimum inhibitory concentrations (MICs), and reduced development of microbial resistance. The combination of thymol and gallic acid demonstrated synergistic inhibition of *Listeria monocytogenes* in meat products (Burt, 2004).

#### 15.7.6.2 Enhanced Antioxidant Defense

Oxidative deterioration leads to rancidity, color changes, and nutrient depletion. The synergistic integration of phenolics with vitamins (C, E), carotenoids, or components of essential oils can more efficiently inhibit lipid peroxidation. The combination of green tea polyphenols and rosemary extract effectively prolongs the oxidation process in vegetable oils (Shan et al., 2005).

#### 15.7.6.3 Shelf-Life Extension of Perishable Foods

Synergistic natural preservatives enhance the microbial and oxidative stability of meat, fish, dairy products, fruits, and baked goods. It minimizes the growth of spoilage organisms (e.g., *Pseudomonas*, *Salmonella*). It prolongs the oxidation of lipids and proteins. It preserves organoleptic properties (flavor, color, texture). Quercetin combined with lactic acid prolongs the shelf life of chicken fillets when stored in refrigeration (Al-Hijazeen et al., 2016).

#### 15.7.6.4 Reduced Need for Synthetic Preservatives

Natural synergistic mixtures can diminish or substitute synthetic substances such as BHA, BHT, sodium benzoate, or nitrites, responding to consumer preferences for "Clean-label" items, environmentally friendly processing, and health-oriented formulations. The combination of cinnamaldehyde and ferulic acid demonstrated preservative effects similar to those of sodium nitrite in cured meat (Holley & Patel, 2005).

##### *15.7.6.4.1 Utilization in Edible Coatings and Packaging*

Phenolics, when combined with various natural antimicrobials, are integrated into edible films/coatings (for instance, chitosan and phenolics) and active packaging materials (such as phenolic-loaded nanofibers). These systems gradually release preservatives, enhancing protection throughout storage and transportation. The film made from chitosan and grape seed extract inhibits microbial growth and preserves the firmness of fresh-cut fruits (Dutta et al., 2009).

##### *15.7.6.4.2 Mechanistic Synergy*

Synergistic preservation arises from complementary mechanisms of action (for instance, membrane disruption combined with pH reduction), sequential effects (such as rapid elimination followed by prolonged inhibition), and enhanced solubility and permeability when used together. The combination of resveratrol and citral enhances the membrane permeability of spoilage bacteria more efficiently than either compound alone (Friedman, Henika, & Mandrell, 2002).

##### *15.7.6.4.3 Future Opportunities in Food Preservation*

Advancement of intelligent packaging that incorporates synergistic phenolic mixtures, application of ML and *in silico* methodologies to forecast ideal combinations, and increased emphasis on dosage standardization and obtaining regulatory approval are the future opportunities in food preservation.

## 15.8 CONCLUSION

The collaborative interactions between phenolic compounds and other natural substances signify a promising area in pharmacology, nutraceuticals, and the development of functional foods. By thoroughly understanding their various sources, chemical characteristics, and biological capabilities, it becomes clear that these interactions significantly enhance therapeutic results, often surpassing the effects of individual compounds. These synergistic effects are evident in a range of bioactivities, including antioxidant, antimicrobial, anticancer, antidiabetic, anti-inflammatory, sun protection, and anti-obesity properties, indicating their broad applicability in addressing complex health issues. From a mechanistic perspective, synergism occurs through several pathways, including redox modulation, complex formation, enzyme inhibition, and biochemical interactions both in vitro and in vivo. Contemporary methodologies—encompassing in vitro, in vivo, and *in silico* techniques—provide strong frameworks for discovering and confirming these interactions, thereby expediting their implementation in practical situations. The ramifications of these discoveries are significant. They range from the creation of more effective medications with fewer side effects to the innovation of new functional foods and natural preservatives. The incorporation of synergistic phenolic systems has the potential to transform strategies for health and disease management. Ongoing interdisciplinary research into these interactions is crucial for realizing the full therapeutic potential of natural compound synergy.

## COMPETING INTERESTS

The authors declare that they have no competing interest.

## DECLARATION OF GENERATIVE AI AND AI-ASSISTED TECHNOLOGIES IN THE WRITING PROCESS

During the preparation of this work, the book chapter author(s) have not taken any assistance from AI, and the book chapter author(s) reviewed and edited the content as needed and take(s) full responsibility for the content of the publication.

## REFERENCES

Adisakwattana, S., Chantarasinlapin, P., Thammarat, H., & Yibchok-Anun, S. (2012). Synergistic inhibition of α-glucosidase by plant-derived phenolic compounds and acarbose: The study in vitro. *Evidence-Based Complementary and Alternative Medicine*, *2012*, Article ID 281436.

Ahmed, T., Gilani, A. H., & Abdollahi, M. (2017). Role of natural products in neuroprotection against Alzheimer's disease. *Neurotoxicity Research*, *31*(4), 560–572.

Akinmoladun, F. O., Akinrinlola, B. L., Komolafe, T. O., Farombi, E. O., & Olaleye, T. M. (2010). Synergistic antioxidant activity of phytochemical combinations. *Journal of Medicinal Plants Research*, *4*(13), 1223–1230.

Al-Hijazeen, M., Lee, E. J., Mendonca, A., & Ahn, D. U. (2016). Effect of tannic acid on lipid and protein oxidation, color, and volatiles of ground chicken meat during storage. *Meat Science*, *117*, 14–20.

Anand, P., Kunnumakkara, A. B., Newman, R. A., & Aggarwal, B. B. (2007). Bioavailability of curcumin: Problems and promises. *Molecular Pharmaceutics*, *4*(6), 807–818.

Bahadoran, Z., Mirmiran, P., & Azizi, F. (2013) Dietary polyphenols as potential nutraceuticals in management of diabetes: a review.J Diabetes Metab Disord.Aug 13;12(1):43. doi: 10.1186/2251-6581-12-43.

Bajpai, V. K., Baek, K. H., & Kang, S. C. (2013). Control of salmonella in foods by using essential oils: A review. *Food Research International*, *45*(2), 722–734.

Balasundram, N., Sundram, K., & Samman, S. (2006). Phenolic compounds in plants and agri-industrial by-products: Antioxidant activity, occurrence, and potential uses. *Food Chemistry*, *99*(1), 191–203.

Banerjee, S., Singh, S. K., Chowdhury, I., & Lillard, J. W. (2012). Curcumin and EGCG combination treatment modulates inflammatory pathways. *Arthritis Research & Therapy*, *14*(2), R45.

Bouarab-Chibane, L., Forquet, V., Lantéri, P., Clément, Y., Léonard-Akkari, L., Oulahal, N., Degraeve, P., & Bordes, C. (2019). Antimicrobial activities of phenolic compounds. *Microorganisms*, *7*(11), 379.

Bourassa, P., Côté, R., Hutchandani, S., Samson, G., & Tajmir-Riahi, H. A. (2016). Binding of polyphenols with human serum albumin. *Biomolecules*, *6*(1), 25.

Burt, S. (2004). Essential oils: Their antibacterial properties and potential applications in foods—A review. *International Journal of Food Microbiology*, *94*(3), 223–253.

Cardona, F., Andrés-Lacueva, C., Tulipani, S., Tinahones, F. J., & Queipo-Ortuño, M. I. (2013). Benefits of polyphenols on gut microbiota and implications in human health. *Journal of Nutritional Biochemistry*, *24*(8), 1415–1422.

Crozier, A., Jaganath, I. B., & Clifford, M. N. (2009). Dietary phenolics: Chemistry, bioavailability, and effects on health. *Natural Product Reports*, *26*(8), 1001–1043.

Daglia, M. (2012). Polyphenols as antimicrobial agents. *Current Opinion in Biotechnology*, *23*(2), 174–181.

Dai, J., & Mumper, R. J. (2010). Plant phenolics: Extraction, analysis and their antioxidant and anticancer properties. *Molecules*, *15*(10), 7313–7352.

Daina, A., Michielin, O., & Zoete, V. (2017). SwissADME: A free web tool to evaluate pharmacokinetics, drug-likeness and medicinal chemistry friendliness of small molecules. *Scientific Reports*, *7*, 42717.

Dinkova-Kostova, A. T., & Kostov, R. V. (2012). Glucosinolates and isothiocyanates in health and disease. *Chemical Biology & Interactions*, *195*(2), 120–128.

Dulloo, A. G., Duret, C., Rohrer, D., Girardier, L., Mensi, N., Fathi, M., Chantre, P., Dutta, P. K., Tripathi, S., Mehrotra, G. K., & Dutta, J. (2009). Perspectives for chitosan-based antimicrobial films in food applications. *Food Chemistry*, *114*(4), 1173–1182.

El-Shitany, N. A., El-Haggar, S. M., & El-Desoky, K. (2014). Synergistic antioxidant effects of quercetin and resveratrol on oxidative stress. *Oxidative Medicine and Cellular Longevity*, *2014*, Article 125846.

Friedman, M., Henika, P. R., & Mandrell, R. E. (2002). Bactericidal activities of plant essential oils and some of their isolated constituents against *Campylobacter jejuni*, *Escherichia coli*, *Listeria monocytogenes*, and *Salmonellenterica*. *Journal of Agricultural and Food Chemistry*, *50*(6), 1639–1643.

Goiris, K., Muylaert, K., Fraeye, I., Foubert, I., De Brabanter, J., & De Cooman, L. (2014).Detection of flavonoids in microalgae from different evolutionary lineages.J Phycol. 2014 Jun;50(3):483-92. doi: 10.1111/jpy.12180.

González-Sarrías, A., García-Villalba, R., Núñez-Sánchez, M. A., Tomás-Barberán, F. A., & Espín, J. C. (2017). Identification of urolithin A as a metabolite produced by human gut microbiota from ellagic acid and its health implications. *Journal of Agricultural and Food Chemistry*, *65*(36), 7817–7825.

Grether-Beck, S., Buettner, R., Krutmann, J., & Ferrara, F. (2012). Non-enzymatic neutralization of singlet oxygen by antioxidants in skin cells: Synergistic protection with α-tocopherol. *Free Radical Biology and Medicine*, *53*(7), 1260–1270.

Gutiérrez, J., Barry-Ryan, C., & Bourke, P. (2008). The antimicrobial efficacy of plant essential oil combinations and interactions with food ingredients. *International Journal of Food Microbiology*, *124*(1), 91–97.

Hemaiswarya, S., Kruthiventi, A. K., & Doble, M. (2008). Synergism between natural products and antibiotics against infectious diseases. *Phytomedicine*, *15*(8), 639–652.

Holley, R. A., & Patel, D. (2005). Improvement in shelf-life and safety of perishable foods by plant essential oils and smoke antimicrobials. *Meat Science*, *71*(1), 100–121.

Hsu, C. H., Liao, Y. L., Lin, S. C., Tsai, T. H., & Huang, C. J. (2008). Does supplementation with green tea extract modify the body weight, blood lipid profile, and glucose homeostasis in obese subjects? A randomized double-blind trial. *Obesity*, *16*(6), 1461–1467.

Hsu, C. L., & Yen, G. C. (2007). Effects of flavonoids and phenolic acids on the inhibition of adipogenesis in 3T3-L1 adipocytes. *Journal of Agricultural and Food Chemistry*, *55*(21), 8404–8410.

Huang, W. Y., Cai, Y. Z., & Zhang, Y. (2010). Natural phenolic compounds from medicinal herbs and dietary plants: Potential use for cancer prevention. *Nutrition and Cancer*, *62*(1), 1–20.

Khan, M. F., Kumar, S., Malik, A., Bhushan, S., & Khan, M. J. (2019). Synergistic anti-inflammatory potential of polyphenols through MD simulation. *Journal of Molecular Graphics and Modelling*, *91*, 162–172.

Khan, N., & Mukhtar, H. (2007). Multitargeted therapy of cancer by green tea polyphenols. *Cancer Letters*, *269*(2), 269–280.

Kumar, S., Prahalathan, P., Raja, B., & Sivasankaran, K. (2013). Assessment of antioxidant synergy between catechin and curcumin. *Food Chemistry*, *138*(2–3), 1041–1049.

Lambert, J. D., Sang, S., Hong, J., Kwon, S. J., Lee, M. J., & Yang, C. S. (2007). Peracetylation as a means of enhancing in vitro bioactivity and bioavailability of (–)- epigallocatechin-3-gallate. *Cancer Letters*, *251*(2), 168–175.

Li, S., Zhang, B., Zhang, N., Chen, H., & Zhu, M. (2014). Network pharmacology in combination therapy discovery: A case study on synergy between EGCG and sulforaphane. *PLoS One*, *9*(4), e93392.

Lin, J. Y., & Lin, C. L. (2005). Photoprotective effect of topical formulations containing vitamin C and E against UV-induced damage in human skin. *Journal of the American Academy of Dermatology*, *52*(6), 867–871.

Liu, Y., Ma, W., Zhang, P., He, S., Huang, D., Wang, M., … & Lu, Y. (2017). Resveratrol and genistein synergistically inhibit prostate cancer growth. *Molecular Cancer Therapeutics*, *16*(12), 2776–2788.

López-Alarcón, C., & Denicola, A. (2013). Evaluating the antioxidant capacity of natural products: A review on chemical and cellular-based assays. *Analytica Chimica Acta*, *763*, 1–10.

Lotito, S. B., & Frei, B. (2006). Consumption of flavonoid-rich foods and increased plasma antioxidant capacity in humans: Cause, consequence, or epiphenomenon? *Free Radical Biology and Medicine*, *41*(12), 1727–1746.

Manach, C., Williamson, G., Morand, C., Scalbert, A., & Rémésy, C. (2004). Bioavailability and bioefficacy of polyphenols in humans. *American Journal of Clinical Nutrition*, *79*(5), 727–747.

Nichols, J. A., & Katiyar, S. K. (2010). Skin photoprotection by natural polyphenols: Anti-inflammatory, antioxidant and DNA repair mechanisms. *Archives of Dermatological Research*, *302*(2), 71–83.

Nutraceuticals in management of diabetes: A review. *Journal of Diabetes & Metabolic Disorders*, 12(1), 43.

Orhan, I., Kartal, M., Naz, Q., Ejaz, A., Yilmaz, G., Kan, Y., & Konuklugil, B. (2007). Antioxidant and anticholinesterase evaluation of selected Turkish Salvia species. *Food Chemistry*, *103*(4), 1247–1254.

Ozdal, T., Capanoglu, E., & Altay, F. (2013). A review on protein–phenolic interactions and associated changes. *Food Research International*, *51*(2), 954–970.

Park, S. Y., Kim, D. S., & Lee, S. K. (2014). Synergistic effects of curcumin and resveratrol on cell viability in human tumor cells. *Journal of Nutritional Biochemistry*, *25*(8), 865–872.

Pérez-Jiménez, J., Neveu, V., Vos, F., & Scalbert, A. (2010). Identification of the 100 richest dietary sources of polyphenols: An application of the Phenol-Explorer database. *British Journal of Nutrition*, *104*(11), 1582–1589.

Pisoschi, A. M., & Pop, A. (2015). The role of antioxidants in the chemistry of oxidative stress: A review. *European Journal of Medicinal Chemistry*, *97*, 55–74.

Rauf, A., Imran, M., Khan, I. A., Ur-Rehman, M., Gilani, S. A., Mehmood, Z., & Mubarak, M. S. (2018). Antibacterial and antifungal activities of polyphenols extracted from medicinal plants: A review. *Biomedicine & Pharmacotherapy*, *103*, 1091–1101.

Rice-Evans, C., Miller, N., & Paganga, G. (1997). Antioxidant properties of phenolic compounds. *Trends in Plant Science*, *2*(4), 152–159.

Rice-Evans, C. A., Miller, N. J., & Paganga, G. (1996). Structure-antioxidant activity relationships of flavonoids and phenolic acids. *Free Radical Biology and Medicine*, *20*(7), 933–956.

Roy, K., Kar, S., & Das, R. N. (2015). QSAR-based prediction of enzyme inhibitory synergy of natural compounds. *Journal of Molecular Modeling*, *21*(4), 100.

Saad, B., Azaizeh, H., Abu-Hijleh, G., & Said, O. (2014). Synergistic inhibition of DNA topoisomerase by phenolics: A molecular docking study. *Computational Biology and Chemistry*, *53*, 281–289.

Sanchez, M., Llorens, A., & Luque-Garcia, J. L. (2011). Phenolic compounds as natural photoprotective agents against UV radiation. *Journal of Photochemistry and Photobiology B: Biology*, *104*(1–2), 1–14.

Santos, C. M. M., Freitas, M., Ferreira, R., & Mateus, N. (2017). Metal chelation by flavonoids: Complexation of flavonoids with $Fe^{3+}$ and $Fe^{2+}$ ions. *Food Chemistry*, *219*, 373–382.

Scalbert, A., Johnson, I. T., & Saltmarsh, M. (2005). Polyphenols: Antioxidants and beyond. *The American Journal of Clinical Nutrition*, *81*(1), 215S–217S.

Sgarbossa, A., Giacomazza, D., & Di Carlo, M. (2015). Polyphenol combinations activate Nrf2 synergistically. *Neurochemistry International*, *89*, 134–142.

Shan, B., Cai, Y. Z., Sun, M., & Corke, H. (2005). Antioxidant capacity of 26 spice extracts and characterization of their phenolic constituents. *Food Chemistry*, *91*(3), 449–455.

Shankar, S., Ganapathy, S., & Srivastava, R. K. (2007). Green tea polyphenols: Biology and therapeutic implications in cancer. *Frontiers in Bioscience*, *12*, 4881–4899.

Shoba, G., Joy, D., Joseph, T., Majeed, M., Rajendran, R., & Srinivas, P. S. (1998). Influence of piperine on the pharmacokinetics of curcumin in animals and human volunteers. *Planta Medica*, *64*(4), 353–356.

Singh, S., Sharma, B., Kanwar, S. S., & Kumar, A. (2020). Lead phytochemicals for anticancer drug development. *Neurochemistry International*, *138*, 104758.

Sun, H., Pan, X., Guan, X., Zhang, J., & Chen, C. (2020). AI-based synergistic screening of natural polyphenols for potential therapeutic agents. *Briefings in Bioinformatics*, *21*(6), 2063–2075.

Tanigawa, S., Fujii, M., & Hou, D. X. (2007). Synergistic activation of the antioxidant response element by phytochemicals. *Biochemical and Biophysical Research Communications*, *362*(2), 575–579.

Tapiero, H., Tew, K. D., Ba, G. N., & Mathé, G. (2004). Polyphenols: Do they play a role in the prevention of human pathologies? *Biomedicine & Pharmacotherapy*, *58*(4), 171–177.

Tsao, R. (2010). Chemistry and biochemistry of dietary polyphenols. *Nutrients*, *2*(12), 1231–1246.

Tundis, R., Loizzo, M. R., & Menichini, F. (2010). Natural products as α-amylase and α-glucosidase inhibitors and their hypoglycaemic potential in the treatment of diabetes: An update. *Mini Reviews in Medicinal Chemistry*, *10*(4), 315–331.

Vauzour, D., Rodriguez-Mateos, A., Corona, G., Oruna-Concha, M. J., & Spencer, J. P. (2010). Polyphenols and human health: Prevention of disease and mechanisms of action. *Nutrients*, *2*(11), 1106–1131.

Vauzour, D., Vafeiadou, K., Rodriguez-Mateos, A., Rendeiro, C., & Spencer, J. P. (2008). The neuroprotective potential of flavonoids: A multiplicity of effects. *Genes & Nutrition*, *3*(3–4), 115–126.

Wang, J., Zou, Y., Yin, Z., Tan, Z., Liu, L., & Peng, J. (2017). A systematic approach for drug repositioning based on clinical side-effects and indications. *Oncotarget*, *8*, 60490–60504.

Wang, W., Sun, C., Mao, L., Ma, P., Liu, F., & Yang, J. (2016). The Biological Activities,chemical stability, metabolism and delivery systems of quercetin: A review, Trends in Food Science & Technology Volume 56, October 2016, pp 21–38

Wink, M. (2010). Introduction: Biochemistry, physiology and ecological functions of secondary metabolites. In Wink M (Ed.). *Annual plant reviews, biochemistry of plant secondary metabolism* (2nd ed.). Wiley-Blackwell, 1–19.

Yallapu, M. M., Khan, S., Maher, D. M., Ebeling, M. C., Sundram, V., Chauhan, N., … & Chauhan, S. C. (2014). Anti-cancer activity of curcumin loaded nanoparticles in prostate cancer. *Molecular Pharmaceutics*, *11*(5), 1594–1611.

Zhou, Y., Zhang, H., Su, J., Xu, F., Kong, Y., Peng, Y., & Yang, Y. (2019). Synergistic hypoglycemic effect of berberine and ferulic acid. *Journal of Ethnopharmacology*, *231*, 261–267.

# 16 Formulation Strategies for Enhancing Phenolic Compound Delivery

*Monica Butnariu**

## 16.1 INTRODUCTION

Polymeric materials, in general, are made up of the basic polymer and additives that favorably replace its physico-chemical characteristics (workability, flexibility, appearance, resistance to oxidation, resistance to the action of microorganisms, etc.). Thus, polymeric materials, depending on their nature and the category of applications for which they are intended, may contain distinct vegetable phenolic substances. It should be noted that the polymer itself is practically inert to the vegetable phenolic substances it comes into contact with due to the large size of its molecules. Instead, vegetable phenolic substances can replace the behavior of the base material in one way or another.

It is not enough that the vegetable phenolic substances used are not toxic but their efficiency about the oxidative decomposition process that takes place in the polymer is also significant [Kaur & Goyal, 2024]. Moreover, many vegetable phenolic substances admitted to the market can become toxic by increasing their concentration or their accumulation in the body. The vegetable phenolic substances used to obtain food packaging must meet certain conditions, something of which are distinctive to the general use of vegetable phenolic substances:

- Compatibility with the polymer, i.e. their molecules must be soluble in the amorphous phase of the polymer;
- Not to affect the color of the finished outcome by damaging the vegetable phenolic substances molecules;,
- Not to be toxic and not to emit unpleasant odors;
- To have great stability;
- Not to emit unpleasant odors, not to be toxic under any conditions, and not to chemically influence the substances;
- The incorporation content should not exceed 0.1–0.2% of the mass of the polymer.

The vegetable phenolic substances antioxidants on the market transform over time due to the oxidation process of the polymer into quinone configurations that are harmful to the human body [Martínková et al., 2016]. The remarkable antioxidant activity of vegetable organism extracts from the first group can be explained by the high content of flavonoids of the type shown in Figure 16.1.

The linear dependence of those values demonstrates the direct involvement of flavonoids in the stabilization of the paraffin substrate [Shi et al., 2021]. The reactions can show the antioxidant protection mechanism of flavonoids (Figure 16.2).

The general conclusion is favorable to the use of natural vegetable organism extracts as stabilizing agents for the thermal decomposition of polymers for their use as packaging in the food industry. From the group of flavonoids, the main constructions with antioxidant indispensable qualities are

* Corresponding author: monicabutnariu@yahoo.com



DOI: 10.1201/9781003527671-16

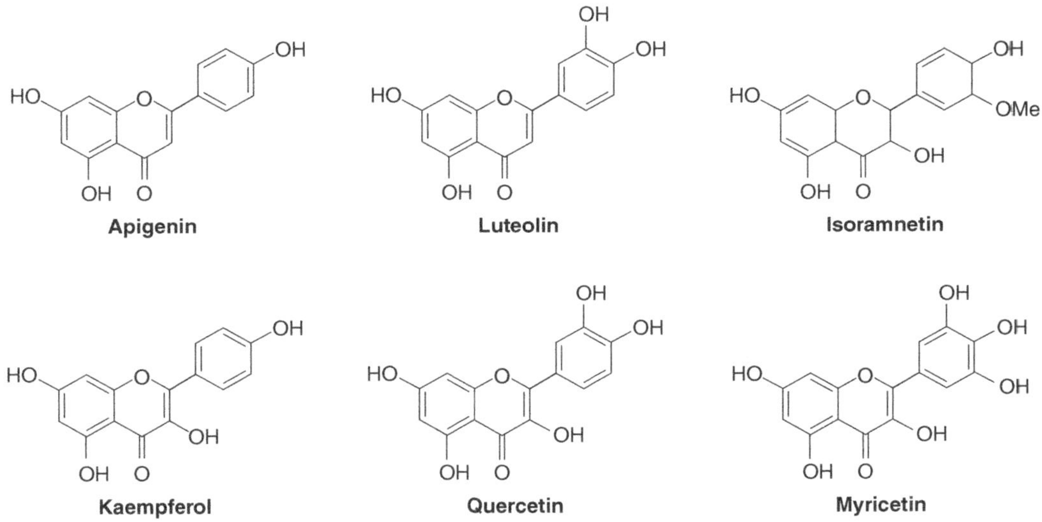


**FIGURE 16.1** The chemical structures of the flavonoids mentioned are as follows: Apigenin, Luteolin, Isoramnetin, Kaempferol, Quercetin and Myricetin.

catechins (colorless), flavonols and falvones (yellow), anthocyanins (colored in shades of red to blue), proanthocyanidols (colorless). In general, flavonoids are found in all vegetable organism organs, even in underground organs (roots, rhizomes), the richest being the leaves, fruits, stems, and barks [Davoudi et al., 2024]. Although polyphenolic compounds possess special, indispensable qualities worthy of being used at an industrial level, their stability to external constituents such as light, atmospheric oxygen, or temperature is relatively limited.

For this reason, vegetable phenolic substances have a major advantage in that the decomposition wares are, in turn, compounds with antioxidant and non-toxic indispensable qualities, and, from another point of view, the compatibility with the human body is perfect. Surveys have shown that

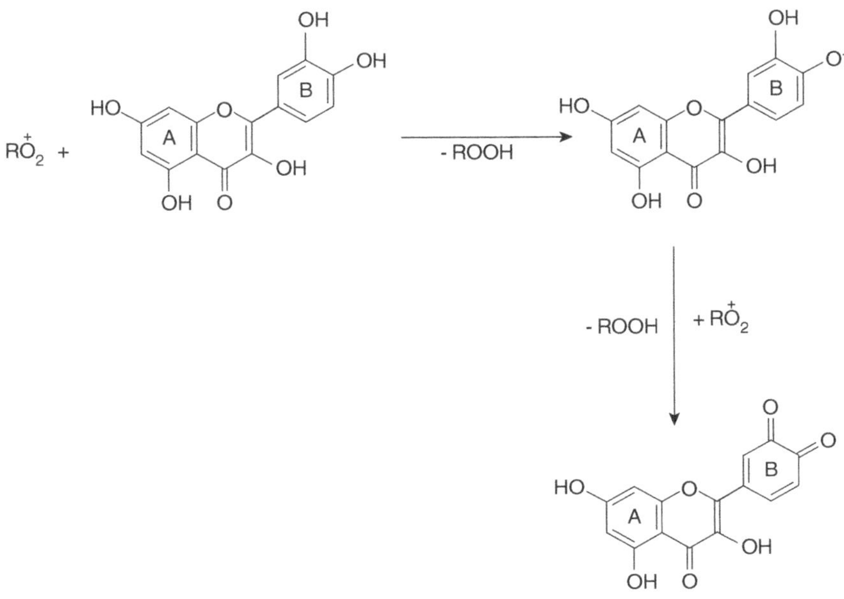


**FIGURE 16.2** The reactions that show the antioxidant protection mechanism of flavonoids.

the incorporation of polyphenolic compounds in polymeric materials or their formulation as nanomaterials not only intensifies the steadiness of vegetable phenolic substances to outside constituents but also associates with ameliorating the biological indispensable qualities of the incorporation material, increasing the value of the outcome obtained.

Thus, biocompounds containing extracts of vegetable phenolic substances from biomass find applicability in electrical and electronic industries in the configuration of metal complexes in circumstance arrangements by removing pollutants from wastewater when formulated as nanoparticles but also in the food industry, allowing the purification of apple juice when included in cryogenics [Turuvekere Vittala Murthy, Agrahari, & Chauhan, 2021].Therefore, the formulation of polyphenolic compounds from biomass extractable into new wares comes with the advantage of increasing the value of the initial compounds, encouraging the use of vegetable waste as raw material sources and thus making the transition to a circular economy with the achievement of the concept of the "zero waste" [Cañadas et al., 2020].

Hydrophilic cyclodextrins (CDs), such as natural cyclodextrins and branched cyclodextrins, are practical in ameliorating the dissolution indispensable qualities of weakly (the hydrophilicity) water-soluble polyphenolic extracts due to their loud dissolving capability and safety profiles.

The instantaneous deliverance of weakly soluble vegetable phenolic substances extracts in water by complexing with hydrophilic CDs is attributed to the intensification of water solubility, the wetting capacity and dispersibility of the polyphenolic extracts, and the reduction of the particle size of the solid complexes.

Vegetable phenolic substances, encompassing flavonoids, represent one of the farthest significant categories of pharmacologically active compounds that have a distinguished biological characteristic.

Their use is constricted by their undersized steadiness and undersized solubility, both in aqueous suspensions and in organic solvents [López-Nicolás, Rodríguez-Bonilla, & García-Carmona, 2014].

To annihilate those inconveniences, flavonoids can be transformed into glycosylated or acylated byproducts by chemical, enzymatic, or chemoenzymatic techniques.

Enzymatic techniques are more effective since they capitalize on the high regioselectivity of these biocatalysts and achieve a distinctive functionalization of the flavonoid.

The transmutation of flavonoids into bioconjugates by acylation with fatty acids gives the likelihood of introducing another pharmacologically active function into their molecule and thus replacing in a positive sense not only physically indispensable qualities such as solubility but also pharmacologically characteristic [Wu et al., 2011].

The CDs are partitioned into the class of the farthest-used receptors in host–guest incorporation chemistry. The facilities of encapsulating a vegetable phenolic substance in CDs are ameliorating bioavailability, increasing steadiness, and reducing side repercussions. Bioconjugates were obtained through the direct enzymatic esterification of flavonoids (aglycones, glycosylated, and flavonolignans) with fatty acids, which were identified, separated, and characterized by chromatographic and spectrometric techniques.

The molecular modeling of the interconnection between the obtained bioconjugates and the CDs was carried out using distinct molecular mechanics programs, and the results obtained regarding the interconnection energies were compared with those resulting from the DSC analysis. The bioconjugate/cyclodextrin incorporation of vegetable phenolic compounds was obtained, and their antioxidant characteristic and transfer through semipermeable membranes were evaluated [Tutunchi et al., 2019].

## 16.2 OBTAINING BIOCONJUGATES

Obtaining bioconjugates through direct enzymatic esterification of flavonoids (aglycones, glycosylated, and flavonolignans) with saturated and unsaturated fatty acids. Molecular modeling of the interconnection of bioconjugates obtained with CDs ($\alpha$-, $\beta$-, and $\gamma$-cyclodextrin); obtaining the incorporation vegetable phenolic compounds of the main CDs (α- and β-cyclodextrin) with flavonoids, fatty acids, and their bioconjugates; by the crystallization method from the ethanol–water suspension; the study of the antioxidant characteristic of pure flavonoids in comparison with their complexes, respectively with bioconjugates and complexes of bioconjugates; evaluation of the transfer through semipermeable membranes of the incorporation vegetable phenolic compounds [Negahdari et al., 2021].

## 16.3 QSAR CORRELATIONS BETWEEN THE STRUCTURE OF POLYPHENOLS AND CYCLODEXTRIN CAPACITY

The modeling of the structure of something bioactive molecules such as betulinic acid, betulin, lupeol, and something structurally related compounds: loganin (iridoid monoterpenoid) [$C_{17}H_{26}O_{10}$], morroniside [$C_{17}H_{26}O_{11}$], oleanolic acid [$C_{30}H_{48}O_3$], ursolic acid [$C_{30}H_{48}O_3$] was achieved by using van der Waals molecular descriptors with the help of which linear quantitative relationships through pharmacologically characteristic are established (beneficial or harmful) of something organic compounds on living organisms and their molecular structure [Harish et al., 2010]. Those molecular descriptors were calculated by a Monte Carlo algorithm accomplished in the IRS computer program. To obtain a succession of van der Waals descriptors, 3D molecular constructions were calculated using the $MM^+$-molecular mechanics and quantum mechanics programs. The chemistry of incorporation complexes is a particularly mediatized field [Gaudreault, van de ven, & Whitehead, 2002].

The importance of incorporation complexes is evident, especially in sustainable agriculture, since the incorporation of compounds of vegetable phenolic substances interest in supramolecular constructions causes dramatic changes in their physical and chemical indispensable qualities. The modifications induced by complexation with CDs include enhancing water solubility, improving stability, and improving bioavailability parameters. The substantial modification of the physical indispensable qualities of vegetable phenolic substances is often exploited to prove their incorporation in the host molecule.

The chapter surveys aspects of complexation in the case of vegetable phenolic substances [Shaddel et al., 2024]. Following their complexation, the change in the vibration spectrum (mid-infrared spectral domain) and the intensification in solubility in aqueous medium are highlighted.

The CDs are used in conditioned configurations to enhance the solubility, stability, and bioavailability of active substance molecules. Although natural CDs have relatively low solubility in both water and organic solvents, a fact that restricts their use in techniques involving vegetable phenolic substances. Distinct byproducts of CDs, such as hydrophilic, water-insoluble, and ionic configurations, have been comprehended in such a way as to extend the physico-chemical indispensable qualities and incorporation capacity differentiated to natural CDs.

CDs are applicable as vegetable phenolic substances vectors to control the speed and/or time profiles of vegetable phenolic substances deliverance.

Hydrophilic CDs can change the deliverance rate of vegetable phenolic substances that are difficult to dissolve in water, a consequence that can be used as a factor to intensify the incorporation of vegetable phenolic substances through biological barriers, serving as vectors of vegetable phenolic substances in instantaneous deliverance techniques [Li et al., 2019].

Amorphous CDs, such as ß-hydroxyalkyl CDs, are practical to inhibit polymorphic transformations and crystallization amount of weakly the hydrophilicity vegetable phenolic substances throughout escalate, which can accordingly keep in existence greater dissolution indispensable qualities and bioavailability of vegetable phenolic substances. From another point of view, water-insoluble CDs can serve as vectors that support the deliverance of the hydrophilicity of vegetable phenolic substances encompassing peptides and protein vegetable phenolic substances.

Delayed-deliverance formulations can be obtained by using enteric-style CDs such as O-carboxymethyl-O-ethyl-ß CD (CME-ß-CD) CDs, which are soluble at physiological pH.

An amalgamated use of distinct styles of CDs and/or additives provides a more balanced bioavailability with prolonged repercussions. The farthest advantageous characteristic of a vegetable phenolic substances vector is the capability to carry the vegetable phenolic substances to a target location [Ishitsuka, Irie, & Matsuo, 2022].

Vegetable phenolic substances derived with CDs can survive the passage through the stomach and small intestine, but the deliverance of vegetable phenolic substances will be triggered by the enzymatic decomposition of the CD circle in the colon; such a byproduct with CDs can be a smart means to design a new class of vegetable phenolic substances oriented for controlled availability. Moreover, cationic polymer byproducts with polydextrins can be new candidates for vectors of vegetable phenolic substances to intensify the gene transfer of plasmid DNA.

The desired characteristics of vegetable phenolic substance vectors in vegetable phenolic substance delivery arrangements are represented by multifunctional indispensable qualities, such as controlled delivery, targeting, and intensified incorporation potentialities. From the point of view of safety, bioadaptability is a must, and quality, cost indispensable qualities, etc., there are supplementary necessities in the case of vegetable phenolic substances vectors [Wang et al., 2019].

The CDs are quite adaptable and absorb and interconnect with distinctive constituents of biomembranes, such as lipids; their macrocyclic constituent perseveres in the body, but they are biodegradable in vegetable organisms, and the multifunctional byproducts of CDs can replace the physico-chemical and incorporation of host compounds [Uekama, 2004]. If the –OH (hydroxyl) substitutes of the CDs are replaced by longer ethyl, acetyl, or acyl substitutes (C1–C12), the solubility of those CDs in water decreases comparatively to the d.s. and/or the extent lengthwise of the alkyl chains. Intensive suspensions of peracylated ß-CD cyclodextrins in organic suspensions are much more viscous and stickier, and after the evaporation of the solvents, the transmutation into a gel takes place.

Those indispensable qualities would be practical for a slow-deliverance vector of the hydrophilicity vegetable phenolic substances. The extended acyl side chains, however, significantly intercede with the capability of peracylated ß-CD cyclodextrins to configuration incorporation complexes [Hu, Tao, & Cao, 2009]. The per-O-butanoyl-ß-CD (TB-ß-CD) has an important retard consequence for the hydrophilicity vegetable phenolic substances, due to its muco-adhesive ownership and due to the corresponding water-insolubility. An X-ray crystallographic study was carried out to obtain an inside view of the interconnection of guest compounds with ß-CD peracylated CDs, and it turned out that the self-alliance of acyl substitutes occurs with increasing chain length; more precisely, a butanoyl chain penetrates the concavity.

The peracylated diclodextrin ß-CD prevents the incorporation of other guest compounds in the central concavity, and the guest compounds can integrate into the matrix through the acyl chains, forming a dispersed monomolecular system. In the case of cyclodextrin per-O-valeryl-ß-CD (TV-ß-CD), an adhesive film was obtained by means of the usual distribution method, when an ethanol suspension containing the compound TV-ß-CD and the vegetable phenolic substances itself was spread on a polyethylene support layer.

The BE7-ß-CD, containing on midpoint seven sulfobutyl ether substitutes per cyclodextrin molecule, is currently used in parenteral preparations since it is highly soluble in water and less toxic than other sulfobutyl ether ß-CD byproducts [Salústio et al., 2011].

Due to the very undersized pKa value of the sulfonic acid substitutes, the BE7-ß-CD molecule possesses multiple negative charges at physiologically compatible pH values. The four-carbon butyl chain, coupled with the repulsion of the negative charges of the last group, allows an extension of the CD concavity. This often leads to stronger bonds with candidate vegetable phenolic substances that are formed with other modified CDs. The use of the BE7-ß-CD compound allows the development of parenteral and oral wares from active compounds too insoluble to be formed by other dissolution techniques.

The amorphous constituent BE7-ß-CD is effective in ameliorating both the solubility and the chemical steadiness of Rivenprost (ONO-AE1-734), a prostaglandin E1 analogue, in contrast to hydrophilic CDs.

The compound BE7-ß-CD also serves as both a solubility modulator and an osmotic pumping agent for controlled-porosity osmotic pump tablets, from which the deliverance rate of both highly soluble and weakly the hydrophilicity vegetable phenolic substances, can be precisely controlled [Wang et al., 2023].

A significant characteristic of CDs consists of the formation of incorporation complexes both in suspensions and in solid states, in which the water-insoluble circumstance of the cyclodextrin concavity surrounds each guest compound.

This fact leads to the alteration of the physico-chemical indispensable qualities of the host molecules, such as chemical steadiness, solubility, and dispersibility., an aspect that can have considerable pharmaceutical potential.

The steadiness constant and the stoichiometry of the incorporation complexes, which depend on the guest/host concentration involved, are practical indices for estimating the binding power of the complex and the changes in the physico-chemical indispensable qualities of the guest compounds in the complex.

Furthermore, consideration should be given to various circumstance constituents, such as temperature, dilution, pH, and additives, when developing practical formulations and routes of administration.

The route of administration presents several challenges related to the incorporation of vegetable phenolic substances, including the first-pass effect, individual variability in bioavailability, and local irritation induced by the vegetable phenolic substances [Petitjean et al., 2022].

Hydrophilic CDs can affect the metabolism of vegetable phenolic substances under the action of CYP enzymes, which exist both in intestinal epithelial cells and hepatocytes.

The CDs were thought not to interconnect with microsomal proteins since CDs cannot enter the cytoplasm.

Thus, CDs can intensify the bioavailability of vegetable phenolic substances in many ways, and the improvement mechanism exerted by CDs on the bioavailability of vegetable phenolic substances can be much more complicated than originally thought.

The influence of CDs on the physical and chemical stability of vegetable phenolic substances in suspensions and in the solid state represents one of the most extensively considered problems regarding the incorporation complexes of CDs.

Vegetable phenolic substances must be stable not only throughout escalation but also in biological fluids since the reactions that generate a less active outcome from a pharmaceutical point of view usually reduce the therapeutic efficiency [Petitjean & Isasi, 2023].

The CDs are known to speed up and slow down distinct styles of reactions, depending on the nature of the complex formed.

In general, when an active site of a vegetable phenolic substance is included in the cyclodextrin concavity, there is a rate-reducing consequence, and the reaction rate is dependent on the concentration of free vegetable phenolic substances resulting from the decomposition of the complex.

From another point of view, if the vegetable phenolic substances are not entirely included in the concavity, leaving the active site sterically fixed near the catalyst, an acceleration of the consequence occurs.

The CDs accelerate or reduce the speed of various reactions, manifesting many kinetic features characteristic of enzyme reactions, such as, for example, the formation of the substrate–catalyst complex, competitive inhibition, saturation, and stereospecific catalysis.

Reactions catalyzed by CDs can be classified into three categories, depending on the style of stimulation:

- The contribution of the –OH (hydroxyl) substitutes of CDs: Alkaline hydrolyzes of phenolic esters are quickening by the supplement of CDs, when the secondary –OH (hydroxyl) substitutes of the CDs (pKa around 12.2) function as a nucleophilic catalyst, acting the carbonyl group of the esters to configuration a covalent middleman, acylated CD, which is gradually hydrolyzed to configuration the last outcome and cyclodextrin.

**TABLE 16.1**
**Examples of Active Substances Derived from Natural Sources**

| Generic Name | Type of Action and Field of Use | Year of Launch | Company | No. of Chiral Centers |
|---|---|---|---|---|
| Ertapenem-Na | Antibacterial | 2002 | Zeneca | 6 |
| Telithromycin | Antibiotic | 2001 | Aventis | 13 |
| Caspofungin acetate | Antifungals | 2001 | Merck | 14 |
| Micafungin-Na | Antifungals | 2002 | Fujisawa | 15 |
| Amrubicin-HCl | Antineoplastic | 2002 | Sumitomo | 5 |
| Pimecrolimus | Immunosuppressant | 2002 | Novartis | 14 |
| Daptomycin | Antibacterial | 2003 | Cubist | 13 |
| Mycophenolate-Na | Transplant rejection | 2003 | Novartis | 0 |
| Everolimus | Transplant rejection | 2004 | Novartis | 15 |
| Mitomycin C | Anticarcinogenic | 2002 | SuperGen | 4 |
| Galantamine | Acetylcholinesterase inhibitor | 2000 | Jansen | 3 |

- The microsolvent consequence of the cyclodextrin water-insoluble concavity: The cyclodextrin concavity has an apolar nature like that of an 80% dioxane/water suspension and prepares a moderately alkaline circumstance since glycosidic ethers surround it. The chemical reactivity of the guest compounds is altered by the change of polarity when they are integrated in such a water-insoluble and alkaline circumstance. The conformational or steric consequence of CDs is that when two reactive parts of the substances are constrained to be in contact by complexing with cyclodextrin, intra- and intermolecular reactions are quickening. From another point of view, reactions are slowed down if the two reactive parts are constrained apart [Petitjean et al., 2020].

Vegetable phenolic substances, derived from pure substances in conjunction with their byproducts and synthetic substances resulting from a precursor of the natural outcome, signify many pharmaceutical sales contemporaries.

Almost all are chiral, with a midpoint of eight chiral sites per molecule [Walsh et al., 2000].

Table 16.1 discloses some of the farthest latterly inaugurated medicines derived from vegetable phenolic substances, and in Figure 16.3 the constructions of the active substances are presented [Dahmer et al., 2022].

## 16.4 ASPECTS OF STEREOCHEMISTRY OF POLYPHENOLS

Vegetable phenolic substances differ from synthetic substances by several indispensable qualities.

Structural, vegetable phenolic substances are very diverse, often exhibiting distinctive pharmacological characteristics and an amazing steric complexity.

This is obviously a consequence of the fact that the biosynthesis details of the enzymes are three-dimensional and chiral.

In addition, based on a constricted range of building blocks used for synthesis, nature must generate new characters by removing ordinary middleman.

Due to their formation by enzymatic pathways, vegetable phenolic substances have shown their capability to bind to enzymes or, in general, to protein targets and to also react with therapeutically relevant target proteins [Gehm et al., 2022].

In numerous statistical surveys, synthetic substances, pharmaceuticals, and plant-derived phenolic compounds are distinguished.

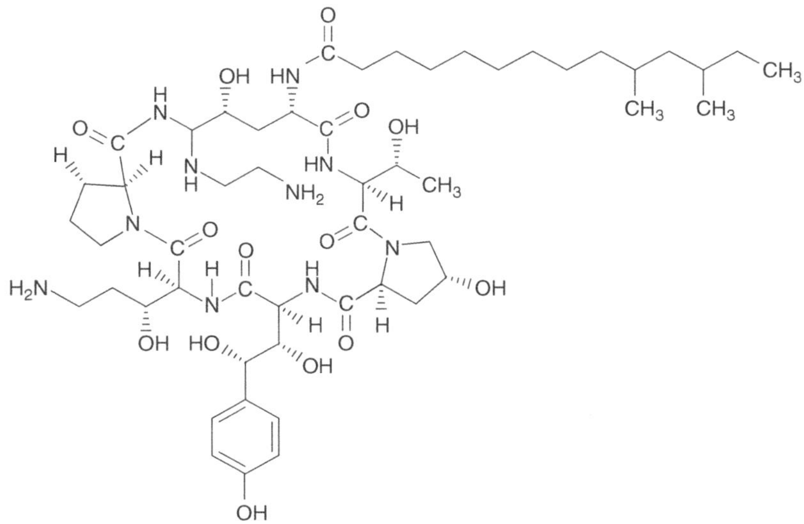

**Caspofungin**

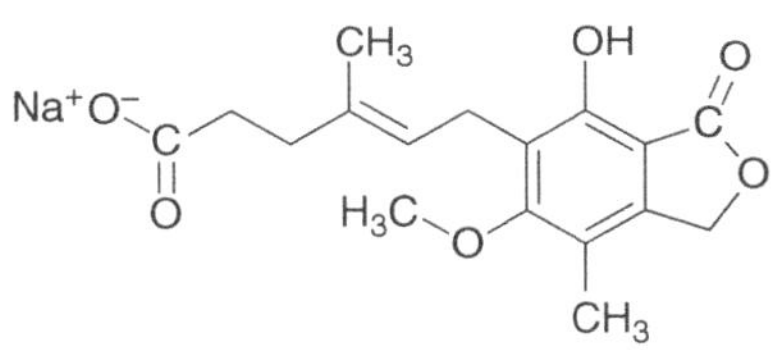

**Myfortic-Na**

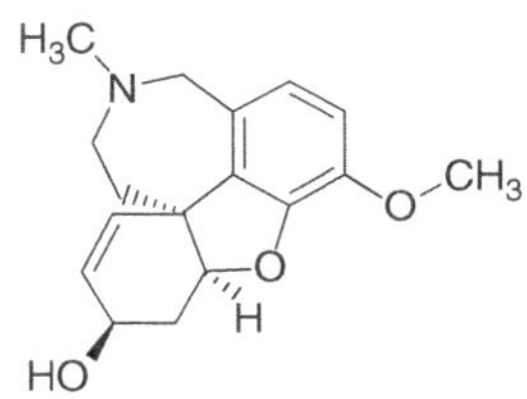

**Galantamina**

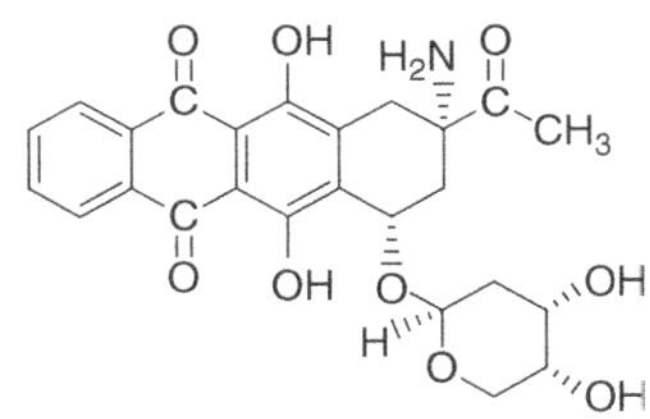

**Amrubicin**

**FIGURE 16.3** The chemical structures of Caspofungin, Myfortic-Na, Galantamina and Amrubicin.

Differentiated from medicamentary and synthetic substances, vegetable phenolic substances contain, on midpoint, less nitrogen, sulfur, or halogen atoms but are considerably richer in oxygen atoms and donors for the formation of hydrogen bonds.

This reflects the fundamental differences between biosynthesis and chemical synthesis.

Many biosynthesis pathways are based on reagents with a similar number of O and N atoms, such as amino acids, or containing more oxygen than nitrogen, as is the case, for example, in the synthesis of polyketides.

Differentiated from synthetic substances, natural wares/vegetable phenolic substances contain significantly more cycles based on sp3 hybridized atoms and chiral sites [Ibrahim, Hermawan, & Sanagi, 2013].

## 16.5 INCREASING SOLUBILITY AND EFFICIENCY OF POLYPHENOLS

In recent years, the scientific interest in amphiphilic polymers has intensified due to their unique associative behavior and their potential applicability in the pharmaceutical industry.

Those polymers mainly consist of hydrophilic and water-insoluble parts [Escobar-Avello et al., 2021].

The first surveys describing associative behavior were published in 1951 and presented a poly (4-vinylpyridine) byproduct with ethyl and dodecyl substituents.

Since then, the interest in those special macromolecules has intensified, as can be seen from the numerous books and articles on this subject.

The structure of such polymers varies enormously, from amphiphilic copolymer blocks to amphiphilic polymer grafts.

It should be known that the balance through intra- and intermolecular interconnections can be easily controlled by adjusting the size and/or the number of water-insoluble substitutes grafted on the chain.

Among amphiphilic polymers, polysaccharides with modified water-insolubility have stood out as promising vehicles for water-insoluble vegetable phenolic substances due to their low toxicity, biocompatibility, and biodegradability [Cai et al., 2024].

The first surveys on polysaccharide-based surfactants began in the 80s. Landoll's pioneering work involved the viscosity, solubility, and surface characteristics of cellulose grafted with alkyl chains.

The structure of amphiphilic polysaccharides was not only modified by grafting distinct hydrophilic substitutes but also by using hydrophilic primary chains (methylcellulose, hydroxyethyl cellulose, etc.).

This new class of polysaccharides possesses unique dissolution properties due to inter- and intramolecular interactions. However, it has been demonstrated that the internal structure of those arrangements can be used as a vehicle for water-insoluble medicamentary.

Indeed, byproducts such as dextran's, cellulose, starch, or alginates can encapsulate water-insoluble molecules such as ciclosporin [$C_{62}H_{111}N_{11}O_{12}$], vitamin B2 [$C_{17}H_{20}N_4O_6$], paclitaxel [$C_{47}H_{51}NO_{14}$], or even proteins [Remígio et al., 2024].

## 16.6 COMPARATIVE CHIRALITY ANALYSIS OF POLYPHENOLS

There are almost no substances that have more than three stereo enters.

In medicines, the number of chiral sites is lower than in vegetable phenolic substances but higher than in synthetic ones.

The midpoint number of chirality sites in vegetable phenolic substances is 2.8 differentiated to 5.2 for vegetable phenolic substances and 0.4 for synthetic wares.

Around 40% of vegetable phenolic substances do not have chiral sites, approx. 20% have a single site, following a wide distribution through 2 and 10 or more stereo enters.

From the point of view of comparing wares of various origins, it is practical to define the notion of *chirality index*.

This index is the ratio through the number of stereo enters in a molecule and the number of distinct hydrogen atoms in the same molecule [Watanabe & Ayugase, 2009].

Analyzing the biological sources of vegetable phenolic substances, certain differences are observed. About 25% of fungal metabolites do not have chiral sites.

The midpoint is 4. Fungi are known as producers of small achiral aromatic substances, such as quinones, xanthones, or pyrones, which are generated by the polyketide pathway.

This may be a reason for the intensification in the percentage of achiral molecules from this source [Čižmáriková, Habala, & Markulia, 2023].

## 16.7 DETERMINATION OF THE RELATIVE AND ABSOLUTE STEREOCHEMISTRY OF PHENOLIC COMPOUNDS

In the case of chemical synthesis, at each step of the synthesis and following each chemical reaction, the structure of the outcome is more or less predictable, and only a constricted number of stereocenters are created.

Starting from molecules with known chirality (carbohydrates, amino acids, hydroxy acids), only the comparative stereochemistry of the new sites must be elucidated since the stereo acquaintanceship of the starting products gives the absolute stereochemistry.

In the case of vegetable phenolic substances, the starting point is distinct.

Initially, nothing is known regarding the stereochemistry and the way the atoms are bonded.

The capability to work without losses and with small amounts of sample is indispensable in this case [Koyanagi et al., 2010].

The NMR (nuclear magnetic resonance spectroscopy) has become the most powerful means of elucidating the structure of vegetable phenolic substances.

Using two-dimensional nuclear magnetic resonance (2D-NMR), such as Heteronuclear Multiple Quantum Coherence (HMQC) or Heteronuclear Multiple Bond Correlation (HMBC), together with modern nuclear magnetic resonance (NMR) equipment (for example, the cryoprobe technique where the sample is cooled to the temperature of liquid helium) and computer-aided techniques, today it is possible to elucidate the structure using a very small amount of natural sample.

However, the determination of comparative and absolute stereochemistry using NMR spectroscopy is constricted.

For a significant number of vegetable phenolic substances, no acquaintanceship regarding their stereochemistry has yet been published [Zhang et al., 2015].

The introduction of the "virtual screening" technique and docking strategies in the design of vegetable phenolic substances requires a rapid acquisition of stereochemical knowledge about these substances.

In recent years, numerous improvements have been made in the standardization of stereochemical determination procedures.

This involves, in many cases, a combination of distinct techniques, such as NMR spectroscopy, chiroptical techniques based on X-ray diffraction, circular vibrational dichroism spectroscopy, total synthesis, and decomposition reactions, which aim to elucidate the three-dimensional molecular structure [Olmo-Cunillera et al., 2020].

## 16.8 NUCLEAR MAGNETIC RESONANCE SPECTROSCOPY (NMR)

The stereochemical analysis of vegetable phenolic substances with defined conformational indispensable qualities is relatively easy to perform.

For cyclic vegetable phenolic substances with small circles, the configuration can be deduced from NMR parameters such as the H–H proton pair and/or the intensity of the vegetable phenolic substances (NOE).

A much more difficult task is to assign the comparative configuration of flexible arrangements, such as macrocyclic vegetable phenolic substances and polysubstituted chains.

On the one hand, this could be solved by studying various byproducts of the original vegetable phenolic substances [Christophoridou & Dais, 2006].

The resulting cycle can be used to determine the comparative stereochemistry of both chiral sites.

From another perspective, the NMR approach based on J-coupling has been comprehensive in determining the stereochemistry of acyclic vegetable phenolic substances.

The coupling constants through the protons separated by the triple bond are directly dependent on their dihedral angles according to the Karplus equation.

Additionally, the vicinal C–H coupling constants ($^3J_{CH}$) respect the Karplus relation and can be practical if the carbon atom has an electronegative substituent, such as an O or halogen atom.

The values of $^{2,3}J_{CH}$ were, for a long time, inaccessible for the elucidation of the molecular vegetable phenolic substances.

Thanks to advances in the 2D-NMR technique, hardware improvements, and high-field magnets, those constants are now measurable even in vegetable phenolic substances with the natural abundance of $^{13}C$ nuclei [Christophoridou & Dais, 2009].

This method is based on the formation of (R)/(S)-MTPA esters (esters of methoxy-alpha-(trifluoromethyl) phenylacetic acid) and the comparison of the NMR data of both diastereomers.

However, the method is constricted to vegetable phenolic substances with functional substitutes that lend themselves to derivatization (for example, the hydroxy group), and in some cases, the conclusions drawn from NMR measurements are unreliable [Christophoridou et al., 2005]

## 16.9 CHIROPTICAL TECHNIQUES

Polarimetry is still used today to characterize the optical characteristics of vegetable phenolic substances and as a purity criterion of enantiomers.

A prerequisite for determining the absolute stereochemistry with this method is the presence of something that references vegetable phenolic substances.

The use of Circular Dichroism – CD (circular dichroism spectroscopy) is practical in the analysis of vegetable phenolic substances.

CD spectroscopy is based on the distinct incorporation of circularly polarized light having opposite directions of circular polarization.

The CD spectra can be calculated geometrically, and the transition moments enable the prediction of absolute stereochemistry [Chen et al., 2022].

In 1970, the method of chiral excitons (Exciton Chirality Circular Dichroism – ECCD) was launched.

If two identical, or almost identical, chromophores are attendant in a chiral vegetable phenolic substance, the symmetric and asymmetric coupling through two oscillating dipoles causes the splitting of the incorporation band in the electronic incorporation spectrum.

The consequence of this splitting of the incorporation band is noticed as a positive or negative Cotton consequence [Schaack et al., 2017].

However, the method also has some limitations.

The assignment of the Cotton consequence retains an empirical constituent when it is applied to new styles of constructions and depends on the presence of chromophores or involves the preparation of byproducts capable of showing the Cotton consequence [Schaack et al., 2019].

## 16.10 X-RAY CRYSTALLOGRAPHY

If the vegetable phenolic substances are in the monocrystalline state, X-ray crystallography is the most efficient way to determine the comparative stereochemistry and, in many cases, even the absolute configuration.

The technique involves the use of X-rays with a wavelength close to the incorporation edge of one of the atoms attendants in the vegetable phenolic substances of the considered molecule.

Throughout the diffraction of X-rays on the single crystal of the considered vegetable phenolic substances, a phase change of the X-rays diffused by those atoms occurs, differentiated to the phase of the X-rays scattered by other atoms of the same molecule.

For this reason, the diffraction image is no longer truly centrosymmetric; the signal pairs (Bijvoet pairs) in the diffraction image, which are in centrosymmetric relation, become unequal in intensity [Loewen et al., 2015].

Starting from an assumed absolute configuration, the comparative intensities of the Bijvoet pairs are calculated for both the R stereoisomer and the S stereoisomer.

In this calculation, a parameter x (Flack parameter) appears; if this parameter takes the value 0, the absolute configuration assumed in the calculation is correct.

If the parameter x has the value 1, the assumed absolute configuration is incorrect.

The phase transmutation generally intensifies with the atomic mass of the element's attendant and the wavelength of the radiation used.

In the past, bulky atoms (such as silicon or even larger atoms) had to be used to employ the method with an acceptable degree of confidence.

Currently, the comprehensive redundancy of high-precision data enables the determination of absolute stereochemistry, starting from constructions containing oxygen as the largest atom in the molecule [Pramanik et al., 2024].

## 16.11 TOTAL SYNTHESIS AND DECOMPOSITION REACTIONS

If no crystalline vegetable phenolic substances are available and spectroscopic techniques are not practical in determining the absolute stereochemistry, total synthesis or decomposition reactions can be practical.

In decomposition reactions, the vegetable phenolic substances are divided into subunits that can be differentiated from molecules whose absolute configuration is known.

If minor quantities of vegetable phenolic substances are available and decomposition reactions are not possible, total synthesis is an alternative for elucidating the absolute stereochemistry.

If the comparative stereochemistry of the vegetable phenolic substances is known, there are only two antipodes that can be considered target molecules; if the comparative stereochemistry is unclear and several enantiomers and diastereomers are possible, the correct synthesis of the stereoisomer can be a difficult task [Magalhães et al., 2010].

For example, the first synthesis of discodermolide led to the unnatural antipode (–)-discodermolide. In some cases, the total synthesis led to the revision of the previously assumed stereochemistry of some vegetable phenolic substances.

In conclusion, there is no general rule for determining the stereochemistry of vegetable phenolic substances.

The most efficient way to solve problems related to stereochemistry depends on the structural class and the availability of techniques [Jing et al., 2023].

## 16.12 MECHANISM OF STEREOCHEMICAL CONTROL OF BIOSYNTHESIS POLYPHENOLS

Vegetable phenolic substances can be differentiated according to their biosynthetic origin.

Characteristics of biosynthesis are the vegetable phenolic substances' building blocks used for the assembly of the basic structure and the categories of enzymes catalyzing the condensation of the formed units, as well as the cyclization and derivatization reactions.

Complex vegetable phenolic substances with a certain number of stereocenters are usually synthesized in an enantiomeric pure configuration.

For example, in DNP, approximately 80% of all wares are unique records, meaning that there are no more molecules with the same connectivity.

In general, stereospecific conformations are essential for the distinctive pharmacological characteristics [Dong et al., 2019].

There are several methods for generating chirality in vegetable phenolic compounds.

- It can be obtained in a stereospecific synthesis step.
- It is introduced from the beginning of biosynthesis using chiral blocks resulting from primary metabolism.
- Enzymes can graft carbohydrate units to form complex constructions, such as the glycosylated antibiotic Vancomycin (Figure 16.4).

Although a growing number of biosynthetic enzymes and their corresponding genes or gene substitutes have been discovered and analyzed in recent years, the schematic detailing of investigations regarding stereochemical control in the biosynthesis of complex vegetable phenolic compounds is rare.

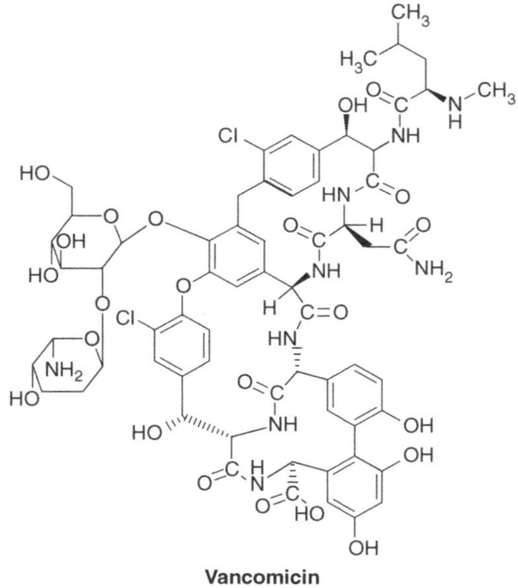


**FIGURE 16.4** The Enzymes can graft carbohydrate units to form complex constructions, such as the glycosylated antibiotic Vancomycin.

Examples may include vegetable phenolic substances typical of microorganisms, such as non-ribosomal peptides, macrolides, aromatic polyketides, and taxol, a terpene of vegetable organism origin [Loll & Axelsen, 2000].

## 16.13 DEPENDENCE OF BIOLOGICAL ACTIVITY OF POLYPHENOLS ON THEIR STEREOCHEMISTRY

Plants organisms use chiral enzymes for the biosynthesis of vegetable phenolic substances. The biochemical reactions carried out by those enzymes are generally enantioselective.

Accordingly, nature normally synthesizes only one stereoisomer of a vegetable phenolic substances.

This fundamental ownership is probably the result of a selection process, comprehensive throughout evolution, to produce, with maximum economy, vegetable phenolic substances with maximum pharmacological characteristics in a chiral circumstance.

Thus, vegetable phenolic substances are a perfect example to illustrate Pfeiffer's rule, which states, "the lower the effective dose of a medicinal substance, the greater the difference through the pharmacological consequence of the optical isomers (enantiomer)."

It means that in a succession of chiral substances, the comparative potential (eudismic ratio) of the stereoisomers intensifies with the intensification of the characteristic of one stereoisomer (eutomer) differentiated from the other stereoisomer (distomer).

This principle can also be demonstrated by observing the fact that stereoisomers can have completely distinct medicinal indispensable qualities, such as the difference between quinidine [$C_{20}H_{24}N_2O_2$] (antiarrhythmic agent) and quinine [$C_{20}H_{24}N_2O_2$] (antimalarial agent) (Figure 16.5), and the fact that the derivatization of particularly active vegetable phenolic substances can lead to substances with profoundly altered pharmacologically characteristic and biological measures with distinct medicamentary interconnections.

In the process of vegetable phenolic substances identification and unfolding events, generally, only the naturally available stereoisomer is tested and promoted for clinical trials [Woiwode, Sievers-Engler, & Lämmerhofer, 2024].

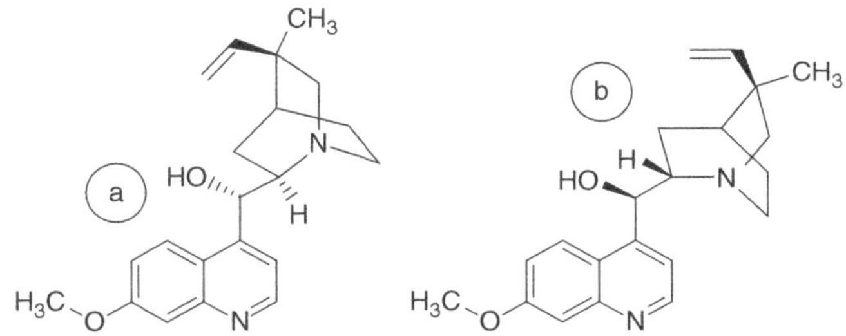


**FIGURE 16.5** Some the stereoisomers can have completely distinct medicinal indispensable qualities, such as the difference between quinidine (antiarrhythmic agent) and quinine (antimalarial agent).

The construction of farthest vegetable phenolic substances is complex and differentiated to synthetic medicinal compounds.

A synthetic approach to obtain both enantiomers or diastereomers usually defined is not feasible. It is outside the scope of medicinal chemistry, despite the spectacular unfolding made in the domain of stereoselective synthesis in the last 20 years.

Currently, in the medicinal development process, it is a standard practice to profile all diastereoisomers of possible medicinal possibilities associated with pairs of enantiomers, especially in the case of synthetic active substances.

The influence of the stereochemical differences of vegetable phenolic substances on biological arrangements has been considered by Pasteur, recognizing the fact that ammonium dextro-tartrate is digested faster by *Penicillium glaucum* than the levo stereoisomer [Kanawati et al., 2011].

Epinephrine (adrenaline) [$C_9H_{13}NO_3$] was the first hormone for which data were published regarding the differential consequence of the two epimers on blood pressure in the early 20th century.

Next was the analysis of the consequences of the optical stereoisomers (mirror image structures) of ephedrine and methylephedrine on heart rate in normal rats and those treated with reserpine to investigate the direct and indirect effects on 1-adrenoceptors.

The ephedrine [$C_{10}H_{15}NO$] are L-ephedrine stereoisomers, and to a less important measurement, D-ephedrine, significantly intensified heart rate [Rice et al., 2018].

## 16.14 GOSSYPOL (CONTAINING HYDROXYL GROUPS), A RACEMIC NATURAL PRODUCT

The gossypol [$C_{30}H_{30}O_8$] is a compound identified from the vegetable organism variety of cotton *Gossypium herbaceum* L., *Gossypium arboreum* L., and/or *Gossypium hirsutum* L. The sesquiterpenoid molecule of the compound discloses axial dissymmetry. This compound has garnered particular interest due to its use as a male antifertility agent, but its effectiveness has been overshadowed by its slow onset of activity and the likelihood of sterility [Wang et al., 2009].The vegetable organism produces a racemic incorporation of gossypol, of which only the conformer significantly, with the (–) enantiomer exhibits antispermatogenic characteristic and cytotoxicity characteristic [Liu et al., 2022]. The (+)-gossypol enantiomer does not affect the sperm count and, at the same time, discloses lower toxicity than the (–) enantiomer (Figure 16.6).

The example of gossypol highlights the fact that it is indispensable to have comprehensive acquaintanceships about the stereochemical steadiness of a chiral site and acquaintanceships about the mode of action of stereoisomers, to start the amplification of a medicamentary based on a pure enantiomer of a racemic incorporation. The configurational steadiness must be controlled throughout industrial operations, in the final escalated configuration, and in the physiological application [Messeha et al., 2020].

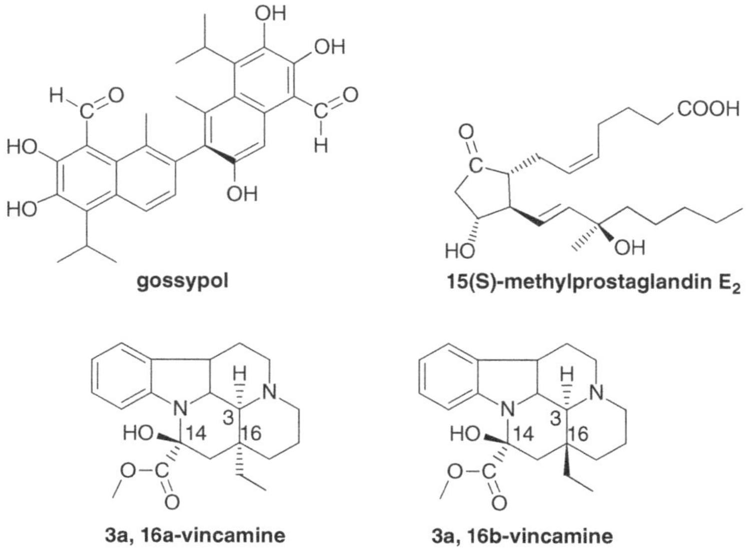


**FIGURE 16.6** The vegetable organism produces a racemic incorporation of gossypol, of which only the conformer significantly, with the (–) enantiomer exhibits antispermatogenic characteristic and cytotoxicity characteristic. The (+)-gossypol enantiomer does not affect the sperm count and, at the same time, discloses lower toxicity than the (–) enantiomer.

## 16.15 CONCLUSION AND FUTURE REMARKS

At the same time, the characteristics and spectrum of action of those vegetable phenolic substances are constricted. When natural vegetable phenolic substances are administered in vivo, the effectiveness of the expected therapeutic consequence is closely correlated with the bioavailability parameter of the active constituent.

Bioavailability is characterized by the fraction of the unchanged dose of the therapeutic compound adsorbed in the systemic circulation of the living organism.

Orally, the amount of vegetable phenolic substances adsorbed decreases considerably due to the influence of the biochemical parameters of the gastrointestinal tract, where they can undergo oxidative chemical changes under the action of enzymes, pH, and the presence of other catalysts, such as transition metals.

Another parameter that influences the therapeutic outcome is the lipophilic nature of polyphenolic substances, which leads to a decrease in the adsorption of vegetable phenolic substances and, consequently, a reduction in their antioxidant and antibacterial properties.

The unfavorable actions of outside constituents, such as humidity, temperature, and the presence of biological agents, can intensify chemical oxidation reactions in the chain and lead to decomposition under the influence of microbiota, affecting their stability and reducing their therapeutic potential. To prevent the decomposition of polyphenolic substances and preserve their beneficial effects when ingested or used in various medical treatments, they can be incorporated into delivery systems or copigmented with other molecules. Accordingly, scientists are striving to find ways to ameliorate the steadiness of polyphenolic substances, their bioavailability, and cellular uptake, as well as to understand the mechanisms behind their biological repercussions and intracellular localization. In the coming decades, we can witness the development of new therapies for melanoma, considering the cytotoxic potential of polyphenolic substances encapsulated in fluorescent nanosystems. This targeted delivery approach could lead to new strategies for treating skin cancer, addressing the issues of frequent recurrence and the limited efficacy of current treatments. In addition, the enhanced stability of cyanidin under various conditions could lead to the formulation of more effective food and pharmaceutical products that retain the beneficial properties of anthocyanins.

Those advances could bring contributions not only to health but also to the food industry by ameliorating the intake of bioactive substances and the appearance/color of food wares based on fruits rich in polyphenolic substances. The conditions of the worrying intensify in the share of synthetic materials and substances, farthest of the time unfriendly about the health of the user, the need to preserve and ameliorate human potential, as well as the introduction into the economic circuit of natural vegetable organisms wares with outstanding chemical and biological qualities, the project proposes the use of vegetable organism extracts to produce multifunctional polymeric materials. Their areas of use are very diverse: manufacturing packaging (especially for the food sector), electrical, thermal, and sound insulation; medical wares (syringes, catheters, bras, probes, prostheses, etc.); and articles of general use, especially toys, the creation of ecological anti-corrosion safeguarding, foils for solariums, recycled materials. The extension of the introduction of these natural polyphenolic compounds is achieved by utilizing the results obtained in this project for manufacturing pharmaceuticals, food, cosmetics, photostabilizers, and treatments for conditions induced by exposure of the human body to ionizing radiation (especially in cancer patients).

Those aspects signify supplementary potential directions for use to take advantage of the capacity of pharmacologically active substances to reduce the consequences of using unhealthy synthetic wares. The chapter attendants a generous give for the use of antioxidant-style vegetable phenolic substances (polyphenolcarboxylic acids, flavonoids, anthocyanins, tannins, phenolic amines), able to satisfy economic necessity (undersized cost price, assortment variety, continuous availability, and abundance of raw material), social (diversity of the wares made, ensuring the health of the population, the takeover by economic agents, through technological transfer, of research results), cultural (broadening the academic horizon through comprehensive research on ameliorating the performance of multifunctional materials and by introducing into the scientific circuit new data on the beneficial repercussions of vegetable organism extracts).

## COMPETING INTERESTS

The authors declare that they have no competing interest.

## DECLARATION OF GENERATIVE AI AND AI-ASSISTED TECHNOLOGIES IN THE WRITING PROCESS

During the preparation of this work, the book chapter author(s) have not taken any assistance from AI, and the book chapter author(s) reviewed and edited the content as needed and take(s) full responsibility for the content of the publication.

## REFERENCES

Cai, D., Wang, X., Wang, Q., Tong, P., Niu, W., Guo, X., Yu, J., Chen, X., Liu, X., Zhou, D., & Yin, F. (2024). β-cyclodextrin inclusion complexes with short-chain phenolipids: An effective formulation for the dual sustained-release of phenolic compounds. *Food Research International (Ottawa, Ont.)*, *187*, 114423. https://doi.org/10.1016/j.foodres.2024.114423

Cañadas, R., González-Miquel, M., González, E. J., Díaz, I., & Rodríguez, M. (2020). Overview of neoteric solvents as extractants in food industry: A focus on phenolic compounds separation from liquid streams. *Food Research International (Ottawa, Ont.)*, *136*, 109558. https://doi.org/10.1016/j.foodres.2020.109558

Chen, L., Chen, N., He, Q., Sun, Q., Gao, M. R., & Zeng, W. C. (2022). Effects of different phenolic compounds on the interfacial behaviour of casein and the action mechanism. *Food Research International (Ottawa, Ont.)*, *162*(Pt B), 112110. https://doi.org/10.1016/j.foodres.2022.112110

Christophoridou, S., & Dais, P. (2006). Novel approach to the detection and quantification of phenolic compounds in olive oil based on 31P nuclear magnetic resonance spectroscopy. *Journal of Agricultural and Food Chemistry*, *54*(3), 656–664. https://doi.org/10.1021/jf058138u

Christophoridou, S., & Dais, P. (2009). Detection and quantification of phenolic compounds in olive oil by high resolution $^{1}H$ nuclear magnetic resonance spectroscopy. *Analytica Chimica Acta*, *633*(2), 283–292. https://doi.org/10.1016/j.aca.2008.11.048

Christophoridou, S., Dais, P., Tseng, L. H., & Spraul, M. (2005). Separation and identification of phenolic compounds in olive oil by coupling high-performance liquid chromatography with postcolumn solid-phase extraction to nuclear magnetic resonance spectroscopy (LC-SPE-NMR). *Journal of Agricultural and Food Chemistry*, *53*(12), 4667–4679. https://doi.org/10.1021/jf040466r

Čižmáriková, R., Habala, L., & Markulia, M. (2023). General intravenous anesthetics – Pharmacodynamics, pharmacokinetics and chiral properties. Celkové intravenózne anestetiká – farmakodynamika, farmakokinetika a chirálne vlastnosti. *Ceska a Slovenska farmacie: casopis Ceske farmaceuticke spolecnosti a Slovenske farmaceuticke spolecnosti*, *72*(4), 155–164.

Dahmer, J., Marangon, P., Adolpho, L. O., Reis, F. L., Maldaner, G., Burrow, R. A., Mostardeiro, M. A., Dalcol, I. I., & Morel, A. F. (2022). Alkaloids from the stem barks of *Scutia buxifolia* Reissek (Rhamnaceae): Structures and antimicrobial evaluation. *Phytochemistry*, *196*, 113071. https://doi.org/10.1016/j.phytochem.2021.113071

Davoudi, Z., Azizi, M. H., Barzegar, M., & Bernkop-Schnürch, A. (2024). Porous starch-inulin loaded quercetin microcapsules: Characterization, antioxidant activity, in-vitro release, and storage stability. *Journal of Pharmaceutical Sciences*, *113*(5), 1228–1238. https://doi.org/10.1016/j.xphs.2023.11.019

Dong, L. B., Zhang, X., Rudolf, J. D., Deng, M. R., Kalkreuter, E., Cepeda, A. J., Renata, H., & Shen, B. (2019). Cryptic and stereospecific hydroxylation, oxidation, and reduction in platensimycin and platencin biosynthesis. *Journal of the American chemical society*, *141*(9), 4043–4050. https://doi.org/10.1021/jacs.8b13452

Escobar-Avello, D., Avendaño-Godoy, J., Santos, J., Lozano-Castellón, J., Mardones, C., von Baer, D., Luengo, J., Lamuela-Raventós, R. M., Vallverdú-Queralt, A., & Gómez-Gaete, C. (2021). Encapsulation of phenolic compounds from a grape cane pilot-plant extract in hydroxypropyl beta-cyclodextrin and maltodextrin by spray drying. *Antioxidants (Basel, Switzerland)*, *10*(7), 1130. https://doi.org/10.3390/antiox10071130

Gaudreault, R., van de ven, T. G., & Whitehead, M. A. (2002). Molecular modeling of poly(ethylene oxide) model cofactors; 1,3,6-tri-O-galloyl-beta- d-glucose and corilagin. *Journal of Molecular Modeling*, *8*(3), 73–80. https://doi.org/10.1007/s00894-001-0070-9

Gehm, A. Z., Cunha, S. B., da Silva, B. W., Mostardeiro, M. A., Bastos, N. R., Burrow, R. A., Caro, M. S. B., Dalcol, I. I., & Morel, A. F. (2022). New cyclopeptide alkaloid of Condalia buxifolia and the absolute stereochemistry of Condaline A. *Fitoterapia*, *159*, 105194. https://doi.org/10.1016/j.fitote.2022.105194

Harish, G., Venkateshappa, C., Mythri, R. B., Dubey, S. K., Mishra, K., Singh, N., Vali, S., & Bharath, M. M. (2010). Bioconjugates of curcumin display improved protection against glutathione depletion mediated oxidative stress in a dopaminergic neuronal cell line: Implications for Parkinson's disease. *Bioorganic & Medicinal Chemistry*, *18*(7), 2631–2638. https://doi.org/10.1016/j.bmc.2010.02.029

Hu, J., Tao, Z., & Cao, S. (2009). Unique structure and property of cyclodextrins and their application in transdermal drug delivery. *Methods and Findings in Experimental and Clinical Pharmacology*, *31*(7), 449–456.

Ibrahim, W. A., Hermawan, D., & Sanagi, M. M. (2013). Cyclodextrin-modified micellar electrokinetic chromatography for enantioseparations. *Methods in molecular biology (Clifton, N.J.)*, *970*, 349–361. https://doi.org/10.1007/978-1-62703-263-6_22

Ishitsuka, Y., Irie, T., & Matsuo, M. (2022). Cyclodextrins applied to the treatment of lysosomal storage disorders. *Advanced Drug Delivery Reviews*, *191*, 114617. https://doi.org/10.1016/j.addr.2022.114617

Jing, S. X., Ferreira, D., Pandey, P., Klein, L. L., Chittiboyina, A. G., McAlpine, J. B., Lankin, D. C., Alania, Y., Reis-Havlat, M., Bedran-Russo, A. K., Chen, S. N., & Pauli, G. F. (2023). Unprecedented benzoquinone motifs reveal post-oligomerizational modification of proanthocyanidins. *The Journal of Organic Chemistry*, *88*(19), 13490–13503. https://doi.org/10.1021/acs.joc.3c00950

Kanawati, B., von Saint Paul, V., Herrmann, C., Schäffner, A. R., & Schmitt-Kopplin, P. (2011). Mass spectrometric stereoisomeric differentiation between α- and β-ascorbic acid 2-O-glucosides. Experimental and density functional theory study. *Rapid Communications in Mass Spectrometry: RCM*, *25*(6), 806–814. https://doi.org/10.1002/rcm.4908

Kaur, H., & Goyal, D. (2024). Lignin extraction from lignocellulosic biomass and its valorization to therapeutic phenolic compounds. *Journal of Environmental Management*, *372*, 123334. https://doi.org/10.1016/j.jenvman.2024.123334

Koyanagi, T., Edler, K. L., Parrott, R. W., Hitchcock, S. R., & Ferrence, G. M.(2010). (6R)-2-tert-Butyl-6-[(4R,5S)-3-isopropyl-4-methyl-5-phenyl-oxazolidin-2-yl]phenol. *Acta Crystallographica. Section E, Structure Reports Online*, *66*(Pt 4), o898–o899. https://doi.org/10.1107/S1600536810009591

Li, R., Bao, R., Yang, Q. X., Wang, Q. L., Adu-Frimpong, M., Wei, Q. Y., Elmurat, T., Ji, H., Yu, J. N., & Xu, X. M.(2019). [6]-Shogaol/β-CDs inclusion complex: Preparation, characterisation, *in vivo* pharmacokinetics, and *in situ* intestinal perfusion study. *Journal of Microencapsulation*, *36*(5), 500–512. https://doi.org/10.1080/02652048.2019.1649480

Liu, Y., Wang, L., Zhao, L., & Zhang, Y. (2022). Structure, properties of gossypol and its derivatives-from physiological activities to drug discovery and drug design. *Natural Product Reports*, *39*(6), 1282–1304. https://doi.org/10.1039/d1np00080b.

Loewen, P. C., Villanueva, J., Switala, J., Donald, L. J., & Ivancich, A. (2015). Unprecedented access of phenolic substrates to the heme active site of a catalase: Substrate binding and peroxidase-like reactivity of *Bacillus pumilus* catalase monitored by x-ray crystallography and EPR spectroscopy. *Proteins*, *83*(5), 853–866. https://doi.org/10.1002/prot.24777

Loll, P. J., & Axelsen, P. H. (2000). The structural biology of molecular recognition by vancomycin. *Annual Review of Biophysics and Biomolecular Structure*, *29*, 265–289. https://doi.org/10.1146/annurev.biophys.29.1.265

López-Nicolás, J. M., Rodríguez-Bonilla, P., & García-Carmona, F. (2014). Cyclodextrins and antioxidants. *Critical Reviews in Food Science and Nutrition*, *54*(2), 251–276. https://doi.org/10.1080/10408398.2011.582544

Magalhães, P. J., Vieira, J. S., Gonçalves, L. M., Pacheco, J. G., Guido, L. F., & Barros, A. A. (2010). Isolation of phenolic compounds from hop extracts using polyvinylpolypyrrolidone: Characterization by high-performance liquid chromatography-diode array detection-electrospray tandem mass spectrometry. *Journal of Chromatography. A*, *1217*(19), 3258–3268. https://doi.org/10.1016/j.chroma.2009.10.068

Martínková, L., Kotik, M., Marková, E., & Homolka, L. (2016). Biodegradation of phenolic compounds by Basidiomycota and its phenol oxidases: A review. *Chemosphere*, *149*, 373–382. https://doi.org/10.1016/j.chemosphere.2016.01.022

Messeha, S. S., Zarmouh, N. O., Mendonca, P., Cotton, C., & Soliman, K. F. A. (2020). Molecular mechanism of gossypol mediating CCL2 and IL-8 attenuation in triple-negative breast cancer cells. *Molecular medicine reports*, *22*(2), 1213–1226. https://doi.org/10.3892/mmr.2020.11240

Negahdari, R., Bohlouli, S., Sharifi, S., Maleki Dizaj, S., Rahbar Saadat, Y., Khezri, K., Jafari, S., Ahmadian, E., Gorbani Jahandizi, N., & Raeesi, S. (2021). Therapeutic benefits of rutin and its nanoformulations. *Phytotherapy Research: PTR*, *35*(4), 1719–1738. https://doi.org/10.1002/ptr.6904

Olmo-Cunillera, A., López-Yerena, A., Lozano-Castellón, J., Tresserra-Rimbau, A., Vallverdú-Queralt, A., & Pérez, M. (2020). NMR spectroscopy: A powerful tool for the analysis of polyphenols in extra virgin olive oil. *Journal of the Science of Food and Agriculture*, *100*(5), 1842–1851. https://doi.org/10.1002/jsfa.10173

Petitjean, M., Aussant, F., Vergara, A., & Isasi, J. R. (2020). Solventless crosslinking of chitosan, xanthan, and locust bean gum networks functionalized with β-cyclodextrin. *Gels (Basel, Switzerland)*, *6*(4), 51. https://doi.org/10.3390/gels6040051

Petitjean, M., & Isasi, J. R. (2023). Preparation of β-cyclodextrin/polysaccharide foams using saponin. *Beilstein Journal of Organic Chemistry*, *19*, 78–88. https://doi.org/10.3762/bjoc.19.7

Petitjean, M., Lamberto, N., Zornoza, A., & Isasi, J. R. (2022). Green synthesis and chemometric characterization of hydrophobic xanthan matrices: Interactions with phenolic compounds. *Carbohydrate Polymers*, *288*, 119387. https://doi.org/10.1016/j.carbpol.2022.119387

Pramanik, C., Jana, A., Brandao, P., Aher, A., Bera, P., Khatua, S., Majumdar, S., Mandal, B., Kumar Manna, S., & Bera, P. (2024). Phenolate-thiazole based reversible "turn-on" chemosensor for the selective detection of carbonate anion: X-ray crystallography, DFT/TDFT, and cell study. *Spectrochimica acta. Part A, Molecular and Biomolecular Spectroscopy*, *315*, 124233. https://doi.org/10.1016/j.saa.2024.124233

Remígio, M. S. D. N., Greco, T., Silva Júnior, J. O. C., Converti, A., Ribeiro-Costa, R. M., Rossi, A., & Barbosa, W. L. R. (2024). Spray-drying microencapsulation of *Bauhinia ungulata* L. var. *Obtusifolia* aqueous extract containing phenolic compounds: A comparative study using different wall materials. *Pharmaceutics*, *16*(4), 488. https://doi.org/10.3390/pharmaceutics16040488

Rice, J., Proctor, K., Lopardo, L., Evans, S., & Kasprzyk-Hordern, B. (2018). Stereochemistry of ephedrine and its environmental significance: Exposure and effects directed approach. *Journal of Hazardous Materials*, *348*, 39–46. https://doi.org/10.1016/j.jhazmat.2018.01.020

Salústio, P. J., Pontes, P., Conduto, C., Sanches, I., Carvalho, C., Arrais, J., & Marques, H. M. (2011). Advanced technologies for oral controlled release: Cyclodextrins for oral controlled release. *AAPS PharmSciTech*, *12*(4), 1276–1292. https://doi.org/10.1208/s12249-011-9690-2

Schaack, C., Arrico, L., Sidler, E., Górecki, M., Di Bari, L., & Diederich, F. (2019). Helicene monomers and dimers: Chiral chromophores featuring strong circularly polarized luminescence. *Chemistry (Weinheim an der Bergstrasse, Germany*, *25*(34), 8003–8007. https://doi.org/10.1002/chem.201901248

Schaack, C., Sidler, E., Trapp, N., & Diederich, F. (2017). Helical threads: Enantiomerically pure carbo[6] helicene oligomers. *Chemistry (Weinheim an der Bergstrasse, Germany, 23*(57), 14153–14157. https://doi.org/10.1002/chem.201703024

Shaddel, R., Akbari-Alavijeh, S., Cacciotti, I., Yousefi, S., Tomas, M., Capanoglu, E., Tarhan, O., Rashidinejad, A., Rezaei, A., Bhia, M., & Jafari, S. M. (2024). Caffeine-loaded nano/micro-carriers: Techniques, bioavailability, and applications. *Critical Reviews in Food Science and Nutrition*, *64*(15), 4940–4965. https://doi.org/10.1080/10408398.2022.2147143

Shi, S., Li, J., Zhao, X., Liu, Q., & Song, S. J. (2021). A comprehensive review: Biological activity, modification and synthetic methodologies of prenylated flavonoids. *Phytochemistry*, *191*, 112895. https://doi.org/10.1016/j.phytochem.2021.112895

Turuvekere Vittala Murthy, N., Agrahari, V., & Chauhan, H. (2021). Polyphenols against infectious diseases: Controlled release nano-formulations. *European Journal of Pharmaceutics and Biopharmaceutics: Official Journal of Arbeitsgemeinschaft fur Pharmazeutische Verfahrenstechnik e.V*, *161*, 66–79. https://doi.org/10.1016/j.ejpb.2021.02.003

Tutunchi, P., Roufegarinejad, L., Hamishehkar, H., & Alizadeh, A. (2019). Extraction of red beet extract with β-cyclodextrin-enhanced ultrasound assisted extraction: A strategy for enhancing the extraction efficacy of bioactive compounds and their stability in food models. *Food chemistry*, *297*, 124994. https://doi.org/10.1016/j.foodchem.2019.124994

Uekama, K. (2004). Design and evaluation of cyclodextrin-based drug formulation. *Chemical & Pharmaceutical Bulletin*, *52*(8), 900–915. https://doi.org/10.1248/cpb.52.900

Walsh, T. J., Viviani, M. A., Arathoon, E., Chiou, C., Ghannoum, M., Groll, A. H., & Odds, F. C. (2000). New targets and delivery systems for antifungal therapy. *Medical Mycology*, *38*(Suppl 1), 335–347.

Wang, P., Huang, S., Hu, Z., Yang, W., Lan, Y., Zhu, J., Hancharou, A., Guo, R., & Tang, B. (2019). In situ formed anti-inflammatory hydrogel loading plasmid DNA encoding VEGF for burn wound healing. *Acta Biomaterialia*, *100*, 191–201. https://doi.org/10.1016/j.actbio.2019.10.004

Wang, X., Howell, C. P., Chen, F., Yin, J., & Jiang, Y. (2009). Gossypol—A polyphenolic compound from cotton plant. *Advances in Food and Nutrition Research*, *58*, 215–263. https://doi.org/10.1016/S1043-4526(09)58006-0

Wang, Y., Li, P., Xu, W., Zhang, D., & Jia, Q. (2023). Hydrophilic magnetic host-guest Ti-phenolic networks: A promising material for the highly sensitive enrichment of glycopeptides and phosphopeptides. *Journal of Materials Chemistry. B*, *11*(22), 4874–4881. https://doi.org/10.1039/d2tb01708c

Watanabe, M., & Ayugase, J. (2009). Chiral separation of catechins in buckwheat groats and the effects of phenolic compounds in mice subjected to restraint stress. *Journal of Agricultural and Food Chemistry*, *57*(14), 6438–6442. https://doi.org/10.1021/jf901281j.

Woiwode, U., Sievers-Engler, A., & Lämmerhofer, M. (2024). Cross-linked polysiloxane-coated stable bond O-9-(2,6-diisopropylphenylcarbamoyl)quinine and quinidine chiral stationary phases as well as application in enantioselective cryo-HPLC. *Electrophoresis*, *45*(11–12), 989–999. https://doi.org/10.1002/elps.202300182

Wu, B., Basu, S., Meng, S., Wang, X., & Hu, M. (2011). Regioselective sulfation and glucuronidation of phenolics: Insights into the structural basis. *Current Drug Metabolism*, *12*(9), 900–916. https://doi.org/10.2174/138920011797470100

Zhang, Y. B., Wu, P., Zhang, X. L., Xia, C., Li, G. Q., Ye, W. C., Wang, G. C., & Li, Y. L. (2015). Phenolic compounds from the flowers of Bombax malabaricum and their antioxidant and antiviral activities. *Molecules (Basel, Switzerland)*, *20*(11), 19947–19957. https://doi.org/10.3390/molecules201119660

# Index

**Note**: Locators in *italics* represent figures and **bold** indicate tables in the text.

## B

## F



## G

## K



## L



## M

## N



## O



## P

## Q

## R



## S

For Product Safety Concerns and Information please contact our EU representative GPSR@taylorandfrancis.com Taylor & Francis Verlag GmbH, Kaufingerstraße 24, 80331 München, Germany